Springer Optimization and Its Applications

Volume 774

Aims and Scope

Optimization has continued to expand in all directions at an astonishing rate. New algorithmic and theoretical techniques are continually developing and the diffusion into other disciplines is proceeding at a rapid pace, with a spot light on machine learning, artificial intelligence, and quantum computing. Our knowledge of all aspects of the field has grown even more profound. At the same time, one of the most striking trends in optimization is the constantly increasing emphasis on the interdisciplinary nature of the field. Optimization has been a basic tool in areas not limited to applied mathematics, engineering, medicine, economics, computer science, operations research, and other sciences.

The series **Springer Optimization and Its Applications (SOIA)** aims to publish state-of-the-art expository works (monographs, contributed volumes, textbooks, handbooks) that focus on theory, methods, and applications of optimization. Topics covered include, but are not limited to, nonlinear optimization, combinatorial optimization, continuous optimization, stochastic optimization, Bayesian optimization, optimal control, discrete optimization, multi-objective optimization, and more. New to the series portfolio include Works at the intersection of optimization and machine learning, artificial intelligence, and quantum computing.

Volumes from this series are indexed by Web of Science, zbMATH, Mathematical Reviews, and SCOPUS.

Lewis Ntaimo

Computational Stochastic Programming

Models, Algorithms, and Implementation

Springer

Lewis Ntaimo
Industrial and Systems Engineering
Texas A&M University
College Station, TX, USA

ISSN 1931-6828 ISSN 1931-6836 (electronic)
Springer Optimization and Its Applications
ISBN 978-3-031-52462-2 ISBN 978-3-031-52464-6 (eBook)
https://doi.org/10.1007/978-3-031-52464-6

This Springer imprint is published by the registered company Springer Nature Switzerland AG
The registered company address is: Gewerbestrasse 11, 6330 Cham, Switzerland

Paper in this product is recyclable.

To past, present, and future students of this subject

Preface

This book is about stochastic programming (SP), a field of optimization that deals with mathematical programming problems involving uncertainty. SP problems are generally difficult to solve and the field has continued to evolve with contributions from various disciplines such as operations research, mathematics, and probability and statistics. SP has a wide range of application areas that include manufacturing, transportation, telecommunications, electricity power generation, health care, agriculture, forestry and wildfire management, mineral, oil and gas exploration, and finance. The purpose of this book is to provide a foundational and thorough treatment of the subject with a focus on models and algorithms and their computer implementation. Therefore, this book is suitable for readers with fundamental knowledge of linear programming, elementary analysis, probability and statistics, and some computer programming background. Specifically, computer programming knowledge is needed for model and algorithm implementation (coding) using available optimization software.

The most important features of this book include a focus on both risk-neutral and risk-averse models, a variety of real-life example applications of SP, decomposition algorithms, detailed illustrative numerical examples of the models and algorithms, and an emphasis on computational experimentation. This book takes a pragmatic approach and places emphasis on both theory and implementation of the models and algorithms for solving practical optimization problems. The benefits readers are expected to derive from this book include learning the following: (a) modeling real-life problems using SP; (b) translating theory into practical algorithms; (c) implementing models and algorithms on a computer; and (d) performing computational experiments in SP. The book is based on the author's hands-on experience with computational implementation of the various models and algorithms for different real-life applications.

This book covers various models of SP with a focus on two-stage recourse models encompassing mean-risk SP and mixed-integer SP. The goal is to provide a focused treatment of the subject with emphasis on modeling and solution methods accompanied by illustrative numerical examples. Therefore, the book starts with definitions and basic concepts in Chap. 1, taking the reader from linear programming

(LP) to SP in Chap. 2, and thereafter delves into mean-risk SP models in Chap. 3. At this juncture, the reader can proceed to Chap. 4, or jump to either Chap. 5 or Chap. 6 based on their background. Chapter 4 provides a sample of example applications of SP while Chap. 5 takes the reader through an adventure of deterministic large-scale LP methods that provide a foundation for the next chapter. Chapter 6 covers classical decomposition algorithms for SP and provides a basis for both Chaps. 7 and 8. Therefore, the reader can proceed to either chapter—Chap. 7 covers mean-risk methods and Chap. 8 focuses on statistic methods for linear SP. Chapter 9 introduces basic properties and algorithms for mixed-integer SP. Finally, Chap. 10 covers the computational aspects of SP, including standard data input formats and computational experimentation.

In essence, this book is appropriate for students and practitioners who are new to this exciting field, and as a reference for the seasoned SP experts. What makes this text different from existing books on the subject is the focus on illustrative numerical examples and an emphasis on computer implementation of the models and algorithms. This focus was borne out of incessant requests from students over the years for numerical examples to help them better understand SP models and algorithms. Readers conversant with SP can simply skip the numerical example illustrations without loss of continuity in the discourse.

This book is not meant to be covered in one course—there is just too much material to fit into one semester. Instead, the book is designed to be self-contained so it can be used in different ways. For example, one option is to use the book for an introductory course on SP for graduate students, including those who want to make SP as their research area. In this case, the course instructor can cover, for example, Chaps. 1, 2, 3, 6, and some selected topics from Chaps. 7, 8, and 9 based on the understanding level of the students. A course focused on model and algorithm implementation should also cover Chap. 10. Some selected topics from both Chaps. 4 and 5 can be used at the discretion of the instructor, especially for students who are not familiar with SP. Advanced graduate students whose research area is SP can focus on Chaps. 7 through 10, while practitioners can use the book as a reference based on their practical needs and interests. Another option is to use the book for an introductory course for advanced undergraduate students with a strong LP background, especially those in operations research, industrial engineering, and related fields. In this case, for example, the course can cover Chaps. 1, 2 (selected topics), 3, 4 (selected example applications), and 5 (selected topics). The course can also include some selected topics from Chap. 6.

I am grateful to my students and several colleagues for the support I have received over the past 20 years of research in this field. My early students worked on different aspects of SP and motivated me to write this book. I am also indebted to several of my colleagues for their support such as Bernardo K. Pagnoncelli and Guy L. Curry. Bernardo inspired me a great deal especially while he was a visiting professor at Texas A&M University and was eager to review the early versions of some of the chapters of this book. He believed such a textbook was needed in both the academic and industry optimization community. Guy, my office neighbor who had been with the department for 50 years at that time, really kept me on my toes. Every time he stopped by my office he would first ask, "how is the book coming along?" Finally, I am deeply grateful to the two people who introduced me to the field of SP, Suvrajeet Sen and Julie L. Higle. Special thanks go to my family as well as my friend Lubinda F. Walubita for their support, and to past, present, and future students of SP to whom I dedicate this book.

College Station, TX, USA
March 2023

Lewis Ntaimo

Contents

Acronyms

ASD	Absolute Semideviation
BAB	Branch-and-Bound
BAC	Branch-and-Cut
BFS	Binary First-Stage
C^3	Common-Cut-Coefficients
CDEV	Central Deviation
CDF	Cumulative Distribution Function
CLT	Central Limit Theorem
CVaR	Conditional Value-at-Risk
D^2	Disjunctive Decomposition
EE	Expected Excess
EP	Excess Probability
FD	Fenchel Decomposition
IID	Independent and Identically Distributed
IP	Integer Programming
LP	Linear Programming
MIP	Mixed-Integer Programming
MPS	Mathematical Programming System
MR-SP	Mean-Risk Stochastic Programming
MR-SLP	Mean-Risk Stochastic Linear Programming
MR-SMIP	Mean-Risk Stochastic Mixed-Integer Programming
PC-SP	Probabilistically Constrained Stochastic Programming
PDF	Probability Density Function
PMF	Probability Mass Function
QDEV	Quantile Deviation
QP	Quadratic Programming
RN-SLP	Risk-Neutral Stochastic Linear Programming
SAA	Sample Average Approximation
SD	Stochastic Decomposition

SLP	Stochastic Linear Programming
SMIP	Stochastic Mixed-Integer Programming
SMPS	Stochastic Mathematical Programming System
SP	Stochastic Programming

Part I
Foundations

Chapter 1
Introduction

1.1 Introduction

Decision-making regarding many interesting systems in business, operations research and management sciences, and engineering often involves modeling the system using mathematical optimization to determine the optimal decisions. Real complex systems involve data which is typically uncertain and evolve with time, requiring important system decisions to be made sequentially prior to observing the entire data stream. In many cases, the decisions have to be made here-and-now in the face of an unknown future, and often the decisions are required to be discrete. Consequently, modeling real-life systems to determine the optimal decisions is challenging. This book provides a pragmatic introduction to the subject of stochastic programming (SP), a vibrant and evolving field of stochastic optimization dealing with decision-making problems involving data uncertainty and risk.

SP is a branch of mathematical programming in which the problem data is allowed to be uncertain and is often characterized using probability distributions. Therefore, SP enables modeling system risk which in simple terms is the probability of loss. Risk comprises two components, the likelihood of an undesirable event or scenario happening and its adverse impact. Therefore, SP seeks optimal solutions in the face of all unforeseen scenarios and risk, which can also include the decision-maker's level of risk. In general, SP seeks optimal policies, i.e., specifications of feasible actions (decisions) to take for a given state of the system.

One of the fundamental motivations of SP stems from the fact that the real world is uncertain, and it is therefore imperative to take into account uncertainty and risk in decision-making. Otherwise, one may end up with suboptimal decisions that do not perform well in the face of uncertainty and risk. It is often pointed out that in the world nothing can be said to be certain except death and taxes. Another fundamental motivation of SP is the sheer number and diverse applications of SP. These applications span all facets of life and industry and include manufacturing, transportation,

L. Ntaimo, *Computational Stochastic Programming*, Springer Optimization and Its Applications 774, https://doi.org/10.1007/978-3-031-52464-6_1

telecommunications, electricity power generation, health care, agriculture, forestry and wildfire management, mineral, oil and gas exploration, and finance.

Manufacturing deals with interesting problems such as supply chain planning under uncertainty in product supply, demand, and prices. Transportation is a vast field and includes classical problems in optimization such as the traveling salesman problem, vehicle routing problem, scheduling, and so on, all under uncertainty. The data uncertainties can be in terms of unknown transportation costs, arc failures, resource availability, and demand, just to name a few. This field also spans the airline, bus, train, and shipping industries. Telecommunications and electricity power generation are traditional application areas of SP. Telecommunications include network design problems under uncertainty in resource demand, arc failure, and node failure. SP has been applied to electricity power generation in areas such as unit commitment and power adequacy planning problems under uncertainty in power generation and load demand.

SP has also seen a lot of interest in recent years in health care, which is still an evolving field. The problems in health care include patient and resource (nurses, physicians, equipment, etc.) scheduling in the different departments of healthcare such as surgery and outpatient clinics. The uncertainty in the health care settings includes unknown patient arrivals and unknown resource availability.

The SP applications of agriculture, forestry, and wildfire management are related. In agriculture, farmers deal with farm planning under uncertainty in weather and climate, which can significantly affect product yield. Forestry not only has a set of problems involving data uncertainties similar to farm planning but also includes wildfire management. Wildfires (forest fires or bushfires) are driven by the condition of the vegetation (fuels) and the prevailing uncertain weather and climate conditions. We have seen in recent years the devastation from wildfires across the globe attributed to weather and climate change. SP has also been applied to the mineral, oil, and gas industries in terms of exploration and production planning. For example, exploration involves field infrastructure planning (e.g., where to drill and set up expensive exploration platforms) under uncertainty in mineral, oil, and gas deposits as well as subsurface geology. Finally, the application of SP to finance deals with problems such as investment planning and portfolio optimization under uncertainty. In this case the data uncertainties include unknown interest rates and returns/yields on investments.

As can be gleaned from the foregoing discussion, the example applications for SP are very diverse and rich. So a natural question to ask is *where does the uncertainty stem from?* In the real world uncertainty and risk can come from different sources. We can classify these into *external* and *internal* sources. External sources of uncertainty can be market related, financial related, technology related, competition related, natural phenomenon related, or simply natural or human-made catastrophic events. Market related sources of uncertainty include product demand and price. For example, the demand and price of crude oil and stock market prices are uncertain at best. Financial related sources of uncertainty include the state of the economy and prevailing political conditions. The introduction of new technology often induces unforeseen uncertainty in the price and demand of new products. An easy to appreciate source of uncertainty that most can relate to on a daily basis is *weather*. Changes in weather can affect, for example, airline reschedules leading to delays and cancellations of flights. Adverse weather and climate can lead to unforeseen catastrophic wildfires, floods, mudslides, etc. Catastrophic events can be potent sources of uncertainty in that they often lead to unexpected impact in our way of life. These include unexpected wars, unforeseen attacks such as the 9/11, and unexpected accidents. Clearly, the 9/11 attacks led to a complete change in the way airlines operate and the kind of modern air travel experience we have today.

All the given example sources of uncertainty make optimal decision-making extremely challenging. In SP, we refer to the external uncertainty as *exogenous uncertainty* and to internal uncertainty as *endogenous uncertainty*. Exogenous uncertainty means that the underlying stochastic data process of the system is not influenced by the decisions being made. For example, weather related uncertainty is external, and the decisions an airline makes regarding rescheduling or canceling flights do not influence or impact the weather in any significant way. Another example of exogenous uncertainty is the demand for a product on the market where the producers' decisions do not influence the price and demand of the product on the market.

On the contrary, endogenous uncertainty is "internal" to the underlying stochastic process and is influenced by the decisions being made. For example, wildfire behavior (e.g., surface rate of spread and direction) is influenced by fire suppression decisions. For example, applying fire retardant on a wildfire affects the fire's rate of spread. Another example of endogenous uncertainty appears in mineral, oil, and gas exploration when the decision of where to drill an exploration hole influences the unveiling of the underling mineral, oil, and gas deposit distribution and the subsurface geology. In general, SP dealing with endogenous uncertainty is significantly more challenging due to the dependency of the probability distribution on the decisions.

Another aspect of SP is the sequential decision-making process. The two-stage SP decision-making process is depicted in Fig. 1.1. In this setting a decision has to be made here-and-now (first-stage) in the face of future uncertainty and risk. Then a recourse or corrective decision (second-stage) is made for each future realization to hedge against costly realizations. In multistage SP, decisions are made sequentially at discrete stages over a time horizon, with the first-stage decision made here-and-now in the face of future uncertainty and risk. Naturally, multistage SP problems are written using dynamic programming, and Bellman equations are employed to capture the initial, intermediate, and terminal conditions of the dynamics involved.

One question that often comes up from students new to SP is that *what characteristics make a stochastic optimization problem suitable to be modeled using SP?* Basically, SP problems are generalizations of deterministic mathematical programming problems in which the problem data is not known with certainty. The *certainty assumption* of deterministic optimization is violated in this case. SP generally assumes that there are no difficulties in specifying the objective function and probability distribution that characterized the future data uncertainty. Also, different statistics can be applied to define the objective function, including the

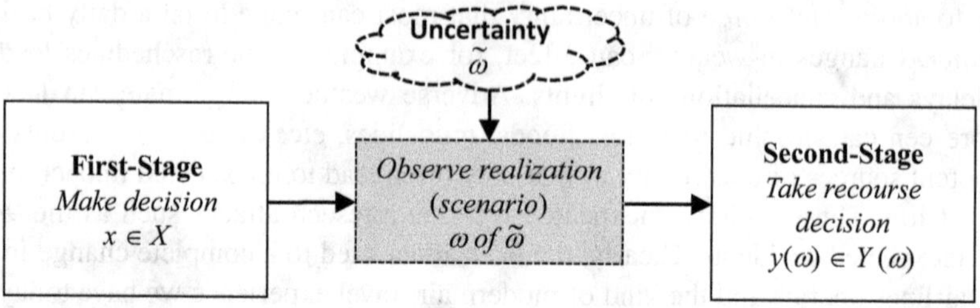

Fig. 1.1 The two-stage decision-making process

expectation (risk-neutral) and mean-risk measures (risk-averse). Emphasis is on finding a solution after a suitable problem statement has been found.

The key features of SP include known or partially known *probability distributions*, discrete time stages for decisions (two-stage or multistage), and a lot of decision variables with many potential values due to the uncertainty involved. The number of constraints is also typically large. Thus, SP problems are typically considered *large scale* due to the sheer size of the resulting mathematical problem. The relative importance of these features contrasts SP with the other stochastic optimization models. SP considers different types of decision variables: *continuous*, *binary*, *general integer*, and *mixed*. SP involving continuous decision variables only falls under stochastic linear programming (SLP), while SP involving binary/integer decision variables falls under stochastic mixed-integer programming (SMIP). These two classes of SP traditionally use expectations in the objective function and are therefore risk-neutral. When mean-risk (MR) measures are used in the objective function, we refer to the resulting risk-averse models as mean-risk stochastic programming (MR-SP) (i.e., MR-SLP and MR-SMIP, respectively) models. In general, SLP and many of the MR-SLP models have nice convexity properties but are still challenging to solve due to multi-dimensional integration and their large-scale nature. On the contrary, both SMIP and MR-SMIP do not have nice convexity properties. Instead, they inherit the undesirable properties of MIPs (e.g., nonconvexity and lower semicontinuity) and are, consequently, extremely difficult to solve.

Even though a lot of progress has been made in SP in the last 20 years, there is still, however, a need for new decomposition methods for SMIP and MR-SMIP and multistage SP. Figure 1.2 gives a pictorial illustration of the deterministic setting of LP/MIP and the stochastic setting of SP. Using the arrows, the figure shows how the different models are related in terms being deterministic or stochastic and possessing convexity or nonconvexity properties. Specifically, the figure shows model transitions from LP to IP, LP to SLP, IP to SMIP, SLP to MR-SLP, and SMIP to MR-SMIP.

One SP modeling framework that is different from the two-stage recourse model is the *probabilistic (chance) constraints* SP (PC-SP) model. Unlike the recourse model framework, PC-SP involves probabilistic (chance or join-chance) constraints,

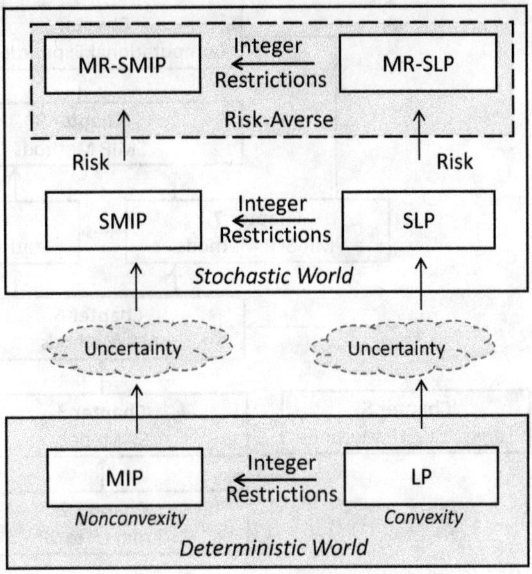

Fig. 1.2 From LP and MIP to MR-SLP and MR-SMIP

i.e., constraints that do not necessarily hold all the time but are allowed to be violated based on a level or risk or reliability. This level of risk is specified by the decision-maker and reflects the decision-maker's level of risk-averseness. PC-SP is a powerful modeling framework for applications where one cannot guarantee a particular set of constraints to hold with certainty. This is common in applications that involve reliability requirements or quality of service constraints.

A quite different branch of SP is *robust optimization* (RO), where a certain measure of robustness is sought against uncertainty that is represented as deterministic variability in the value of the problem parameters and/or its solution. In SP, uncertain is captured using probability distributions, whereas in RO the uncertainty values are defined using a continuous set called the "uncertainty set." RO assumes no distributional knowledge about the underlying uncertainty other than its support. The model minimizes the worst-case cost over the uncertainty set. Relatively recently, RO has been extended to *distributionally robust optimization* (DRO). The uncertain parameters in DRO are governed by any probability distribution from within an "ambiguity set," and a decision is sought that minimizes a cost function under the most adverse outcome of the uncertainty.

Another class of SP, referred to as *stochastic dominance constraints* (SDC), has been developed in finance. The SDC modeling framework allows measurement of risk by comparing distributions of two random variables. This modeling setting is in the context of managing risk by selecting options that are preferable to a random benchmark with a goal to maximize expected profits.

The focus of this book is on two-stage SP with recourse, encompassing SLP, MR-SLP, SMIP, and MR-SMIP models. The aim is to provide a focused treatment of the subject with emphasis on modeling and solution methods for SP with recourse. The

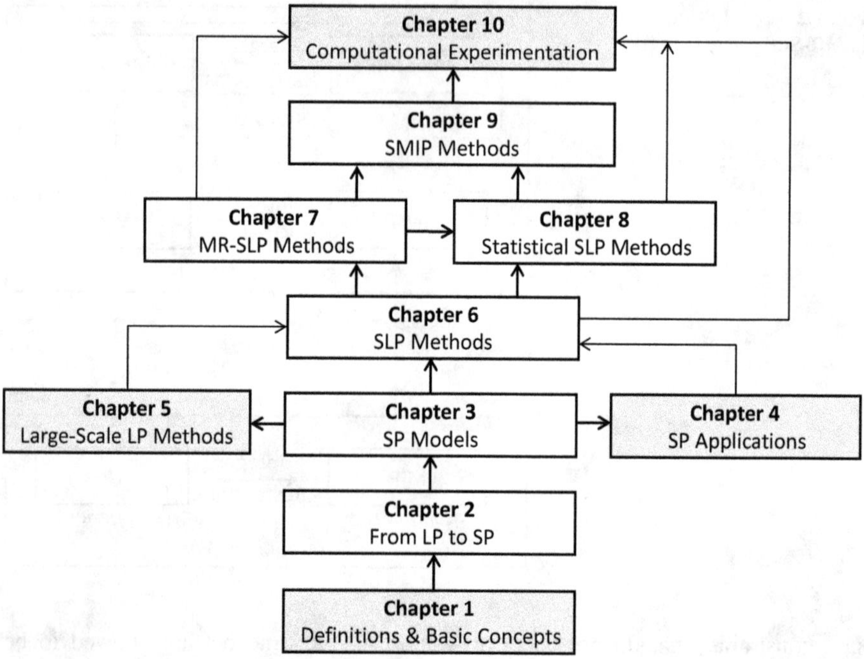

Fig. 1.3 Organization of the book

book takes the student on an adventure from deterministic large-scale LP models and algorithms, on to SP models and decomposition algorithms. In essence, this text introduces SP to both students and practitioners who are new to this exciting field. What makes this text different from existing books on the subject is the focus on numerical examples and emphasis on the computer implementation of the models and algorithms. This focus was borne out of student requests over the years for numerical examples to help them understand SP. Figure 1.3 provides a summary of how the chapters of this textbook are organized.

Chapter 1 provides definitions and basic concepts that are used throughout the textbook, with more advanced concepts introduced and defined in each chapter as needed. The chapter reviews the deterministic LP model and introduces different approaches for dealing with uncertainty and risk in SP. Chapter 2 provides a formal and rigorous definition of SP models, while Chap. 3 provides detailed numerical example illustrations of different SP models. Chapter 4 provides motivational example modeling applications of SP. Chapter 5 covers models and algorithms for large-scale deterministic LP, which motivate the algorithms for SP covered in Chaps. 6, 7, and 8. Students conversant with the material covered in Chap. 5 can skip this chapter without loss in continuity of the topics covered. Chapter 6 covers selected classical decomposition methods for SLP and provides detailed example illustrations for each method. From Chap. 6 one can move on either to MR-

SLP models and methods in Chap. 7 or to sampling methods for SLP in Chap. 8. Chapter 9 covers the basic properties of SMIP, introduces some three pragmatic algorithms for SMIP, and gives example illustrations of each. Finally, Chap. 10 provides an overview of how to perform computational experiments in SP. This chapter can be covered right after completing any of the Chaps. 6, 7, 8, or 9.

This book does not cover PC-SP, RO/DRO, and multistage SP. These topics deserve their own special treatment of a similar kind to this text. Students and practitioners with SP background may find other textbooks more suitable for their needs. Examples of textbooks on SP include the following: Birge and Louveaux [7, 8], Shapiro and Ruszczyński [32], Wallace and Ziemba [37], Shapiro, Dentcheva and Ruszczyński [34], King and Wallace [17], and Prékopa [28].

1.2 Preliminaries

We are now in a position to review some fundamental mathematical concepts that we believe are useful in later chapters that the reader should be familiar with. The reader conversant with these concepts can skip this section and/or chapter without loss of continuity in the discourse.

1.2.1 Basic Notations

Different notations and mathematical symbols are used in SP. In this book, we chose some specific notations that are commonly used. In Table 1.1 we list selected notations, i.e., mathematical symbols and their meaning, to provide a starting point for the reader. The rest of the notations are introduced in each section or chapter as needed. Here we simply provide a list of notations that appear throughout the book.

1.2.2 Vectors and Matrices

Definition 1.1 (Vector) A real n-dimensional vector is an ordered set of n real numbers $\{a_1, a_2, \ldots, a_n\}$ and is usually written in the form

$$a = \begin{pmatrix} a_1 \\ a_2 \\ \vdots \\ a_n \end{pmatrix}$$

Table 1.1 Selected mathematical symbols and their meaning

Symbol	Meaning
Min	Minimization operation
Max	Maximization operation
min	Minimum operator
max	Maximum operator
$\mathbb{R}$	Set of numbers, real line
$\bar{\mathbb{R}} = \mathbb{R} \cup \{-\infty\} \cup \{+\infty\}$	Extended real line
$\mathbb{R}^n_+$	Set of nonnegative numbers in n-dimensional space
$\mathbb{Z}$	Set of integers
$\mathbb{Z}^n_+$	Set of nonnegative integers in n-dimensional space
$\mathbb{E}$	Expectation operator
$\mathbb{D}$	Risk measure
$\in$	Set membership, belongs to
$\forall$	For all
$\ni, \mid$ or $:$	Such that, subject to
$\exists$	There exists
$\lvert \cdot \rvert$	Absolute value, cardinality of a set
$\lVert \cdot \rVert$	Vector norm
$d(x, y)$	Euclidean distance between $x \in \mathbb{R}^n$ and $y \in \mathbb{R}^n$
$(a)_+ = \max\{a, 0\}, a \in \mathbb{R}$	Largest nonnegative value
$A \in \mathbb{R}^{m_1 \times n_1}, T \in \mathbb{R}^{m_2 \times n_1}, W \in \mathbb{R}^{m_2 \times n_2}$	Matrices

or

$$a = (a_1, a_2, \ldots, a_n)^\top$$

(column vector) or $a = (a_1, a_2, \ldots, a_n)^\top$(row vector). The numbers $a_1, a_2, \ldots, a_n$ are called the components of a.

The set of all n-dimensional vectors is called *Euclidean n-space* and is usually denoted by $\mathbb{R}^n$. For clarity, unless otherwise specified, by a vector we will mean the column vector.

Definition 1.2 (Matrix) A *real matrix* is a rectangular array of real numbers composed of rows and columns. We write

$$A = [a_{ij}]_{m \times n} = \begin{bmatrix} a_{11} & a_{12} & \cdots & a_{1n} \\ a_{21} & a_{22} & \cdots & a_{2n} \\ \vdots & \vdots & \ddots & \vdots \\ a_{m1} & a_{m2} & \cdots & a_{mn} \end{bmatrix}$$

for a matrix of m rows and n columns, and we say that the matrix A is of *order* $m \times n$.

We will deal with real matrices (i.e., $A \in \mathbb{R}^{m \times n}$), and $m \times n$ will always denote *rows* $\times$ *columns*.

Definition 1.3 (Matrix Transpose) Given a matrix $A \in \mathbb{R}^{m \times n}$, the *transpose* of A is an $n \times m$ matrix whose rows are the columns of A. The *transpose* matrix of A is denoted by $A^\top$.

Example 1.1 Given

$$A = \begin{bmatrix} a_{11} & a_{12} & \cdots & a_{1n} \\ a_{21} & a_{22} & \cdots & a_{2n} \\ \vdots & \vdots & \ddots & \vdots \\ a_{m1} & a_{m2} & \cdots & a_{mn} \end{bmatrix},$$

we write

$$A^\top = \begin{bmatrix} a_{11} & a_{21} & \cdots & a_{m1} \\ a_{12} & a_{22} & \cdots & a_{m2} \\ \vdots & \vdots & \ddots & \vdots \\ a_{1n} & a_{2n} & \cdots & a_{mn} \end{bmatrix}.$$

Definition 1.4 (Vector Norm) A *norm* is the generalization of the intuitive notion of length (distance) in the real world to real vector spaces. A norm is a real-valued function defined on the vector space, $x \mapsto \|x\|$, and has the following properties:

(a) $\|x\| \geq 0$ for every vector x.
(b) $\|x\| = 0$ implies $x = 0$.
(c) For every vector x and every scalar α, $\|\alpha x\| = |\alpha| \, \|x\|$.
(d) For every vectors x and y, $\|x + y\| \leq \|x\| + \|y\|$.

Condition (a) means that a norm is nonnegative and (b) means that it is positive on nonzero vectors. Part (c) shows how a norm is proportional to $|\alpha|$, while part (d) is the so-called *triangle inequality*.

A norm induces a distance given by $d(x, y) = \|y - x\|$. This is called its *(norm) induced metric*.

Throughout this book, we shall make use of the ℓ^2-norm for $x \in \mathbb{R}^n$, denoted $\|x\|_2$ and defined as follows:

$$\|x\|_2 = \left\{ \sum_{i=1}^{n} |x_i|^2 \right\}^{\frac{1}{2}}.$$

1.2.3 Convex Sets and Functions

One of the prerequisites of optimization is to understand the structure and properties of mathematical sets and functions. The feasible region of an optimization problem is characterized using sets via constraints, while the objective is described by means of a function, i.e., the objective function. Therefore, in this subsection we review convex sets and functions, limit points, epigraphs, level sets, and separation hyperplanes. All these basic mathematical concepts are used in later chapters.

Definition 1.5 (Polyhedral Set) A set $C \subseteq \mathbb{R}^n$ is polyhedral if it can be written as

$$C = \{x \in \mathbb{R}^n \mid Ax \geq b\}$$

for some matrix $A \in \mathbb{R}^{m \times n}$ and a vector $b \in \mathbb{R}^m$.

In other words, a polyhedron $C \subseteq \mathbb{R}^n$ is the set of all points $x \in \mathbb{R}^n$ that satisfy a finite set of linear inequalities. A polyhedron can be expressed in other equivalent ways, e.g.,

$$C = \{x \in \mathbb{R}^n \mid Ax \leq b\} \text{ or } C = \{x \in \mathbb{R}^n \mid Ax = b,\ x \geq 0\}.$$

If a polyhedron is bounded, it is called a *polytope*, i.e., it can be enclosed in a ball of finite radius.

Definition 1.6 (Convex Set) A set $C \subseteq \mathbb{R}^n$ is a convex set if for all $x_1, x_2 \in C$ we have

$$\lambda x_1 + (1 - \lambda)x_2 \in C,\ \forall \lambda \in [0, 1].$$

Figure 1.4 shows examples of two convex sets. Observe that all points, $x_\lambda = \lambda x_1 + (1 - \lambda)x_2$ for all $\lambda \in [0, 1]$, on the line segment are contained within the set. In contrast, not all points on the line segment are contained in the sets shown in Fig. 1.5. Therefore, those sets are not convex. We will encounter different types of feasible regions in SP characterized as sets. Figure 1.6 gives an example convex feasible set for an LP/SLP, while Fig. 1.7 gives an example of a nonconvex feasible set for an MIP/SMIP. It is desirable to have convex feasible regions as optimizing over nonconvex sets can pose challenges as we shall in later chapters.

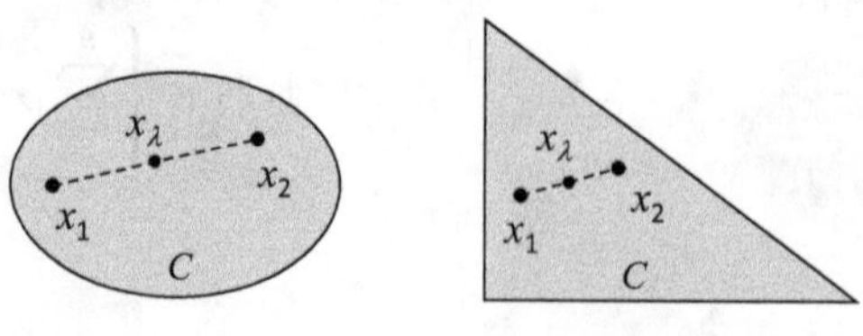

Fig. 1.4 Examples of convex sets

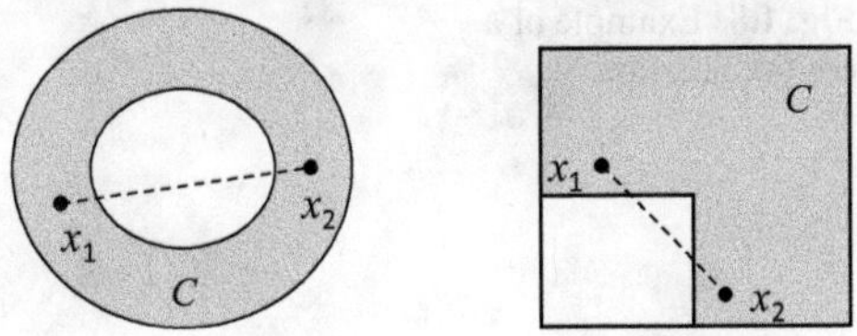

Fig. 1.5 Examples of nonconvex sets

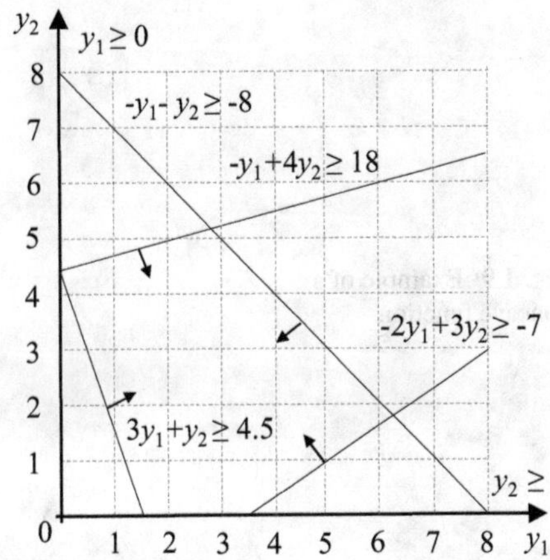

Fig. 1.6 Example of a convex set: feasible region of an LP/SLP

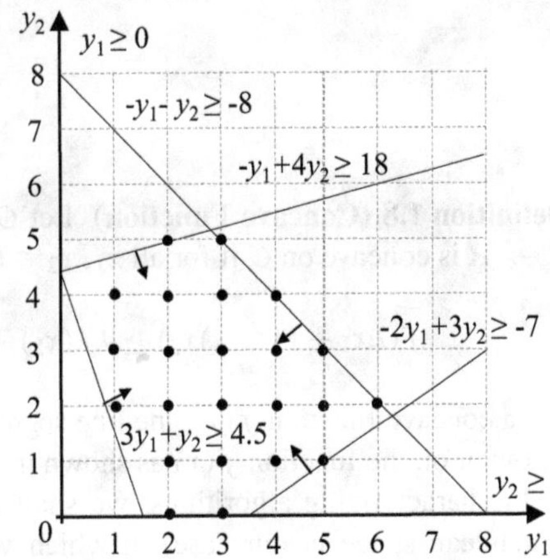

Fig. 1.7 Example of a nonconvex set: feasible region of an MIP/SMIP

Definition 1.7 (Convex Function) Let $C \subseteq \mathbb{R}^n$ be a convex set. A function $f : C \to \mathbb{R}$ is convex on C if for all $x_1, x_2 \in C$ we have

$$f(\lambda x_1 + (1 - \lambda)x_2) \leq \lambda f(x_1) + (1 - \lambda) f(x_2), \ \forall \lambda \in [0, 1].$$

As shown in Fig. 1.8, for a convex function $f(x)$, the line segment joining the two points $x_1, x_2 \in C$ is above or on the function $f(x)$.

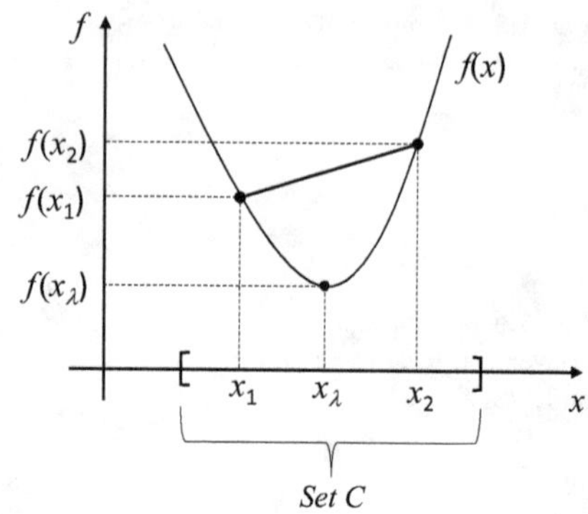

Fig. 1.8 Example of a convex function

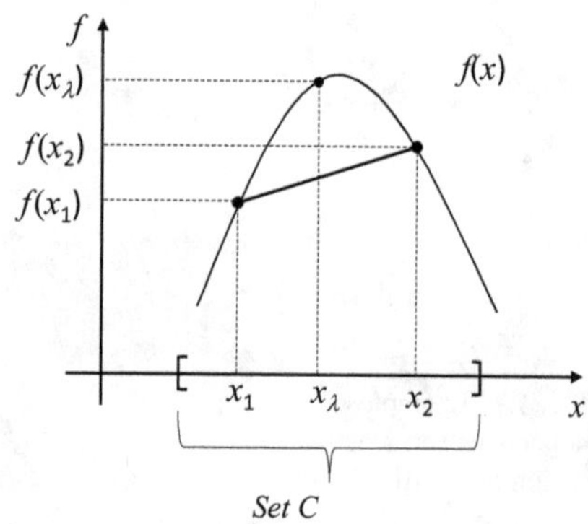

Fig. 1.9 Example of a concave function

Definition 1.8 (Concave Function) Let $C \subseteq \mathbb{R}^n$ be a convex set. A function $f : C \to \mathbb{R}$ is concave on C if for all $x_1, x_2 \in C$ we have

$$f(\lambda x_1 + (1-\lambda)x_2) \geq \lambda f(x_1) + (1-\lambda)f(x_2), \ \forall \lambda \in [0, 1].$$

For a concave function $f(x)$, the line segment joining the two points $x_1, x_2 \in C$ is below or on the function $f(x)$ as shown in Fig. 1.9.

In characterizing algorithms, we shall use the concept of a topological space (Euclidean space in our case) in which we will define *closeness*. The elements of Euclidean space are a set of points, and its topology can be defined as a set of neighborhoods for each point. The neighborhoods will satisfy some axioms formalizing the concept of closeness. With a topological space, we can define the notion of open/closed sets, limits, continuity, and connectedness.

Definition 1.9 (Limit Point) Let C be a subset of a topological space X. A point $y \in X$ is a *limit point* or an *accumulation point* of the set C if every neighborhood of y contains at least one point of C different from y itself. More specifically, y is a *limit (accumulation) point* of a set C if for all $\varepsilon > 0$, there exists $x \in C$ such that $\|x - y\| < \varepsilon$.

Let us now review several other basic concepts related to limit points: *closed set*, *open set*, *compact set*, and *covers*. A *closed* set C contains all its *limit points*. In other words, a closed set is a set whose complement is an *open* set. Also, a set is closed if and only if it contains all of its boundary points. A set that is both closed and bounded is called a *compact* set. A key thing about a compact set is that it contains its limit points. In other words, from any subsequence of elements, say $c_1, c_2, \cdots$, of the set, a subsequence can always be extracted which tends to some *limit elements* of the set. We should mention that $\mathbb{R}^n$ is closed but is *not* bounded. Another concept we should point out is that of a *cover*. A cover of a set C is a collection of sets whose union includes X as a subset. An *open cover* of a real line with respect to Euclidean topology is the set of all open intervals $(-n, n)$, where $n \in \mathbb{Z}$. For example, the set of all intervals $(\frac{1}{n}, 1)$, where $n \in \mathbb{Z} \setminus \{0\}$, is an open cover of the interval $(0, 1)$. An open cover contains a finite subcover.

Two concepts that are similar to *minimum* and *maximum* but are more useful because they characterize special sets which may have no minimum or maximum are *infimum* and *supremum*. We shall denote these concepts by inf and sup, respectively. The infimum of a subset C of a partially ordered set X, denoted inf C, is the greatest element in X that is less than or equal to each element of C, if such an element exists. In other words, inf C is the greatest lower bound of subset C. Conversely, the supremum of a subset C of a partially ordered set X, denoted sup C, is the least element in X that is greater than or equal to each element of C, if such an element exists. In other words, sup C is the greatest lower bound of subset C.

We are now in a position to review the concept of an *epigraph* of a function over the set C, which is the set of all points in $C \times \mathbb{R}$ lying on or above its graph.

Definition 1.10 (Epigraph of a Function) The *epigraph* of a function $f : C \to \mathbb{R}$, denoted *epi-f*, is defined as

$$epi\text{-}f := \big\{(x, \alpha) \in \mathbb{R}^{n+1} : \alpha \geq f(x)\big\}.$$

In convex analysis, epigraphs are used to provide geometrical interpretations of a convex function's properties and to help formulate or prove hypotheses. Figure 1.10 gives an illustration of *epi-f*. Consider $\big(x, f(x)\big) \in \mathbb{R}^{n+1}$. If f is convex, then

$$\begin{aligned}\lambda\big(x_1, f(x_1)\big) + (1-\lambda)\big(x_2, f(x_2)\big) &= \big(\lambda x_1 + (1-\lambda)x_2, \lambda f(x_1) + (1-\lambda) f(x_2)\big)\\ &\geq \big(\lambda x_1 + (1-\lambda)x_2, f(\lambda x_1 + (1-\lambda)x_2)\big).\end{aligned}$$

Let $(x, \alpha_1), (x, \alpha_2) \in epi\text{-}f$ such that $\alpha_1 \geq f(x_1)$ and $\alpha_2 \geq f(x_2)$ as shown in Fig. 1.11. Then

$$\begin{aligned}\lambda(x_1, \alpha_1) + (1-\lambda)(x_2, \alpha_2) &\geq \big(\lambda x_1 + (1-\lambda)x_2, f(\lambda x_1 + (1-\lambda)x_2)\big)\\ &= \big(x_\lambda, f(x_\lambda)\big).\end{aligned}$$

Therefore, it follows from the above that if f is a convex function, then *epi-f* is a convex set. The concept of an epigraph bridges convex sets and convex functions,

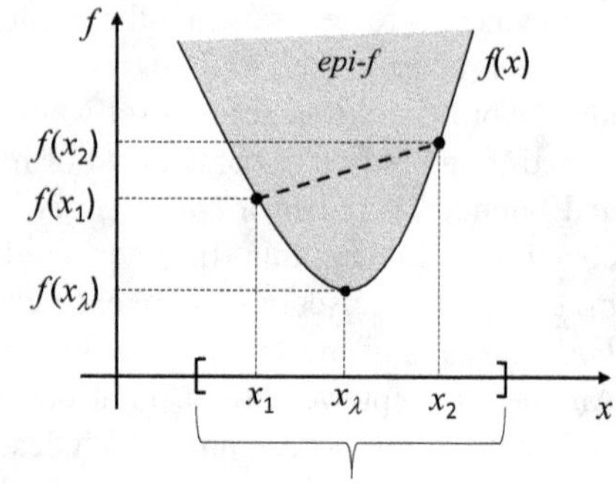

Fig. 1.10 An epigraph of function f, epi-f

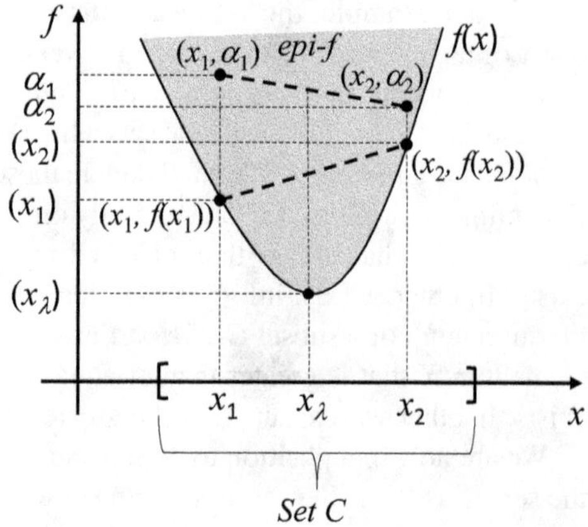

Fig. 1.11 An epigraph of function f, epi-f

and we sometimes use epigraphs in formulating convex optimization problems as we shall see in later chapters of this book. Specifically, we can interchange the object function and the constraints while maintaining the same convex optimization problem.

Another important set related to f is its *level set*. Next we define the *lower level* set of a function.

Definition 1.11 (Lower Level Set) Given $\alpha \in \mathbb{R}$, the *lower level* (*sublevel*) set of function $f : C \to \mathbb{R}$ is defined as

$$X_\alpha^- = \{x : f(x) \leq \alpha\}.$$

If f is convex, then its level set will have certain properties. Figure 1.12 gives an illustration of a level set. Observe that if f is convex, then $X_\alpha^- \subseteq C$ is a convex set. Consider a convex function f and $x_1, x_2 \in X_\alpha$ for some level α. Then we have

$$\begin{aligned} f(\lambda x_1 + (1-\lambda)x_2) &\leq \lambda f(x_1) + (1-\lambda) f(x_2) \\ &\leq \lambda\alpha + (1-\lambda)\alpha = \alpha, \end{aligned}$$

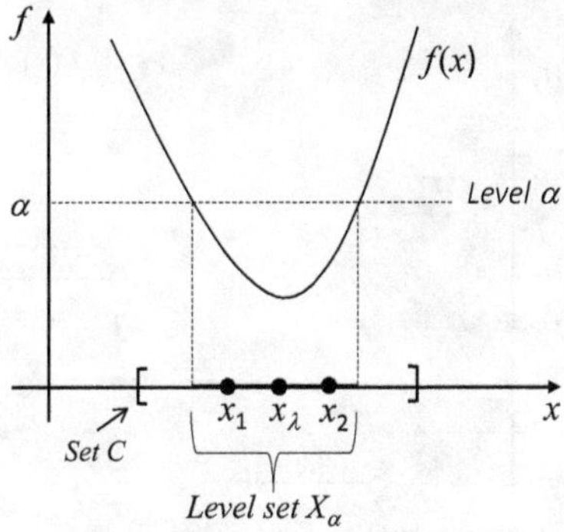

Fig. 1.12 A lower level set X_α^- for function f

implying that $\lambda x_1 + (1-\lambda)x_2 \in X_\alpha^-$. The upper level (superlevel) set of function f, denoted X_α^+, is defined as

$$X_\alpha^+ = \{x : f(x) \geq \alpha\}.$$

If f is convex, then $X_\alpha^+ \subseteq C$ is not a convex set.

Another important set of properties of functions besides convexity, which we encounter when we study MIP and SMIP, is *discontinuity* and *semicontinuity* of the value functions of this class of SP.

Definition 1.12 (Lower Semicontinuity) A function $f : X \to \overline{\mathbb{R}}$ is called *lower semicontinuous* at a point $x_0 \in X$ if for every real $\alpha < f(x_0)$ there exists a neighborhood $\mathscr{B}$ of x_0 such that $f(x) > \alpha$ for all $x \in \mathscr{B}$. Equivalently, f is *lower semicontinuous* at x_0 if and only if

$$\liminf_{x \to x_0} f(x) \geq f(x_0),$$

where lim inf is the limit inferior of the function f at point x_0.

Figure 1.13 gives two illustrations of lower semicontinuous functions, both value functions of SMIP (to be studied in Chap. 9). Observe that both functions satisfy the definition of lower semicontinuity but also, both functions have discontinuities. In Fig. 1.13a we see discontinuities in the function values at $x = 1, 2, 3, 4$, while in Fig. 1.13b the discontinuities occur at $x = 1, 2, 3$. Alternatively, a function $f : X \to \overline{\mathbb{R}}$ is lower semicontinuous if it satisfies any of the following equivalent conditions:

(a) The function is lower semicontinuous at every point of its domain.
(b) With $\alpha \in \mathbb{R}$ and $(\alpha, \to) = \{t \in \overline{\mathbb{R}} : t > \alpha\}$, all sets $f^{-1}((\alpha, \to)) = \{x \in X : f(x) > \alpha\}$ are *open* in X.
(c) All *lower level* sets $\{x \in X : f(x) \leq \alpha\}$ with $\alpha \in \mathbb{R}$ are closed in X.
(d) The *epigraph* $\{(x, t) \in X \times \mathbb{R} : t \geq f(x)\}$ is closed in $X \times \mathbb{R}$.

Upper semicontinuity can be defined in a similar manner.

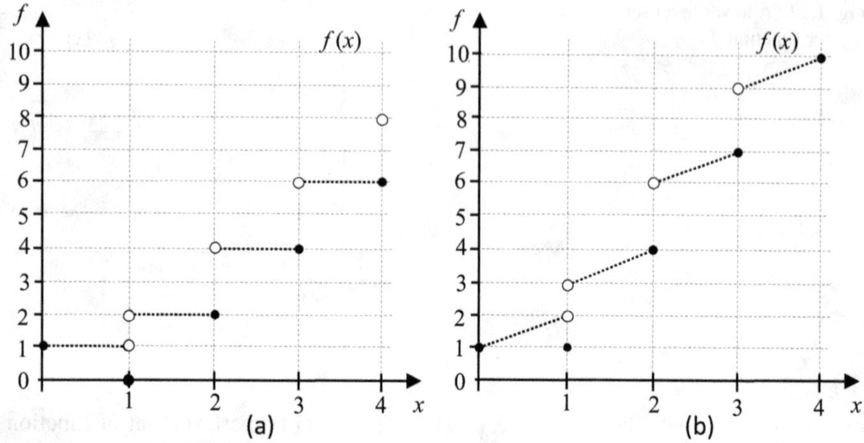

Fig. 1.13 Example of lower semicontinuous functions: (a) $f(x) = \max\{2\lceil x-1\rceil, \lceil x-1\rceil\}$ and (b) $f(x) = x + \max\{2\lceil x-1\rceil, \lceil x-1\rceil\}$

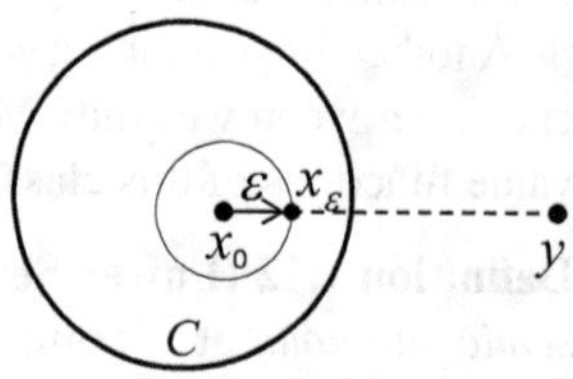

Fig. 1.14 Distance between a closed convex set and a point outside the set

1.2.4 Separation Hyperplanes

Separation hyperplane theorems characterize disjoint convex sets in $\mathbb{R}^n$ having a hyperplane in between them, and there are several rather similar versions. We are interested in separation hyperplanes between a convex set and a point outside the set. We also review a supporting hyperplane theorem which involves a hyperplane that has one point on the boundary of the set, while the set is entirely contained on one side of the hyperplane. These concepts are essential in deriving cutting-plane decomposition methods for large-scale LP and for SP covered in later chapters.

Let us start with characterizing the distance $d(x_0, y)$ between a point $x_0 \in C$ in a closed convex set C and a point $y \notin C$ (see Fig. 1.14). For $x_0 \in \mathbb{R}^n$ and $\varepsilon \in \mathbb{R}_+$, let $\mathscr{B}_\varepsilon(x_0) = \{x \in \mathbb{R}^n : \|x - x_0\| \leq \varepsilon\}$ denote the closed ε-ball centered at x_0.

Lemma 1.1 (Distance Between a Convex set and a Point) *Suppose that C is a closed convex set, and there exists a point $y \notin C$. Then $d(x_0, y) \leq d(x, y)$ for all $x \in C$ for some x_0 on the boundary of C.*

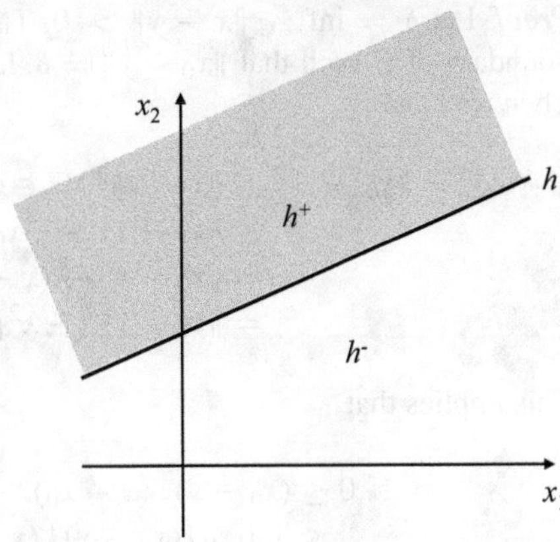

Fig. 1.15 Example hyperplane h and its half-spaces h^+ and h^-

Proof (By Contradiction) Suppose there exists $x_0 \in C$ such that $d(x_0, y) \leq d(x, y)\, \forall x \in C$, but x_0 is not on the boundary of C. Then there exists $\varepsilon > 0$ such that $\mathscr{B}_\varepsilon(x_0) \subset C$ implying that the line $\lambda x_0 + (1-\lambda)y$ intersects the ball $\mathscr{B}_\varepsilon(x_0)$ at some $x_\varepsilon \in C$. Note that $d(x_0, x_\varepsilon) = \varepsilon$, and

$$\begin{aligned} d(x_0, y) &= d(x_0, x_\varepsilon) + d(x_\varepsilon, y) \\ &= \varepsilon + d(x_\varepsilon, y) > d(x_\varepsilon, y). \end{aligned}$$

This is a contradiction, and therefore, $d(x_0, y) \leq d(x, y)$ for all $x \in C$ and x_0 must be on the boundary of C.

Definition 1.13 (Hyperplane) A *hyperplane* h in $\mathbb{R}^n$ is a collection of points of the form $\{x \mid a^\top x = \delta\}$, where $a \in \mathbb{R}^n \setminus \{\mathbf{0}\}$ and $\delta \in \mathbb{R}$ is a scalar.

A hyperplane h defines two *closed half-spaces*

$$h^+ = \{x \mid a^\top x \geq \delta\} \text{ and } h^- = \{x \mid a^\top x \leq \delta\}$$

as shown in Fig. 1.15. A closed half-space is the half-space that includes the points within the hyperplane. In addition to having two closed half-spaces, h has two *open half-spaces*

$$\{x \mid a^\top x > \delta\} \text{ and } \{x \mid a^\top x < \delta\}.$$

Therefore, any point in $\mathbb{R}^n$ lies in h^+, in h^-, or in both. A hyperplane h and the corresponding half-spaces can be written in reference to a fixed point, say, $x_0 \in h$. If $x_0 \in h$, then $a^\top x_0 = \delta$ and any point $x \in h$ must satisfy $a^\top x - a^\top x_0 = \delta - \delta = 0$. Consequently, $h^+ = \{x \mid a^\top (x - x_0) \geq 0\}$ and $h^- = \{x \mid a^\top (x - x_0) \leq 0\}$.

Theorem 1.1 (Separation Hyperplane) *Let $C \subset \mathbb{R}^n$ be closed and convex, and let $y \notin C$. There exists $a \in \mathbb{R}^n$ such that $a^\top y \leq a^\top x$ for all $x \in C$.*

Proof Let $\delta = \inf_{x \in C} \|x - y\| > 0$. Then by Lemma 1.1 there exists x_0 on the boundary of C such that $\|x_0 - y\| = \delta$. Let $a = (x_0 - y)$, and let $x \in C$ be given. Then we have

$$\begin{aligned}\delta^2 = \|x_0 - y\|^2 &\le \|x - y\|^2 \ \forall x \in C \\ &\le \|\lambda x + (1 - \lambda)x_0 - y\|^2 \ \ \forall \lambda \in [0, 1] \\ &= \|x_0 - y + \lambda(x - x_0)\|^2 \\ &= \|x_0 - y\|^2 + \lambda^2 \|x - x_0\|^2 + 2\lambda(x - x_0)^\top (x_0 - y).\end{aligned}$$

This implies that

$$\begin{aligned}0 &\le (x_0 - y)^\top (x - x_0) + \tfrac{\lambda}{2}\|x - x_0\|^2 \ \ \forall \lambda \in (0, 1] \\ &\le \lim_{\lambda \downarrow 0} \left((x_0 - y)^\top (x - x_0) + \tfrac{\lambda}{2}\|x - x_0\|^2\right) \\ &\le (x_0 - y)^\top (x - x_0) \\ &\le (x_0 - y)^T x - (x_0 - y)^T x_0.\end{aligned}$$

Rearranging the left hand side, we have

$$\begin{aligned}(x_0 - y)^T x_0 &\le (x_0 - y)^T x \\ \underbrace{(x_0 - y)^T}_{a} y + \underbrace{(x_0 - y)^\top (x_0 - y)}_{\|x_0 - y\|^2 = \delta^2} &\le \underbrace{(x_0 - y)^T}_{a} x.\end{aligned}$$

Substituting for $a = (x_0 - y)$ and $\delta^2 = \|x_0 - y\|^2$, we get $a^T y \le a^T x - \delta^2$. Since $\delta^2 \ge 0$, we can drop it to get

$$a^T y \le a^T x, \ \forall x \in C.$$

This completes the proof.

Theorem 1.1 can be interpreted as follows: $a^T y \le a^T x$ is a hyperplane that contains set C entirely on one side and is thus a *separation hyperplane*. If y is on the boundary of C, we can take a sequence of points $\{y^k\} \to y$, where $y^k \notin C \, \forall k$. Then there exists a limit point $\bar{a}$ to a sequence $\{a^k\}$ such that $\bar{a}^\top y \le \bar{a}^\top x$ for all $x \in C$. A hyperplane that contains C in one half-space and intersects it at the boundary is called a *supporting hyperplane* (see Fig. 1.16 for an example). A supporting hyperplane has the following two properties:

(a) Set C is entirely contained in one of the two closed half-spaces bounded by the hyperplane.
(b) Set C has at least one boundary point on the hyperplane.

In reference to Fig. 1.16, recall that a point on the boundary of *epi*-f has the form $(x_0, f(x_0)) \in \mathbb{R}^{n+1}$, and if f is a convex function, then *epi*-f is a convex set.

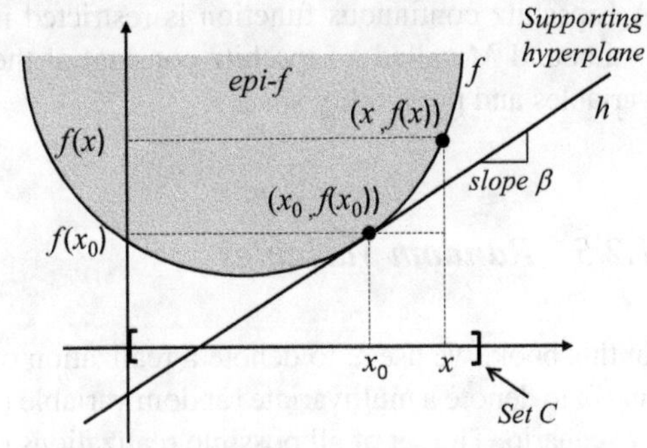

Fig. 1.16 Supporting hyperplane h supporting $epi\text{-}f$ at x_0

Suppose that we have a supporting hyperplane h with slope $\beta \in \mathbb{R}^n$ at x_0 supporting $epi\text{-}f$. Then we have

$$f(x) - f(x_0) \geq \beta^\top (x - x_0), \ \forall x \in C$$
$$f(x) \geq \beta^\top x + f(x_0) - \beta^\top x_0, \ \forall x \in C.$$

Setting $\alpha_0 := f(x_0) - \beta^\top x_0$ and assuming that f is differentiable at x_0 with gradient β, then we can clearly see that h has the form $f(x) = \beta^\top x + \alpha_0$, which is the equation of a straight line or an affine function. Such supporting hyperplanes will correspond to cutting-planes that we shall employ to approximate convex functions in later chapters.

To analyze mathematical functions, we are often interested in how functions behave with respect to changes in the argument, i.e., function *continuity*. Thus, a function can be characterized as *continuous* or *discontinuous*. A continuous function is one such that a continuous change of its argument induces a continuous change in the value of the function. This means that there is a change without jumps or abrupt changes in the function value, known as *discontinuities*. A discontinuous function is not continuous, and we shall encounter such functions when we study value functions of SMIP problems. Another concept is that of *uniform continuity* of functions.

For a uniformly continuous function f, there is a $\delta > 0$ such that

$$|f(x) - f(y)| < \epsilon$$

at any x and y in any function interval of size δ. A strong form of uniform continuity is *Lipschitz continuity*.

Definition 1.14 (Lipschitz Continuity) A real-valued function $f : R \to R$ is called *Lipschitz continuous* if there exists a positive real constant M such that, for all $x_1, x_2 \in \mathbb{R}^n$,

$$|f(x_1) - f(x_2)| \leq M|x_1 - x_2|.$$

A Lipschitz continuous function is restricted in how fast it can change, i.e., there is a bound M called a Lipschitz constant of the function. Next, we review random variables and probability spaces.

1.2.5 Random Variables

In this book, we use ω to denote a realization of a random process or "experiment" and $\tilde{\omega}$ to denote a multivariate random variable (vector). We refer to ω as an outcome or scenario. The set of all possible realizations (or sample space) will be denoted by Ω. A different notation is used in the SP literature, but we shall use this one since it is commonly used. Another common notation is ξ and $\boldsymbol{\xi}$ to denote a realization and a random variable, respectively. In this case, the sample space is denoted Ξ. Let us denote by $\mathscr{A}$ a collection of random realizations (events) of Ω. In probability theory and in mathematical analysis, this collection is called a *σ-algebra* or *σ-field*. Examples of events for an experiment involving rolling a standard die include the following: 1, 2, 3, 4, 5, 6, odd number, even number, number less than 2, number less than 3, etc. For each $A \in \mathscr{A}$, there is a probability measure or *distribution* $\mathbb{P}$ that returns the probability with which $A \in \mathscr{A}$ occurs and satisfies the following axioms:

(a) $0 \le \mathbb{P}(A) \le 1$.
(b) $\mathbb{P}(\Omega) = 1, \mathbb{P}(\emptyset) = 0$.
(c) $\mathbb{P}(A_1 \cup A_2) = \mathbb{P}(A_1) + \mathbb{P}(A_2)$ if $A_1 \cap A_2 = \emptyset$.

Thus, $\mathbb{P}$ is a function $\mathbb{P} : \mathscr{A} \to [0, 1]$. Impossible events have probability zero, and the event that happens almost surely or with almost total certainty has probability one.

The triple $(\Omega, \mathscr{A}, \mathbb{P})$ is called a *probability space* or *probability triple* in probability theory. It is a mathematical construct that gives a formal model of a random process or experiment. A random variable $\tilde{\omega}$ on a probability space $(\Omega, \mathscr{A}, \mathbb{P})$ is a real-valued function $\tilde{\omega}(\omega), \omega \in \Omega$ such that $\{\omega | \tilde{\omega}(\omega) \le z\}$ is an event for all finite z. For the random variable $\tilde{\omega}$, we define its *cumulative distribution function* (CDF) by $G(\tilde{\omega}) = \mathbb{P}(\tilde{\omega} \le z)$.

A discrete random variable takes on a finite number of values $\omega^s, s = 1, \cdots, S$, $S = |\Omega|$, with associated probabilities $g(\omega^s) = \mathbb{P}(\tilde{\omega} = \omega^s)$ with

$$\sum_{s=1}^{S} g(\omega^s) = 1.$$

A discrete random variable is described by a *probability mass function* (PMF) $g(\tilde{\omega})$. Unlike discrete random variables, continuous random variables take an infinite number of possible values. Such a random variable is described by *probability density function* (PDF) $g(\tilde{\omega})$. The probability of $\tilde{\omega}$ being in an interval $[a, b]$ is calculated as follows:

$$\begin{aligned}\mathbb{P}(a \leq \tilde{\omega} \leq b) &= \int_a^b g(\tilde{\omega})d\tilde{\omega} \\ &= \int_a^b dG(\tilde{\omega}) \\ &= G(b) - G(a).\end{aligned}$$

Contrary to the discrete case, $\mathbb{P}(\tilde{\omega} = z) = 0$.

Example 1.2 Consider an experiment involving one toss of a fair coin, where the outcome is either heads (H) or tails (T), i.e., the sample space is $\Omega = \{H, T\}$. The σ-algebra $\mathscr{A} = \{\{\}, \{H\}, \{T\}, \{H, T\}\}$. The null (empty) set, $\{\}$, means neither heads nor tails, $\{H\}$ means heads, $\{T\}$ means tails, and $\{H, T\}$ means either heads or tails. Since it is a fair coin, there is a 50% chance of tossing heads and 50% chance of tossing tails. Therefore, the probability measure is

$$\mathbb{P}(\{\}) = 0, \ \mathbb{P}(\{H\}) = 0.5, \ \mathbb{P}(\{T\}) = 0.5, \ \mathbb{P}(\{H, T\}) = 1.$$

Next, we need to characterize how to measure a random variable since it takes on different values. From probability and statistics, we can compute an *expected value* (mean or average) of a random variable as well as its *variance*. The expected value μ of a discrete random variable $\tilde{\omega}$ is given as

$$\mu = \mathbb{E}[\tilde{\omega}] = \sum_{s=1}^{S} \omega^s g(\omega^s)$$

and that of a continuous random is given as

$$\mu = \mathbb{E}[\tilde{\omega}] = \int_{-\infty}^{\infty} \tilde{\omega} g(\tilde{\omega})d\tilde{\omega} = \int_{-\infty}^{\infty} \tilde{\omega} dG(\tilde{\omega}).$$

The variance, denoted σ^2, of a random variable $\tilde{\omega}$ is given as

$$\sigma^2 = \mathbb{E}[(\tilde{\omega} - \mu)^2].$$

Remark 1.1 We shall introduce other *statistics* or risk measures in Chap. 2 for defining the objective function involving random variables. In particular, we consider quantile and deviation measures for SP involving risk measures in the objective. The quantile measures include conditional value-at-risk and quantile deviation, while deviation measures include expected excess and absolute semideviation.

1.3 Deterministic to Stochastic Programming

In this section we introduce a basic linear programming (LP) model and extend it to the stochastic setting. We focus on providing a conceptual understanding of transitioning from LP to SP and give several example models that one can consider, including the SP with recourse model. These models are covered in more depth in later chapter. The main aim here is to introduce the models to the reader in a relatively simple way. Nevertheless, we assume that the reader is conversant with LP; otherwise, we recommend to review LP. Consider an LP of the following form:

$$\text{Min } c^\top x \tag{1.1a}$$
$$\text{s.t. } Ax \geq b \tag{1.1b}$$
$$Tx \geq r \tag{1.1c}$$
$$x \geq 0, \tag{1.1d}$$

where

$x \in \mathbb{R}^{n_1}_+$ is the decision variable vector.
$c \in \mathbb{R}^{n_1}$ is the cost vector.
$A \in \mathbb{R}^{m_1 \times n_1}$ is the constraint matrix.
$b \in \mathbb{R}^{m_1}$ is the right hand side vector.
$T \in \mathbb{R}^{m_2 \times n_1}$ is the technology matrix.
$r \in \mathbb{R}^{m_2}$ is another right hand side vector.

We intentionally designed Problem (1.1) to have two sets of constraints and the reason for this will become apparent later. The problem data (c, A, b, T, r) captures the underlying physical aspects of the problem. In particular, matrix T will typically represent the "technology" being used and is thus called the *technology matrix*. In the deterministic setting of LP, we assume the *certainty principle*, which means that the LP data (c, A, b, T, r) is known with certainty. We are simply concerned with determining the optimal solution x^*, i.e., a solution such that $x^* \in \{x \in \mathbb{R}^{n_1}_+ \mid Ax \geq b,\ Tx \geq r\}$ with $c^\top x^* \leq c^\top \bar{x}$ for all $\bar{x} \in \{x \in \mathbb{R}^{n_1}_+ \mid Ax \geq b,\ Tx \geq r\}$.

In LP, we deal with two fundamental concepts: *feasibility* and *optimality*. Feasibility ensures that

$$x^*, \bar{x} \in \{x \in \mathbb{R}^{n_1}_+ \mid Ax \geq b,\ Tx \geq r\},$$

while optimality guarantees that

$$c^\top x^* \leq c^\top \bar{x} \quad \forall \bar{x} \in \{x \in \mathbb{R}^{n_1}_+ \mid Ax \geq b,\ Tx \geq r\}.$$

The two concepts are very clear, and in fact LP duality and sensitivity analysis deals with these both concepts. We are interested in the case when the certainty principle is violated, i.e., suppose the problem data (T, r) contains random variables whose values are not known and the data is given as $(T(\tilde{\omega}), r(\tilde{\omega}))$. We can now write a generic stochastic problem as follows:

$$\begin{aligned} \text{Min } & c^\top x \\ \text{s.t. } & Ax \geq b \\ & T(\tilde{\omega})x \geq r(\tilde{\omega}) \\ & x \geq 0, \end{aligned} \tag{1.2}$$

where

$T(\tilde{\omega}) \in \mathbb{R}^{m_2 \times n_1}$ is the *random* technology matrix.
$r(\tilde{\omega}) \in \mathbb{R}^{m_2}$ is the *random* right hand side vector.

Problem (1.2) is not well-defined yet since $(T(\tilde{\omega}), r(\tilde{\omega}))$ is a random variable. The question is how should we restate this formulation (in the face of uncertainty) so that it is well-defined? To be more concrete, we shall let $\tilde{\omega}$ be defined on a probability space $(\Omega, \mathscr{A}, \mathbb{P})$. Also, to ensure that the problem at hand is tractable, we make the following assumption:

(A1) The multivariate random variable $\tilde{\omega}$ is discretely distributed with finitely many realizations (scenarios) $\omega \in \Omega$, each with the probability of occurrence $p(\omega)$.

Note that with assumption (A1), it also means that $\sum_{\omega \in \Omega} p(\omega) = 1$. We shall continue to make assumption (A1) throughout this chapter. A *scenario* $\omega \in \Omega$ defines the realization of the random problem data, i.e., $\omega := (T(\omega), r(\omega))$. Uncertainty is expressed using a multivariate probability distribution (or "scenarios"):

$$\mathbb{P}(T(\tilde{\omega}), r(\tilde{\omega})) : \mathbb{P}(T(\omega), r(\omega)) = p(\omega), \; \omega \in \Omega.$$

With the proliferation of data for many applications in operations research, science, and engineering, probability distributions can be estimated or may be partially known. In some cases experts and forecasting/prediction models can provide the needed probability distributions. The decision-making problem under uncertainty we are faced with is that:

Find an optimal solution x^* here-and-now without knowing the real value of $(T(\tilde{\omega}), r(\tilde{\omega}))$ but knowing its probability distribution.

We should point out that $T(\tilde{\omega})x \geq r(\tilde{\omega})$ can be interpreted as a goal constraint to be specified more precisely once information is unveiled. This is what makes this problem interesting, and therefore, in the next subsections we introduce several approaches to addressing this problem.

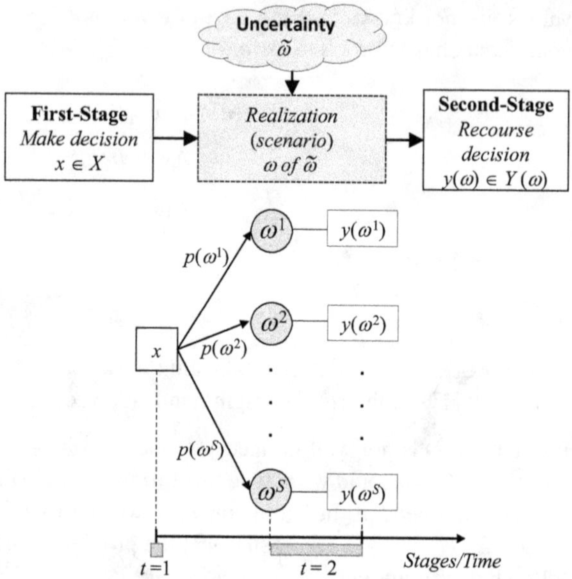

Fig. 1.17 Two-stage SP with recourse decision-making process and corresponding scenario tree

1.3.1 Scenario Trees

In SP, we use scenario trees to depict the evolution of the underlying stochastic data process for a given SP problem. Invoking assumption (A1), let $S = |\Omega|$ denote the total number of scenarios and re-index the scenarios $\omega \in \Omega$ as $\omega^s, s = 1, \cdots, S$, each with the probability of occurrence $p(\omega^s)$. We can create a two-stage *scenario tree* shown to depict the underlying stochastic process. The two-stage decision-making process, depicted in the top part of Fig. 1.17, involves a decision tree shown in the bottom part of Fig. 1.17. In the figure, time progresses from left to right, with the first-stage here-and-now decision event (square) at the root node. The random events (circles) occur at terminal nodes at which time/stage information is revealed, and then a recourse decision is made (rectangles). The root node is joined to the terminal node by an arc to indicate a (sample) path leading to the scenario at the terminal node. Therefore, the number of terminal nodes is equal to the number of scenarios.

1.3.2 Expected Value Solution

One straightforward way to address Problem (1.2) is to use the average or expectation of the random variable. This approach is widely used and is referred to as the *expected value* (EV) solution approach. What we do in this case is to simply replace $T(\tilde{\omega})x \geq r(\tilde{\omega})$ in Problem (1.2) with $\mathbb{E}[T(\tilde{\omega})]x \geq \mathbb{E}[r(\tilde{\omega})]$. Letting

$$\bar{T} = \sum_{\omega \in \Omega} p(\omega)T(\omega) \text{ and } \bar{r} = \sum_{\omega \in \Omega} p(\omega)r(\omega),$$

Problem (1.2) can be rewritten as follows:

$$\begin{aligned} z^{EV} := \text{Min } & c^\top x \\ \text{s.t. } & Ax \geq b \\ & \bar{T}x \geq \bar{r} \\ & x \geq 0, \end{aligned} \tag{1.3}$$

where z^{EV} is the EV optimal objective value. We shall denote by x_{EV} the optimal solution to Problem (1.3) (EV solution) with the corresponding objective value denoted by $z^{EV}(x_{EV})$. The advantage of Problem (1.3) is that it is a simple model, i.e., it is a deterministic LP. The disadvantage is clearly that this model does not take risk into account, and all the scenario information is condensed into average values. This means that we lose out on the variability in the data, and in fact, the constraint $T(\omega)x \geq r(\omega)$ may only hold for some scenarios only. This means we need to consider other alternative approaches. What else can you do here?

1.3.3 Scenario Analysis Solution

An alternative approach to EV is to consider data for each scenario $\omega \in \Omega$ separately and then form and solve a scenario LP. This approach is called a *scenario analysis* (SA) solution or a wait-and-see (WS) solution [19]. Let $x(\omega)$ be the decision variable for scenario $\omega \in \Omega$. We will now replace $T(\tilde{\omega})x \geq r(\tilde{\omega})$ in Problem (1.2) with $T(\omega)x(\omega) \geq r(\omega)$ for a given ω. The SA problem for $\omega \in \Omega$ can be written as follows:

$$\begin{aligned} z^{SA}(\omega) := \text{Min } & c^\top x(\omega) \\ \text{s.t. } & Ax(\omega) \geq b \\ & T(\omega)x(\omega) \geq r(\omega) \\ & x(\omega) \geq 0, \end{aligned} \tag{1.4}$$

where $z^{SA}(\omega)$ is the optimal objective value for scenario ω. The optimal solution to Problem (1.4) (SA solution) is denoted by $x_{SA}(\omega)$. The advantage of the SA approach is that each scenario problem is a simple deterministic LP and offers an improvement over the EV approach. Thus, SA can be an attractive approach, but the issue is how to find an overall solution based on the scenario solutions from each scenario LP (1.4). The WS solution is the expected value of the SA optimal values, i.e.,

$$z^{WS} := \sum_{\omega \in \Omega} p(\omega) z^{SA}(\omega).$$

1.3.4 Extreme Event Solution

Another approach to address Problem (1.2) is to simply consider all the scenarios simultaneously. Replace $T(\tilde{\omega})x \geq r(\tilde{\omega})$ in Problem (1.2) with $T(\omega)x \geq r(\omega),\ \forall \omega \in \Omega$. Then the new problem can be given as follows:

$$\begin{aligned} \text{Min}\ & c^\top x \\ \text{s.t.}\ & Ax \geq b \\ & T(\omega)x \geq r(\omega),\ \forall \omega \in \Omega \\ & x \geq 0. \end{aligned} \tag{1.5}$$

We shall call this approach the *extreme event solution* (EES). This name is fitting because the solution to this model is usually based on an "extreme" scenario. This advantage of this model is just that it is a deterministic LP, although it may be large scale. The disadvantage is that this model can be relatively conservative or restrictive and is often *infeasible*. In summary, the EES yields overly conservative solutions driven by extreme events, no matter how rare.

1.3.5 Two-Stage Recourse Model

The idea of the two-stage recourse problem (RP) formulation involves introducing recourse or corrective decisions $y(\tilde{\omega}) \in \mathbb{R}_+^{n_2}$ and replacing constraint $T(\tilde{\omega})x \geq r(\tilde{\omega})$ in Problem (1.2) with the following constraint:

$$T(\tilde{\omega})x + Wy(\tilde{\omega}) \geq r(\tilde{\omega}).$$

The key idea is to penalize the corrective actions to force the constraint $T(\tilde{\omega})x \geq r(\tilde{\omega})$ to hold, if possible; otherwise, we incur a penalty. Therefore, we introduce the second-stage objective cost $q(\tilde{\omega}) \in \mathbb{R}^{n_2}$. Consequently, the values the recourse decision $y(\tilde{\omega})$ takes depend on the realization (scenario) $\omega \in \Omega$ of $\tilde{\omega}$. A scenario ω defines the following problem data: $\omega := (q(\omega), T(\omega), r(\omega))$. What we want now is to minimize the here-and-now (first-stage) cost plus the total future cost. The RP model takes a risk-neutral approach and uses the *expectation* in computing the total future cost. We shall extend this model to the risk-averse setting in Chap. 2. We are now ready to state a two-stage SP with recourse as follows:

$$\begin{aligned} z^{RP} := \text{Min}\ & c^\top x + \mathbb{E}[\varphi(x, \tilde{\omega})] \\ \text{s.t.}\ & Ax \geq b \\ & x \geq 0, \end{aligned} \tag{1.6}$$

where for a given $x \in X := \{Ax \geq b, x \geq 0\}$ and realization $\omega \in \Omega$ of $\tilde{\omega}$

$$\begin{aligned} \varphi(x, \omega) = \text{Min}\ & q(\omega)^\top y(\omega) \\ \text{s.t.}\ & Wy(\omega) \geq r(\omega) - T(\omega)x \\ & y(\omega) \geq 0. \end{aligned} \tag{1.7}$$

It should be clear that x in the second-stage Problem (1.7) is data and not a decision variable. Thus, it is moved to the right hand side in the constraints. Because for a given ω we have a specification of the stochastic data $(q(\omega), T(\omega), r(\omega))$, the recourse decision $y(\omega)$ in formulation (1.7) is often written as y with an understanding that it depends on ω.

Let us now introduce another assumption and some standard SP terminology. To ensure that Problem (1.6–1.7) is well-defined, in addition to assumption (A1), we make the following assumption:

(A2) $\{Wy(\omega) \geq r(\omega) - T(\omega)x,\ y(\omega) \geq 0\} \neq \emptyset$ for all $x \in X$.

Recall that assumption (A1) stated earlier makes the problem tractable. Assumption (A2) is the *relatively complete recourse* assumption in SP. It guarantees the feasibility of the second-stage problem for every $x \in X$. We say that Problem (1.6–1.7) has relatively complete recourse if $\varphi(x, \omega) < \infty$ with probability one for all $x \in X$. Also, we say the problem has *complete recourse* if $\varphi(x, \omega) < \infty$ with probability one for all $x \in \mathbb{R}^{n_1}$. If assumption (A2) cannot be guaranteed, then by convention we have that $\varphi(x, \omega) = +\infty$ if the problem is *infeasible* and $\varphi(x, \omega) = -\infty$ if it is *unbounded*.

In terms of SP terminology, $\varphi(x, \tilde{\omega})$ is referred to as the *recourse function*, while $\mathbb{E}[\varphi(x, \tilde{\omega})]$ is called the *expected recourse function*. The matrix $W(\omega)$ is called the *recourse matrix*. We say Problem (1.6–1.7) has *fixed recourse* if $W(\omega) = W$ for all $\omega \in \Omega$. Otherwise, we say the problem has *random recourse*. A special case of fixed recourse is when $W = [I, -I]$, where I is the identity matrix. This is referred to as *simple recourse*, and it is the structure possessed by the classical newsvendor problem.

Based on assumption (A1), let us consider all the scenarios $\omega \in \Omega$ and let $S = |\Omega|$ denote the total number of scenarios. We can re-index the scenarios as ω^s, $s = 1, \cdots, S$, each with the probability of occurrence $p(\omega^s)$. We can now write the deterministic equivalent problem (DEP) for the RP model as follows:

$$z^{RP} := \text{Min}\ c^\top x + \sum_{s=1}^{S} p(\omega^s) q(\omega^s)^\top y(\omega^s) \tag{1.8a}$$

$$\text{s.t.}\ Ax \geq b \tag{1.8b}$$

$$T(\omega^s)x + Wy(\omega^s) \geq r(\omega^s),\ s = 1, \ldots, S \tag{1.8c}$$

$$x \geq 0,\ y(\omega^s) \geq 0,\ s = 1, \ldots, S. \tag{1.8d}$$

The advantage of the RP model is that the scenarios are taken care of explicitly, albeit in a risk-neutral setting. However, even though $S < \infty$, S can be very large making the DEP large scale and challenging to solve. For example, consider ten independent random variables each with six realizations. Then $S = 6^{10} = 60, 466, 176$, which is a large number of scenarios. Assuming matrix A is $(m_1 \times n_1)$ and W is $(m_2 \times n_2)$, then the total number of decision variables and constraints in the DEP is $(n_1 + n_2 S)$ and $(m_1 + m_2 S)$, respectively. Clearly, the DEP can be too large to even read into an optimization solver for large S. This motivates and calls for decomposition methods, which we explore in later chapters.

1.3.6 Relationships Among EV, SA, and RP Models

Recall that x_{EV} is the EV solution to Problem (1.3) with optimal value z^{EV} and $x_{SA}(\omega)$ is the SA solution to Problem (1.4) with optimal value $z^{SA}(\omega)$. Now let z^{RP} be the optimal value to RP (1.8) with the corresponding optimal solution x_{RP}. We refer to x_{RP} as the *here-and-now* solution. Given $z^{SA}(\omega)$ for all $\omega \in \Omega$, we can compute the WS solution , i.e., the expected value of the SA optimal values, as follows:

$$z^{WS} := \mathbb{E}[z^{SA}(\tilde{\omega})]. \tag{1.9}$$

The difference between the RP and WS solutions is called the *expected value of perfect information* (EVPI) and is given as

$$EVPI := z^{RP} - z^{WS}. \tag{1.10}$$

EVPI is a measure of how much to pay for perfect information regarding the uncertainty. Given the EV solution x_{EV}, let z^{EEV} be the optimal value to RP with $x := x_{EV}$ in Problem (1.8), i.e., x is fixed to x_{EV}. The value z^{EEV} is the expectation of the EV solution. The difference between the RP solution and z^{EEV} is called the *value of the stochastic solution* (VSS) and is given as

$$VSS := z^{RP} - z^{EEV}. \tag{1.11}$$

VSS can be thought of as a measure of the benefit of using the RP model over the EV model. In other words, VSS measures how bad the EV solution x_{EV} is relative to the RP solution x_{RP}. A positive VSS value indicates that it is necessary to use the RP model.

1.3.7 *Probabilistic (Chance) Constraints Model*

In certain applications it may not be possible to satisfy constraint $T(\tilde{\omega})x \geq r(\tilde{\omega})$ all the time. Probabilistic (chance) constraints allow for modeling *qualitative* measures of violation of the constraint. This is done by replacing $T(\tilde{\omega})x \geq r(\tilde{\omega})$ with $\mathbb{P}(T(\tilde{\omega})x \geq r(\tilde{\omega})) \geq \alpha$, for some specified reliability level $\alpha \in (0.5, 1)$. Then the problem can be written as follows:

$$\begin{aligned} \text{Min } & c^\top x \\ \text{s.t. } & Ax \geq b && (1.12a) \\ & \mathbb{P}(T(\tilde{\omega})x \geq r(\tilde{\omega})) \geq \alpha && (1.12b) \\ & x \geq 0. \end{aligned}$$

Constraint (1.12b) is referred to as the probabilistic constraint, and it can be interpreted as requiring $T(\tilde{\omega})x \geq r(\tilde{\omega})$ to hold with at least α probability. This parameter has to be specified by the modeler. Given $T(\tilde{\omega}) \in \mathbb{R}^{m_1 \times n_1}$ and $r(\tilde{\omega}) \in \mathbb{R}^{m_1}$, if $m_1 = 1$, we have a single probabilistic constraint (1.12b), and this is referred to as a *chance* constraint. If $m_1 \geq 2$, constraint (1.12b) is referred to as a *joint-chance* constraint.

The advantage of this model over the other models is that risk is taken care of explicitly, with $1 - \alpha$ being the maximal acceptable risk. The disadvantage of this model is that it is generally computationally difficult to solve. This is because the model is *nonconvex* in general. In fact, with discrete distributions Problem (1.12) can be written in extensive form (DEP) as an MIP. Let us continue to assume that $\tilde{\omega}$ has a discrete distribution as per assumption (A1) with the realizations (scenarios) denoted ω^s, $s = 1, \cdots, S$, each with the probability of occurrence $p(\omega^s)$. The total number of scenarios is $S = |\Omega|$. Also, for scenario s let $M(\omega^s)$ be an appropriately sized scalar and let e be an appropriately dimensioned vector of ones. For all $s = 1, \cdots, S$, define binary decision variables z^s: $z^s = 1$ if under scenario s at least one of the inequalities in the joint-chance constraint is violated, and $z^s = 0$ otherwise. Then Problem (1.12) is equivalent to the following DEP formulation:

$$\begin{aligned} \text{Min } & c^\top x \\ \text{s.t. } & Ax \geq b \\ & T(\omega^s)x + M(\omega^s)ez^s \geq r(\omega^s), \forall s = 1, \cdots, S \end{aligned}$$

$$\sum_{s=1}^{S} p(\omega^s) z^s \leq 1 - \alpha \tag{1.13}$$

$$x \geq 0,\ z^s \in \{0, 1\},\ \forall s = 1, \cdots, S.$$

The total probability of violating the probabilistic constraint (1.12b) is

$$\mathbb{P}(\tilde{T}x \ngeq \tilde{r}) \leq \sum_{s=1}^{S} p(\omega^s) z^s \leq 1 - \alpha.$$

When z^s=1, it means that scenario s is excluded from the formulation. Thus, we can assume that $p(\omega^s) \leq \alpha,\ \forall s = 1, \cdots, S$ so that the knapsack constraint (1.13) has a well-defined subset of scenarios that can be excluded from the formulation without exceeding the risk level $1 - \alpha$. Observe that DEP (1.13) is an MIP with a "big-M" and generally tends to have a very weak LP relaxation. Thus, it can be large scale and hard to solve and requires decomposition methods to solve efficiently.

1.3.8 *Integrated-Chance Constraints*

Probabilistic (chance) constraints allow for modeling *qualitative* measures of violation. However, in certain cases it is necessary to determine the exact amount of violation and then to bound it. Modeling *quantitative* measures of violation can be accomplished via *integrated-chance constraints* (ICC). In this case, we replace constraint (1.2), $T(\tilde{\omega})x \geq r(\tilde{\omega})$, in our generic stochastic problem as follows: Let a constant $d \in \mathbb{R}$ be given; then an ICC is defined as follows:

$$\mathbb{E}[\max_i (r_i(\tilde{\omega}) - T_i(\tilde{\omega})x)_+] \leq d,$$

where $T_i(\tilde{\omega})$ is the i-th row of matrix $T(\tilde{\omega})$ and $r_i(\tilde{\omega})$ is the i-th component of vector $r(\tilde{\omega})$. The ICC problem can be given as follows:

$$\begin{aligned} &\text{Min } c^\top x \\ &\text{s.t. } Ax \geq b \\ &\qquad \mathbb{E}[\max_i (r_i(\tilde{\omega}) - T_i(\tilde{\omega})x)_+] \leq d \\ &\qquad x \geq 0. \end{aligned} \tag{1.14}$$

What is the interpretation of the ICC constraint in Problem (1.14)? Given constraint $\mathbb{P}(T(\tilde{\omega})x \geq r(\tilde{\omega})) \geq 1 - \alpha$, we want to avoid positive shortages. However, it may not be possible to avoid positive shortages for all ω due to the randomness induced by $\tilde{\omega}$. The quantity $\mathbb{E}[\max_i (r_i(\tilde{\omega}) - T_i(\tilde{\omega})x)_+]$ in the ICC captures the expected positive shortage, which is limited by the amount d. This model offers the advantage that, like the probabilistic (chance) constraints model, risk is explicitly taken care of. In addition, the ICC model may be relatively easier to solve compared to the probabilistic constraints model.

1.3.9 Multistage Model

The two-stage decision-making process we introduced can be extending to the multistage setting where decisions have to be made sequential over a planning horizon H. Again, we graphically represent the underlying stochastic data process using scenario trees, which aid in keeping track of the interplay between decisions and random events. Recall that a scenario tree captures the evolution of the stochastic data process and represents a (finite) number of possible future outcomes. We show an example of a scenario tree with H stages in Fig. 1.18. As in the two-stage setting, time progresses from left to right with decision events (squares) and random events (circles) occurring at discrete points in time. A random event is a point at which information is revealed or provided. At stage (time) $t = 1$ we have one root node ω_1, and at stage $t = 2$ we have nodes equal to the number of the different realizations of data that may occur. Each is joined to the root node by an arc.

A generic node at time $t = 2$ is denoted as ω_2 and at time t is denoted ω_t. The set of all nodes at stage $t = 1, \cdots, H$ is denoted by Ω_t. At time t we can view a node $\omega_t \in \Omega_t$ as a *state* of the system. If we are at node (state) ω_t, we can talk about child nodes of a node $\omega_t \in \Omega_t$—these are nodes at the next stages $t + 1, \cdots, H$. Then a *scenario* is a particular realization (sample path) of the data process, i.e., a sequence $\omega_1, \cdots, \omega_H$ of nodes such that $\omega_t \in \Omega_t$ and $\omega_{t+1} \in \Omega_{t+1}$ is a child of the node ω_t. The process $\omega_1, \cdots, \omega_H$ will have a probabilistic structure, which we ignore for now. To capture the history of the process, for $1 \leq s \leq t \leq H$ let $\omega_{[s,t]} := \omega_s, \cdots, \omega_t$. In particular, $\omega_{[1,t]}$ represents the history of the process from the root node to node ω_t.

We should point out that scenario trees help keep track of the interplay between decisions and random events. Decisions that follow the data being revealed can adapt to it, while decisions that precede it cannot. Thus, our decisions should be adapted to the time structure of the information (data) process described by the scenario tree following the form depicted in Fig. 1.19. At stage t the values of the decision vector $x_t \in \mathbb{R}^{n_t}$ selected may depend on the information (data) available up to time t, but not on future observations. This is the basic requirement of *nonanticipativity* in SP, which simply means that decisions $x_t = x_t(\omega_{[1,t]})$ are functions of the history $\omega_{[1,t]}$ of the information process up to time t. However, for the first-stage $t = 1$, we have only one (root) node, and the decision x_1 is independent of the data.

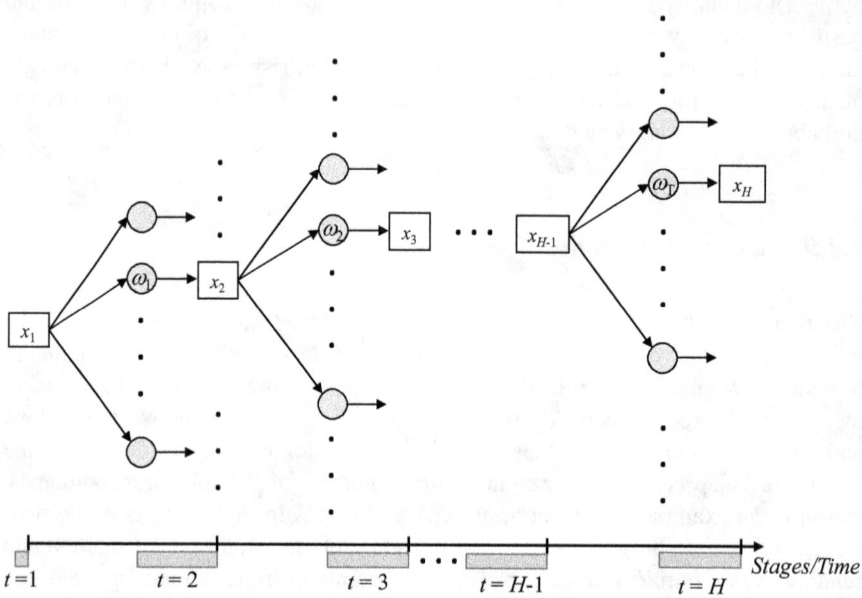

Fig. 1.18 An example scenario tree

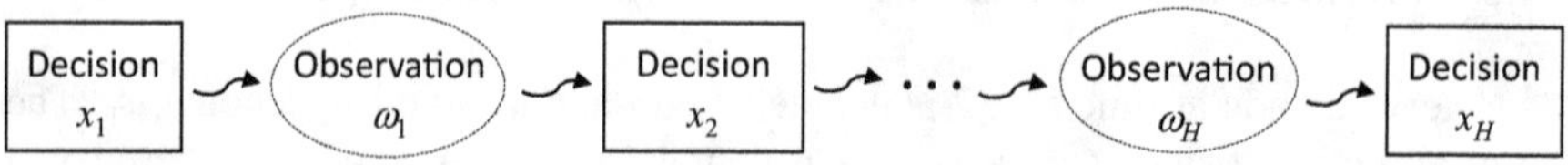

Fig. 1.19 The multistage decision-making process

More formally, we have a stochastic process $\tilde{\omega} = (\tilde{\omega}_1, \tilde{\omega}_2, \cdots, \tilde{\omega}_H)$ and a decision process $x = (x_1, x_2, \cdots, x_H)$. The component x_1 is a non-random vector-valued decision variable and $\tilde{\omega}_1$ is deterministic. The remaining components $x_2, \cdots, x_H$ of x and $\tilde{\omega}_2, \cdots, \tilde{\omega}_H$ of $\tilde{\omega}$ are random vectors (not necessarily of the same dimension) defined on $(\Omega, \mathscr{A}, \mathbb{P})$. The sequential decision and stochastic data process depicted in Fig. 1.19 is as follows:

$$x_1, \tilde{\omega}_2, x_2(x_1, \tilde{\omega}_2), \cdots, x_H(x_{H-1}, \tilde{\omega}_2, \cdots, \tilde{\omega}_H).$$

Because the decision process is *nonanticipative*, the decisions made at a given stage do not depend on future outcomes of the stochastic data or on future decisions. Mathematically, we have $\mathscr{A}_t \subseteq \mathscr{A}$ is a σ-field generated by $\omega_{[t]} := (\omega_1, \cdots, \omega_t)$ of the stochastic process $\tilde{\omega}$, which includes the stochastic data up to stage t. Thus, $\mathscr{A}_1 = \{\emptyset, \Omega\}$ is the trivial σ-field. Since the decision x_t at stage t depends only on the available information, it means that x_t is $\mathscr{A}_t$-measurable. We have $x_{[t]} := (x_1, \cdots, x_t)$ to be the sequence of decisions at stages $1, \cdots, t$ and $\mathbb{P}_t$ the distribution of ω_t. The stochastic process is considered *stagewise independent* if the process $\tilde{\omega}_t$ is independent of $\omega_{[t-1]}$, for all $t = 2, \ldots, H$.

We are now ready to write a multistage formulation of the problem using Bellman equations. Let $x_t(\omega_t) \in \mathbb{R}_+^{n_t}$ and $c_t \in \mathbb{R}^{n_t}$ be decision vector and the corresponding cost vector at stage t. Denote by $X_t(x_{t-1}, \omega_t) \subseteq \mathbb{R}^{n_t}$ the set of feasible decisions at stage $t \in \{2, \ldots, H-1\}$ for realization ω_t. Then a multistage SP with $H \geq 2$ stages can be formally stated as follows:

$$\operatorname*{Min}_{x_1 \in X_1} \quad c_1^\top x_1 + \mathbb{E}[f_2(x_1, \tilde{\omega}_2)] \tag{1.15}$$

where $x_1 \in \mathbb{R}_+^{n_1}$ and $c_1 \in \mathbb{R}^{n_1}$ are the first-stage decision and cost vectors, respectively, and $X_1 \subseteq \mathbb{R}^{n_1}$ is the first-stage feasible set. The real cost variable $f_t(x_{t-1}, \omega_{[t]})$ at stage t is defined recursively as

$$f_t(x_{t-1}, \omega_{[t]}) := \operatorname*{Min}_{x_t(\omega_t) \in X_t(x_{t-1}, \omega_{[t]})} \quad c_t^\top x_t(\omega_t) + \mathbb{E}[f_{t+1}(x_t, \tilde{\omega}_{[t+1]}) \,|\omega_{[t]}], \tag{1.16}$$

where $\mathbb{E}[. \,|\omega_{[t]}]$ is the conditional expectation. For the final stage $t = H$, given a solution $x_{H-1} \in X_{H-1}(x_{H-2}, \omega_{H-1})$, the real cost variable $f_H(x_{H-1}, \omega_{[H]})$ is given as

$$f_H(x_{H-1}, \omega_{[H]}) := \operatorname*{Min}_{x_H(\omega_H) \in X_H(x_{H-1}, \omega_{[H]})} \quad c_H^\top x_H(\omega_H). \tag{1.17}$$

To make sure that Problem (1.15–1.17) is well-defined and has an optimal solution, several assumptions have to be made. For example, the first-stage feasible set X_1 and Ω can be required to be compact. Also, it is desirable that the problem has suitable properties such as convexity to allow the formulation to be amenable to decomposition methods.

Bibliographic Notes

SP with recourse was initiated by Dantzig [11] and can be differentiated from other stochastic models in several respects. For example, statistical decision theory (SDT) introduced by Wald [36] focuses on determining best levels of variables $x \in X$ that affect the realization of an experiment involving a random variable $\tilde{\omega}$ with associated distribution $\mathbb{P}(\tilde{\omega})$ and reward $r(x, \tilde{\omega})$. The basic form of the problem can be given as follows:

$$\operatorname*{Max}_{x \in X} \mathbb{E}[r(x, \tilde{\omega}) \mid \mathbb{P}] = \operatorname*{Max}_{x \in X} \int_{\tilde{\omega}} r(x, \tilde{\omega}) d\mathbb{P}(\tilde{\omega}). \tag{1.18}$$

The emphasis in SDT is on the changes in $\mathbb{P}$ to some distribution $\hat{\mathbb{P}}_x$ that depends on a partial choice of x and some observations of ω. SDT assumes that X is a relatively small finite set that can be enumerated. A closely related field to SDT is decision analysis (DA) developed by Raiffa [29] whose emphasis is on the acquisition of information about possible realization, determining the utility associated with possible realizations and defining scenario trees to form a limited set of possible realizations.

Multistage SP (MSP) provides a framework for modeling problems involving sequential decision-making under uncertainty over time and extends the two-stage SP with recourse which was first formulated by [1, 11]. Decomposition approaches for solving MSP based on the L-shaped method of [35] were first proposed by Birge [6]. Later methods include the progressive hedging method MSP [30] and the stochastic dual dynamic programming (SDDP) algorithm [22], which was developed for solving hydrothermal scheduling problems.

Dynamic programming (DP), introduced by Bellman [2], and Markov decision processes (MDPs), pioneered by Ross [31], emphasize searching for optimal *actions* (decisions) to take at discrete points in time. These actions are influenced by random realizations and carry on from some *state* at some stage to another state at the next stage. DP has been extended to approximate DP (ADP) by [23, 24]. ADP considers strategies for tackling the curse of dimensionality in large-scale multi-period optimization problems. The emphasis in MPD is on identifying finite or low-dimensional state and action spaces while assuming some Markovian structure (e.g., actions and realizations depend only on the current state). Thus, Bellman equations are employed to state the problem, and backward recursion is used to determine an optimal policy. For infinite horizon problems, discounting is used with an emphasis placed on finding stationary policies. Optimal stochastic control is another field that is closely related to SP. In this field the models are similar to SP, but the problem dimensions are relatively lower, and emphasis is on determining control rules. Also, there are typically more restrictive constraint assumptions in optimal stochastic control.

Probabilistic (chance) constraint stochastic programming (PC-SP) was also introduced in the 1950s beginning with the work of Charnes, Cooper, and Symonds [9, 10] and extended by [20]. PC-SP was developed later with the work of Prékopa [25–27]. Closely related to PC-SP is integrated-chance constraint (ICC), which was introduced by [16, 18] but can be traced back to [27].

Robust optimization (RO) also goes back to the 1950s in the establishment of modern decision theory and the use of worst-case analysis and Wald's maximin model as a tool for the treatment of severe uncertainty [36]. Around the same time, distributionally robust optimization (DRO) was introduced by Scarf [33], and both fields have evolved over time [3–5, 15, 38]. RO involves uncertain parameters that belong to an uncertainty set, while DRO involves uncertainty parameters that are governed by probability distributions from within an ambiguity set. In both RO and DRO, the goal is to find a decision that minimizes a cost function under the most adverse outcome of the uncertainty.

Stochastic dominance constraint (SDC) [12, 13, 21] is a relatively new field and enables modeling risk by comparing distributions of two random variables. For example, given a concave nondecreasing function $u(.)$, then a random variable Z_1 dominates another random variable Z_2 in the second order, i.e., $Z_1 \succeq_{(2)} Z_2$, if $\mathbb{E}[u(Z_1)] \geq \mathbb{E}[u(Z_2)]$ for every $u(.)$ for which these expected values are finite. SDC has been applied to portfolio optimization problems in finance [14].

Problems

1.1 (Convex Sets) Determine whether each of the following statements is True or False. In either case, provide a proof or a counterexample.

(a) All polyhedral sets are convex.
(b) The intersection of convex sets is convex.
(c) The union of convex sets is convex.

1.2 (Convex and Concave Functions) Determine whether each of the following statements is True or False. In either case, provide a proof or a counterexample.

(a) The maximum of *convex* functions is a convex function.
(b) The minimum of *concave* functions is a convex function.
(c) The maximum of *affine* functions is a piecewise linear and convex function.
(c) The minimum of *affine* functions is a piecewise linear and convex function.

1.3 (Level Sets) Consider the problem, Min $\{f(x) \mid x \in X\}$, where

$$f(x) = (x-4)^2 + 2 \text{ and } X = \{2 \leq x \leq 6\}.$$

(a) Plot the function $f(x)$ and the feasible set X.
(b) Graph, X_α^-, the *lower level* set of $f(x)$ at level $\alpha := 3$, on your plot in part (a).
(c) What is the relation between the sets X_α^- and X.
(d) Graph, X_α^+, the *upper level* set of $f(x)$ at level $\alpha := 3$, on your plot in part (b).
(e) What is the relation between the sets X_α^+ and X.

1.4 (Level Sets and Convexity) Determine whether each of the following statements is True or False. In either case, provide a proof or a counterexample.

(a) The *lower level* set of a *concave* function is a *convex* set.
(b) The *upper level* set of a *concave* function is a *convex* set.
(c) The *upper level set* of a *convex* function is a *convex* set.
(d) The level set X_α^- in Problem 1.3 is *convex*.
(e) The level set X_α^+ in Problem 1.3 is *convex*.

1.5 (Epigraphs) Determine whether each of the following statements is True or False. In either case, provide a proof or a counterexample.

(a) The *epigraph* of a *concave* function is a *convex* set.
(b) The *epigraph* of a *concave* function is a *convex* set.

1.6 (Function Properties) Determine whether each of the following statements is True or False. In either case, provide a proof or a counterexample.

(a) An *affine* function is both *convex* and *concave*.
(b) The function $f(x) = \max\ \{\alpha_t + \beta_t^\top x \mid t = 0, 1, \cdots, k\}$ is *concave*.
(c) The function $f(x) = \text{Max}\ \{q^\top y \mid Wy \leq r - Tx,\ y \geq 0\}$ is *convex* (assume that this LP is dual feasible).
(d) The function $f(x) = \text{Min}\ \{q^\top y \mid Wy \geq r - Tx,\ y \geq 0\}$ is *convex* (assume that this LP is dual feasible).

1.7 (Function Properties) Consider the function

$$f(x) = \max\{2\lceil x - 2\rceil, \lceil x - 2\rceil\},$$

where $0 \leq x \leq 3,\ x \in \mathbb{Z}_+$.

(a) Plot the function $f(x)$.
(b) Show that the function $f(x)$ is *not* convex.
(c) What type of function is $f(x)$? Describe the function and state where the *discontinuities* are located if any.

1.8 (Separation Hyperplanes) Consider the function given in Problem 1.3, and assume that the function is differentiable. Answer the following questions:

(a) What is the supporting hyperplane at $x = 2$.
(b) What is the supporting hyperplane at $x = 3$.

1.9 (Random Variables) Consider an experiment involving two tosses of a fair coin, where the outcome for each toss is either heads (H) or tails (T). The outcomes (first toss and second toss) are ordered pairs, e.g., HH means heads followed by heads, and TH means tails followed by heads. Answer the following questions:

(a) What is the sample space Ω?
(b) How many elements are in the σ-algebra $\mathscr{A}$?
(c) Write out explicitly the σ-algebra $\mathscr{A}$.
(d) What is the probability measure $\mathbb{P}(\{\})$?

1.10 (Multistage Model) Consider a multistage SP with recourse and planning horizon $H = 3$ (three stages). Let the t-stage problem data be denoted $c^t(\omega_t) \in \mathbb{R}^{n_t}$, $T_t(\omega_t) \in \mathbb{R}^{m_t \times n_t}$, $W_t(\omega_t) \in \mathbb{R}^{m_t \times n_t}$, and $r_t(\omega_t) \in \mathbb{R}^{m_t}$ for $t = 1, 2, 3$. Answer the following questions:

(a) Write the t-stage constraints $X_t(x_{t-1}, \omega_{[t]})$ for $t = 1, 2, 3$ explicitly and state any assumptions you wish to make.
(b) Using your definition of $X_t(x_{t-1}, \omega_{[t]})$ for $t = 1, 2, 3$ in part (a), write the Bellman equations for all the three stages.
(c) Write the explicit extensive form or DEP formulation.

References

1. E.M.L. Beal. On minimizing a convex function subject to linear inequalities. *Journal of the Royal Statistical Society*, B17:173–184, 1955.
2. R. Bellman. *Dynamic Programming*. Princeton Univ. Press, New York, 1957.
3. A. Ben-Tal, L. El Ghaoui, and A. Nemirovski. *Robust Optimization*. Princeton Series in Applied Mathematics. Princeton University Press, 2009.
4. D. Bertsimas, V. Gupta, and N. Kallus. Data-driven robust optimization. *Mathematical Programming*, 167(2):235–292, 2018.
5. D. Bertsimas and M. Sim. The price of robustness. *Operations Research*, 52(1):35–53, 2004.
6. J.R. Birge. Decomposition and partitioning methods for multistage stochastic linear programs. *Operations Research*, 33.5:989–1007, 1985.
7. J.R. Birge and F.V. Louveaux. *Introduction to Stochastic Programming*. Springer, New York, 1997.
8. J.R. Birge and F.V. Louveaux. *Introduction to stochastic programming*. Springer Science & Business Media, Second edition, 2011.
9. A. Charnes and W.W. Cooper. Chance constrained programming. *Management Science*, 5:73–79, 1959.
10. A. Charnes, W.W. Cooper, and G.H. Symonds. Cost horizons and certainty equivalents: An approach to stochastic programming of heating oil. *Management Science*, 4:235–263, 1958.
11. G.B. Dantzig. Linear programming under uncertainty. *Management Science*, 1(3–4):197–206, 1955. Republished in the 50th anniversary issue of Management Science 50(12):1764–1769, 2004.
12. D. Dentcheva and A. Ruszczyński. Optimization with stochastic dominance constraints. *SIAM Journal on Optimization*, 14(2):548–566, 2003.
13. D. Dentcheva and A. Ruszczyński. Semi-infinite probabilistic optimization: First order stochastic dominance constraints. *Optimization*, 53(5–6):583–601, 2004.
14. D. Dentcheva and A. Ruszczyński. Portfolio optimization with stochastic dominance constraints. *Journal of Banking & Finance*, 30(2):433–451, 2006. Risk Management and Optimization in Finance.
15. J. Goh and M. Sim. Distributionally robust optimization and its tractable approximations. *Operations Research*, 58(4-part-1):902–917, 2010.
16. W.K.K. Haneveld and M.H. van der Vlerk. Integrated chance constraints: Reduced forms and an algorithm. *Computational Management Science*, 3:245–269, 2006.
17. A.J. King and S.W. Wallace. *Modeling with Stochastic Programming*. Springer, New York, NY, USA, 2012.
18. W. K. Klein Haneveld. On integrated chance constraints. In F. Archetti, G. Di Pillo, and M. Lucertini, editors, *Stochastic Programming*, pages 194–209. Springer Berlin Heidelberg, Berlin, Heidelberg, 1986.
19. A. Madansky. Inequalities for stochastic linear programming problems. *Management Science*, 6(2):197–204, 1960.
20. B.L. Miller and H.M. Wagner. Chance constrained programming with joint constraints. *Operations Research*, 13:930–945, 1965.
21. A. Müller and D. Stoyan. *Comparison Methods for Stochastic Models and Risks*. John Wiley & Sons, Chichester, UK, 2002.
22. M.V. Pereira and L.M. Pinto. Multi-stage stochastic optimization applied to energy planning. *Mathematical Programming*, 52.1–3:359–375, 1991.
23. W.B. Powell. *Approximate Dynamic Programming: Solving the Curses of Dimensionality*. John Wiley & Sons, Hoboken, NJ, 2nd edition, 2011.
24. W.B. Powell, H.P. Simao, and B. Bouzaiene-Ayari. Approximate dynamic programming in transportation and logistics: a unified framework. *European Journal of Transportation Logistics*, 1:237–284, 2012.

25. A. Prékopa. On probabilistic constrained programming. In Princeton University Press, editor, *Proceedings of the Princeton Symposium on Mathematical Programming*, 1970.
26. A. Prékopa. Logarithmic concave measures with application to stochastic programming. *Acta Scientiarium Mathematiarum (Szeged)*, 32:301–316, 1971.
27. A. Prékopa. Contributions to the theory of stochastic programming. *Mathematical Programming*, 4:202–221, 1973.
28. A. Prékopa. *Stochastic programming*, volume 324. Springer Science & Business Media, 2013.
29. H. Raiffa. *Decision analysis: introductory lectures on choices under uncertainty*. Addison-Wesley, 1968.
30. R.T. Rockafellar and R.J-B. Wets. Scenarios and policy aggregation in optimization under uncertainty. *Mathematics of Operations Research*, 16:119–147, 1991.
31. S. Ross. *Introduction to Stochastic processes*. Academic Press, New York, 1983.
32. A. Ruszczyn'ski and A. Shapiro, editors. *Stochastic Programming*, volume 10 of *Handbooks in Operations Research and Management Science*. Elsevier, 2003.
33. H. Scarf. A min-max solution of an inventory problem. In *Studies in the mathematical theory of inventory and production*. The RAND Corporation, Santa Monica, California, 1958.
34. A. Shapiro, D. Dentcheva, and A. Ruszczyński. *Lectures on Stochastic Programming: Modeling and Theory*. SIAM, Philadelphia, PA., 2009.
35. R. Van Slyke and R.-B. Wets. L-shaped linear programs with application to optimal control and stochastic programming. *SIAM Journal on Applied Mathematics*, 17:638–663, 1969.
36. A. Wald. *Statistical Decision Functions*. John Wiley and Sons; Chapman and Hall, London, 1950.
37. S.W. Wallace and W.T. Ziemba. *Applications of stochastic programming*. SIAM, 2005.
38. W. Wiesemann, D. Kuhn, and M. Sim. Distributionally robust convex optimization. *Operations Research*, 62(6):1358–1376, 2014.

Chapter 2
Stochastic Programming Models

2.1 Introduction

One of the greatest challenges in decision-making in engineering is to balance cost, risks, and results. Risk is the probability of loss and involves the likelihood and impact. It is the likelihood of an undesirable event occurring and the impact of that event if it occurs. An undesirable event is one that has adverse consequences when it occurs. For example, when demand for a product turns out to be lower than expected resulting in a huge loss (impact) for a company. In general, reducing risk is a continual process throughout the life-cycle of a system, and it has to be managed. Risk management is the methodology employed to identify and minimize risk in a system. In this chapter we shall focus on how to make decisions that not only minimize cost but also minimize risk by taking into account the possibility of undesirable events happening in the future. To model this mathematically, we introduce the concepts of a *risk measure* and a *risk function*, which can be incorporated into the objective of the stochastic program. A *risk measure*, for its classical meaning, is a countably additive set function. In our stochastic programming context, however, we shall first focus our attention on *risk functions*. We consider risk functions that can take values in the extended real line $\bar{\mathbb{R}} = \mathbb{R} \cup \{-\infty\} \cup \{+\infty\}$. Therefore, a *risk function* $\mathbb{D}$ is a functional from some set of real random variables $\mathscr{F}$ to the extended real numbers, where the random variables $Z \in \mathscr{F}$ represent some uncertain value (usually monetary).

Definition 2.1 A *risk function* is a functional $\mathbb{D} : \mathscr{F} \to \mathbb{R}$, which assigns to a random variable $Z \in \mathscr{F}$, a real value $\mathbb{D}(Z)$.

We assume that Ω is a certain space of elementary events whose realization is represented by the function $Z : \Omega \to \mathbb{R}$. We will let $\mathscr{F}$ be a linear space of real random variables on some probability space $(\Omega, \mathscr{A}, \mathbb{P})$. In other words, Ω is a measurable space equipped with σ-algebra $\mathscr{A}$ of subsets of Ω, $\mathbb{P}$ is a probability measure on $\mathscr{A}$, and $\mathbb{F}$ is a linear space of $\mathscr{A}$-measurable functions $Z : \Omega \to \mathbb{R}$. In

L. Ntaimo, *Computational Stochastic Programming*, Springer Optimization and Its Applications 774, https://doi.org/10.1007/978-3-031-52464-6_2

this chapter, we restrict ourselves to *one-period* risk measures, and we shall assume that *smaller* realizations for the Z values (e.g., cost) are *preferred* to larger ones. For $Z_1, Z_2 \in \mathscr{F}$, the notation $Z_1 \succeq Z_2$ will mean that $Z_1(\omega) \geq Z_2(\omega)$ for all $\omega \in \Omega$. Thus if $Z_1 \succeq Z_2$, then $\mathbb{D}(Z_1) \geq \mathbb{D}(Z_2)$ (monotonicity). We also assume that $\mathbb{D}(\emptyset) = 0$, which means that there is zero risk in taking no position.

In the literature on finance, researchers have developed a class of risk measures called *coherent risk measures* [2]. A coherent risk measure is simply a function that satisfies properties of *translation invariance*, *positive homogeneity*, *monotonicity*, and *subadditivity*. The combination of *positive homogeneity* and *subadditivity* is *sublinearity*, implying *convexity*. These properties are motivated by portfolio optimization to provide a formal way of describing common sense ideas about risk. We will use the following definition to define a coherent risk measure:

Definition 2.2 An extended real-valued function $\mathbb{D} : \mathscr{F} \mapsto \bar{\mathbb{R}}$ is said to be *coherent* if it satisfies the following conditions:

A1. *Translation invariance:* If $a \in \mathbb{R}$ and $Z \in \mathscr{F}$, then $\mathbb{D}(Z + a) = a + \mathbb{D}(Z)$.
A2. *Positive homogeneity:* If $c > 0$ and $Z \in \mathscr{F}$, then $\mathbb{D}(cZ) = c\mathbb{D}(Z)$.
A3. *Monotonicity:* If $Z_1, Z_2 \in \mathscr{F}$ and $Z_1 \leq Z_2$, then $\mathbb{D}(Z_1) \leq \mathbb{D}(Z_2)$.
A4. *Convexity:* If $Z_1, Z_2 \in \mathscr{F}$ and $\beta \in [0, 1]$, then $\mathbb{D}(\beta Z_1 + (1 - \beta)Z_2) \leq \beta\mathbb{D}(Z_1) + (1 - \beta)\mathbb{D}(Z_2)$.

A function is *convex* and *positive homogeneous* if and only if it is *subadditive*. Therefore, property (A4) can be replaced by *subadditivity*: $\mathbb{D}(Z_1 + Z_2) \leq \mathbb{D}(Z_1) + \mathbb{D}(Z_2)$. *Subadditivity* means that having a position in two different assets in a portfolio can only decrease the risk of the portfolio (diversification). *Positive homogeneity* captures the notion that doubling a position on an asset, for example, doubles your risk. *Translation invariance* means that adding cash to a portfolio, for example, only diminishes the risk. And lastly, *monotonicity* captures the notion that if one risk always has greater losses than another risk, the risk measure should always be greater. Consequently, a risk measure is said to be coherent if and only if it satisfies all the four properties.

Coherence is a convention motivated by the fact that in finance, a typical investor expects these properties to hold for a risk measure. However, for engineering applications in general, whether or not one should use a coherent risk measure should be determined in the context of the application itself. For example, in finance the focus could be to minimize expected losses above a certain target amount or quantile, whereas in an engineering design application one may be interested in minimizing deviations from a specific design target or quantile. In the case of financial problems, it becomes natural to consider a coherent risk measure that minimizes expected deviations above (one-sided) a target or quantile. In the engineering design example, however, one may want to minimize deviations both above and below the target (two-sided) and thus employ a non-coherent risk measure. In any event, the decision-maker needs to understand which of the coherence properties the risk measure being used satisfies and which ones it violates.

Turning to the risk-averse SP setting, we shall introduce mean-risk stochastic programming (MR-SP) which includes a weighted mean and dispersion statistic in the objective. We will introduce several risk measures commonly used in risk-averse stochastic programming. So from hereon, the real random variables Z will be represented by a random cost function $f(x, \tilde{\omega})$, which we define below. Throughout this chapter, we define for $a \in \mathbb{R}$, $(a)_+ = \max\{a, 0\}$, where max is the *maximum* operator. We shall denote by $|a|$ the absolute value of a, i.e., $|a| = a$ if $a \geq 0$ and $|a| = -a$ if $a < 0$.

Before we mathematically define an MR-SP, let us first define the classical risk-neutral two-stage stochastic programming (RN-SP) with recourse (recourse model), which can be written as follows:

$$\underset{x \in X}{\text{Min}}\ \mathbb{E}[f(x, \tilde{\omega})], \tag{2.1}$$

where $x \in \mathbb{R}_+^{n_1}$ is a vector of decision variables, $X := \{Ax \geq b, x \geq 0\}$ is a nonempty polyhedron that defines the set of first-stage feasible solutions, $A \in \mathbb{R}^{m_1 \times n_1}$, and $b \in \mathbb{R}^{m_1}$. The family of real random cost variables $\{f(x, \tilde{\omega})\}_{x \in X} \subseteq \mathscr{F}$ is defined on a probability space $(\Omega, \mathscr{A}, \mathbb{P})$. The mathematical operator $\mathbb{E} : \mathscr{F} \mapsto \mathbb{R}$ denotes the expected value, where $\mathscr{F}$ is the space of all real random cost variables $f : \Omega \mapsto \mathbb{R}$ satisfying $\mathbb{E}[|f(x, \tilde{\omega})|] < +\infty$ (finite first moment). For a given $x \in X$, the real random cost variable $f(x, \tilde{\omega})$ is given by

$$f(x, \tilde{\omega}) := c^\top x + \varphi(x, \tilde{\omega}). \tag{2.2}$$

For a realization (*scenario*) ω of $\tilde{\omega}$, the *recourse function* $\varphi(x, \omega)$ is given by

$$\begin{aligned} \varphi(x, \omega) := &\text{Min } q(\omega)^\top y \\ &\text{s.t. } Wy \geq r(\omega) - T(\omega)x \\ &\quad\ \ y \geq 0. \end{aligned} \tag{2.3}$$

In the above formulation $q \in \mathbb{R}^{n_2}$ is the second-stage cost vector and $y \in \mathbb{R}_+^{n_2}$ is the recourse decision. The matrix $W \in \mathbb{R}^{m_2 \times n_2}$ is the recourse matrix, $T(\omega) \in \mathbb{R}^{m_2 \times n_1}$ is the technology matrix, and $r(\omega) \in \mathbb{R}^{m_2}$ is the right hand side (RHS) vector. A *scenario* ω defines the realization of the stochastic problem data, i.e., $\omega := (q(\omega), T(\omega), r(\omega))$. If X includes binary or integer restrictions, the problem becomes a stochastic mixed-integer programming (SMIP) problem. In this chapter, we restrict our attention to SLP and deal with SMIP in a later chapter.

Modeling problems using only the expectation in the objective makes the formulation risk-neutral. In this chapter, we first state the properties of the risk-neutral stochastic programs and then present the properties for the risk-averse cases. To introduce risk, a *risk measure* $\mathbb{D} : \mathscr{F} \mapsto \mathbb{R}$ is added to Problem (2.1) resulting in the following MR-SP:

$$\text{MR-SP} : \underset{x \in X}{\text{Min}} \ \mathbb{E}[f(x, \tilde{\omega})] + \lambda \mathbb{D}[f(x, \tilde{\omega})], \tag{2.4}$$

where $\lambda > 0$ is a suitable weight factor that quantifies the trade-off between expected cost and risk. Risk measure $\mathbb{D}$ is chosen so that the problem remains a *convex optimization problem*, allowing it to have access to the readily available convex optimization methods. The two main classes of mean-risk measures in the literature are *quantile* and *deviation* risk measures. Quantile risk measures, as the name implies, are based on some quantile, while deviation risk measures are based on the deviation from some target. For computational purposes, it is desirable for the risk measures to be convexity preserving. Furthermore, we can make the following additional assumptions to ensure that Problem (2.4) is well-defined:

(A1) The multivariate random variable $\tilde{\omega}$ is discretely distributed with finitely many scenarios $\omega \in \Omega$, each with the probability of occurrence $p(\omega)$.
(A2) $\{Wy(\omega) \geq r(\omega) - T(\omega)x, \ y(\omega) \geq 0\} \neq \emptyset$ for all $x \in X$.

Assumption (A1) is needed to make the problem tractable, while assumption (A2) is the *relatively complete recourse* assumption that guarantees the feasibility of the second-stage problem for every $x \in X$. In other words, Problem (2.4) has relatively complete recourse if $\varphi(x, \omega) < \infty$ with probability one for all $x \in X$. Problem (2.4) is said to have *complete recourse* if $\varphi(x, \omega) < \infty$ with probability one for all $x \in \mathbb{R}^{n_1}$. If assumption (A2) cannot be guaranteed, then by convention we have $\varphi(x, \omega) = +\infty$ if Problem (2.3) is infeasible and $\varphi(x, \omega) = -\infty$ if Problem (2.3) is unbounded.

2.2 Risk-Neutral Models

The two main ways of formulating the two-stage RN-SP are *stage-wise* and *scenario-wise* formulation. The stage-wise formulation states the problem in terms of the first stage with decision variable x and the second stage with decision variable $y(\omega)$ for a given scenario $\omega \in \Omega$. The scenario-wise formulation approach on the other hand introduces the first-stage decision variable $x(\omega)$ and the second-stage decision variable $y(\omega)$ for a given scenario $\omega \in \Omega$. To enforce the requirement that all the first-stage decisions x(ω) are the same, *nonanticipativity* constraints are added to the formulation. Next, we state the stage-wise formulation and then present some important structural properties in the next subsection. We present the scenario formulation in the subsequent subsection.

Let $Q(x) := \mathbb{E}[\varphi(x, \tilde{\omega})]$ denote the *expected recourse function* so that we can rewrite Problem (2.1) as follows:

$$\begin{aligned}\text{RN-SLP : Min } & c^\top x + \mathcal{Q}(x) \\ \text{s.t. } & Ax \geq b \\ & x \geq 0,\end{aligned} \tag{2.5}$$

where for each realization $\omega \in \Omega$ of $\tilde{\omega}$, the recourse function $\varphi(x, \omega)$ is given by

$$\begin{aligned}\varphi(x, \omega) ::= \text{ Min } & q(\omega)^\top y(\omega) \\ \text{s.t. } & Wy(\omega) \geq r(\omega) - T(\omega)x \quad \leftarrow \quad \pi(\omega) \\ & y(\omega) \geq 0.\end{aligned} \tag{2.6}$$

Assuming assumptions A1–A3 hold, let $\pi(\omega)$ denote the dual multipliers associated with the constraints of subproblem (2.6) for scenario ω. Then by LP *strong duality*, at optimality we have

$$\begin{aligned}\varphi(x, \omega) := \text{ Max } & \pi(\omega)^\top (r(\omega) - T(\omega)x) \\ \text{s.t. } & W^\top \pi(\omega) \leq q(\omega) \\ & \pi(\omega) \geq 0.\end{aligned} \tag{2.7}$$

Let $\Pi(\omega)$ denote the dual feasible set, defined as follows:

$$\Pi(\omega) := \{\pi(\omega) \mid W^\top \pi(\omega) \leq q(\omega)\}.$$

Observe that if W has fixed recourse and $q(\omega) = q$ for all $\omega \in \Omega$, the set $\Pi(\omega) = \Pi$ and remains fixed regardless of the choice of x.

2.2.1 Structural Properties

Let us now look at some important structural properties of the RN-SLP formulation. These properties will be used in later chapters for devising solution methods for RN-SLP. We begin with the following important convexity result:

Theorem 2.1 *The objective function $F(x)$ of a two-stage SLP is convex over its effective domain $X_E := \{x \in X \mid \varphi(x, \tilde{\omega}) < \infty\}$.*

Proof Let $x^1, x^2 \in X_E$, and for any $\lambda \in [0, 1]$ define $x_\lambda = \lambda x^1 + (1 - \lambda)x^2$. Then

$$\begin{aligned}F(x_\lambda) &= c^\top x_\lambda + \mathbb{E}[\varphi(x_\lambda, \tilde{\omega}] \\ &= c^\top x_\lambda + \sum_{\omega \in \Omega} p(\omega)\varphi(x_\lambda, \omega)\end{aligned}$$

$$
\begin{aligned}
&= c^\top x_\lambda + \sum_{\omega\in\Omega} p(\omega) \operatorname*{Max}_{\pi(\omega)\in\Pi(\omega)} \left(\pi(\omega)^\top (r(\omega) - T(\omega)x_\lambda)\right) \\
&= \lambda c^\top x_1 + (1-\lambda)c^\top x_2 + \\
&\qquad \sum_{\omega\in\Omega} p(\omega) \operatorname*{Max}_{\pi(\omega)\in\Pi(\omega)} (\lambda\pi(\omega)^\top (r(\omega) - T(\omega)x^1) + \\
&\qquad (1-\lambda)\pi(\omega)^\top (r(\omega) - T(\omega)x^2)) \\
&\le \lambda c^\top x_1 + (1-\lambda)c^\top x_2 + \\
&\qquad \lambda \sum_{\omega\in\Omega} p(\omega) \operatorname*{Max}_{\pi(\omega)\in\Pi(\omega)} \left(\pi(\omega)^\top (r(\omega) - T(\omega)x^1)\right) \\
&\qquad + (1-\lambda) \sum_{\omega\in\Omega} p(\omega) \operatorname*{Max}_{\pi(\omega)\in\Pi(\omega)} \left(\pi(\omega)^\top (r(\omega) - T(\omega)x^2)\right) \\
&= \lambda(c^\top x_1 + \mathbb{E}[\varphi(x^1,\tilde\omega)]) + (1-\lambda)(c^\top x_2 + \mathbb{E}[\varphi(x^2,\tilde\omega)]) \\
\Rightarrow \quad & F(x_\lambda) \le \lambda F(x^1) + (1-\lambda)F(x^2).
\end{aligned}
$$

■

Theorem 2.1 means that we can solve Problem (2.5) by approximating the convex *expected recourse function* using supporting hyperplanes (see Chap. 4).

The next important property of Problem (2.5) is *Lipschitz continuity* of the expected recourse function $\mathcal{Q}(x)$. Let the *extreme points* (vertices) of the set of dual feasible multipliers $\Pi(\omega)$ be denoted by $\Pi_E(\omega)$.

Theorem 2.2 *Suppose that Problem* (2.5) *has relatively complete recourse, i.e., $\varphi(x,\tilde\omega) < \infty$ with probability one for all $x \in X$ and that $\mathbb{E}[\|T(\tilde\omega)\|] < \infty$. Furthermore, suppose that the set $\Pi(\omega)$ is nonempty for all $\omega \in \Omega$. Then*

(a) For every $\omega \in \Omega$ outside a set of probability zero, there exists $M(\omega) < \infty$, such that $\mathbb{E}[M(\tilde\omega)] < \infty$, and $|\varphi(x^1,\omega) - \varphi(x^2,\omega)| \le M(\omega)\|x^1 - x^2\|$.

(b) There exists $M < \infty$ such that $|\varphi(x^1,\omega) - \varphi(x^2,\omega)| \le M\|x^1 - x^2\| \quad \forall x^1, x^2 \in X$.

Proof

(a) Our supposition ensures that both the primal and dual subproblems are feasible for all $x \in X$ with probability one. Therefore, it follows that for all realizations ω outside a set of $\mathbb{P}$-measure zero

$$\varphi(x,\omega) = \operatorname*{Max}_{\pi(\omega)\in\Pi(\omega)} \pi(\omega)^\top (r(\omega) - T(\omega)x), \quad \forall x \in X.$$

Given $x^1,\ x^2 \in X$, let

$$\pi^i(\omega) \in \operatorname{argmax}\{\pi(\omega)^\top (r(\omega) - T(\omega)x^i) \mid \pi(\omega) \in \Pi_E(\omega)\},\ i = 1, 2.$$

Given that the recourse function $\varphi(., \omega)$ is convex, then the following *subgradient inequalities* hold:

$$\varphi(x^1, \omega) - \varphi(x^2, \omega) \leq \pi^1(\omega)^\top T(\omega)(x^2 - x^1)$$

and

$$\varphi(x^2, \omega) - \varphi(x^1, \omega) \leq \pi^2(\omega)^\top T(\omega)(x^1 - x^2).$$

Combining these two inequalities, we obtain

$$-\pi^2(\omega)^\top T(\omega)(x^1 - x^2) \leq \varphi(x^1, \omega) - \varphi(x^2, \omega) \leq -\pi^1(\omega)^\top T(\omega)(x^1 - x^2). \tag{2.8}$$

Thus if we define

$$M(\omega) = \max\{\|\pi(\omega)\| \mid \pi(\omega) \in \Pi_E(\omega)\}\|T(\omega)\|, \tag{2.9}$$

it follows that $\mathbb{E}[M(\tilde{\omega})] = \mathbb{E}[\max\{\|\pi(\tilde{\omega})\| \mid \pi(\tilde{\omega}) \in \Pi_E(\tilde{\omega})\}]\mathbb{E}[\|T(\tilde{\omega})\|] < \infty$. Hence

$$\begin{aligned}
|\varphi(x^1, \omega) - \varphi(x^2, \omega)| &\leq \max\{\pi^1(\omega)^\top T(\omega)(x^1 - x^2),\ \pi^2(\omega)^\top T(\omega)(x^1 - x^2)\} \\
&\leq \max\{\|\pi(\omega)^\top T(\omega)\| \mid \pi(\omega) \in \Pi_E(\omega)\}\|x^1 - x^2\| \\
&\leq \max\{\|\pi(\omega)\| \mid \pi(\omega) \in \Pi_E(\omega)\}\|T(\omega)\|\|x^1 - x^2\| \\
&= M(\omega)\|x^1 - x^2\|
\end{aligned}$$

almost everywhere with respect to the measure $\mathbb{P}$.

(b) Since $\mathcal{Q}(x) = \mathbb{E}[\varphi(x, \tilde{\omega})]$, the inequality (2.8) yields

$$-\mathbb{E}[\pi^2(\tilde{\omega})^\top T(\omega)(x^1 - x^2)] \leq \varphi(x^1) - \varphi(x^2) \leq -\mathbb{E}[\pi^1(\tilde{\omega})^\top T(\omega)(x^1 - x^2)]. \tag{2.10}$$

Setting $M = \mathbb{E}[M(\tilde{\omega})] \equiv \mathbb{E}[\max\{\|\pi(\tilde{\omega})\| \mid \pi(\tilde{\omega}) \in \Pi_E(\tilde{\omega})\}]\mathbb{E}[\|T(\tilde{\omega})\|]$, the result follows from part (a).

■

Another important result in risk-neutral two-stage SLP with recourse is *Jensen's inequality* [10] applied to the expected recourse function. Jensen's inequality states that for any convex function, say $h(\tilde{\omega})$ of $\tilde{\omega}$, $\mathbb{E}[h(\tilde{\omega})] \geq h(\mathbb{E}[\tilde{\omega}])$. Let $q(\omega) = q$ for all $\omega \in \Omega$ so that the set of dual feasible multipliers $\Pi(\omega) = \Pi$. In our context, *Jensen inequality* shows that replacing the random variables $r(\tilde{\omega})$ and $T(\tilde{\omega})$ by the expectations, denoted $\bar{r}$ and $\bar{T}$, respectively, results in *underestimating* the expected recourse function. This is known as *Jensen's lower bound* and can be formally stated as follows:

Theorem 2.3 $Q(x) = \mathbb{E}[\varphi(x, \tilde{\omega})] \geq \mathrm{Max}_{\pi\in\Pi}\pi(\bar{r} - \bar{T}x)$ *for all* x.

Proof We first consider two trivial cases. If $\Pi = \emptyset$, by convention the right hand side of the inequality is $-\infty$, and hence the inequality holds. Also, if $\mathbb{E}[\varphi(x, \tilde{\omega})] = +\infty$, then the inequality holds trivially. Therefore, let us assume that $\Pi \neq \emptyset$ and that $\mathbb{E}[\varphi(x, \tilde{\omega})] < \infty$. Then using LP duality, we have

$$
\begin{aligned}
\mathbb{E}[\varphi(x, \tilde{\omega})] &= \sum_{\omega\in\Omega} p(\omega) \operatorname*{Max}_{\pi\in\Pi} \pi(\bar{r} - \bar{T}x) \\
&\geq \sum_{\omega\in\Omega} p(\omega)\pi(\bar{r} - \bar{T}x), \quad \forall\pi \in \Pi.
\end{aligned}
$$

Hence, it follows that

$$
\begin{aligned}
\mathbb{E}[\varphi(x, \tilde{\omega})] &\geq \operatorname*{Max}_{\pi\in\Pi} \sum_{\omega\in\Omega} p(\omega)\pi(\bar{r} - \bar{T}x) \\
&= \operatorname*{Max}_{\pi\in\Pi} \pi(\bar{r} - \bar{T}x).
\end{aligned}
$$

■

By imposing further assumptions on the stochastic problem data, we obtain the following useful case of *Jensen's lower bound*:

Corollary 2.1 *If* $r(\omega)$ *and* $T(\omega)$ *are linear function of* ω, *we can write* $\bar{r} = r(\mathbb{E}[\tilde{\omega}])$ *and* $\bar{T} = T(\mathbb{E}[\tilde{\omega}])$, *and then it follows that* $Q(x) = \mathbb{E}[\varphi(x, \tilde{\omega})] \leq \varphi(x, \mathbb{E}[\tilde{\omega}])$.

This means that replacing the random variables $r(\tilde{\omega})$ and $T(\tilde{\omega})$ by their expected values will, in general, result in *underestimation* of the expected recourse function. Next, we present the scenario formulation of the RN-SLP.

2.2.2 *Scenario Formulation*

To write the scenario formulation of the RN-SLP, let us introduce copies of the first-stage decision variable x, $x(\omega)$ for each $\omega \in \Omega$, and then add *nonanticipativity* constraints to guarantee that $x(\omega)$ is the same for all $\omega \in \Omega$. There are three main ways of stating the *nonanticipativity* constraints. The first way is to simply add the following set of constraints:

$$x(\omega) = x, \ \forall\omega \in \Omega.$$

The second is the so-called *cyclic* nonanticipativity constraints. In this case index the ω's as $\omega^1, \omega^2, \cdots, \omega^{|\Omega|}$ and then add the following constraints:

$$x(\omega^1) = x(\omega^2),\ x(\omega^2) = x(\omega^3),\ \cdots,\ x(\omega^{|\Omega|-1}) = x(\omega^{|\Omega|}).$$

The third way is the *expectation* nonanticipativity constraints given as follows:

$$x(\omega) = \sum_{\varpi \in \Omega} p(\varpi)x(\varpi), \ \forall \omega \in \Omega.$$

The type of which nonanticipativity constraints to use is the modeler's choice and often depends on the solution method being employed.

Using *expectation nonanticipativity* constraints, RN-SLP (2.5) can be reformulated using the scenario formulation as follows:

$$\begin{aligned}
\text{Min} \quad & \sum_{\omega \in \Omega} p(\omega)\left(c^\top x(\omega) + q(\omega)^\top y(\omega)\right) \\
\text{s.t.} \quad & Ax(\omega) \geq b, \ \forall \omega \in \Omega \\
& T(\omega)x(\omega) + Wy(\omega) \geq r(\omega), \ \forall \omega \in \Omega \\
& x(\omega) = \sum_{\varpi \in \Omega} p(\varpi)x(\varpi), \ \forall \omega \in \Omega \\
& x, x(\omega), \ y(\omega) \geq 0, \ \forall \omega \in \Omega.
\end{aligned} \tag{2.11}$$

Observe that the nonanticipativity constraints (2.11) link all scenarios and are thus considered *complicating* constraints. Thus decomposition methods for problem are based on *Lagrangian relaxation* or dual decomposition, which involves relaxing the linking constraints by placing them in the objective and penalizing them. We cover this topic in Chap. 5. Next, we turn to the risk-averse setting.

2.3 Mean-Risk Models

Let us now turn to the risk-averse setting and define some of the common risk measures used for MR-SPs. We first deal with quantile risk measures and then move on to deviation risk measures.

2.3.1 *Quantile Risk Measures*

There are several quantile risk measures that have been introduced in the literature on stochastic programming. However, we shall focus only on the following commonly used quantile risk measures for MR-SPs: excess probability (EP), quantile deviation (QDEV), and conditional value-at-risk (CVaR).

Excess Probability

Given a target level $\eta \in \mathbb{R}$ and $x \in X$, excess EP [30] reflects the probability of exceeding the target η. It is given by

$$\phi_{EP_\eta}(x) := \mathbb{P}(\omega \in \Omega : f(x, \omega) > \eta).$$

The function $\phi_{EP_\eta}(x)$ gives the sum of the probabilities for all ω's exceeding the target η. Substituting $\mathbb{D} := \phi_{EP_\eta}$ in Problem (2.4) with $\lambda \geq 0$, we obtain the following MR-SP with *excess probability*:

$$\operatorname*{Min}_{x \in X} \mathbb{E}[f(x, \tilde{\omega})] + \lambda \phi_{EP_\eta}(x). \tag{2.12}$$

Quantile Deviation

Given $x \in X$ and $\alpha \in (0, 1)$, QDEV can be defined as follows [21]:

$$\phi_{QDEV_\alpha}(x) := \mathbb{E}[(1-\alpha)(\kappa_\alpha[f(x, \tilde{\omega})] - f(x, \tilde{\omega}))_+ + \alpha(f(x, \tilde{\omega}) - \kappa_\alpha[f(x, \tilde{\omega})])_+],$$

where $\kappa_\alpha[f(x, \tilde{\omega})]$ is the α-quantile of the cumulative distribution of $f(x, \tilde{\omega})$. This means that $\mathbb{P}(f(x, \tilde{\omega}) \leq \kappa_\alpha[f(x, \tilde{\omega})]) \geq \alpha$ and $\mathbb{P}(f(x, \tilde{\omega}) \geq \kappa_\alpha[f(x, \tilde{\omega})]) \geq 1-\alpha$. Therefore, for constants $\varepsilon_1 > 0$ and $\varepsilon_2 > 0$, we have $\alpha = \varepsilon_2/(\varepsilon_1 + \varepsilon_2) = 1 - \varepsilon_1/(\varepsilon_1 + \varepsilon_2)$. Risk measure QDEV is two-sided and reflects the expectation of the deviation above and below the α-quantile of the cumulative distribution of $f(x, \tilde{\omega})$. The given expression for $\phi_{QDEV_\alpha}(x)$ is *not* amenable to optimization, and we thus need to rewrite it in a form involving the α-quantile of the cumulative distribution of $f(x, \tilde{\omega})$. An equivalent expression of ϕ_{QDEV_α} can now be given as follows [21, 28]:

$$\phi_{QDEV_\alpha}(x) \equiv \phi_{QDEV_{\varepsilon_1,\varepsilon_2}}(x) := \operatorname*{Min}_{\eta \in \mathbb{R}} \mathbb{E}[\varepsilon_1(\eta - f(x, \tilde{\omega}))_+ + \varepsilon_2(f(x, \tilde{\omega}) - \eta)_+].$$

Note that the minimum is attained at some η which is the α-quantile of the cumulative distribution of $f(x, \tilde{\omega})$. We can now write an MR-SP with $\mathbb{D} := \phi_{QDEV_\alpha}$ and $\lambda \geq 0$ as follows:

$$\operatorname*{Min}_{x \in X} \mathbb{E}[f(x, \tilde{\omega})] + \lambda \phi_{QDEV_\alpha}(x). \tag{2.13}$$

Conditional Value-at-Risk

Given $x \in X$ and $\alpha \in (0, 1)$, CVaR [26] can be expressed as follows:

$$\phi_{CVaR_\alpha}(x) := \operatorname*{Min}_{\eta \in \mathbb{R}} \{\eta + \frac{1}{1-\alpha} \mathbb{E}[(f(x, \tilde{\omega}) - \eta)_+]\}.$$

CVaR reflects the expectation of the $(1-\alpha).100\%$ worst realizations for a given probability level $\alpha \in (0, 1)$. Alternatively, CVaR can be expressed in terms of $\phi_{QDEV_{\varepsilon_1,\varepsilon_2}}(x)$, and we can derive the expression as follows: First, rewrite the expression for $\phi_{CVaR_\alpha}(x)$ as follows:

$$\phi_{CVaR_\alpha}(x) := \underset{\eta \in \mathbb{R}}{\text{Min}} \{\mathbb{E}[\eta + \frac{1}{1-\alpha}(f(x,\tilde{\omega}) - \eta)_+]\}.$$

Observe that $\eta = f(x,\tilde{\omega}) + (\eta - f(x,\tilde{\omega}))_+ - (f(x,\tilde{\omega}) - \eta)_+$. Second, substitute this expression of η in the above expression for $\phi_{CVaR_\alpha}(x)$ to get the following result:

$$\begin{aligned}
\phi_{CVaR_\alpha}(x) := &\underset{\eta \in \mathbb{R}}{\text{Min}} \{\mathbb{E}[f(x,\tilde{\omega}) + (\eta - f(x,\tilde{\omega}))_+ - (f(x,\tilde{\omega}) - \eta)_+ \\
&+ \frac{1}{1-\alpha}(f(x,\tilde{\omega}) - \eta)_+]\} \\
= &\underset{\eta \in \mathbb{R}}{\text{Min}} \{\mathbb{E}[f(x,\tilde{\omega}) + (\eta - f(x,\tilde{\omega}))_+ + \frac{\alpha}{1-\alpha}(f(x,\tilde{\omega}) - \eta)_+]\} \\
= &\mathbb{E}[f(x,\tilde{\omega})] + \underset{\eta \in \mathbb{R}}{\text{Min}} \{\mathbb{E}[(\eta - f(x,\tilde{\omega}))_+ + \frac{\alpha}{1-\alpha}(f(x,\tilde{\omega}) - \eta)_+]\}.
\end{aligned}$$

Since we have $\alpha = \frac{\varepsilon_2}{\varepsilon_1+\varepsilon_2}$, then $\frac{\alpha}{1-\alpha} = \frac{\varepsilon_2}{\varepsilon_1}$. Substituting this in the above expression, we get the following:

$$\begin{aligned}
\phi_{CVaR_\alpha}(x) := &\mathbb{E}[f(x,\tilde{\omega})] + \underset{\eta \in \mathbb{R}}{\text{Min}} \{\mathbb{E}[(\eta - f(x,\tilde{\omega}))_+ + \frac{\varepsilon_2}{\varepsilon_1}(f(x,\tilde{\omega}) - \eta)_+]\} \\
= &\mathbb{E}[f(x,\tilde{\omega})] + \frac{1}{\varepsilon_1} \underset{\eta \in \mathbb{R}}{\text{Min}} \{\mathbb{E}[\varepsilon_1(\eta - f(x,\tilde{\omega}))_+ + \varepsilon_2(f(x,\tilde{\omega}) - \eta)_+]\}.
\end{aligned}$$

CVaR can now be expressed as follows [28]:

$$\phi_{CVaR_\alpha}(x) := \mathbb{E}[f(x,\tilde{\omega})] + \frac{1}{\varepsilon_1}\phi_{QDEV_{\varepsilon_1,\varepsilon_2}}(x). \tag{2.14}$$

For any $\lambda \geq 0$, an MR-SP with $\mathbb{D} := \phi_{CVaR_\alpha}$ can be given as follows:

$$\underset{x \in X}{\text{Min}}\ \mathbb{E}[f(x,\tilde{\omega})] + \lambda\phi_{CVaR_\alpha}(x). \tag{2.15}$$

An alternative to this formulation, for $0 \leq \lambda \leq 1$, is given as follows:

$$\underset{x \in X}{\text{Min}}\ (1-\lambda)\mathbb{E}[f(x,\tilde{\omega})] + \lambda\phi_{CVaR_\alpha}(x). \tag{2.16}$$

Notice that this model has a *convex combination* of the mean and the CVaR dispersion statistic. Unlike model (2.15), we show in the next subsection that model (2.16) is coherent.

2.3.2 Deviation Risk Measures

Several deviation risk measures have been defined in the literature. The commonly used deviation risk measures for MR-SPs are expected excess (EE), absolute semideviation (ASD), and central deviation (CDEV). We will study each of these risk measures in detail next, starting with EE.

Expected Excess

Given $x \in X$, a target $\eta \in \mathbb{R}$, and $\lambda \geq 0$, EE [17] is given as

$$\phi_{EE_\eta}(x) := \mathbb{E}[f(x, \tilde{\omega}) - \eta]_+.$$

It reflects the expected value of the excess over the target $\eta \in \mathbb{R}$. Substituting $\mathbb{D} := \phi_{EE_\eta}$ in Problem (2.4), we obtain an MR-SP with EE as follows:

$$\operatorname*{Min}_{x \in X} \; \mathbb{E}[f(x, \tilde{\omega})] + \lambda \phi_{EE_\eta}(x). \tag{2.17}$$

The mean-risk objective function for EE can be expressed as follows:

$$\begin{aligned}
&\mathbb{E}[f(x, \tilde{\omega})] + \lambda \phi_{EE}(x) \\
&\quad = \mathbb{E}[f(x, \tilde{\omega})] + \lambda \mathbb{E}[f(x, \tilde{\omega}) - \eta]_+ \\
&\quad = \mathbb{E}[f(x, \tilde{\omega})] + \lambda \mathbb{E}[\max\{f(x, \tilde{\omega}), \eta\}] - \lambda \eta.
\end{aligned} \tag{2.18}$$

Absolute Semideviation

ASD is defined in the same way as EE but with the target replaced by the mean value $\mathbb{E}[f(x, \tilde{\omega})]$, with $0 \leq \lambda \leq 1$. It can be written as follows [21]:

$$\phi_{ASD}(x) := \mathbb{E}[[f(x, \tilde{\omega}) - \mathbb{E}[f(x, \tilde{\omega})]]_+].$$

It reflects the expected value of the excess over the mean value. By setting $\mathbb{D} := \phi_{ASD}$ in Problem (2.4), we obtain the following MR-SP with ASD:

$$\operatorname*{Min}_{x \in X} \; \mathbb{E}[f(x, \tilde{\omega})] + \lambda \phi_{ASD}(x). \tag{2.19}$$

The mean-risk objective function for ASD can be written in the following form:

$$\begin{aligned}
&\mathbb{E}[f(x, \tilde{\omega})] + \lambda \phi_{ASD}(x) \\
&\quad = \mathbb{E}[f(x, \tilde{\omega})] + \lambda \mathbb{E}[[f(x, \tilde{\omega}) - \mathbb{E}[f(x, \tilde{\omega})]]_+].
\end{aligned}$$

$$= \mathbb{E}[f(x, \tilde{\omega})] + \lambda(\mathbb{E}[\max\{f(x, \tilde{\omega}), \mathbb{E}[f(x, \tilde{\omega})]\}] - \mathbb{E}[f(x, \tilde{\omega})])$$
$$= (1 - \lambda)\mathbb{E}[f(x, \tilde{\omega})] + \lambda\mathbb{E}[\max\{f(x, \tilde{\omega}), \mathbb{E}[f(x, \tilde{\omega})]\}]. \quad (2.20)$$

Central Deviation

CDEV is a two-sided deviation measure with the mean value as the target like in ASD. Given $x \in X$ and $0 \le \lambda \le \frac{1}{2}$, CDEV is given as follows [17]:

$$\phi_{CDEV}(x) := \mathbb{E}[|f(x, \tilde{\omega}) - \mathbb{E}[f(x, \tilde{\omega})]|].$$

CDEV reflects the expectation of the sum of the excess above the mean value and the shortfall below the mean value. By setting $\mathbb{D} := \phi_{CDEV}$ in Problem (2.4), we obtain the following MR-SP with CD:

$$\underset{x \in X}{\text{Min}}\ \mathbb{E}[f(x, \tilde{\omega})] + \lambda\phi_{CDEV}(x). \quad (2.21)$$

The mean-risk objective function for CDEV can be expressed as follows:

$$\begin{aligned}
&\mathbb{E}[f(x, \tilde{\omega})] + \lambda\phi_{CDEV}(x) \\
&\quad = \mathbb{E}[f(x, \tilde{\omega})] + \lambda\mathbb{E}[|f(x, \tilde{\omega}) - \mathbb{E}[f(x, \tilde{\omega})]|] \\
&\quad = \mathbb{E}[f(x, \tilde{\omega})] + \lambda(\mathbb{E}[[f(x, \tilde{\omega}) - \mathbb{E}[f(x, \tilde{\omega})]]_+] + \mathbb{E}[[\mathbb{E}[f(x, \tilde{\omega})] \\
&\qquad - f(x, \tilde{\omega})]_+]) \\
&\quad = \mathbb{E}[f(x, \tilde{\omega})] + \lambda(\mathbb{E}[\max\{f(x, \tilde{\omega}), \mathbb{E}[f(x, \tilde{\omega})]\}] - \mathbb{E}[f(x, \tilde{\omega})] + \\
&\qquad \mathbb{E}[\max\{\mathbb{E}[f(x, \tilde{\omega})], f(x, \tilde{\omega})\} - f(x, \tilde{\omega})]) \\
&\quad = (1 - 2\lambda)\mathbb{E}[f(x, \tilde{\omega})] + 2\lambda\mathbb{E}[\max\{f(x, \tilde{\omega}), \mathbb{E}[f(x, \tilde{\omega})]\}].
\end{aligned}$$

We illustrate how to prove the coherence properties of the risk measures in the next section.

2.4 Checking Coherence Properties of a Risk Measure

Given a risk measure, it is important to know its properties before it is used in an MR-SP. In particular, it is desirable for computational purposes to have a risk measure that is convex so that the resulting optimization problem is convex. This means we can draw from a wide range of convex optimization approaches that are available. We would also like to know which coherence properties our risk measure has and which ones it violates. In this subsection, we use one quantile measure (CVaR) and one deviation measure (EE) to illustrate how to prove whether or not the risk measure is coherent. We also prove coherence properties for MR-SPs with CVaR and with EE. Proving coherence for the rest of the quantile and deviation risk measures is left as an exercise for the reader and is given at the end of this chapter.

2.4.1 Example: Conditional Value-at-Risk

Example 2.1 (CVaR) Recall that for a given $x \in X$ and $\alpha \in (0, 1)$, CVaR is given as

$$\phi_{CVaR_\alpha}(x) := \underset{\eta \in \mathbb{R}}{\text{Min}} \{\eta + \frac{1}{1-\alpha}\mathbb{E}[(f(x, \tilde{\omega}) - \eta)_+]\}.$$

A1. Translation invariance Given $a \in \mathbb{R}$, we have

$$\underset{\eta \in \mathbb{R}}{\text{Min}} \{\eta + \frac{1}{1-\alpha}\mathbb{E}[(f(x, \tilde{\omega}) + a - \eta)_+]\}.$$

Now performing a decision variable substitution $\eta' = \eta - a$ in this expression, we get

$$\begin{aligned}
&\underset{\eta \in \mathbb{R}}{\text{Min}} \{\eta + \frac{1}{1-\alpha}\mathbb{E}[(f(x, \tilde{\omega}) + a - \eta)_+]\\
&= \underset{\eta' \in \mathbb{R}}{\text{Min}} \{\eta' + a + \frac{1}{1-\alpha}\mathbb{E}[(f(x, \tilde{\omega}) - \eta')_+]\}\\
&= a + \underset{\eta' \in \mathbb{R}}{\text{Min}} \{\eta' + \frac{1}{1-\alpha}\mathbb{E}[(f(x, \tilde{\omega}) - \eta')_+]\}.
\end{aligned}$$

Observe that the last term is equivalent to

$$\underset{\eta \in \mathbb{R}}{\text{Min}} \{\eta + \frac{1}{1-\alpha}\mathbb{E}[(f(x, \tilde{\omega}) - \eta)_+]\},$$

showing that CVaR satisfies the translation invariance property.

A2. Positive homogeneity Given $c > 0$, we have

$$\begin{aligned}
\underset{\eta \in \mathbb{R}}{\text{Min}} \{\eta + \frac{1}{1-\alpha}\mathbb{E}[(cf(x, \tilde{\omega}) - \eta)_+]\} &= \underset{\eta \in \mathbb{R}}{\text{Min}} \{\eta + \frac{c}{1-\alpha}\mathbb{E}[(f(x, \tilde{\omega}) - \frac{\eta}{c})_+]\}.\\
&= c\,\underset{\eta \in \mathbb{R}}{\text{Min}} \{\frac{\eta}{c} + \frac{1}{1-\alpha}\mathbb{E}[(f(x, \tilde{\omega}) - \frac{\eta}{c})_+]\}.\\
&= c\,\underset{\eta' \in \mathbb{R}}{\text{Min}} \{\eta' + \frac{1}{1-\alpha}\mathbb{E}[(f(x, \tilde{\omega}) - \eta')_+]\},
\end{aligned}$$

where $\eta' = \frac{\eta}{c}$ is simply a change of variable, thus showing that CVaR satisfies the property of positive homogeneity.

A3. Monotonicity Let $f(x, \tilde{\omega}_1), f(x, \tilde{\omega}_2) \in \mathscr{F}$ such that $f(x, \tilde{\omega}_1) \leq f(x, \tilde{\omega}_2)$. Then $f(x, \tilde{\omega}_1) - \eta \leq f(x, \tilde{\omega}_2) - \eta$, which implies that

$$\underset{\eta \in \mathbb{R}}{\text{Min}} \{\eta + \frac{1}{1-\alpha}\mathbb{E}[(f(x, \tilde{\omega}_1) - \eta)_+]\} \leq \underset{\eta \in \mathbb{R}}{\text{Min}} \{\eta + \frac{1}{1-\alpha}\mathbb{E}[(f(x, \tilde{\omega}_2) - \eta)_+]\},$$

showing that CVaR satisfies the monotonicity property.

A4. Convexity Let $f(x, \tilde{\omega}_1), f(x, \tilde{\omega}_2) \in \mathscr{F}$, and for any $\beta \in [0, 1]$ define $f(x, \tilde{\omega}_\beta) = \beta f(x, \tilde{\omega}_1) + (1 - \beta) f(x, \tilde{\omega}_2)$. Then
$\text{Min}_{\eta \in \mathbb{R}} \{\eta + \frac{1}{1-\alpha}\mathbb{E}[(f(x, \tilde{\omega}_\beta) - \eta)_+]\}$

$$= \underset{\eta \in \mathbb{R}}{\text{Min}} \{\eta + \frac{1}{1-\alpha}\mathbb{E}[(\beta f(x, \tilde{\omega}_1) + (1 - \beta) f(x, \tilde{\omega}_2) - \eta)_+]\}$$

$$= \underset{\eta \in \mathbb{R}}{\text{Min}} \{\eta + \beta\eta - \beta\eta + \frac{1}{1-\alpha}\mathbb{E}[(\beta f(x, \tilde{\omega}_1) - \beta\eta + (1-\beta) f(x, \tilde{\omega}_2) - \eta + \beta\eta)_+]\}$$

$$= \underset{\eta \in \mathbb{R}}{\text{Min}} \{\beta\eta + (1 - \beta)\eta + \frac{1}{1-\alpha}\mathbb{E}[(\beta(f(x, \tilde{\omega}_1) - \eta) + (1 - \beta)(f(x, \tilde{\omega}_2) - \eta)_+]\}$$

$$\leq \beta \underset{\eta \in \mathbb{R}}{\text{Min}} \{\eta + \frac{1}{1-\alpha}\mathbb{E}[(f(x, \tilde{\omega}_1) - \eta)_+]\} + (1 - \beta) \underset{\eta \in \mathbb{R}}{\text{Min}} \{\eta +$$

$$\frac{1}{1-\alpha}\mathbb{E}[(f(x, \tilde{\omega}_2) - \eta)_+]\}.$$

This shows that CVaR satisfies the convexity property. Overall, we have just proved that CVaR is a coherent risk measure as it satisfies all the four axioms for a coherent risk measure.

2.4.2 Example: Mean-Risk Conditional Value-at-Risk

Example 2.2 (Mean-Risk with CVaR) Let us now consider the MR-SP with CVaR (2.15) for any $\lambda \geq 0$:

$$\underset{x \in X}{\text{Min}} \ \mathbb{E}[f(x, \tilde{\omega})] + \lambda \phi_{CVaR_\alpha}(x),$$

where

$$\phi_{CVaR_\alpha}(x) := \underset{\eta \in \mathbb{R}}{\text{Min}} \{\eta + \frac{1}{1-\alpha}\mathbb{E}[(f(x, \tilde{\omega}) - \eta)_+]\}.$$

A1. Translation invariance Given $a \in \mathbb{R}$, we have

$$\mathbb{E}[a + f(x, \tilde{\omega})] + \lambda \operatorname*{Min}_{\eta \in \mathbb{R}} \{\eta + \frac{1}{1-\alpha}\mathbb{E}[(f(x, \tilde{\omega}) + a - \eta)_+]\}.$$

Now performing a decision variable substitution $\eta' = \eta - a$ in the second term of the above expression, we get

$$a + \mathbb{E}[f(x, \tilde{\omega})] + \lambda \operatorname*{Min}_{\eta' \in \mathbb{R}} \{\eta' + a + \frac{1}{1-\alpha}\mathbb{E}[(f(x, \tilde{\omega}) - \eta')_+]\}$$

$$= (1+\lambda)a + \mathbb{E}[f(x, \tilde{\omega})] + \lambda \operatorname*{Min}_{\eta' \in \mathbb{R}} \{\eta' + \frac{1}{1-\alpha}\mathbb{E}[(f(x, \tilde{\omega}) - \eta')_+]\}$$

$$= (1+\lambda)a + \mathbb{E}[f(x, \tilde{\omega})] + \lambda \operatorname*{Min}_{\eta \in \mathbb{R}} \{\eta + \frac{1}{1-\alpha}\mathbb{E}[(f(x, \tilde{\omega}) - \eta)_+]\}.$$

When $a \neq 0$ or $\lambda \neq 0$, the last result is not equal to

$$a + \mathbb{E}[f(x, \tilde{\omega})] + \lambda \operatorname*{Min}_{\eta \in \mathbb{R}} \{\eta + \frac{1}{1-\alpha}\mathbb{E}[(f(x, \tilde{\omega}) - \eta)_+]\}.$$

This proves that a mean-risk objective function with CVaR *does not* necessarily satisfy *translation invariance* despite CVaR being a coherent risk measure.

A2. Positive homogeneity Given $c > 0$, we have

$$\mathbb{E}[f(x, \tilde{\omega})] + \lambda \operatorname*{Min}_{\eta \in \mathbb{R}} \{\eta + \frac{1}{1-\alpha}\mathbb{E}[(cf(x, \tilde{\omega}) - \eta)_+]\}$$

$$= \mathbb{E}[f(x, \tilde{\omega})] + \lambda \operatorname*{Min}_{\eta \in \mathbb{R}} \{\eta + \frac{c}{1-\alpha}\mathbb{E}[(f(x, \tilde{\omega}) - \frac{\eta}{c})_+]\}$$

$$= \mathbb{E}[f(x, \tilde{\omega})] + c\lambda \operatorname*{Min}_{\eta \in \mathbb{R}} \{\frac{\eta}{c} + \frac{1}{1-\alpha}\mathbb{E}[(f(x, \tilde{\omega}) - \frac{\eta}{c})_+]\}$$

$$= \mathbb{E}[f(x, \tilde{\omega})] + c\lambda \operatorname*{Min}_{\eta' \in \mathbb{R}} \{\eta' + \frac{1}{1-\alpha}\mathbb{E}[(f(x, \tilde{\omega}) - \eta')_+]\},$$

where $\eta' = \eta/c$ is simply a change of variable, thus showing that the mean-risk objective function with CVaR satisfies the property of positive homogeneity.

A3. Monotonicity Let $f(x, \tilde{\omega}_1), f(x, \tilde{\omega}_2) \in \mathscr{F}$ such that $f(x, \tilde{\omega}_1) \leq f(x, \tilde{\omega}_2)$. Then $f(x, \tilde{\omega}_1) - \eta \leq f(x, \tilde{\omega}_2) - \eta$, which implies that

$$\mathbb{E}[f(x, \tilde{\omega}_1)] + \lambda \operatorname*{Min}_{\eta \in \mathbb{R}} \{\eta + \frac{1}{1-\alpha}\mathbb{E}[(f(x, \tilde{\omega}_1) - \eta)_+]\}$$

$$\leq \mathbb{E}[f(x, \tilde{\omega}_2)] + \lambda \operatorname*{Min}_{\eta \in \mathbb{R}} \{\eta + \frac{1}{1-\alpha}\mathbb{E}[(f(x, \tilde{\omega}_2) - \eta)_+]\}. \tag{2.22}$$

This shows that the mean-risk objective function with CVaR satisfies the monotonicity property.

A4. Convexity Let $f(x, \tilde{\omega}_1), f(x, \tilde{\omega}_2) \in \mathscr{F}$, and for any $\beta \in [0, 1]$ define $f(x, \tilde{\omega}_\beta) = \beta f(x, \tilde{\omega}_1) + (1 - \beta) f(x, \tilde{\omega}_2)$. Then
$\mathbb{E}[f(x, \tilde{\omega}_\beta)] + \lambda \text{Min}_{\eta \in \mathbb{R}} \{\eta + \frac{1}{1-\alpha}\mathbb{E}[(f(x, \tilde{\omega}_\beta) - \eta)_+]\}$

$$
\begin{aligned}
&= \mathbb{E}[\beta f(x, \tilde{\omega}_1) + (1 - \beta) f(x, \tilde{\omega}_2)] + \\
&\qquad \lambda \underset{\eta \in \mathbb{R}}{\text{Min}} \{\eta + \frac{1}{1-\alpha}\mathbb{E}[(\beta f(x, \tilde{\omega}_1) + (1 - \beta) f(x, \tilde{\omega}_2) - \eta)_+]\} \\
&= \mathbb{E}[\beta f(x, \tilde{\omega}_1)] + (1 - \beta)\mathbb{E}[f(x, \tilde{\omega}_2)] + \\
&\quad \lambda \underset{\eta \in \mathbb{R}}{\text{Min}} \{\eta + \beta\eta - \beta\eta + \frac{1}{1-\alpha}\mathbb{E}[(\beta f(x, \tilde{\omega}_1) - \beta\eta + (1 - \beta) f(x, \tilde{\omega}_2) \\
&\qquad - \eta + \beta\eta)_+]\} \\
&= \mathbb{E}[\beta f(x, \tilde{\omega}_1)] + (1 - \beta)\mathbb{E}[f(x, \tilde{\omega}_2)] + \\
&\qquad \lambda \underset{\eta \in \mathbb{R}}{\text{Min}} \{\beta\eta + (1 - \beta)\eta + \frac{1}{1-\alpha}\mathbb{E}[(\beta(f(x, \tilde{\omega}_1) - \eta) + (1 - \beta)(f(x, \tilde{\omega}_2) \\
&\qquad - \eta)_+]\} \\
&\leq \beta \left(\mathbb{E}[f(x, \tilde{\omega}_1)] + \lambda \underset{\eta \in \mathbb{R}}{\text{Min}} \{\eta + \frac{1}{1-\alpha}\lambda\mathbb{E}[(f(x, \tilde{\omega}_1) - \eta)_+]\}\right) + \\
&\qquad (1 - \beta)\left(\mathbb{E}[f(x, \tilde{\omega}_2)] + \underset{\eta \in \mathbb{R}}{\text{Min}} \{\eta + \frac{1}{1-\alpha}\mathbb{E}[(f(x, \tilde{\omega}_2) - \eta)_+]\}\right).
\end{aligned}
$$

This shows that the mean-risk objective function with CVaR satisfies the convexity property. Overall, we have just proved that the mean-risk objective function with CVaR is *not* a coherent risk measure because it violates translation invariance.

2.4.3 Example: Alternative Mean-Risk Conditional Value-at-Risk

Example 2.3 (Alternative Mean-Risk with CVaR) In this example we prove that the alternative MR-SP with CVaR (2.16) satisfies *translation invariance*. Recall that for $0 \leq \lambda \leq 1$, we have

$$
\underset{x \in X}{\text{Min}} \; (1 - \lambda)\mathbb{E}[f(x, \tilde{\omega})] + \lambda\phi_{CVaR_\alpha}(x),
$$

where

$$\phi_{CVaR_\alpha}(x) := \underset{\eta \in \mathbb{R}}{\text{Min}} \{\eta + \frac{1}{1-\alpha}\mathbb{E}[(f(x, \tilde{\omega}) - \eta)_+]\}.$$

The objective function is given as follows:

$$(1-\lambda)\mathbb{E}[f(x, \tilde{\omega})] + \lambda \underset{\eta \in \mathbb{R}}{\text{Min}} \{\eta + \frac{1}{1-\alpha}\mathbb{E}[(f(x, \tilde{\omega}) - \eta)_+]\}.$$

To prove translation invariance, given an $a \in \mathbb{R}$, we have

$$(1-\lambda)\mathbb{E}[a + f(x, \tilde{\omega})] + \lambda \underset{\eta \in \mathbb{R}}{\text{Min}} \{\eta + \frac{1}{1-\alpha}\mathbb{E}[(a + f(x, \tilde{\omega}) - \eta)_+]\}.$$

Performing a decision variable substitution $\eta' = \eta - a$, we get the following:

$$(1-\lambda)\mathbb{E}[a + f(x, \tilde{\omega})] + \lambda \underset{\eta' \in \mathbb{R}}{\text{Min}} \{\eta' + a + \frac{1}{1-\alpha}\mathbb{E}[(f(x, \tilde{\omega}) - \eta')_+]\}$$

$$= (1-\lambda)a + (1-\lambda)\mathbb{E}[f(x, \tilde{\omega})] + \lambda a + \lambda \underset{\eta' \in \mathbb{R}}{\text{Min}} \{\eta' + \frac{1}{1-\alpha}\mathbb{E}[(f(x, \tilde{\omega}) - \eta')_+]\}$$

$$= a + (1-\lambda)\mathbb{E}[f(x, \tilde{\omega})] + \lambda \underset{\eta \in \mathbb{R}}{\text{Min}} \{\eta + \frac{1}{1-\alpha}\mathbb{E}[(f(x, \tilde{\omega}) - \eta)_+]\}.$$

This shows that a mean-risk objective function with CVaR *does* indeed satisfy translation invariance. In fact, a mean-risk objective function with CVaR satisfies the rest of the coherence properties and is therefore coherent. The next example illustrates the expected excess risk measure.

2.4.4 Example: Expected Excess

Example 2.4 (Expected Excess) In this example we prove the coherence properties for the EE risk measure. Recall that given $x \in X$, EE is given as

$$\phi_{EE_\eta}(x) := \mathbb{E}[(f(x, \tilde{\omega}) - \eta)_+].$$

for a target $\eta \in \mathbb{R}$.

A1. Translation invariance Let $a \in \mathbb{R}$ and define $\eta' = \eta - a$. Then

$$\mathbb{E}[(f(x, \tilde{\omega}) + a - \eta)_+] = \mathbb{E}[(f(x, \tilde{\omega}) - \eta')_+].$$

Thus $\mathbb{E}[(f(x, \tilde{\omega}) + a - \eta)_+]$ is not necessarily equal to $a + \mathbb{E}[(f(x, \tilde{\omega}) - \eta)_+]$, showing that EE *does not* necessarily satisfy the *translation invariance* property.

We can also use a *counterexample* to illustrate this result. Suppose $\tilde{\omega}$ has two equiprobable realizations ω^1 and ω^2, i.e., $\mathbb{P}(\tilde{\omega} = \omega^1) = \mathbb{P}(\tilde{\omega} = \omega^2) = 1/2$. Let for some $x \in X$ the corresponding value $f(x, \omega^1) = 2$ and $f(x, \omega^2) = 6$. Let the target $\eta = 3$. Then we have

$$\mathbb{E}[(f(x, \tilde{\omega}) - \eta)_+] = \frac{1}{2}(2-3)_+ + \frac{1}{2}(6-3)_+ = 1.5.$$

Now let $a > 1$. Then

$$\begin{aligned}\mathbb{E}[(f(x, \tilde{\omega}) + a - \eta)_+] &= \frac{1}{2}[2+a-3]_+ + \frac{1}{2}(6+a-3)_+ \\ &= \frac{1}{2}(-1+a) + \frac{1}{2}(3+a) \\ &= 1+a.\end{aligned}$$

Observe that

$$\mathbb{E}[(f(x, \tilde{\omega}) + a - \eta)_+] \neq a + \mathbb{E}[(f(x, \tilde{\omega}) - \eta)_+].$$

Therefore, EE does *not* satisfy the translation invariance property.

A2. Positive homogeneity Given $c > 0$, we have

$$\begin{aligned}\mathbb{E}[(cf(x, \tilde{\omega}) - \eta)_+] &= c\mathbb{E}[(f(x, \tilde{\omega}) - \frac{\eta}{c})_+] \\ &\neq c\mathbb{E}[(f(x, \tilde{\omega}) - \eta)_+].\end{aligned}$$

This shows that EE does *not* satisfy the positive homogeneity property. Using the previous counterexample, we further show that EE indeed does *not* satisfy positive homogeneity. Let $c \geq 2$, then

$$\mathbb{E}[(f(x, \tilde{\omega}) - \eta)_+] = \frac{1}{2}(2-3)_+ + \frac{1}{2}(6-3)_+ = 1.5$$

$$\begin{aligned}\mathbb{E}[(cf(x, \tilde{\omega}) - \eta)_+] &= \frac{1}{2}(2c-3)_+ + \frac{1}{2}(6c-3)_+ \\ &= \frac{1}{2}(2c-3) + \frac{1}{2}(6c-3) \\ &= 4c-3 \\ &\neq 1.5 \text{ for } c \geq 2.\end{aligned}$$

A3. Monotonicity Let $f(x, \tilde{\omega}_1), f(x, \tilde{\omega}_2) \in \mathscr{F}$ such that $f(x, \tilde{\omega}_1) \leq f(x, \tilde{\omega}_2)$. Then $f(x, \tilde{\omega}_1) - \eta \leq f(x, \tilde{\omega}_2) - \eta$, which implies that $(f(x, \tilde{\omega}_1) - \eta)_+ \leq (f(x, \tilde{\omega}_2) - \eta)_+$. Therefore,

$$\mathbb{E}[(f(x, \tilde{\omega}_2) - \eta)_+] \geq \mathbb{E}[(f(x, \tilde{\omega}_1) - \eta)_+],$$

showing that EE satisfies the monotonicity property.

A4. Convexity Let $f(x, \tilde{\omega}_1), f(x, \tilde{\omega}_2) \in \mathscr{F}$, and for any $\beta \in [0, 1]$ define $f(x, \tilde{\omega}_\beta) = \beta f(x, \tilde{\omega}_1) + (1 - \beta) f(x, \tilde{\omega}_2)$. Then

$$\begin{aligned}
\mathbb{E}[(f(x, \tilde{\omega}_\beta) - \eta)_+] &= \mathbb{E}[(\beta f(x, \tilde{\omega}_1) + (1 - \beta) f(x, \tilde{\omega}_2) - \eta)_+] \\
&= \mathbb{E}[(\beta f(x, \tilde{\omega}_1) - \beta\eta + (1 - \beta) f(x, \tilde{\omega}_2) - \eta + \beta\eta)_+] \\
&\leq \beta\mathbb{E}[(f(x, \tilde{\omega}_1) - \eta)_+] + (1 - \beta)\mathbb{E}[(f(x, \tilde{\omega}_2) - \eta)_+].
\end{aligned}$$

This shows that EE satisfies the convexity property. In summary, we have just shown that EE does not satisfy two of the four coherence properties, namely, translation invariance and positive homogeneity. Nevertheless, EE satisfies monotonicity and convexity.

2.4.5 Example: Mean-Risk Expected Excess

Example 2.5 (Mean-Risk Expected Excess) Let us now turn to proving coherence properties for the mean-risk with EE risk measure (2.17). Recall that given $x \in X$, a target $\eta \in \mathbb{R}$, and $\lambda \geq 0$, an MR-SP with EE is given as follows:

$$\operatorname*{Min}_{x \in X} \mathbb{E}[f(x, \tilde{\omega})] + \lambda \phi_{EE_\eta}(x),$$

where

$$\phi_{EE_\eta}(x) := \mathbb{E}[f(x, \tilde{\omega}) - \eta]_+.$$

The objective function can be expressed as

$$\mathbb{E}[f(x, \tilde{\omega})] + \lambda\mathbb{E}[f(x, \tilde{\omega}) - \eta]_+.$$

A1. Translation invariance Let $a \in \mathbb{R}$ and define $\eta' = \eta - a$. Then

$$\begin{aligned}
&\mathbb{E}[a + f(x, \tilde{\omega})] + \lambda\mathbb{E}[(a + f(x, \tilde{\omega}) - \eta)_+] \\
&= \mathbb{E}[a + f(x, \tilde{\omega})] + \lambda\mathbb{E}[(f(x, \tilde{\omega}) - \eta')_+] \\
&= a + \mathbb{E}[f(x, \tilde{\omega})] + \lambda\mathbb{E}[(f(x, \tilde{\omega}) - \eta')_+].
\end{aligned}$$

Observe that the last result does not necessarily equal

$$a + \mathbb{E}[f(x, \tilde{\omega})] + \lambda\mathbb{E}[(f(x, \tilde{\omega}) - \eta)_+],$$

thus proving that EE *does not* satisfy translation invariance.

Consider the previous *counterexample* we used for EE with $\tilde{\omega}$ having two equiprobable realizations ω^1 and ω^2 with corresponding values $f(x, \omega^1) = 2$ and $f(x, \omega^2) = 6$ for some $x \in X$. Let the target $\eta = 3$ and $\lambda = 1$. Then we have

$$\begin{aligned}
&\mathbb{E}[f(x, \tilde{\omega})] + \lambda\mathbb{E}[(f(x, \tilde{\omega}) - \eta)_+] \\
&= \frac{1}{2}(2) + \frac{1}{2}(6) + \frac{1}{2}[2 - 3]_+ + \frac{1}{2}[6 - 3]_+ = 5.5.
\end{aligned}$$

Now let $a > 1$. Then

$$\begin{aligned}
&\mathbb{E}[a + f(x, \tilde{\omega})] + \lambda\mathbb{E}[(a + f(x, \tilde{\omega}) - \eta)_+] \\
&= \frac{1}{2}(a + 2) + \frac{1}{2}(a + 6) + \frac{1}{2}[2 + a - 3]_+ + \frac{1}{2}[6 + a - 3]_+ \\
&= (a + 4) + \frac{1}{2}(-1 + a) + \frac{1}{2}(3 + a) \\
&= 2a + 5.
\end{aligned}$$

Observe that

$$\begin{aligned}
&\mathbb{E}[a + f(x, \tilde{\omega})] + \lambda\mathbb{E}[(f(x, \tilde{\omega}) + a - \eta)_+] \\
&\neq a + \mathbb{E}[(f(x, \tilde{\omega}) - \eta)_+].
\end{aligned}$$

Therefore, we can conclude that the mean-risk SLP with EE risk does *not* satisfy the translation invariance property.

A2. Positive homogeneity Given $c > 0$, we have

$$\begin{aligned}
\mathbb{E}[cf(x, \tilde{\omega})] + \lambda\mathbb{E}[(cf(x, \tilde{\omega}) - \eta)_+] &= c(\mathbb{E}[f(x, \tilde{\omega})] + \lambda\mathbb{E}[(f(x, \tilde{\omega}) - \frac{\eta}{c})_+]) \\
&\neq c(\mathbb{E}[f(x, \tilde{\omega})] + \lambda\mathbb{E}[(f(x, \tilde{\omega}) - \eta)_+]).
\end{aligned}$$

This shows that the mean-risk with EE risk measure does *not* satisfy the positive homogeneity property.

A3. Monotonicity Let $f(x, \tilde{\omega}_1), f(x, \tilde{\omega}_2) \in \mathscr{F}$ such that $f(x, \tilde{\omega}_1) \leq f(x, \tilde{\omega}_2)$. Then $f(x, \tilde{\omega}_1) - \eta \leq f(x, \tilde{\omega}_2) - \eta$, which implies that $(f(x, \tilde{\omega}_1) - \eta)_+ \leq (f(x, \tilde{\omega}_2) - \eta)_+$. Therefore,

$$\mathbb{E}[f(x, \tilde{\omega}_1)] + \lambda\mathbb{E}[(f(x, \tilde{\omega}_1) - \eta)_+] \leq \mathbb{E}[f(x, \tilde{\omega}_2)] + \lambda\mathbb{E}[(f(x, \tilde{\omega}_2) - \eta)_+],$$

proving that the mean-risk with EE risk measure satisfies the monotonicity property.

A4. Convexity Let $f(x, \tilde{\omega}_1), f(x, \tilde{\omega}_2) \in \mathscr{F}$, and for any $\beta \in [0, 1]$ define $f(x, \tilde{\omega}_\beta) = \beta f(x, \tilde{\omega}_1) + (1 - \beta) f(x, \tilde{\omega}_2)$. Then

$$
\begin{aligned}
&\mathbb{E}[f(x, \tilde{\omega}_\beta)] + \lambda \mathbb{E}[(f(x, \tilde{\omega}_\beta) - \eta)_+] \\
&= EE[\beta f(x, \tilde{\omega}_1) + (1 - \beta) f(x, \tilde{\omega}_2)] + \\
&\lambda \mathbb{E}[(\beta f(x, \tilde{\omega}_1) + (1 - \beta) f(x, \tilde{\omega}_2) - \eta)_+] \\
&= EE[\beta f(x, \tilde{\omega}_1)] + (1 - \beta) EE[f(x, \tilde{\omega}_2)] + \\
&\lambda \mathbb{E}[(\beta f(x, \tilde{\omega}_1) - \beta \eta + (1 - \beta) f(x, \tilde{\omega}_2) - \eta + \beta \eta)_+] \\
&\leq \beta (EE[f(x, \tilde{\omega}_1)] + \mathbb{E}[(f(x, \tilde{\omega}_1) - \eta)_+]) + \\
&(1 - \beta) (EE[f(x, \tilde{\omega}_2)] + \mathbb{E}[(f(x, \tilde{\omega}_2) - \eta)_+]) .
\end{aligned}
$$

This shows that the mean-risk with EE risk measure satisfies the convexity property. Overall, we see that like the EE risk measure, a mean-risk SLP with EE does not satisfy translation invariance and positive homogeneity. It satisfies monotonicity and convexity.

2.5 Deterministic Equivalent Problem Formulations

Let us assume that $\tilde{\omega}$ is discrete with finitely many scenarios $\omega \in \Omega$, each with corresponding probability $p(\omega)$. Then we can write deterministic equivalent problem (DEP) formulations of the stochastic programs defined in the previous sections. Due to the problem size, which grows with the number of scenarios, $|\Omega|$, the problems become too large for direct solvers. This motivates the study of decomposition methods to solve these problems. Decomposition methods for RS-SPs and MR-SPs include subgradient optimization based methods [1] and Lagrangian dual decomposition [17]. These methods are covered in later chapters.

2.5.1 *Risk-Neutral Case*

The RN-SLP (2.1) defined in Sect. 2.1 is equivalent to Problem (2.5) if $\lambda := 0$. The DEP formulation for RN-SLP is defined as follows [34]:

$$
\begin{aligned}
\text{Min} \quad & c^\top x + \sum_{\omega \in \Omega} p(\omega) q(\omega)^\top y(\omega) \qquad (2.23) \\
\text{s.t.} \quad & T(\omega) x + W y(\omega) \geq r(\omega), \ \forall \omega \in \Omega \\
& x \in X, \ \ y(\omega) \in \mathbb{R}^{n_2}_+, \ \forall \omega \in \Omega .
\end{aligned}
$$

Problem (2.23) is an LP with a dual block angular structure and is suitable for standard SLP decomposition methods. The first-stage decision variable is x, and the second-stage decision variable is $y(\omega)$.

2.5.2 Excess Probability

Let $\lambda \geq 0$ and choose $\eta \in \mathbb{R}$. If X is bounded, then there exists a constant $M > 0$ such that Problem (2.12) is equivalent to the following problem [29, 30]:

$$\text{Min} \quad c^\top x + \sum_{\omega\in\Omega} p(\omega)q(\omega)^\top y(\omega) + \lambda . \sum_{\omega\in\Omega} p(\omega)\gamma(\omega) \tag{2.24a}$$

$$\text{s.t.} \quad T(\omega)x + Wy(\omega) \geq r(\omega),\ \forall\omega \in \Omega \tag{2.24b}$$

$$-c^\top x - q(\omega)^\top y(\omega) + M.\gamma(\omega) \geq -\eta,\ \forall\omega \in \Omega \tag{2.24c}$$

$$x \in X,\ \ y(\omega) \in \mathbb{R}_+^{n_2},\ \gamma(\omega) \in \{0, 1\},\ \forall\omega \in \Omega. \tag{2.24d}$$

Since X is bounded, the constant M can be selected as $\sup\{f(x, \omega) : x \in X, \forall\omega \in \Omega\}$. Observe that (2.24) has a dual block angular structure: Consider x as the first-stage variable and $y(\omega)$ and $\gamma(\omega)$ as the second-stage decision variables. Then second-stage decision variables for different ω's never occur in the same constraint but are linked through the first-stage decision variables only. However, Problem (2.24) is a mixed 0-1 LP due to the introduction of the $\gamma(\omega)$ variable. Thus decomposition methods for SMIP are needed to solve (2.24).

2.5.3 Quantile Deviation

Given $\lambda \in [0, 1/\varepsilon_1]$, Problem (2.13) is equivalent to the following LP [1]:

$$\text{Min} \quad (1-\lambda\varepsilon_1)c^\top x + \lambda\varepsilon_1\eta + (1-\lambda\varepsilon_1)\sum_{\omega\in\Omega} p(\omega)q(\omega)^\top y(\omega)$$

$$+\lambda(\varepsilon_1+\varepsilon_2)\sum_{\omega\in\Omega} p(\omega)\upsilon(\omega) \tag{2.25a}$$

$$\text{s.t.} \quad T(\omega)x + Wy(\omega) \geq r(\omega),\ \forall\omega \in \Omega \tag{2.25b}$$

$$-c^\top x - q(\omega)^\top y(\omega) + \eta + \upsilon(\omega) \geq 0,\ \forall\omega \in \Omega \tag{2.25c}$$

$$x \in X,\ \eta \in \mathbb{R},\ y(\omega) \in \mathbb{R}_+^{n_2},\ \upsilon(\omega) \in \mathbb{R}_+, \forall\omega \in \Omega. \tag{2.25d}$$

Problem (2.25) is an LP, has a dual block angular structure, and is amenable to standard decomposition methods for two-stage SLP such as the L-shaped method [32]. Now x and η are the first-stage decision variables, and $y(\omega)$ and $v(\omega)$ are the second-stage decision variables.

2.5.4 Conditional Value-at-Risk

Given $\lambda \geq 0$, Problem (2.15) is equivalent to the following LP [31]:

$$\begin{aligned}
\text{Min} \quad & c^\top x + \sum_{\omega \in \Omega} p(\omega) q(\omega)^\top y(\omega) + \lambda.\eta + \frac{\lambda}{1-\alpha} \sum_{\omega \in \Omega} p(\omega) v(\omega) && \text{(2.26a)}\\
\text{s.t.} \quad & T(\omega)x + W y(\omega) \geq r(\omega), \ \forall \omega \in \Omega && \text{(2.26b)}\\
& -c^\top x - q(\omega)^\top y(\omega) + \eta + v(\omega) \geq 0, \ \forall \omega \in \Omega && \text{(2.26c)}\\
& x \in X, \ \eta \in \mathbb{R}, \ y(\omega) \in \mathbb{R}_+^{n_2}, \ v(\omega) \in \mathbb{R}_+, \forall \omega \in \Omega. && \text{(2.26d)}
\end{aligned}$$

Problem (2.26) is also an LP and has a dual block angular structure. The first-stage variables are x and η, while $y(\omega)$ and $v(\omega)$ are the second-stage variables. Problem (2.26) can also be solved using standard SLP decomposition methods.

2.5.5 Expected Excess

Given $\lambda \geq 0$ and a target level $\eta \in \mathbb{R}$, Problem (2.17) is equivalent to the following formulation [17]:

$$\begin{aligned}
\text{Min} \quad & c^\top x + \sum_{\omega \in \Omega} p(\omega) q(\omega)^\top y(\omega) + \lambda. \sum_{\omega \in \Omega} p(\omega) v(\omega) && \text{(2.27a)}\\
\text{s.t.} \quad & T(\omega)x + W y(\omega) \geq r(\omega), \ \forall \omega \in \Omega && \text{(2.27b)}\\
& -c^\top x - q(\omega)^\top y(\omega) + v(\omega) \geq -\eta, \ \forall \omega \in \Omega && \text{(2.27c)}\\
& x \in X, \ y(\omega) \in \mathbb{R}_+^{n_2}, \ v(\omega) \in \mathbb{R}_+, \ \forall \omega \in \Omega. && \text{(2.27d)}
\end{aligned}$$

Problem (2.27) is also an LP with a dual block angular structure and is suitable for standard SLP decomposition methods. The first-stage decision variable is x, and the second-stage decision variables are $y(\omega)$ and $v(\omega)$. Problem (2.27) can also be solved using standard SLP decomposition methods.

2.5.6 *Absolute Semideviation*

Given $\lambda \in [0, 1]$, then Problem (2.19) is equivalent to the following formulation [1, 12]:

$$\text{Min} \quad (1-\lambda)c^\top x + (1-\lambda)\sum_{\omega\in\Omega} p(\omega)q(\omega)^\top y(\omega) \ + \lambda. \sum_{\omega\in\Omega} p(\omega)v(\omega) \tag{2.28a}$$

$$\text{s.t.} \quad T(\omega)x + Wy(\omega) \geq r(\omega), \ \forall \omega \in \Omega \tag{2.28b}$$

$$-c^\top x - q(\omega)^\top y(\omega) + v(\omega) \geq 0, \ \forall \omega \in \Omega \tag{2.28c}$$

$$-c^\top x - \sum_{\varpi\in\Omega} p(\varpi)q(\varpi)^\top y(\varpi) + v(\omega) \geq 0, \ \forall \omega \in \Omega \tag{2.28d}$$

$$x \in X, \ \ y(\omega) \in \mathbb{R}^{n2}_{+}, \ v(\omega) \in \mathbb{R}, \ \forall \omega \in \Omega. \tag{2.28e}$$

Unlike the rest of the formulations presented above, Problem (2.28) does *not* have a dual block angular structure due to the "complicating" or "linking" constraints (2.28d). Observe that these constraints link all the scenarios. Therefore, standard SLP methods such as the L-shaped method cannot be used directly to solve SD. However, a *subgradient optimization* approach or column generation can be applied.

2.5.7 *Central Deviation*

Given $\lambda \in [0, \frac{1}{2}]$, then Problem (2.21) is equivalent to the following formulation [17]:

$$\text{Min} \quad (1-2\lambda)c^\top x + (1-2\lambda)\sum_{\omega\in\Omega} p(\omega)q(\omega)^\top y(\omega) \ + 2\lambda. \sum_{\omega\in\Omega} p(\omega)v(\omega) \tag{2.29a}$$

$$\text{s.t.} \quad T(\omega)x + Wy(\omega) \geq r(\omega), \ \forall \omega \in \Omega \tag{2.29b}$$

$$-c^\top x - q(\omega)^\top y(\omega) + v(\omega) \geq 0, \ \forall \omega \in \Omega \tag{2.29c}$$

$$-c^\top x - \sum_{\varpi\in\Omega} p(\varpi)q(\varpi)^\top y(\varpi) + v(\omega) \geq 0, \ \forall \omega \in \Omega \tag{2.29d}$$

$$x \in X, \ \ y(\omega) \in \mathbb{R}^{n2}_{+}, \ v(\omega) \in \mathbb{R}, \ \forall \omega \in \Omega. \tag{2.29e}$$

Like the formulation for ASD, Problem (2.29) does *not* have a dual block angular structure due to the scenario linking constraints (2.29d) and is also amenable to *subgradient optimization* or column generation.

2.6 Probabilistically Constrained Models

In certain applications in science and engineering, the occurrence of undesirable random realizations may be unavoidable. For example, an electricity power blackout may be unavoidable due to a hurricane, or flight delays for an airline may be unavoidable due to inclement weather. In such cases, one may consider making decisions that take into account this fact so as to allow for such undesirable random events to happen but with a specified level of "risk." This *probabilistic thinking* is very natural and different from the mean-risk aversion mindset of making "recourse" actions and leads to the so-called *chance* or *probabilistic* constraints approach for modeling uncertainty. In this approach, the decision-maker or modeler imposes their level of reliability $\alpha \in (0, 1)$ or level of acceptable risk $(1 - \alpha)$. This means that the given probabilistic constraint(s) must hold $\alpha \times 100\%$ of the time, i.e., it can be violated up to $(1 - \alpha) \times 100\%$ of the time. In practice, the typical values for α are usually between 0.9 and 1. Since their introduction by [5], probabilistic (chance) constraints stochastic programs have been extensively studied in the literature (e.g., [3, 4, 6, 13–16, 18, 22, 24, 27, 33]).

2.6.1 Probabilistically Constrained Models

Formally, a generic linear probabilistically constrained stochastic program (PC-SP) can be given as follows:

$$
\begin{aligned}
\text{Min } & f(x) \\
\text{s.t. } & x \in X \\
& \mathbb{P}(T(\tilde{\omega})x \geq r(\tilde{\omega})) \geq \alpha \qquad (2.30)\\
& x \geq 0,
\end{aligned}
$$

where $x \in \mathbb{R}^{n_1}$ is the decision variable vector, $T(\tilde{\omega}) \in \mathbb{R}^{m_1 \times n_1}$ is the technology matrix, and $r(\tilde{\omega}) \in \mathbb{R}^{m_1}$ is the right hand side vector. We assume that the function $f(x)$ is convex, and in many applications it is considered to be linear. Constraint (2.30) is the probabilistic or *joint-chance* constraint and requires that all the constraints hold jointly. If any one of the constraints is violated, then the entire set of constraints is violated. PC-SPs are generally difficult to solve. This is partly because the feasible set defined by probabilistic constraints is generally nonconvex. Furthermore, reformulation of the problem results in an MIP, which is also difficult to solve in general.

2.6.2 Single-Chance Constrained Models

A single-chance PC-SP can be given as follows:

$$
\begin{aligned}
\text{Min } & f(x) \\
\text{s.t. } & x \in X \\
& \mathbb{P}(t(\tilde{\omega})x \geq r(\tilde{\omega})) \geq \alpha \qquad (2.31) \\
& x \geq 0,
\end{aligned}
$$

where $x \in \mathbb{R}^{n_1}$ is the decision variable vector, $t(\tilde{\omega}) \in \mathbb{R}^{1\times n_1}$ is the technology matrix, and $r(\tilde{\omega}) \in \mathbb{R}^{m_1}$ is the right hand side vector. As in the general case, the function $f(x)$ is convex in x and is often linear. Constraint (2.31) is the *single-chance* or *single-probabilistic* constraint. In this case, it may be able to express the chance constraint in a relatively simpler way if the distribution of $\tilde{\omega}$ is "nice." For example, if $\tilde{\omega}$ follows a multivariate normal distribution, the chance constraint can be converted to a second-order conic constraint. Thus the problem can be solved using an off-the-shelf optimization solver. Since their introduction by [5], PC-SPs have been extensively studied in the literature (e.g., [3, 6, 18, 22, 24, 27]). In the next subsection, we define the deterministic equivalent formulation for the generic PC-SP model.

2.6.3 Deterministic Equivalent Problem Formulation

Assuming that $\tilde{\omega}$ is discrete with finitely many scenarios $\omega \in \Omega$, each with corresponding probability $p(\omega)$, we can derive an extensive form or deterministic equivalent problem (DEP) formulation for the PC-SP. This leads to the study of decomposition approaches for PC-SPs. These include MIP approaches [13, 16, 27, e.g.,], p-efficient points method [15, 24, e.g.,], and IIS branch-and-cut [33].

Let $M(\omega)$ be an appropriately sized scalar for scenario ω, and let e be an appropriately dimensioned vector of ones. Let us define a binary decision variable $z(\omega)$ as follows: $z(\omega)$=1 if under scenario ω at least one of the inequalities in the probabilistic constraint is violated, and $z(\omega)$=0 otherwise. Then a DEP formulation for Problem (2.30) can be written as follows:

$$\begin{aligned} &\text{Min } c^\top x \\ &\text{s.t. } Ax \geq b \\ &\quad T(\omega)x + M(\omega)ez(\omega) \geq r(\omega), \forall \omega \in \Omega \\ &\quad \sum_{\omega \in \Omega} p(\omega)z(\omega) \leq 1 - \alpha \\ &\quad x \geq 0,\ z(\omega) \in \{0, 1\},\ \forall \omega \in \Omega. \end{aligned} \tag{2.32}$$

The total probability of violating the probabilistic constraint is given by $\mathbb{P}(T(\tilde{\omega})x \not\geq r(\tilde{\omega})) \leq \sum_{\omega \in \Omega} p(\omega)z(\omega) \leq 1 - \alpha$. When $z(\omega) = 1$, it means that scenario ω is excluded from the PC-SP formulation. Thus we can assume that $p(\omega) \leq \alpha,\ \forall \omega \in \Omega$, so that the knapsack constraint (2.32) has a well-defined subset of scenarios that can be excluded from the formulation without exceeding the risk/reliability level $1 - \alpha$. The parameter α can be thought of as a budget and that the knapsack is a budget constraint for constraint removal.

Observe that DEP formulation (2.32) is a mixed-binary program and is therefore nonconvex. In fact, for a small or moderate number of scenarios, one can attack this problem using an off-the-shelf MIP solver. However, Problem (2.32) involves the big-M and generally tends to have a very weak LP relaxation. Thus this calls for decomposition methods to solve this problem. These include MIP approaches (e.g., [13, 16, 27]), the p-efficient points approach (e.g., [15, 24]), and the IIS branch-and-cut method [33], among others.

2.7 Other Models

There are several other approaches in the literature to capture and include a measure of risk in optimization models. These include *integrated-chance constraints* (ICCs) [9, 11, 23], *stochastic dominance constraints* (SDCs) [7, 8], and *economic expected utility* [19]. We shall briefly describe the basic idea of each of these approaches and leave it to the interested reader to study these topics in the cited references.

The use of probabilistic constraints also allows for incorporating other measures of violation. One type of measure of violation one can use is the ICC. Given a constant $d \in \mathbb{R}$, an ICC can be defined as follows:

$$\mathbb{E}[\max_i (r_i(\tilde{\omega}) - T_i(\tilde{\omega})x)_+] \leq d,$$

where $T_i(\tilde{\omega})$ is the i-th row of matrix $T(\tilde{\omega})$ and $r_i(\tilde{\omega})$ is the i-th component of vector $r(\tilde{\omega})$. The interpretation of the ICC is as follows: Given the probabilistic constraint $\mathbb{P}(T(\tilde{\omega})x \geq r(\tilde{\omega})) \geq 1-\alpha$, the constraint $T(\tilde{\omega})x \geq r(\tilde{\omega})$ reflects the idea of wanting to avoid positive shortage, but in general it may be impossible to avoid positive shortages for all ω due to the random nature induced by $\tilde{\omega}$. Therefore, in the

ICC the quantity $\mathbb{E}[\max_i (r_i(\tilde{\omega}) - T_i(\tilde{\omega})x)_+]$ captures the expected positive shortage, which must be limited by the amount d. For example, in power generation planning this measure of positive shortage is the loss of load probability, where $r_i(\tilde{\omega})$ is the power demand on a given day for a given area, while $T_i(\tilde{\omega})$ is the total generating capacity.

SDCs measure risk by comparing distributions of two random variables. The original context for SDC was in managing risk by selecting options that are preferable to a random benchmark with a goal of maximizing expected profits. The basic concept of stochastic dominance can be explained as follows: Given $\eta \in \mathbb{R}$, a random variable $Z_1 \in \mathscr{F}$, and its cumulative distribution $F_1(Z_1; \eta) = \mathbb{P}(Z_1 \leq \eta)$, define recourse functions for $n \geq 2$

$$F_n(Z_1; \eta) := \int_{-\infty}^{\eta} F_{n-1}(Z_1; t)dt,$$

assuming that the $n-1$ first moments are finite. It is said that random variable Z_1 dominates another random variable $Z_2 \in \mathscr{F}$ in n-th order, denoted $Z_1 \succeq_{(n)} Z_2$, if

$$F_n(Z_1; \eta) \leq F_n(Z_2; \eta) \text{ for all } \eta \in \mathbb{R}.$$

By ensuring that $F_n(Z_1; \eta) < \infty$ for all η, the $n-1$ first moments of Z_1 are finite for $n \geq 2$. This is captured by the following equivalence relation [20]:

$$F_i(Z_1; \eta) = \frac{1}{(j-1)!}\mathbb{E}[((\eta - z_1)_+)^{i-1}].$$

Now we can state an optimization model with SDC as follows:

$$\begin{aligned} &\text{Min } g(x) \\ &\text{s.t. } G(x, \tilde{\omega}) \succeq_{(n)} Z_2 \\ &\quad\;\; x \in X, \end{aligned}$$

where $g(x)$ and $G(x, \tilde{\omega})$ are the objective and constraint functions. When $n = 1$, Problem (2.33) is nonconvex and has similar computational difficulties as in the probabilistic constraint setting. When $n = 2$, however, Problem (2.33) is convex as long as $g(x)$ is convex, $G(x, \tilde{\omega})$ is concave, and the set X is convex. Using the equivalence relation above, we can write the problem using expected value constraints as the following convex program:

$$\begin{aligned} &\text{Min } g(x) \\ &\text{s.t. } \mathbb{E}[(\eta - G(x, \tilde{\omega}))_+] \leq \mathbb{E}[(\eta - Z_2)_+] \quad \forall \eta \in \mathbb{R} \\ &\quad\;\; x \in X. \end{aligned}$$

An important observation to make here is that Problem (2.34) has uncountably many constraints, one for each $\eta \in \mathbb{R}$. However, if the random variable Z_2 has finitely many realizations, then it suffices to write constraints only for each realization resulting in a problem with finitely many expected value constraints. Furthermore, assuming the distribution of $\tilde{\omega}$ has finite support, the expectations can be written as sums leading to a deterministic problem.

An approach that is related to stochastic dominance is the *expected utility theory*. Expected utility theory contends that a rational decision-maker has a utility function U belonging to a certain set of functions $\mathscr{U}$ such that the random realization Z_1 is preferred to Z_2 if $\mathbb{E}[U(Z_1)] \geq \mathbb{E}[U(Z_2)]$. This set of functions is determined by the level of risk averseness of the decision-maker. For example, a risk-averse decision-maker would be characterized by having a utility function that is nondecreasing and concave. Actually, the decision-maker's exact utility function U is unknown, and in such a case it is said that Z_1 is preferred over Z_2 if $\mathbb{E}[U(Z_1)] \geq \mathbb{E}[U(Z_2)]$ for all $U \in \mathscr{U}$. Let $\mathscr{U}_1$ be the set of *nondecreasing* functions $U : \mathbb{R} \mapsto \mathbb{R}$ and $\mathscr{U}_2$ be the set of *nondecreasing concave* functions $U : \mathbb{R} \mapsto \mathbb{R}$. Then the notions of stochastic dominance are related to expected utility theory through the following well-known result: Assuming that the expectations exist, we have

$$Z_1 \succeq_{(n)} Z_2 \Leftrightarrow \mathbb{E}[U(Z_1)] \geq \mathbb{E}[U(Z_2)], \quad \forall u \in \mathscr{U}_n, \; n \in \{0, 1\}.$$

Finally, we should point out that stochastic dominance is also related to the notion of *stochastic ordering*, where, for example, $\mathbb{E}[U(Z_1)] \geq \mathbb{E}[U(Z_2)]$ for all $U \in \mathscr{U}_2$ is referred to as the stochastic increasing concave order.

Problems

2.1 Excess Probability and Central Deviation Risk Measures

Prove coherence properties of *translation invariance*, *positive homogeneity*, *monotonicity*, and *convexity* for each of the following risk measures:

(a) Excess probability (EP)
(b) Central deviation (CDEV)

2.2 Proving Subadditivity

Prove or disprove the property of *subadditivity* for each of the two risk measures in Problem 2.1.

2.3 Quantile Deviation Risk Measure

Prove that quantile deviation (QDEV) is not coherent by showing that it does not satisfy the properties of *translation invariance* and *monotonicity* but satisfies *positive homogeneity* and *convexity*.

2.4 Absolute Semideviation Risk Measure
Prove that absolute semideviation (ASD) is not coherent by showing that it does not satisfy the properties of *translation invariance* and *monotonicity* but satisfies *positive homogeneity* and *convexity*.

2.5 Mean-Semideviation Risk Measure
Consider the following mean-semideviation risk function for some $p \in [1, +\infty)$ and $\lambda \geq 0$:

$$\phi_{MS}(x) := \left(\mathbb{E}[(f(x, \tilde{\omega}) - \mathbb{E}[f(x, \tilde{\omega})])_+^p]\right)^{\frac{1}{p}}.$$

Notice that for $p = 1$ this upper semideviation risk measure $\phi_{MS}(x)$ is simply the absolute semideviation (ASD). Prove coherence properties of *translation invariance, positive homogeneity, monotonicity*, and *convexity* for $p = 2$.

2.6 Mean-Deviation Risk Measure
Consider the following mean-deviation risk function for some $p \in [1, +\infty)$:

$$\phi_{MD}(x) := \left(\mathbb{E}[|f(x, \tilde{\omega}) - \mathbb{E}[f(x, \tilde{\omega})]|^p]\right)^{\frac{1}{p}}.$$

Observe that for $p = 1$ this risk measure $\phi_{MD}(x)$ is simply the central deviation (CDEV). Prove coherence properties of *translation invariance, positive homogeneity, monotonicity*, and *convexity* for $p = 2$.

2.7 Mean-Variance Risk Measure
Consider the following mean-deviation risk function for some $p \in [1, +\infty)$:

$$\phi_{MV}(x) := \left(\mathbb{E}[(f(x, \tilde{\omega}) - \mathbb{E}[f(x, \tilde{\omega})])^p]\right)^{\frac{1}{p}}.$$

Prove whether or not $\phi_{MV}(x)$ satisfies the *convexity* property for $p = 2$.

2.8 A Hybrid Mean-Risk and Probabilistically Constrained Model
Write a formulation for a mean expected excess stochastic program with joint-chance constraints. Derive the DEP formulation and prove the coherence properties for the model.

2.9 Conditional Value-at-Risk Deviation
Given $x \in X$ and $\alpha \in (0, 1)$, conditional value-at-risk deviation (CVaR-D) [25] is expressed as follows:

$$\phi_{CVaR\text{-}D_\alpha}(x) := \operatorname*{Min}_{\eta \in \mathbb{R}} \{\eta + \frac{1}{1-\alpha}\mathbb{E}[((f(x, \tilde{\omega}) - \mathbb{E}[f(x, \tilde{\omega})]) - \eta)_+]\}.$$

For any λ, an MR-SP with CVaR-D can be given as follows:

$$\operatorname*{Min}_{x \in X} \mathbb{E}[f(x, \tilde{\omega})] + \lambda \phi_{CVaR\text{-}D_\alpha}(x). \tag{2.33}$$

(a) Prove coherence properties of *translation invariance*, *positive homogeneity*, *monotonicity*, and *convexity* for $\phi_{CVaR\text{-}D_\alpha}(x)$.
(b) Prove coherence properties of *translation invariance*, *positive homogeneity*, *monotonicity*, and *convexity* for the objective function of Problem (2.33).

2.10 Deterministic Equivalent Formulation
Derive the deterministic equivalent formulation for the MR-SP with CVaR-D (2.33) defined in Problem (2.9).

References

1. S. Ahmed. Convexity and decomposition of mean-risk stochastic programs. *Mathematical Programming*, 106(3):433–446, 2006.
2. P. Artzner, F. Delbean, J.M Eber, and D. Heath. Coherent measures of risk. *Mathematical Finance*, 9:203–228, 1999.
3. P. Beraldi and A. Ruszczyński. A branch and bound method for stochastic integer programming under probabilistic constraints. *Optimization methods and Software*, 17:359–382, 2001.
4. A. Shapiro B.K. Pagnoncelli, S. Ahmed. Sample average approximation method for chance constrained programming: Theory and applications. *Journal of Optimization Theory and Applications*, 142:399–416, 2009.
5. A. Charnes, W.W. Cooper, and G.H. Symonds. Cost horizons and certainty equivalents: An approach to stochastic programming of heating oil. *Management Science*, 4:235–263, 1958.
6. D. Dentcheva, A. Prékopa, and A. Ruszczyński. Bounds for probabilistic integer programming problems. *Discrete Applied Mathematics*, A 89:55–65, 2002.
7. D. Dentcheva and A. Ruszczyński. Optimization with stochastic dominance constraints. *SIAM Journal on Optimization*, 14(2):548–566, 2003.
8. D. Dentcheva and A. Ruszczyński. Semi-infinite probabilistic optimization: First order stochastic dominance constraints. *Optimization*, 53(5–6):583–601, 2004.
9. W.K.K. Haneveld and M.H. van der Vlerk. Integrated chance constraints: Reduced forms and an algorithm. *Computational Management Science*, 3:245–269, 2006.
10. J.L. Jensen. Sur les fonctions convexes et les inégualités entre les valeurs moyennes. *Acta Mathematica*, 30:175–193, 1906.
11. W.K. Klein Haneveld. On integrated chance constraints. In F. Archetti, G. Di Pillo, and M. Lucertini, editors, *Stochastic Programming*, pages 194–209. Springer Berlin Heidelberg, Berlin, Heidelberg, 1986.
12. H. Konno and H. Yamazaki. Mean-absolute deviation portfolio optimization model and its applications to Tokyo stock market. *Management Science*, 37(5):519–531, 1991.
13. S. Kücükyavuz. On mixing sets arising in chance-constrained programming. *Mathematical Programming*, 132(1–2):31–56, 2012.
14. M.A. Lejeune. Pattern-based modeling and solution of probabilistically constrained optimization problems. *Operations Research*, 60(6):1356–1372, 2010.
15. M.A. Lejeune and N. Noyan. Mathematical programming approaches for generating p-efficient points. *European Journal of Operational Research*, 207:590–600, 2010.
16. J. Luedtke, S. Ahmed, and G. Nemhauser. An integer programming approach for linear programs with probabilistic constraints. *Mathematical Programming*, 122(2):247–272, 2010.
17. A. Markert and R. Schultz. On deviation measures in stochastic integer programming. *Operations Research Letters*, 33(5):441–449, 2005.
18. B.L. Miller and H.M. Wagner. Chance constrained programming with joint constraints. *Operations Research*, 13:930–945, 1965.

19. John Von Neumann and Oskar Morgenstern, editors. *Theory of Games and Economic Behavior*. Princeton University Press, Princeton, NJ., 1953.
20. W. Ogryczak and A. Ruszczyński. From stochastic dominance to mean-risk model: Semideviations as risk measures. *European Journal of Operational Research*, 116:33–50, 1999.
21. W. Ogryczak and A. Ruszczyński. Dual stochastic dominance and related mean-risk models. *SIAM Journal on Optimization*, 13:60–78, 2002.
22. A. Prékopa. On probabilistic constrained programming. In Princeton University Press, editor, *Proceedings of the Princeton Symposium on Mathematical Programming*, 1970.
23. A. Prékopa. Contributions to the theory of stochastic programming. *Mathematical Programming*, 4:202–221, 1973.
24. A. Prékopa, B. Vizvari, and T. Badics. Programming under probabilistic constraint with discrete random variable. In F. Giannessi, S. Komlósi, and T. Rapcsák, editors, *New Trends in Mathematical Programming*, pages 235–255. Springer, Boston, MA, USA, 1998.
25. R. Tyrrell Rockafellar, S. Uryasev, and M. Zabarankin. Generalized deviations in risk analysis. *Finance and Stochastics*, 10:51–74, 2006.
26. R.T. Rockafellar and S. Uryasev. Optimization of conditional value-at-risk. *Journal of Risk*, 2:21–41, 2000.
27. A. Ruszczyński. Probabilistic programming with discrete distributions and precedence constrained knapsack polyhedra. *Mathematical Programming*, 93:195–215, 2002.
28. A. Ruszczyński and A. Shapiro. Optimization of convex risk functions. *Mathematics of Operations Research*, 31(3):433–452, 2006.
29. R. Schultz. Risk aversion in two-stage stochastic integer programming. In G. Infanger, editor, *International Series in Operations Research & Management Science*, volume 150, pages 165–187. Springer Science+Business Media, New York, 2011.
30. R. Schultz and S. Tiedemann. Risk aversion via excess probabilities in stochastic programs with mixed-integer recourse. *SIAM J. on Optimization*, 14(1):115–138, 2003.
31. R. Schultz and S. Tiedemann. Conditional value-at-risk in stochastic programs with mixed-integer recourse. *Mathematical Programming*, 105:365–386, 2006.
32. R. Van Slyke and R.-B. Wets. L-shaped linear programs with application to optimal control and stochastic programming. *SIAM Journal on Applied Mathematics*, 17:638–663, 1969.
33. M. Tanner and L. Ntaimo. IIS branch-and-cut for joint chance-constrained stochastic programs and application to optimal vaccine allocation. *European Journal of Operational Research*, 207(1):290–296, 2010.
34. R. J-B. Wets. Stochastic programs with fixed recourse: The equivalent deterministic problem. *SIAM Review*, 16:309–339, 1974.

Part II
Modeling and Example Applications

Chapter 3
Modeling and Illustrative Numerical Examples

3.1 Introduction

To begin, let us restate the two-stage mean-risk stochastic linear programming (MR-SLP) model from Chap. 2:

$$\underset{x \in X}{\text{Min}} \; \mathbb{E}[f(x, \tilde{\omega})] + \lambda \mathbb{D}[f(x, \tilde{\omega})], \tag{3.1}$$

where $\mathbb{E} : \mathscr{F} \mapsto \mathbb{R}$ denotes the expected value, $\mathbb{D} : \mathscr{F} \mapsto \mathbb{R}$ is the *risk measure*, and $\lambda \geq 0$ is a suitable weight factor that quantifies the trade-off between expected cost and risk. The problem is risk-neutral if $\lambda := 0$. We assume that risk measure $\mathbb{D}$ is chosen so that the problem remains a *convex optimization problem*, allowing it to be solved using convex optimization methods. The set $X = \{Ax \geq b, \; x \geq 0\}$ is a nonempty polyhedron that defines the set of first-stage feasible solutions. The matrix $A \in \mathbb{R}^{m_1 \times n_1}$ and vector $b \in \mathbb{R}^{m_1}$ are the first-stage matrix and right hand side (RHS) vector, respectively. The family of real random cost variables $\{f(x, \tilde{\omega})\}_{x \in X} \subseteq \mathscr{F}$ is defined on $(\Omega, \mathscr{A}, \mathbb{P})$, where $\mathscr{F}$ is the space of all real random cost variables $f : \Omega \mapsto \mathbb{R}$ satisfying $\mathbb{E}[|f(\tilde{\omega})|] < \infty$. For a given $x \in X$, the real random cost variable $f(x, \tilde{\omega})$ is given by

$$f(x, \tilde{\omega}) := c^\top x + \varphi(x, \tilde{\omega}). \tag{3.2}$$

For a given realization ω of $\tilde{\omega}$, the *recourse function* $\varphi(x, \omega)$ is given by

$$\begin{aligned} \varphi(x, \omega) :=& \text{Min } q(\omega)^\top y(\omega) \\ & \text{s.t. } Wy(\omega) \geq r(\omega) - T(\omega)x \\ & \quad y(\omega) \geq 0, \end{aligned} \tag{3.3}$$

L. Ntaimo, *Computational Stochastic Programming*, Springer Optimization and Its Applications 774, https://doi.org/10.1007/978-3-031-52464-6_3

where $q(\omega) \in \mathbb{R}^{n_2}$ is the second-stage cost vector and $y(\omega) \in \mathbb{R}_+^{n_2}$ is the recourse decision. The matrix $W \in \mathbb{R}^{m_2 \times n_2}$ is the *recourse matrix*, $T(\omega) \in \mathbb{R}^{m_2 \times n_1}$ is the *technology matrix*, and $r(\omega) \in \mathbb{R}^{m_2}$ is the RHS vector. By *scenario* ω we mean the realization of the stochastic problem data, i.e., $\omega := (q(\omega), T(\omega), r(\omega))$.

To ensure that Problem (3.1) is well-defined for computational purposes, we make several assumptions listed in Chap. 2. Here we shall focus on illustrating how to model a specific problem using MR-SLP and other approaches. Next, we give a formal statement of the numerical example.

3.2 Motivating Example

We begin with a simplified production planning problem that we shall use as a motivating example application throughout the book. We refer to this problem as the *abc-Production Planning Problem* or *abc*-PPP for short. This example problem is suitable for illustration because it embeds many key aspects of stochastic programming. We use it to illustrate how to formulate a given problem first as a deterministic linear program (LP) and then show how to extend it to the stochastic setting using different models of stochastic programming. In particular, we not only focus on MR-SLP but also include other risk models such as probabilistic (chance) constrained SP (PC-SP) discussed in Chap. 2.

Example 3.1 (The *abc*-Production Planning Problem—*abc*-PPP) Consider the problem of monthly production planning under demand uncertainty for a manufacturer that produces three products: *Product-a*, *Product-b*, and *Product-c*. The manufacture of each of these products requires raw material, *Material-abc*, and three manufacturing processes: *Process-1*, *Process-2*, and *Process-3*. The production process uses machines/equipment for each process and thus takes time and incurs costs. In this setting involving demand uncertainty, the materials and/or products and the processing time have to be acquired/manufactured *before* product demand becomes known. In other words, the resource acquisition decisions have to be made here-and-now before demand is realized. The amounts of each resource needed, cost, and maximum amount that can be acquired to make each of the products are given in Table 3.1. The table also shows the *testing and inspection cost* for each product after it is made and the planned demand amounts that are used in a deterministic setting.

(a) **Deterministic Setting.** Assume that the selling prices, resource costs, and demand (planned) are known and are as given in Table 3.1. Formulate problem *abc*-PPP as a linear program (LP) to maximize profit.
(b) **Stochastic Setting.** Under the stochastic setting assume that the selling prices and resource costs are known, but *product demand* is random, denoted $\tilde{\omega}$. Also, for simplicity, assume that there are five product demand realizations (scenarios)

Table 3.1 Monthly resource requirements for *abc*-PPP

Item	Product-a	Product-b	Product-c	Unit cost ($)	Available
Material-abc (units)	6	8	10	50	300
Process-1 (hrs.)	20	25	28	30	700
Process-2 (hrs.)	12	15	18	15	600
Process-3 (hrs.)	8	10	14	10	500
Total time available (hrs.)					1600
Testing and inspection cost ($)	50	75	100		
Planned demand (number)	23	18	20		
Selling price ($)	1200	1600	2000		

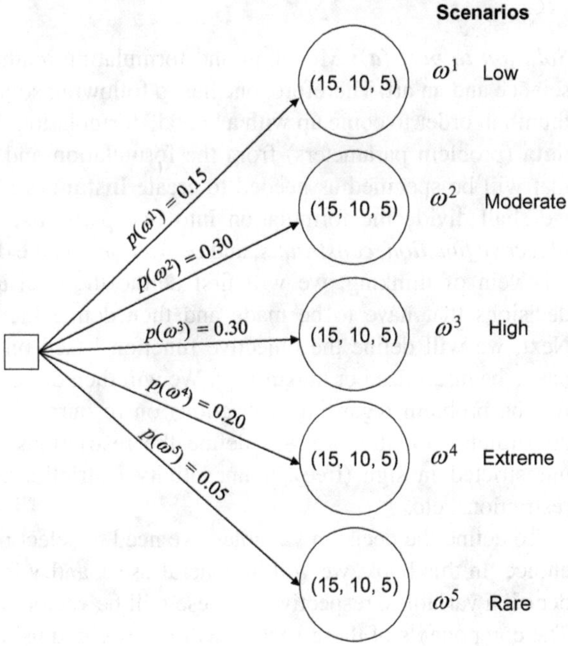

Fig. 3.1 Demand scenario tree for *abc*-PPP

of $\tilde{\omega}$, denoted $\omega^s := (d_a(\omega^s), d_b(\omega^s), d_c(\omega^s)),\ s = 1, \cdots, 5$. These scenarios correspond to *Low*, *Moderate*, *High*, *Extreme*, and *Rare* demand realizations.

The product demand realizations are depicted in the scenario tree in Fig. 3.1. The demand outcomes are given in Table 3.2. The collection of demand scenarios and the corresponding probabilities form a *multivariate probability distribution*.

Assuming that material and processing time are acquired before the beginning of the production period after which demand becomes known and then products are made, formulate problem *abc*-PPP as an MR-SLP to maximize profit.

We provide the solution to parts (a) and (b) in the next two subsections.

Table 3.2 Product demand scenarios for *abc*-PPP

	Scenario	Demand	Probability	
s	ω^s	$(d_a(\omega^s), d_b(\omega^s), d_c(\omega^s))$	$p(\omega^s)$	Description
1	ω^1	(15, 10, 5)	0.15	*Low*
2	ω^2	(20, 15, 15)	0.30	*Moderate*
3	ω^3	(25, 20, 25)	0.30	*High*
4	ω^4	(30, 25, 30)	0.20	*Extreme*
5	ω^5	(10, 10, 10)	0.05	*Rare*

3.2.1 Deterministic Setting

***Solution to part (a)*:** Modeling and formulating mathematical problems is both a science and an art. Therefore, one has to following some basic principles or rules of thumb in order to come up with a "good" formulation. To that end, we will separate **data** (problem parameters) from the formulation and treat the data as something that will be specified as needed to create **instances** of the problem. Furthermore, we shall divide the formulation into five parts: *decision variables*, *parameters*, *objective function*, *constraints*, and *restrictions* on the decision variables. Following this vein of thinking, we will first define the decision variables to capture the decisions that have to be made and then define the problem parameters (data). Next, we will define the objective function based on what performance measure has to be minimized or maximized. We will then discover and define the constraints for the problem regarding restrictions on resources, including money (budgetary constraints). Finally, we will define the restrictions on the decision variables—unrestricted in sign (free), nonnegativity restrictions, binary restrictions, integer restrictions, etc.

To define the decision variables, we need to select mathematical symbols of our choice. In this book, we will in general use x and y for the first- and second-stage decision variables, respectively. These will be vectors in $\mathbb{R}^{n_1}$ and $\mathbb{R}^{n_2}$, respectively. The components of these vectors will be specified using subscripts, e.g., x_i and y_j. We will generally use uppercase letters to denote mathematical sets. Returning to the problem at hand, let us define the decision variables as follows:

Decision Variables:

x_1 : Number of units of Material-abc purchased
x_2 : Number of hours of Process-1 purchased
x_3 : Number of hours of Process-2 purchased
x_4 : Number of hours of Process-3 purchased
y_1 : Number of units of Product-a produced
y_2 : Number of units of Product-b produced
y_3 : Number of units of Product-c produced

In vector notation, the decision variables can be given as $x = (x_1, x_2, x_3, x_4)^\top$ and $y = (y_1, y_2, y_3)^\top$.

Parameters:

m_a : Units of Material-abc needed to produce a unit of Product-a
m_b : Units of Material-abc needed to produce a unit of Product-b
m_c : Units of Material-abc needed to produce a unit of Product-c
m : Units of Material-abc available
c_1 : Cost of Material-abc ($/unit)
h_{1a} : Hours of Process-1 needed to produce a unit of Product-a
h_{1b} : Hours of Process-1 needed to produce a unit of Product-b
h_{1c} : Hours of Process-1 needed to produce a unit of Product-c
h_{2a} : Hours of Process-2 needed to produce a unit of Product-a
h_{2b} : Hours of Process-2 needed to produce a unit of Product-b
h_{2c} : Hours of Process-2 needed to produce a unit of Product-c
h_{3a} : Hours of Process-3 needed to produce a unit of Product-a
h_{3b} : Hours of Process-3 needed to produce a unit of Product-b
h_{3c} : Hours of Process-3 needed to produce a unit of Product-c
h_1 : Hours of Process-1 available
h_2 : Hours of Process-2 available
h_3 : Hours of Process-3 available
c_2 : Unit cost of hours for Process-1
c_3 : Unit cost of hours for Process-2
c_4 : Unit cost of hours for Process-3
H : Total hours available for all processes
t_a : Testing and inspection cost for Product-a ($ unit)
t_b : Testing and inspection cost for Product-b ($ unit)
t_c : Testing and inspection cost for Product-c ($ unit)
d_a : Demand for Product-a (units)
d_b : Demand for Product-b (units)
d_c : Demand for Product-c (units)
s_a : Selling price for Product-a ($)
s_b : Selling price for Product-b ($)
s_c : Selling price for Product-c ($)

Now that we have defined the decision variables and parameters, we can derive the objective function. In this case we want to maximize profit, which is simply total revenue from selling the products minus total production (material and processing) cost and testing and inspection cost.

Objective Function:

$$
\begin{aligned}
f(x, y) &= c^\top x + q^\top y \\
&= c_1x_1 + c_2x_2 + c_3x_3 + c_4x_4 + q_1y_1 + q_2y_2 + q_3y_3.
\end{aligned}
$$

Setting $q_1 = -(s_a - t_a)$, $q_2 = -(s_b - t_b)$, and $q_3 = -(s_c - t_c)$, we get

$$f(x, y) = c_1x_1 + c_2x_2 + c_3x_3 + c_4x_4 - (s_a - t_a)y_1 - (s_b - t_b)y_2 - (s_c - t_c)y_3.$$

Constraints: In this problem, we have three main types of constraints to derive: capacity, production, and demand constraints. The capacity constraints limit the amount of material and the number of processing hours by the available amounts. Production constraints link material processing to products, while demand constraints limit the number of products up to the demand.

We are now in a position to state the general LP formulation for *abc*-PPP as follows:

$$
\begin{aligned}
\text{Min } & c_1x_1 + c_2x_2 + c_3x_3 + c_4x_4 - q_1y_1 - q_2y_2 - q_3y_3 && \text{(3.4a)} \\
\text{s.t. } & -x_1 \geq -m && \text{(3.4b)} \\
& -x_2 \geq -h_1 && \text{(3.4c)} \\
& -x_3 \geq -h_2 && \text{(3.4d)} \\
& -x_4 \geq -h_3 && \text{(3.4e)} \\
& -x_2 - x_3 - x_4 \geq -H && \text{(3.4f)} \\
& x_1 - m_ay_1 - m_by_2 - m_cy_3 \geq 0 && \text{(3.4g)} \\
& x_2 - h_{1a}y_1 - h_{1b}y_2 - h_{1c}y_3 \geq 0 && \text{(3.4h)} \\
& x_3 - h_{2a}y_1 - h_{2b}y_2 - h_{2c}y_3 \geq 0 && \text{(3.4i)} \\
& x_4 - h_{3a}y_1 - h_{3b}y_2 - h_{3c}y_3 \geq 0 && \text{(3.4j)} \\
& -y_1 \geq -d_a && \text{(3.4k)} \\
& -y_2 \geq -d_b && \text{(3.4l)} \\
& -y_3 \geq -d_c && \text{(3.4m)} \\
& x_1,\ x_2,\ x_3,\ x_4,\ y_1,\ y_2,\ y_3 \geq 0. && \text{(3.4n)}
\end{aligned}
$$

Notice that we have intentionally written the problem as a *minimization* problem and therefore have reversed the objective function to be the total cost minus total revenue. So the negative of the objective function value is equal to the profit, which is what we want to maximize. The objective function is given in (3.4a). Constraints (3.4b–3.4f) are the capacity constraints on the available Material-abc, Process-1 hours, Process-2 hours, Process-3 hours, and the total available hours for processing, respectively. Constraints (3.4g–3.4j) are the production linking constraints and constraints (3.4k–3.4m) correspond to the product demand constraints. Lastly, constraint (3.4n) imposes the nonnegativity restrictions on all the decision variables.

We should point out that this problem can be formulated as a mixed-integer programming (MIP) problem by imposing integer requirements on all decision involving integrality. To write an *instance* of Problem (3.4), we need to specify data for all the problem parameters. Using the data given in Table 3.1, we can write an LP instance of the deterministic *abc*-PPP as follows:

$$
\begin{aligned}
\text{Min } & 50x_1 + 30x_2 + 15x_3 + 10x_4 - 1150y_1 - 1525y_2 - 1900y_3 \\
\text{s.t. } & -x_1 \geq -300 \\
& -x_2 \geq -700 \\
& -x_3 \geq -600 \\
& -x_4 \geq -500 \\
& -x_2 - x_3 - x_4 \geq -1600 \\
& x_1 - 6y_1 - 8y_2 - 10y_3 \geq 0 \\
& x_2 - 20y_1 - 25y_2 - 28y_3 \geq 0 \\
& x_3 - 12y_1 - 15y_2 - 18y_3 \geq 0 \\
& x_4 - 8y_1 - 10y_2 - 14y_3 \geq 0 \\
& -y_1 \geq -23 \\
& -y_2 \geq -18 \\
& -y_3 \geq -20 \\
& x_1,\ x_2,\ x_3,\ x_4,\ y_1,\ y_2,\ y_3 \geq 0.
\end{aligned}
\tag{3.5}
$$

3.2.2 *Stochastic Setting*

***Solution to Part (b)*:** In the two-stage stochastic setting, we have to make here-and-now decisions in the first stage and then take corrective actions in the second stage. Therefore, we can divide the decisions into the first- and second-stage decision variables, i.e., $x = (x_1, x_2, x_3, x_4)^\top$ and $y = (y_1, y_2, y_3)^\top$, respectively. For the sake of illustrating modeling, the decision variables are restated below as follows:

First-stage:

x_1: Number of units of material acquired.
x_2: Number of Process-1 hours acquired.

x_3: Number of Process-2 hours acquired.
x_4: Number of Process-3 hours acquired.

Second-stage:

y_1: Number of Product-a produced.
y_2: Number of Product-b produced.
y_3: Number of Product-c produced.

We are now in a position to state a two-stage MR-SLP formulation of problem *abc*-PPP as follows:

$$\text{Min } \mathbb{E}[f(x,\tilde{\omega})] + \lambda\mathbb{D}[f(x,\tilde{\omega})] \tag{3.6a}$$
$$\text{s.t.} \quad -x_1 \geq -m \tag{3.6b}$$
$$-x_2 \geq -h_1 \tag{3.6c}$$
$$-x_3 \geq -h_2 \tag{3.6d}$$
$$-x_4 \geq -h_3 \tag{3.6e}$$
$$-x_2 - x_3 - x_4 \geq -H \tag{3.6f}$$
$$x_1,\ x_2,\ x_3,\ x_4 \geq 0, \tag{3.6g}$$

where for a given first-stage solution $x = (x_1, x_2, x_3, x_4)^\top$ and for a realization (scenario) ω^s of $\tilde{\omega}$, $s = 1, \cdots, 5$, the function

$$f(x,\omega^s) := c^\top x + \varphi(x,\omega^s).$$

The first-stage cost function is

$$c^\top x = c_1x_1 + c_2x_2 + c_3x_3 + c_4x_4,$$

while the recourse function $\varphi(x,\omega^s)$ is given by

$$\varphi(x,\omega^s) := \text{Min } -q_1y_1 - q_2y_2 - q_3y_3 \tag{3.7a}$$
$$\text{s.t.} \quad -m_ay_1 - m_by_2 - m_cy_3 \geq -x_1 \tag{3.7b}$$
$$-h_{1a}y_1 - h_{1b}y_2 - h_{1c}y_3 \geq -x_2 \tag{3.7c}$$
$$-h_{2a}y_1 - h_{2b}y_2 - h_{2c}y_3 \geq -x_3 \tag{3.7d}$$
$$-h_{3a}y_1 - h_{3b}y_2 - h_{3c}y_3 \geq -x_4 \tag{3.7e}$$
$$-y_1 \geq -d_a(\omega^s) \tag{3.7f}$$
$$-y_2 \geq -d_b(\omega^s) \tag{3.7g}$$
$$-y_3 \geq -d_c(\omega^s) \tag{3.7h}$$
$$y_1,\ y_2,\ y_3 \geq 0. \tag{3.7i}$$

Notice that we have suppressed the dependency of y on ω since solving Problem (3.7) for a given scenario ω specifies a solution for that scenario. Otherwise, one can replace y_1, y_2, and y_3 with $y_1(\omega^s), y_2(\omega^s)$, and $y_3(\omega^s)$, respectively. Consequently, the second-stage cost function can be explicitly expressed as

$$\begin{aligned} q^\top y(\omega^s) &= q_1 y_1(\omega^s) + q_2 y_2(\omega^s) + q_3 y_3(\omega^s) \\ &= -(s_a - t_a) y_1(\omega^s) - (s_b - t_b) y_2(\omega^s) - (s_c - t_c) y_3(\omega^s). \end{aligned}$$

Because the first-stage solution $x = (x_1, x_2, x_3, x_4)^\top$ is data in Problem (3.7), we put it on the RHS in constraints (3.7b–3.7e). Also, notice that the randomness (demand) appears only on the RHS of constraints (3.7f–3.7h).

We are now in a position to specify an instance of the MR-SLP for *abc*-PPP under different risk measures $\mathbb{D}$. Based on the given data, we can write an instance of the problem as follows:

$$\begin{array}{llllll} \text{Min} & \mathbb{E}[f(x,\tilde{\omega})] & +\lambda\mathbb{D}[f(x,\tilde{\omega})] & & & \\ \text{s.t.} & -x_1 & & & \geq & -300 \\ & & -x_2 & & \geq & -700 \\ & & & -x_3 & \geq & -600 \\ & & & & -x_4 \geq & -500 \\ & & -x_2 & -x_3 & -x_4 \geq & -1600 \\ & x_1, & x_2, & x_3, & x_4 \geq & 0, \end{array}$$

where

$$f(x,\tilde{\omega}) := 50x_1 + 30x_2 + 15x_3 + 10x_4 + \varphi(x,\tilde{\omega}),$$

and for a given realization ω^s of $\tilde{\omega}$, $s = 1, \cdots, 5$, the second-stage subproblem is given as follows:

$$\begin{array}{lllllll} \varphi(x,\omega^s) = \text{Min} & -1150y_1 & -1525y_2 & -1900y_3 & & \\ \text{s.t.} & -6y_1 & -8y_2 & -10y_3 & \geq & -x_1 \\ & -20y_1 & -25y_2 & -28y_3 & \geq & -x_2 \\ & -12y_1 & -15y_2 & -18y_3 & \geq & -x_3 \\ & -8y_1 & -10y_2 & -14y_3 & \geq & -x_4 \\ & -y_1 & & & \geq & -d_a^s(\omega^s) \\ & & -y_2 & & \geq & -d_b^s(\omega^s) \\ & & & -y_3 & \geq & -d_c^s(\omega^s) \\ & y_1, & y_2, & y_3 & \geq & 0. \end{array}$$

The problem data can be summarized into first- and second-stage data as follows:

First-Stage:

$$c = (50, 30, 15, 10)^\top,\ A = \begin{bmatrix} -1 & 0 & 0 & 0 \\ 0 & -1 & 0 & 0 \\ 0 & 0 & -1 & 0 \\ 0 & 0 & 0 & -1 \\ 0 & -1 & -1 & -1 \end{bmatrix},$$

$b = (-300, -700, -600, -500, -1600)^\top.$

Second-Stage:

Number of scenarios $|\Omega| = 5$.

Scenario probabilities $p(\omega^s),\ s = 1, \cdots, 5$:

$p(\omega^1) = 0.15,\ p(\omega^2) = 0.30, p(\omega^3) = 0.30, p(\omega^4) = 0.20, p(\omega^5) = 0.05.$

Objective coefficient vector $q(\omega^s),\ s = 1, \cdots, 5$:

$q(\omega^1) = q(\omega^2) = \cdots = q(\omega^5) = (-1150, -1525, -1900)^\top.$

Recourse matrix:

$$W = \begin{bmatrix} -6 & -8 & -10 \\ -20 & -25 & -28 \\ -12 & -15 & -18 \\ -8 & -10 & -14 \\ -1 & 0 & 0 \\ 0 & -1 & 0 \\ 0 & 0 & -1 \end{bmatrix}.$$

Technology matrix:

$$T(\omega^1) = T(\omega^2) = \cdots = T(\omega^5) = T = \begin{bmatrix} 1 & 0 & 0 & 0 \\ 0 & 1 & 0 & 0 \\ 0 & 0 & 1 & 0 \\ 0 & 0 & 0 & 1 \\ 0 & 0 & 0 & 0 \\ 0 & 0 & 0 & 0 \\ 0 & 0 & 0 & 0 \end{bmatrix}.$$

RHS vectors $r(\omega^s), s = 1, \cdots, 5$:

$$r(\omega^1) = (0, 0, 0, 0, -15, -10, -5)^\top.$$

$$r(\omega^2) = (0, 0, 0, 0, -20, -15, -15)^\top.$$

$$r(\omega^3) = (0, 0, 0, 0, -25, -20, -25)^\top.$$

$$r(\omega^4) = (0, 0, 0, 0, -30, -25, -30)^\top.$$

$$r(\omega^5) = (0, 0, 0, 0, -10, -10, -10)^\top.$$

Expressing the data this way in vector and matrix form is necessary for decomposition algorithms, which are discussed in later chapters of this book. For now, we just want the reader to be conversant with how to do this.

3.3 Risk-Neutral Approaches

We shall now explore different solution approaches to the numerical example. We start with deterministic models and then move on to the risk-neutral recourse approach. We end the chapter by exploring risk-averse approaches, which include the mean-risk and probabilistic (join-chance) constrained models. Next, we start with examples of deterministic models.

3.3.1 *Linear Programming and Simple Profit Analysis*

To solve the deterministic version of *abc*-PPP in Example 3.1 (a) is to formulate the problem as an LP as given in Problem (3.5). Using an LP solver to solve this LP, we get an optimal value of −3280.00 and the following solution:

$x_1 = 244.8$
$x_1 = 700.0$
$x_3 = 444.0$
$x_4 = 336.0$
$y_1 = 0.$
$y_2 = 5.6$
$y_3 = 20.0.$

The LP model provides a profit of \$3280.00.

Another way to solve the deterministic *abc*-PPP is *simple per item profit analysis*. This is illustrated in Table 3.3. For each product we first compute the unit production cost (e.g., Product-a production cost is 6(\$50) + 20(\$30) + 12(\$15) + 8(\$10) + \$50 = \$1210) plus the testing and inspection cost. We then subtract the unit selling price to get the profit per item, which if positive indicates that we can produce the product for a profit. Then starting with the most profitable item (Product-c), we determine how many units (of Product-b) can be produced (Table 3.4) and then calculate the resource amounts as shown in Table 3.5. Because Product-a is not profitable (−\$10/unit), we do not produce any. We produce 20 units of Product-c (most profitable) and 5.6 units of Product-b (next profitable) at a profit of \$3280.00, which is the amount we determined using LP.

Table 3.3 Production solution for the *abc*-Production Planning instance

Item	Prod. Cost ($/unit)	Selling price ($)	Profit ($/unit)	Demand
Product-a	1210	1200	−10	23
Product-b	1550	1600	50	18
Product-c	1850	2000	150	20

Table 3.4 Amount of Product-b to produce after producing Product-c

	Material-abc	Process-1	Process-2	Process-3
Amount for producing Product-c	200	560	360	280
Amount remaining	100	140	240	220
Amount of Product-b to produce (min)	12.5	5.6	16	22

Table 3.5 Simple per item profit analysis solution

Resource	Decision	Amount
Material-abc (units)	x_1	244.8
Process-1 (hrs)	x_2	700.0
Process-2 (hrs)	x_3	444.0
Process-3 (hrs)	x_4	336.0
Product-a	y_1	0
Product-b	y_2	5.6
Product-c	y_3	20.0
Profit ($)		3280.00

Table 3.6 *abc*-Production Planning product demand scenarios

	Scenarios				
	Low	*Moderate*	*High*	*Extreme*	*Rare*
	1	2	3	4	5
Product-a	15	20	25	30	10
Product-b	10	15	20	25	10
Product-c	5	15	25	30	10
Probability	0.15	0.30	0.30	0.20	0.05

3.3.2 Expected Value Solution

A common approach in practice is to determine a solution based on the average of the random outcomes. Recall from Chap. 1 that this approach is referred to as the *expected value* (EV) solution. It involves replacing the random variables (e.g., $q(\omega)$, $r(\tilde{\omega})$, and $T(\tilde{\omega})$) by their expected values. This affords the advantage of having a relatively smaller problem to solve in terms of size (number of decision variables and number of constraints). However, the EV approach ignores the uncertainty as well as the risk involved and in general results in *underestimation* of the expected recourse function. For convenience, we re-state the *abc*-PPP instance demand scenarios in Table 3.6.

The expected demand for each product can be calculated as follows:

Product-a: $0.15 \times 15 + 0.3 \times 20 + 0.3 \times 25 + 0.2 \times 30 + 0.05 \times 10 = 22.25$.
Product-b: $0.15 \times 10 + 0.3 \times 15 + 0.3 \times 20 + 0.2 \times 25 + 0.05 \times 10 = 17.50$.
Product-c: $0.15 \times 5 + 0.3 \times 15 + 0.3 \times 25 + 0.2 \times 30 + 0.05 \times 10 = 19.25$.

Notice that the planned demand in Table 3.1 is actually rounded up values of the expected demand. The formulation for the EV problem can be written as follows:

$$
\begin{aligned}
z^{EV} := \text{Min } & 50x_1 + 30x_2 + 15x_3 + 10x_4 - 1150y_1 - 1525y_2 - 1900y_3 \\
\text{s.t.} \quad & -x_1 \geq -300 \\
& -x_2 \geq -700 \\
& -x_3 \geq -600 \\
& -x_4 \geq -500 \\
& -x_2 - x_3 - x_4 \geq -1600 \\
& x_1 - 6y_1 - 8y_2 - 10y_3 \geq 0 \\
& x_2 - 20y_1 - 25y_2 - 28y_3 \geq 0 \\
& x_3 - 12y_1 - 15y_2 - 18y_3 \geq 0 \\
& x_4 - 8y_1 - 10y_2 - 14y_3 \geq 0 \\
& -y_1 \geq -22.25 \\
& -y_2 \geq -17.50 \\
& -y_3 \geq -19.25 \\
& x_1,\ x_2,\ x_3,\ x_4,\ y_1,\ y_2,\ y_3 \geq 0.
\end{aligned}
\tag{3.8}
$$

Solving the EV problem, we get an optimal objective value of $z^{EV} := -3209.50$ and the solution with:

$x_1 = 244.02$
$x_2 = 700.00$
$x_3 = 443.10$
$x_4 = 333.90$
$y_1 = 0$
$y_2 = 6.44$
$y_3 = 19.25$.

Therefore, the EV provides a profit of \$3209.50, which is very close to the LP solution. We should point out that the data for many deterministic models in practice often represent the average case.

3.3.3 Scenario Analysis Solution

Another approach is the *scenario analysis* (SA) solution also referred to as the *wait-and-see (WS) solution* . This involves solving the deterministic LP problem for each scenario separately and then implementing one of the solutions based on some criteria. In this case we create copies of decision variables for each scenario. For convenience, let us re-index each decision variable $x_1, x_2, x_3, x_4, y_1, y_2$, and y_3 by adding scenario indices $s = 1, 2, \cdots, 5$ as follows: $x_1^s, x_2^s, x_3^s, x_4^s, y_1^s, y_2^s$, and y_3^s. Similarly, let us index the product demand realization for each scenario as follows: (d_a^s, d_b^s, d_c^s), $s = 1, 2, \cdots, 5$. The SA problem for scenario s is written as follows:

$$
\begin{aligned}
z_s^{SA} := \text{Min } & 50x_1^s + 30x_2^s + 15x_3^s + 10x_4^s - 1150y_1^s - 1525y_2^s - 1900y_3^s \\
\text{s.t. } & -x_1^s \geq -300 \\
& -x_2^s \geq -700 \\
& -x_3^s \geq -600 \\
& -x_4^s \geq -500 \\
& -x_2^s - x_3^s - x_4^s \geq -1600 \\
& x_1^s - 6y_1^s - 8y_2^s - 10y_3^s \geq 0 \\
& x_2^s - 20y_1^s - 25y_2^s - 28y_3^s \geq 0 \\
& x_3^s - 12y_1^s - 15y_2^s - 18y_3^s \geq 0 \\
& x_4^s - 8y_1^s - 10y_2^s - 14y_3^s \geq 0 \\
& -y_1^s \geq -d_a^s \\
& -y_2^s \geq -d_b^s \\
& -y_3^s \geq -d_c^s \\
& x_1^s, x_2^s, x_3^s, x_4^s, y_1^s, y_2^s, y_3^s \geq 0.
\end{aligned}
\tag{3.9}
$$

Solving the SA DEP (3.9) for each scenario, we get the solutions reported in Table 3.7. The WS solution z^{RP} is simply the expectation of the SA solutions, i.e., $z^{RP} := \sum_{s=1}^{5} p_s z_s^{SA} = \$3005.50$.

We should note that the basic structure of the SA formulation is the same as that of the deterministic LP as well as the EV problem. The only change is that the product demand is replaced by the amounts for each scenario. All the scenario solutions show that Product-a should not be produced. This is in agreement with the deterministic LP and EV solutions, which are based on the assumption of perfect information. However, in reality, the occurrence of each scenario is based on the given probabilities, and this information is being ignored.

Table 3.7 Scenario analysis (SA) solution for individual demand scenarios

		Scenarios s				
		Low	*Moderate*	*High*	*Extreme*	*Rare*
Item	Decision	1	2	3	4	5
Material-abc (units)	x_1^s	130.0	239.6	250.0	250.0	180.0
Process-1 (hrs)	x_2^s	390.0	700.0	700.0	700.0	530.0
Process-2 (hrs)	x_3^s	240.0	438 .0	450.0	450.0	330.0
Process-3 (hrs)	x_4^s	170.0	322.0	350.0	350.0	240.0
Product-a	y_1^s	0	0	0	0	0
Product-b	y_2^s	10	11.2	0	0	10
Product-c	y_3^s	5	15	25	25	10
Objective value, z_s^{SA}		−1250.00	−2810.00	−3750.00	−3750.00	−2000.00
Profit ($)		1250.00	2810.00	3750.00	3750.00	2000.00

For the SA approach, the question remains: which of the scenario solutions should one implement? Well, there is no easy answer to this question, and the choice is left to the decision-maker. For example, if the decision-maker is risk-neutral, they can pick a solution that is the closest to the risk-neutral case (Scenario 2, with the probability of occurrence of 0.3). However, there is a 70% chance that Scenario 2 will not occur. If Scenario 3 occurs, for example, *abc*-Production will produce 11.2 units of Product-b and 15 units of Product-c and will not satisfy the demand of 20 units of Product-b and 25 units of Product-c. In addition, instead of the $3750.00 profit that the model suggests getting under Scenario 3, *abc*-Production will realize a profit of $2810.00, occurring with a 30% chance.

Alternatively, the decision-maker may choose the solution suggested by scenarios 3 and 4 since both scenarios have the same solution. However, since the combined probability of occurrence of the scenarios is 0.5, there is still a 50% chance that any of the other three scenarios will occur. In summary, we should point out that the scenario solutions are best for one particular demand scenario. In general, it is desirable to get a solution that balances the impact of various scenarios.

3.3.4 Extreme Event Solution

Another approach is to consider all the outcomes in one formulation while keeping the decisions as in the deterministic LP or EV Problem (3.8). We refer to this approach as the "extreme event solution" (EES) since the model is driven by the "extreme" event or outcome. The formulation is written as follows:

$$
\begin{aligned}
z^{EES} := \text{Min } & 50x_1 + 30x_2 + 15x_3 + 10x_4 - 1150y_1 - 1525y_2 - 1900y_3 \\
\text{s.t. } & -x_1 \geq -300 \\
& -x_2 \geq -700 \\
& -x_3 \geq -600 \\
& -x_4 \geq -500 \\
& -x_2 - x_3 - x_4 \geq -1600 \\
& x_1 - 6y_1 - 8y_2 - 10y_3 \geq 0 \\
& x_2 - 20y_1 - 25y_2 - 28y_3 \geq 0 \\
& x_3 - 12y_1 - 15y_2 - 18y_3 \geq 0 \\
& x_4 - 8y_1 - 10y_2 - 14y_3 \geq 0 \\
& -y_1 \geq -15 \\
& -y_2 \geq -10 \\
& -y_3 \geq -5 \\
& -y_1 \geq -20 \\
& -y_2 \geq -15 \\
& -y_3 \geq -15 \\
& -y_1 \geq -25 \\
& -y_2 \geq -20 \\
& -y_3 \geq -25 \\
& -y_1 \geq -30 \\
& -y_2 \geq -25 \\
& -y_3 \geq -30 \\
& -y_1 \geq -10 \\
& -y_2 \geq -10 \\
& -y_3 \geq -10 \\
& x_1,\ x_2,\ x_3,\ x_4,\ y_1,\ y_2,\ y_3 \geq 0.
\end{aligned}
\tag{3.10}
$$

Solving the EES Problem (3.10), we obtain the solution summarized in Table 3.8. Notice that this solution corresponds to the SA solution for Scenario 1. The solution is driven by the demand scenario "Low" with a 15% chance of happening and is "very conservative." It is also interesting to realize that regardless of what scenario

Table 3.8 Extreme event solution (EES)

Resource	Decision	Amount
Material-abc (units)	x_1	130.0
Process-1 (hrs)	x_2	390.0
Process-2 (hrs)	x_3	240.0
Process-3 (hrs)	x_4	170.0
Product-a	y_1	0
Product-b	y_2	10
Product-c	y_3	5
Objective value, z^{EES}		−1250.00
Profit ($)		1250.00

occurs (there is an 85% chance that Scenario 1 will not occur), *abc*-Production will make the decisions given in Table 3.8.

3.3.5 Two-Stage Risk-Neural Recourse Model

A special case of the MR-SLP instance is the risk-neutral case, i.e., when $\lambda := 0$. We shall refer to this as the *recourse problem* (RP). The RP instance is written as follows:

$$
\begin{array}{rlllll}
z^{RP} := \text{Min} & 50x_1 & +30x_2 & +15x_3 & +10x_4 & +\sum_{s=1}^{5} p(\omega^s)\varphi(x,\omega^s) \\
\text{s.t.} & -x_1 & & & & \geq -300 \\
 & & -x_3 & & & \geq -700 \\
 & & & -x_4 & & \geq -600 \\
 & & & & -x_4 & \geq -500 \\
 & & -x_3 & -x_4 & -x_4 & \geq -1600 \\
 & x_1, & x_2, & x_3, & x_4 & \geq 0,
\end{array}
$$

where for a given scenario ω^s, $s = 1, \cdots, 5$, the second-stage subproblem is given as follows:

$$
\begin{array}{rllll}
\varphi(x,\omega^s) = \text{Min} & -1150y_1 & -1525y_2 & -1900y_3 & \\
\text{s.t.} & -6y_1 & -8y_2 & -10y_3 \geq & -x_1 \\
 & -20y_1 & -25y_2 & -28y_3 \geq & -x_2 \\
 & -12y_1 & -15y_2 & -18y_3 \geq & -x_3 \\
 & -8y_1 & -10y_2 & -14y_3 \geq & -x_4 \\
 & -y_1 & & \geq & -d_a^s(\omega^s) \\
 & & -y_2 & \geq & -d_b^s(\omega^s) \\
 & & & -y_3 \geq & -d_c^s(\omega^s) \\
 & y_1, & y_2, & y_3 \geq & 0.
\end{array}
$$

Let us denote the probability of occurrence of scenario ω^s by $p(\omega^s) = p^s$. Recall from Chap. 2 that we can write the deterministic equivalent problem (DEP) formulation in which we include all scenarios as a large-scale LP as follows:

$$
\begin{aligned}
z^{RP} := \text{Min } & 50x_1 + 30x_2 + 15x_3 + 10x_4 - \sum_{s=1}^{5} p^s(1150y_1^s + 1525y_2^s + 1900y_3^s) \\
\text{s.t. } & -x_1 \geq -300 \\
& -x_2 \geq -700 \\
& -x_3 \geq -600 \\
& -x_4 \geq -500 \\
& -x_2 - x_3 - x_4 \geq -1600 \\
& x_1 - 6y_1^s - 8y_2^s - 10y_3^s \geq 0, \ \forall s = 1, \cdots, 5 \\
& x_2 - 20y_1^s - 25y_2^s - 28y_3^s \geq 0, \ \forall s = 1, \cdots, 5 \\
& x_3 - 12y_1^s - 15y_2^s - 18y_3^s \geq 0, \ \forall s = 1, \cdots, 5 \\
& x_4 - 8y_1^s - 10y_2^s - 14y_3^s \geq 0, \ \forall s = 1, \cdots, 5 \\
& -y_1^s \geq -d_a^s, \ \forall s = 1, \cdots, 5 \\
& -y_2^s \geq -d_b^s, \ \forall s = 1, \cdots, 5 \\
& -y_3^s \geq -d_c^s, \ \forall s = 1, \cdots, 5 \\
& x_1, x_2, x_3, x_4, y_1^s, y_2^s, y_3^s \geq 0, \ \forall s = 1, \cdots, 5.
\end{aligned}
\tag{3.11}
$$

Problem (3.11) can be challenging to solve for a large number of scenarios. Solving this problem, we obtain the RP solution given in Table 3.9. Notice that, unlike the SA, the RP solution commits the first-stage here-and-now decisions (x_1, x_2, x_3, and x_4) regardless of what scenario occurs. Now we see that contrary to all the models we have looked at so far, the RP solution suggests producing Product-a if Scenario 1 or 5 occurs. Furthermore, the RP model suggests an expected profit of \$2268.50, which is lower than \$3209.50 suggested by the EV model. Clearly, the EV model provides a more optimistic profit. This is in fact expected as we pointed out earlier that, in general, the EV model underestimates the expected recourse function.

3.3.6 Putting Everything Together

In Table 3.10 we provide a summary of the output from all the models. We can now make several observations from this table. Notice that there is no meaningful

Table 3.9 Risk-neutral RP solution

			Scenarios s				
			Low	*Moderate*	*High*	*Extreme*	*Rare*
Resource	Decision	Amount	1	2	3	4	5
Material-abc (units)	x_1	236.4					
Process-1 (hrs)	x_2	690.0					
Process-2 (hrs)	x_3	432.0					
Process-3 (hrs)	x_4	318.0					
Product-a	y_1^s		15	0	0	0	8
Product-b	y_2^s		10	10.8	10.8	10.8	10
Product-c	y_3^s		5	15	15	15	10
Objective value, z^{RP}		−2268.50					
Profit ($)		2268.50					

relationship in terms of resource quantities between the solution to the RP and the solutions to the scenario analysis (SA) problems. Similarly, we do not see any connection between the solutions of the EV and that of SA. In contrast, there is a relationship between the extreme event solution (EES) and the scenario problem solutions: The SA solution corresponds to one of the scenario solutions, in this case, Scenario 1 (Low).

We should point out that the objective value of the RP corresponds to the *expected profit*, which is not the case for the EV and SA problems. Furthermore, it is interesting to see that the RP model is the only model that suggests that Product-a should be produced under demand Scenario 1 (Low) and Scenario 5 (Rare). This illustrates the flexibility of the RP model—it acquires enough resources to satisfy product demand in Low and Rare scenarios. When the product demand scenario is Low or Rare, the RP model is able to recover some of the resource expense by producing Product-a, which is preferable to producing and not being able to sell Product-b and Product-c. Furthermore, the RP model uses up all the resources purchased in the first stage, and in all scenarios, all the products produced are sold. The solutions provided by the SA model are optimistic in the sense that they are all based on each scenario and assume perfect information.

In Table 3.11, we compare the objective value associated with the solutions of the EV, EES, and SA models to the corresponding *expected profits* in the RP setting. We can clearly see that the RP model provides the highest expected profit of $2268.50. The lowest expected profit of $1250.00 is provided by the EES model and SA model for the Low scenario. Recall from Chap. 1 the *expected value of perfect information* (EVPI) and the *value of the stochastic solution* (VSS). EVPI is the difference between the SA solution (WS) and the RP solution (Equation 1.10). It is the amount the decision-maker/modeler is willing to pay for perfect information. On the other hand, VSS is the difference between the expectation of the EV value and RP value (Equation 1.11). It is a measure of the benefit of using the RP model

Table 3.10 Output summary for all models

			SA (WS) Scenarios s					RP				
Decision	EV	EES	1	2	3	4	5					
x_1	244.02	130.0	130.0	239.6	250.0	250.0	180.0	236.4				
x_2	700.0	390.0	390.0	700.0	700.0	700.0	530.0	690.0				
x_3	443.1	240.0	240.0	438.0	450.0	450.0	330.0	432.0				
x_4	333.9	170.0	170.0	350.0	350.0	350.0	240.0	318.0				
								Scenarios s				
								1	2	3	4	5
y_1^s	0	0	0	0	0	0	0	15	0	0	0	8
y_2^s	6.44	10	10	11.2	0	0	10	10	10.8	10.8	10.8	10
y_3^s	19.25	5	5	15	25	25	10	5	15	15	15	10
Profit ($)	3209.5	1250.0	1250.0	2810.0	3750.0	3750.0	2000.0	2268.5				

Table 3.11 Expected profit associated with each model's solution

Model			Profit suggested by model ($)	Expected profit ($)
RP				2268.50
EV			3209.50	2186.75
EES			1250.0,0	1250.00
SA			$3005.50	
	Scenario *s*			
	Low	1	1250.00	1250.00
	Moderate	2	2810.00	2195.25
	High	3	3750.00	2175.25
	Extreme	4	3750.00	2175.25
	Rare	5	2000.00	1782.50

over the EV model where a positive VSS value indicates that it is necessary to use the RP model. In this case,

$$EVPI := z^{RP} - z^{WS} = z^{RP} - \sum_{s=1}^{5} p_s z_s^{SA}$$

$$= \$2268.50 - \$3005.50$$

$$= -\$737.00.$$

This means that the decision-maker should be willing to pay $737.00 for perfect information on the product demand forecast. In terms of VSS, we have

$$VSS = \$2268.50 - \$2186.75$$

$$= \$81.75.$$

This indicates that it is necessary to use the RP model in this case.

3.4 Risk-Averse Approaches

Let us now turn to the risk-averse approaches for solving *abc*-PPP. We encourage the reader to review the definitions of the risk measures in Chap. 2. In this section, we simply restate the DEP formulations for the risk measures we consider assuming the reader knows their mathematical definitions. We then illustrate how to formulate *abc*-PPP for each risk measure. Specifically, we provide illustrations for two quantile measures, excess probability (EP) and conditional value-at-risk (CVaR) model, and one deviation risk measure, expected excess (EE). The rest of the risk measures are left as exercise problems at the end of the chapter; these include quantile deviation (QDEV) and absolute semideviation (ASD). These risk measures were originally introduced and defined in the following works: CVaR[3], QDEV [2, 4], ASD [2], EP [5], and EE [1].

Example 3.2 (The abc-Production Planning Problem—abc-PPP) Consider the stochastic version of problem abc-PPP described in Example 3.1, but now with risk considerations regarding making profit (or loss). The task at hand is to formulate and solve the problem as an MR-SLP for the following cases:

(a) Formulate the DEP for MR-SLP with the excess probability (EP) risk measure. Create and solve two separate instances, one with a profit target $\eta := 0$ and $\lambda := 1000$ and the other with $\eta := -2268.5$ and $\lambda := 1000$. How do the two solutions compare to each other and to the risk-neutral case (when $\lambda := 0$)?
(b) Formulate the DEP for MR-SLP with the conditional value-at-risk (CVaR) measure. Create and solve two separate instances, one with a profit target $\alpha := 0.85$ and $\lambda := 100$ and the other with $\alpha := 0.80$ and $\lambda := 100$. How do the two solutions compare to each other and to the risk-neutral case (when $\lambda := 0$)?
(c) Formulate the DEP for MR-SLP with the expected excess (EE) risk measure. Create and solve two separate instances, one with a profit target $\eta := 0$ and $\lambda := 1000$ and the other with $\eta := -2268.5$ and $\lambda := 100$. How do the two solutions compare to each other and to the risk-neutral case (when $\lambda := 0$)?

We will provide the solution to this example problem in the next subsections.

3.4.1 Excess Probability Model

Given $\lambda \geq 0$, target $\eta \in \mathbb{R}$, and bounded X, then there exists a constant $M > 0$ such that the DEP for EP can be given as follows:

$$
\begin{aligned}
\text{Min} \quad & c^\top x + \sum_{\omega \in \Omega} p(\omega) q(\omega)^\top y(\omega) + \lambda \sum_{\omega \in \Omega} p(\omega) \gamma(\omega) \\
\text{s.t.} \quad & x \in X \\
& T(\omega)x + Wy(\omega) \geq r(\omega), \ \forall \omega \in \Omega \\
& -c^\top x - q(\omega)^\top y(\omega) + M.\gamma(\omega) \geq -\eta, \ \forall \omega \in \Omega \\
& y(\omega) \in \mathbb{R}_+^{n_2}, \ \gamma(\omega) \in \{0, 1\}, \ \forall \omega \in \Omega.
\end{aligned}
\tag{3.12}
$$

Notice that M has be chosen so that it is large enough ("big M") to exceed the excess above the target η. Problem abc-PPP can be formulated as an MR-SLP with EP in the form of DEP (3.12) as follows:

$$
\begin{aligned}
\text{Min } & 50x_1+30x_2+15x_3+10x_4-\sum_{s=1}^{5} p^s(1150y_1^s+1525y_2^s+1900y_3^s)+\lambda\sum_{s=1}^{5} p^s\gamma^s \\
\text{s.t. } & -x_1 \geq -300 \\
& -x_2 \geq -700 \\
& -x_3 \geq -600 \\
& -x_4 \geq -500 \\
& -x_2 - x_3 - x_4 \geq -1600 \\
& x_1 - 6y_1^s - 8y_2^s - 10y_3^s \geq 0, \ \forall s = 1, \cdots, 5 \\
& x_2 - 20y_1^s - 25y_2^s - 28y_3^s \geq 0, \ \forall s = 1, \cdots, 5 \\
& x_3 - 12y_1^s - 15y_2^s - 18y_3^s \geq 0, \ \forall s = 1, \cdots, 5 \\
& x_4 - 8y_1^s - 10y_2^s - 14y_3^s \geq 0, \ \forall s = 1, \cdots, 5 \\
& -y_1^s \geq -d_a^s, \ \forall s = 1, \cdots, 5 \\
& -y_2^s \geq -d_b^s, \ \forall s = 1, \cdots, 5 \\
& -y_3^s \geq -d_c^s, \ \forall s = 1, \cdots, 5 \\
& -50x_1-30x_2-15x_3-10x_4 + 1150y_1^s + 1525y_2^s + 1900y_3^s + M.\gamma^s \geq -\eta, \\
& \qquad \forall s = 1, \cdots, 5 \\
& x_1, x_2, x_3, x_4, y_1^s, y_2^s, y_3^s, \gamma^s \in \{0, 1\}, \ \forall s = 1, \cdots, 5.
\end{aligned}
\tag{3.13}
$$

Notice that DEP (3.13) is an MIP and has to be solved using an MIP solver. Substituting for $\eta \in \{0, -2268.5\}$, $\lambda := 1000$, and $M := 50{,}000$, we solve the DEP to obtain the solutions given in Table 3.12. For the profit target $\eta := 0$ and $\lambda := 1000$ in part (a), the EP model provides an optimal solution with a profit of \$2166.40. This profit is lower than the risk-neutral RP profit of \$2268.50 shown in part (c). EP quantifies the excess probability for scenarios above the target profit η. Thus, we see that in this case $\gamma^s := 0$ for all s, meaning that no scenario is considered in the excess probability. In part (b), for $\eta := -2268.5$ and $\lambda := 1000$, the model provides an increased profit of \$2268.50, which is the same as for the RP case. In this case, $\gamma^1 = \gamma^5 := 1$, meaning that Scenarios 1 and 5 are considered in the excess probability. Overall, we see that the first- and second-stage decisions vary in both cases (a) and (b), while case (b) provides the same solution as the RP case.

Table 3.12 MR-SLP with EP solution

			Scenarios *s*				
			Low	*Moderate*	*High*	*Extreme*	*Rare*
Resource	Decision	Amount	1	2	3	4	5
(a) $\eta := 0$ and $\lambda := 1000$							
Material-abc (units)	x_1	234.5					
Process-1 (hrs)	x_2	690.0					
Process-2 (hrs)	x_3	429.8					
Process-3 (hrs)	x_4	312.9					
Product-a	y_1		15	0	0	0	8
Product-b	y_2		10	12.9	12.9	12.9	10
Product-c	y_3		5	13.2	13.2	13.2	10
Scenario variable	γ^s		0	0	0	0	0
Objective value		−2166.4					
Expected Profit ($)		2166.40					
(b) $\eta := -2268.5$ and $\lambda := 1000$							
Material-abc (units)	x_1	236.4					
Process-1 (hrs)	x_2	690.0					
Process-2 (hrs)	x_3	432.0					
Process-3 (hrs)	x_4	318.0					
Product-a	y_1		15	0	0	0	8
Product-b	y_2		10	10.8	10.8	10.8	10
Product-c	y_3		5	15	15	15	10
Scenario variable	γ^s		1	0	0	0	1
Objective value		−2068.5					
Expected profit ($)		2268.50					
(c) $\eta := -2268.5$ and $\lambda := 0$, Risk-Neutral RP							
Material-abc (units)	x_1	236.4					
Process-1 (hrs)	x_2	690.0					
Process-2 (hrs)	x_3	432.0					
Process-3 (hrs)	x_4	318.0					
Product-a	y_1		15	0	0	0	8
Product-b	y_2		10	10.8	10.8	10.8	10
Product-c	y_3		5	15	15	15	10
Scenario variable	γ^s		1	1	1	1	1
Objective value		−2268.5					
Expected profit ($)		2268.50					

3.4.2 Conditional Value-at-Risk Model

Given $\lambda \geq 0$, the DEP for CVaR can be given as follows:

$$\begin{aligned}
\text{Min} \quad & c^\top x + \sum_{\omega \in \Omega} p(\omega) q(\omega)^\top y(\omega) + \lambda \eta + \frac{\lambda}{1-\alpha} \sum_{\omega \in \Omega} p(\omega) v(\omega) \\
\text{s.t.} \quad & x \in X \\
& T(\omega)x + W y(\omega) \geq r(\omega), \ \forall \omega \in \Omega \\
& -c^\top x - q(\omega)^\top y(\omega) + \eta + v(\omega) \geq 0, \ \forall \omega \in \Omega \\
& \eta \in \mathbb{R}, \ y(\omega) \in \mathbb{R}_+^{n_2}, \ v(\omega) \in \mathbb{R}_+, \forall \omega \in \Omega.
\end{aligned} \tag{3.14}$$

Problem *abc*-PPP can be formulated as an MR-SLP with CVaR in the form of DEP (3.14) as follows:

$$\begin{aligned}
\text{Min } & 50x_1 + 30x_2 + 15x_3 + 10x_4 - \sum_{s=1}^{5} p^s (1150 y_1^s + 1525 y_2^s + 1900 y_3^s) + \\
& \lambda \eta + \frac{\lambda}{1-\alpha} \sum_{s=1}^{5} p^s \nu^s \\
\text{s.t. } & -x_1 \geq -300 \\
& -x_2 \geq -700 \\
& -x_3 \geq -600 \\
& -x_4 \geq -500 \\
& -x_2 - x_3 - x_4 \geq -1600 \\
& x_1 - 6y_1^s - 8y_2^s - 10y_3^s \geq 0, \ \forall s = 1, \cdots, 5 \\
& x_2 - 20y_1^s - 25y_2^s - 28y_3^s \geq 0, \ \forall s = 1, \cdots, 5 \\
& x_3 - 12y_1^s - 15y_2^s - 18y_3^s \geq 0, \ \forall s = 1, \cdots, 5 \\
& x_4 - 8y_1^s - 10y_2^s - 14y_3^s \geq 0, \ \forall s = 1, \cdots, 5 \\
& -y_1^s \geq -d_a^s, \ \forall s = 1, \cdots, 5 \\
& -y_2^s \geq -d_b^s, \ \forall s = 1, \cdots, 5 \\
& -y_3^s \geq -d_c^s, \ \forall s = 1, \cdots, 5 \\
& -50x_1 - 30x_2 - 15x_3 - 10x_4 + 1150 y_1^s + 1525 y_2^s + 1900 y_3^s + \eta + \nu^s \geq 0, \\
& \qquad \forall s = 1, \cdots, 5 \\
& x_1, \ x_2, \ x_3, \ x_4, \ y_1^s, \ y_2^s, \ y_3^s, \nu^s \geq 0, \eta \text{ free }, \ \forall s = 1, \cdots, 5.
\end{aligned} \tag{3.15}$$

Table 3.13 MR-SLP with CVaR solution

			Scenarios s				
			Low	*Moderate*	*High*	*Extreme*	*Rare*
Resource	Decision	Amount	1	2	3	4	5
(a) $\alpha := 0.85$ and $\lambda := 1$							
Material-abc (units)	x_1	130.0					
Process-1 (hrs)	x_2	390.0					
Process-2 (hrs)	x_3	240.0					
Process-3 (hrs)	x_4	170.0					
Product-a	y_1		0	0	0	0	0
Product-b	y_2		10	10	10	10	10
Product-c	y_3		5	5	5	5	5
Scenario variable	ν^s		0	0	0	0	0
VaR	η	−1250.0					
Objective value		−2500.0					
Expected profit ($)		1250.00					
(b) $\alpha := 0.80$ and $\lambda := 1$							
Material-abc (units)	x_1	180.0					
Process-1 (hrs)	x_2	530.0					
Process-2 (hrs)	x_3	330.0					
Process-3 (hrs)	x_4	240.0					
Product-a	y_1		7	0	0	0	0
Product-b	y_2		10	10	10	10	10
Product-c	y_3		5	10	10	10	10
Scenario variable	ν^s		1450	0	0	0	0
VaR	η	−2000.0					
Objective value		−2695.0					
Expected profit ($)		1782.50					
(c) $\alpha := 0.85$ and $\lambda := 0$, Risk-Neutral RP							
Material-abc (units)	x_1	236.4					
Process-1 (hrs)	x_2	690.0					
Process-2 (hrs)	x_3	432.0					
Process-3 (hrs)	x_4	318.0					
Product-a	y_1		15	0	0	0	8
Product-b	y_2		10	10.8	10.8	10.8	10
Product-c	y_3		5	15	15	15	10
Scenario variable	ν^s		1	0	0	0	0
VaR	η	−1250.0					
Objective value		−2268.5					
Expected profit ($)		2268.50					

Substituting for $\alpha \in \{0.85, 0.95\}$ and $\lambda := 100$, we solve DEP (3.15) to get the solutions given in Table 3.13. We see that with a quantile of $\alpha := 0.85$ and $\lambda := 1$ in part (a), the model provides an optimal solution with a profit of \$1250.00.

This profit is lower than the risk-neutral RP profit of \$2268.50 shown in part (c). CVaR quantifies the average profit of unlikely scenarios beyond the α confidence level and can be very conservative. Thus, we see that the value-at-risk (VaR) in this case is $\eta := -1250.00$. In part (b), for $\eta := 0.80$ and $\lambda := 1$, VaR increases to $\eta := -2000.00$, and the model provides an increased profit of \$1782.50. It is worth noting how both the first- and second-stage decisions vary in both cases (a) and (b) and how they differ from the risk-neutral case.

3.4.3 Expected Excess Model

Given $\lambda \geq 0$ and a target level $\eta \in \mathbb{R}$, a general DEP for EE can be formulated as follows:

$$
\begin{aligned}
\text{Min} \quad & c^\top x + \sum_{\omega \in \Omega} p(\omega) q(\omega)^\top y(\omega) + \lambda \sum_{\omega \in \Omega} p(\omega) v(\omega) \\
\text{s.t.} \quad & x \in X \\
& T(\omega)x + W y(\omega) \geq r(\omega), \ \forall \omega \in \Omega \\
& -c^\top x - q(\omega)^\top y(\omega) + v(\omega) \geq -\eta, \ \forall \omega \in \Omega \\
& y(\omega) \in \mathbb{R}_+^{n_2}, \ v(\omega) \in \mathbb{R}_+, \ \forall \omega \in \Omega.
\end{aligned}
\tag{3.16}
$$

Problem *abc*-PPP can be formulated as an MR-SLP with EE in the form of DEP (3.12) as follows:

$$
\begin{aligned}
\text{Min } & 50x_1 + 30x_2 + 15x_3 + 10x_4 - \sum_{s=1}^{5} p^s (1150 y_1^s + 1525 y_2^s + 1900 y_3^s) + \lambda \sum_{s=1}^{5} p^s v^s \\
\text{s.t.} \quad & -x_1 \geq -300 \\
& -x_2 \geq -700 \\
& -x_3 \geq -600 \\
& -x_4 \geq -500 \\
& -x_2 - x_3 - x_4 \geq -1600 \\
& x_1 - 6y_1^s - 8y_2^s - 10y_3^s \geq 0, \ \forall s = 1, \cdots, 5 \\
& x_2 - 20y_1^s - 25y_2^s - 28y_3^s \geq 0, \ \forall s = 1, \cdots, 5 \\
& x_3 - 12y_1^s - 15y_2^s - 18y_3^s \geq 0, \ \forall s = 1, \cdots, 5 \\
& x_4 - 8y_1^s - 10y_2^s - 14y_3^s \geq 0, \ \forall s = 1, \cdots, 5 \\
& -y_1^s \geq -d_a^s, \ \forall s = 1, \cdots, 5
\end{aligned}
$$

$$
\begin{aligned}
& - y_2^s \geq -d_b^s, \ \forall s = 1, \cdots, 5 \\
& - y_3^s \geq -d_c^s, \ \forall s = 1, \cdots, 5 \\
& - 50x_1 - 30x_2 - 15x_3 - 10x_4 + 1150y_1^s + 1525y_2^s + 1900y_3^s + \nu^s \geq -\eta, \\
& \qquad \forall s = 1, \cdots, 5 \\
& x_1, \ x_2, \ x_3, \ x_4, \ y_1^s, \ y_2^s, \ y_3^s, \nu^s \geq 0, \ \forall s = 1, \cdots, 5.
\end{aligned}
\tag{3.17}
$$

Substituting for $\eta \in \{0, -2268.5\}$ and $\lambda := 100$, we solve DEP (3.17) to obtain the solution given in Table 3.14. We see that with a profit target of $\eta := 0$ and $\lambda := 100$ in part (a), the model provides an optimal solution with a profit of \$2166.40. This profit is lower than the risk-neutral RP profit of \$2268.50 shown in part (c). In part (b), for a profit target of $\eta := -2268.5$ and $\lambda := 100$, the model provides a profit of \$1956.30, which is even lower. This illustrates how the risk-averse model allows the generation of first-stage solutions that hedge against extreme scenarios in the second-stage based on the target η as can be seen by the different scenario solutions for different values of η and λ. It is also interesting to note how the decisions differ from the risk-neutral (RP) case.

3.4.4 *Probabilistic (Chance) Constraints Model*

The last risk-averse method we are going to illustrate is probabilistically constrained (joint-chance) SP (PC-SP). In PC-SP we want the stochastic constraint to hold $\alpha.100\%$ of the time based on the decision-maker's level of risk $(1 - \alpha)$. This means that we allow the stochastic constraints to be violated by up to $(1 - \alpha).100\%$ of the time. This is a different mindset from the mean-risk measures. We want to formulate problem *abc*-PPP as a PC-SP considering a joint-chance constraint on satisfying the demand for each scenario $\alpha.100\%$ of the time. Specifically, we shall illustrate PC-SP for two cases:

Formulate and solve the DEP for PC-SP with $\alpha = 0.85$ and for $\alpha = 0.80$. How do the solutions compare to each other and to the risk-neutral case?

Recall from Chap. 2 that a generic linear PC-SP can be given as follows:

$$
\begin{aligned}
\text{Min } & f(x) \\
\text{s.t. } & x \in X \\
& \mathbb{P}(T(\tilde{\omega})x \geq r(\tilde{\omega})) \geq \alpha \\
& x \geq 0,
\end{aligned}
\tag{3.18}
$$

Table 3.14 MR-SLP with EE solution

			Scenarios s				
			Low	*Moderate*	*High*	*Extreme*	*Rare*
Resource	Decision	Amount	1	2	3	4	5
(a) $\eta := 0$ and $\lambda := 100$							
Material-abc (units)	x_1	234.5					
Process-1 (hrs)	x_2	690.0					
Process-2 (hrs)	x_3	429.8					
Process-3 (hrs)	x_4	312.9					
Product-a	y_1		15	0	0	0	8
Product-b	y_2		10	12.9	12.9	12.9 10	
Product-c	y_3		5	13.2	13.2	13.2	10
Scenario variable	ν^s		0	0	0	0	0
Objective value		−2166.4					
Expected Profit ($)		2166.40					
(b) $\eta := -2268.5$ and $\lambda := 100$							
Resource	Decision	Amount	1	2	3	4	5
Material-abc (units)	x_1	221.2					
Process-1 (hrs)	x_2	658.5					
Process-2 (hrs)	x_3	407.2					
Process-3 (hrs)	x_4	291.7					
Product-a	y_1		13.4	0	0	0	6.4
Product-b	y_2		10	15	15	15	10
Product-c	y_3		5	10.1	10.1	10.1	10
Scenario variable	ν^s		1923.3	0	0	0	473.3
Objective value		2925.9					
Expected Profit ($)		1956.30					
(c) $\eta := 0$ and $\lambda := 0$, Risk-Neutral RP							
Material-abc (units)	x_1	236.4					
Process-1 (hrs)	x_2	690.0					
Process-2 (hrs)	x_3	432.0					
Process-3 (hrs)	x_4	318.0					
Product-a	y_1		15	0	0	0	8
Product-b	y_2		10	10.8	10.8	10.8	10
Product-c	y_3		5	15	15	15	10
Scenario variable	ν^s		1	0	0	0	0
Objective value		−2268.5					
Expected Profit ($)		2268.50					

where $x \in \mathbb{R}^{n_1}$ is the decision variable vector, $T(\tilde{\omega}) \in \mathbb{R}^{m_1 \times n_1}$ is the technology matrix and $r(\tilde{\omega}) \in \mathbb{R}^{m_1}$ is the right hand side vector.

Let $M(\omega)$ be an appropriately sized scalar for scenario ω and let e be an appropriately dimensioned vector of ones. Let us define a binary decision variable $z(\omega)$ as follows: $z(\omega) = 1$ if under scenario ω at least one of the inequalities in the probabilistic constraint is violated, and $z(\omega) = 0$ otherwise. Then a DEP formulation for Problem (3.18) can be written as follows:

$$\begin{aligned} \text{Min } & c^\top x \\ \text{s.t. } & x \in X \\ & T(\omega)x + M(\omega)ez(\omega) \geq r(\omega), \forall \omega \in \Omega \\ & \sum_{\omega \in \Omega} p(\omega)z(\omega) \leq 1 - \alpha \\ & z(\omega) \in \{0, 1\}, \ \forall \omega \in \Omega. \end{aligned} \tag{3.19}$$

Notice that the DEP for PC-SP is an MIP. Problem *abc*-PPP can be formulated as a PC-SP in the form of DEP (3.19) as follows:

$$\begin{aligned} \text{Min } & 50x_1 + 30x_2 + 15x_3 + 10x_4 - 1150y_1 - 1525y_2 - 1900y_3 \\ \text{s.t. } & -x_1 \geq -300 \\ & -x_2 \geq -700 \\ & -x_3 \geq -600 \\ & -x_4 \geq -500 \\ & -x_2 - x_3 - x_4 \geq -1600 \\ & x_1 - 6y_1 - 8y_2 - 10y_3 \geq 0 \\ & x_2 - 20y_1 - 25y_2 - 28y_3 \geq 0 \\ & x_3 - 12y_1 - 15y_2 - 18y_3 \geq 0 \\ & x_4 - 8y_1 - 10y_2 - 14y_3 \geq 0 \\ & -y_1 + M^s z^s \geq -d_a^s, \ \forall s = 1, \cdots, 5 \\ & -y_2 + M^s z^s \geq -d_b^s, \ \forall s = 1, \cdots, 5 \\ & -y_3 + M^s z^s \geq -d_c^s, \ \forall s = 1, \cdots, 5 \\ & 0.15z^1 + 0.30z^2 + 0.30z^3 + 0.20z^4 + 0.05z^5 \leq 1 - \alpha \\ & x_1, x_2, x_3, x_4, y_1, y_2, y_3 \geq 0, \ z^s \in \{0, 1\}, \ \forall s = 1, \cdots, 5. \end{aligned} \tag{3.20}$$

Table 3.15 PC-SP solution

			Scenarios *s*				
			Low	*Moderate*	*High*	*Extreme*	*Rare*
Resource	Decision	Amount	1	2	3	4	5
(a) $\alpha := 0.85$							
Material-abc (units)	x_1	180.0					
Process-1 (hrs)	x_2	530.0					
Process-2 (hrs)	x_3	330.0					
Process-3 (hrs)	x_4	240.0					
Product-a	y_1	0					
Product-b	y_2	10					
Product-c	y_3	10					
Scenario variable	z^s		1	0	0	0	0
Objective value		−2000.0					
Profit ($)		2000.00					
(b) $\alpha := 0.80$							
Material-abc (units)	x_1	239.6					
Process-1 (hrs)	x_2	700.0					
Process-2 (hrs)	x_3	438.0					
Process-3 (hrs)	x_4	322.0					
Product-a	y_1	0					
Product-b	y_2	11.2					
Product-c	y_3	15					
Scenario variable	z^s		1	0	0	0	1
Objective value		−2810.0					
Profit ($)		2810.00					

Substituting for $\alpha \in \{0.85, 0.80\}$ and $M^s := 10$, for all $s = 1, \cdots, 5$, we solve DEP (3.20) to get the solutions given in Table 3.15. Unlike the MR-SLP model, the PC-SP model provides here-and-now solutions for both the first- and second-stage, i.e., specifies the (x, y)-decision. In Table 3.15, we see that with $\alpha = 0.85$ the model provides an optimal solution in which scenario $s = 1$ (with the probability of occurrence 0.15) is violated as indicated by $z^1 = 1$. This satisfies $1 - \alpha = 0.15$ as expected. By taking the risk of violating demand constraints for scenario $s = 1$, the profit is \$2000.00, which is below the (risk-neutral) RP profit of \$2268.50. However, increasing the risk level by setting $\alpha = 0.80$, the model provides an optimal solution that results in an optimal profit of \$2810.00, which is higher than the RP value. This time, however, two scenarios are excluded, Scenarios 1 and 5 (as indicated by $z^1 = z^5 = 1$). The total probability of the two scenarios sums to $1 - \alpha = 0.20$, which is the level of risk for the instance.

Table 3.16 Product demand scenarios

	Scenario	Demand	Probability	
s	ω^s	$(a(\omega^s), d_b(\omega^s), d_c(\omega^s))$	$p(\omega^s)$	Description
1	ω^1	(15, 10, 5)	0.10	*Low*
2	ω^2	(20, 15, 15)	0.30	*Moderate*
3	ω^3	(25, 20, 25)	0.30	*High*
4	ω^4	(30, 25, 30),	0.20	*Extreme*
5	ω^5	(10, 10, 10)	0.10	*Rare*

Problems

3.1 (Risk-Neutral SLP with Recourse) Consider the stochastic version of problem *abc*-PPP described in Example 3.1. You are given the demand distribution shown in Table 3.16.

(a) Formulate the recourse problem (RP), i.e., the risk-neural SLP with recourse, to maximize profit.
(b) Solve the problem and report your optimal solution. Denote the optimal value by z^*_{RP}.

3.2 (Deterministic Models) Consider problem *abc*-PPP given in Problem 3.1.

(a) Formulate and solve the expected value (EV) solution problem.
(b) Formulate and solve the extreme event solution (EES) problem.
(c) Formulate and solve the scenario analysis (SA) solution problem.
(d) Compute and compare the objective value associated with the solutions of the EV, EES, and SA models to the corresponding *expected profits* in the RP setting as in Table 3.11. Discuss your findings. Is it necessary to use the RP model in this case? Justify your answer.

3.3 (Excess Probability) Formulate the DEP for MR-SLP with the excess probability (EP) risk measure for problem *abc*-PPP given in Problem 3.1. Create and solve two separate instances, one with a profit target $\eta := 0$ and $\lambda := 1000$ and the other with $\eta = z^*_{RP}$ (from Problem 3.1) and $\lambda := 1000$. How do the two solutions compare to each other and to the risk-neutral case (when $\lambda := 0$)?

3.4 (Conditional Value-at-Risk) Formulate the DEP for MR-SLP with the Conditional Value-at-Risk (CVaR) measure for problem *abc*-PPP given in Problem 3.1. Create and solve two separate instances, one with a profit target $\alpha := 0.85$ and $\lambda := 100$ and the other with $\alpha := 0.95$ and $\lambda := 100$. How do the two solutions compare to each other and to the risk-neutral case (when $\lambda := 0$)?

3.5 (Expected Excess) Formulate the DEP for MR-SLP with the expected excess (EE) risk measure for problem *abc*-PPP given in Problem 3.1. Create and solve two separate instances, one with a profit target $\eta := 0$ and $\lambda := 1000$ and the other with

$\eta = z^*_{RP}$ (from Problem 3.1) and $\lambda := 100$. How do the two solutions compare to each other and to the risk-neutral case (when $\lambda := 0$)?

3.6 (Probabilistically (Joint-Chance) Constrained SP) Formulate the DEP for probabilistically (joint-chance) constrained stochastic program (PC-SP) for problem *abc*-PPP given in Problem 3.1. Create and solve two separate instances, one with a profit target $\alpha := 0.90$ and the other with $\alpha := 0.80$. How do the two solutions compare to each other and to the risk-neutral case (when $\lambda := 0$)?

3.7 (Quantile Deviation) Consider the stochastic version of problem *abc*-PPP described in Example 3.1. Formulate the DEP for MR-SLP with the quantile deviation (QDEV) risk measure. Create and solve two separate instances, one with a profit target $\alpha := 0.85$ and the other with $\alpha := 0.95$. How do the two solutions compare to each other and to the risk-neutral case (when $\lambda := 0$)? Note: $\alpha = \varepsilon_2/(\varepsilon_1 + \varepsilon_2)$, $\varepsilon_1 > 0$ and $\varepsilon_2 > 0$. Choose ϵ_1 and ϵ_2 to meet this condition and set λ values to satisfy $\lambda \in \{0, \frac{1}{\epsilon_1}\}$ as required.

3.8 (Expected Excess) Consider the stochastic version of problem *abc*-PPP described in Example 3.1. Formulate the DEP for MR-SLP with expected excess (EE) risk measure. Create and solve two separate instances, one with a profit target $\eta := 0$ and $\lambda := 100$ and the other with $\eta := -2268.5$ and $\lambda := 100$. How do the two solutions compare to each other and to the risk-neutral case (when $\lambda := 0$)?

3.9 (Absolute Semideviation) Consider the stochastic version of problem *abc*-PPP described in Example 3.1. Formulate the DEP for MR-SLP with the absolute semideviation (ASD) risk measure and solve for $\lambda \in \{0.5, 1\}$. Compare the solutions for each λ value relative to the risk-neutral solution. How do the solutions compare to each other and to the risk-neutral case (when $\lambda := 0$)?

3.10 (Quantile Deviation) Consider the stochastic version of problem *abc*-PPP described in Example 3.1. Formulate the DEP for MR-SLP with the QDEV risk measure and solve for $\alpha = \varepsilon_2/(\varepsilon_1 + \varepsilon_2) = 0.80$, $\varepsilon_1 > 0$ and $\varepsilon_2 > 0$. Choose $\epsilon_1 = 1$ and $\epsilon_2 = 4$, and $\lambda \in \{0, 0.5, 1\}$. Notice that these λ values satisfy $\lambda \in \{0, \frac{1}{\epsilon_1}\}$ as required. How do the solutions compare to each other and to the risk-neutral case (when $\lambda := 0$)?

3.11 (Risk Measures) Compare and contrast the different risk measures (see Chap. 2). Discuss when it is appropriate to use the risk-neutral solution versus using the risk-averse solution. Give examples of applications where you think each risk measure is appropriate and explain why.

References

1. A. Markert and R. Schultz. On deviation measures in stochastic integer programming. *Operations Research Letters*, 33(5):441–449, 2005.
2. W. Ogryczak and A. Ruszczyński. Dual stochastic dominance and related mean-risk models. *SIAM Journal on Optimization*, 13:60–78, 2002.

3. R.T. Rockafellar and S. Uryasev. Optimization of conditional value-at-risk. *Journal of Risk*, 2:21–41, 2000.
4. A. Ruszczyński and A. Shapiro. Optimization of convex risk functions. *Mathematics of Operations Research*, 31(3):433–452, 2006.
5. R. Schultz and S. Tiedemann. Risk aversion via excess probabilities in stochastic programs with mixed-integer recourse. *SIAM J. on Optimization*, 14(1):115–138, 2003.

Chapter 4
Example Applications of Stochastic Programming

4.1 Introduction

In the previous chapter, we introduced and illustrated different stochastic programming (SP) models, including risk-neutral and risk- averse models. One of the fundamental motivations of using SP is its variety of applications. Typical applications of SP involve data uncertainty and/or decision-maker risk-averseness. In this chapter, we preview a variety of example applications of SP to both motivate and expose the student to different application areas of SP. We consider a mix of classical and recent applications of SP. These include flexible manufacturing production planning, facility location, supply chain planning, fuel treatment planning, healthcare appointment scheduling, airport time slot allocation, air traffic flow management, satellite constellation scheduling, wildfire response planning, and vaccine allocation for epidemics. The example applications were carefully selected so as to exemplify different SP approaches, i.e., stochastic linear programming (SLP), stochastic mixed-integer programming (SMIP), and probabilistically constrained stochastic programming (PC-SP).

Because formulating SP problems is both a science and an art, we place emphasis on the modeling aspects with a goal of exposing the student to different ways of formulating SP problems. In particular, for each application we define the decision variables as well as problem data (model parameters). We then state the formulation, which includes the objective function, constraints, and restrictions on the decision variables. Historically, several applications have been modeled using SP. We describe in some level of detail a subset of these classical applications that are used as SP problem test instances in Chap. 10. Here we simply provide a brief description of each test instance in terms of its application area. SP test instances are typically named using acronyms, and they include gbd, pgp2, LandS, ssn, storm, cep1, 20term, SIZES, SEMI, DCAP, SSLP, MPTSP, SMKP, VACCINE, and PROBPORT.

L. Ntaimo, *Computational Stochastic Programming*, Springer Optimization and Its Applications 774, https://doi.org/10.1007/978-3-031-52464-6_4

Test instance *gbd* is an aircraft allocation problem described in [11] and deals with the optimal allocation of different types of aircraft to routes with uncertain demand. Test instance *pgp2* is described by [22] and appears in [14]. It is a power generation planning problem involving electrical capacity expansion with unknown load forecasts. The goal is to select the minimum cost strategy for investing in electricity generated from gas-fired, coal-fired, and nuclear power plants. *LandS* is described in [19] and deals with electrical investment planning, while *ssn* is a telecommunications network design problem under uncertain demand and is described in [31]. Test instance *storm* is described in [21] and deals with the allocation of aircraft routes by the US military during the Gulf War of 1991. Instance *cep1* is a two-stage machine capacity expansion planning (CEP) problem for a flexible manufacturing facility and is described in [14]. The instance *20term* deals with freight carrier operations planning and is described in [20] . The applications described so far are modeled using SLP.

The next set of applications is modeled using SMIP. The SIZES test instance is a two-stage multi-period SMIP problem of a product substitution application and is described in [16, 17]. SEMI is a two-stage multi-period SMIP problem for semiconductor tool purchases planning and is described in [6]. DCAP is a two-stage SMIP dynamic capacity acquisition and allocation problem under uncertainty and is described in [1, 2]. SSLP is a two-stage SMIP problem for server location under uncertainty and is described in [25], while MPTSP is a two-stage SMIP city logistics problem and is described in [26, 32]. SMKP is a two-stage stochastic multiple knapsack problem with a random second-stage vector and is described in [4]. The last two test cases, VACCINE and PROBPORT, are both PC-SP problems. VACCINE is a chance-constrained vaccine allocation problem in epidemiology and is described in [33, 34]. PROBPORT is a chance-constrained portfolio optimization problem and is described in [30].

In the next section, we use CEP to illustrate how to formulate an SLP problem for a specific application. Similarly, in Sect. 4.3 we provide a detailed formulation of SSLP. In each case, we state and describe the two-stage formulation followed by the deterministic equivalent problem (DEP) formulation. For the remainder of the chapter, we describe each example application of SP including reference(s) for further reading and give a simplified generic formulation of the problem. We introduce a stochastic supply chain planning problem in Sect. 4.4, fuel treatment planning in Sect. 4.5, appointment scheduling in nuclear medicine in Sect. 4.6, airport time slot allocation under uncertainty in Sect. 4.7, stochastic air traffic flow management in Sect. 4.8, satellite constellation scheduling under uncertainty in Sect. 4.9, wildfire response planning in Sect. 4.10, and vaccine allocation for epidemics in Sect. 4.11.

4.2 Capacity Expansion Problem (CEP)

We recall the *abc*-production planning problem (*abc*PPP) from Chap. 3 which involves monthly production planning under demand uncertainty for a manufacturer

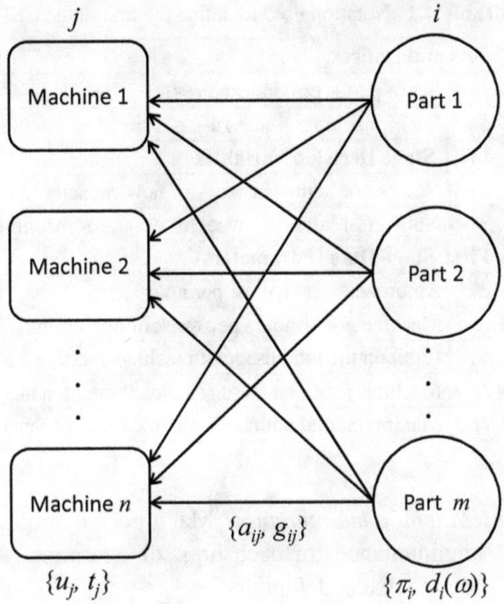

Fig. 4.1 An illustration of CEP: assignment of parts to machines

that produces multiple products. The manufacture of each of the products requires raw materials and a series of different manufacturing processes. The manufacturing process requires different machines for each process, which takes time to complete and incurs a cost. The stochastic setting of *abc*PPP involves product demand uncertainty such that the raw materials and manufacturing time have to be acquired before product demand becomes known. The resource acquisition decisions have to be made here-and-now before demand is realized. Related to *abc*PPP is CEP [14], which involves the task of planning for the expansion of productive capacity in a flexible manufacturing facility that produces different types of parts.

CEP can be defined as follows: Suppose you are given the task of planning for increasing productive capacity of a flexible manufacturing facility that produces m types of parts on n machines (see Fig. 4.1). Because the facility is a flexible manufacturing system, each of the machines under consideration is flexible in nature, and thus each part can be produced on any one of the n machines. The part types are indexed $i = 1, \cdots, m$, while the machines are indexed $j = 1, \cdots, n$. Machine j is currently available for h_j hours of operation per week, and additional hours may be acquired at an amortized weekly cost of c_j per hour.

The following information is also provided:

- *Production rate:* Part type i may be produced on machine j at rate a_{ij} with an associated cost of g_{ij} per hour.
- *Machine usage:* The total usage of machine j is constrained by an upper limit of u_j.

Table 4.1 Notation used to define the first-stage CEP model

Sets and Indices	
$\mathbb{I}$	Set of part types, indexed $i \in \mathbb{I} := \{i = 1, \cdots, m\}$
$\mathbb{J}$	Set of machines, indexed $j \in \mathbb{J} := \{j = 1, \cdots, n\}$
First-Stage Decision Variables	
x_j	Number of hours per week of new capacity that is acquired for machine $j \in J$
z_j	Number of hours per week of usage for machine $j \in J$. Define vector $z = (z_1, \cdots, z_n)^\top$
First-Stage Data (Parameters)	
c_j	Amortized weekly cost per hour for machine $j \in J$
h_j	Machine $j \in J$ hours per week of operation
u_j	Limit on the total usage of machine $j \in J$
t_j	Machine $j \in J$ hours of scheduled maintenance for each hour of operation
H	Maximum total hours of scheduled maintenance

- *Machine maintenance:* Machine j is required to undergo t_j hours of scheduled maintenance for each hour of operation, and the total scheduled maintenance cannot exceed T hours.
- *Production plan:* Weekly production plans are determined by the demand for parts, which varies on a weekly basis. Thus, the demand for part type i in any given week is represented by the random variable $\tilde{\omega}$.
- *Part demand:* Weekly demands are independent and identically distributed (IID) random variables. Upon learning the demand profile for a given week, the allocation of parts to machines is done on a least cost basis.
- *Inventory:* Management has recommended that there should not be any inventory, and furthermore, there is a penalty π_i for each unit of unsatisfied demand for part type i. The cost π_i may be thought of as either a penalty for a lost sale or the additional cost required to satisfy demand by outsourcing.

The problem at hand is to formulate CEP as a two-stage SLP with recourse to minimize the amortized expansion cost plus the expected weekly production costs. Production planning is usually postponed until better information (demand) is available. We define the notation we use to formulate CEP in Tables 4.1 and 4.2.

We are now in a position to write the formulation for CEP. We should note that for a fixed set of capacity values $(x_1, \ldots, x_n)$, the utilization y_{ij} and the corresponding unmet demand s_i vary according to weekly demand realization. The two-stage SLP formulation for CEP can be written as follows:

$$\text{Min} \sum_{j \in J} c_j x_j + \mathbb{E}[f(z, \tilde{\omega})] \tag{4.1a}$$

$$\text{s.t.} \; -x_j + z_j \leq h_j \qquad \forall j \in J \tag{4.1b}$$

$$\sum_{j \in J} t_j z_j \leq H \tag{4.1c}$$

Table 4.2 Notation used to define the second-stage CEP model

Sets and Indices	
Ω	Set of outcomes (scenarios) for parts demand, indexed $\omega \in \Omega$
Second-Stage Decision Variables	
y_{ij}	Number of hours per week that machine $j \in J$ is devoted to the production of part $i \in I$
s_i	Unsatisfied demand for part $i \in I$
Second-Stage Data (Parameters)	
$\tilde{\omega}$	Multivariate random variable whose outcome (scenario) ω specifies the part type demand, i.e., $\omega := \{d_i^\omega\}_{i \in \mathbb{I}},\ \omega \in \Omega$
p_ω	Probability of occurrence of scenario $\omega \in \Omega$
d_i^ω	Amount of part type $i \in \mathbb{I}$ demand realized under scenario $\omega \in \Omega$
g_{ij}	Hourly cost of production of part $i \in I$ on machine $j \in J$
a_{ij}	Hourly rate of production of part $i \in I$ on machine $j \in J$
π_i	Penalty for each unit of unsatisfied demand for part $i \in I$

$$z_j \le u_j \qquad \forall j \in J \tag{4.1d}$$

$$x_j,\ z_j \ge 0 \qquad \forall j \in J, \tag{4.1e}$$

where for each outcome (scenario) $\omega \in \Omega$ of $\tilde{\omega}$, the second-stage problem is given by

$$f(z, \omega) = \text{Min} \sum_{i \in I} \sum_{j \in J} g_{ij} y_{ij} + \sum_{i \in I} \pi_i s_i \tag{4.2a}$$

$$\text{s.t.} \sum_{j \in J} a_{ij} y_{ij} + s_i \ge d_i^\omega \quad \forall i \in I \tag{4.2b}$$

$$z_j - \sum_{i \in I} y_{ij} \ge 0 \qquad \forall j \in J \tag{4.2c}$$

$$y_{ij},\ s_i \ge 0 \qquad \forall i \in I,\ j \in J. \tag{4.2d}$$

The objective function (4.1a) minimizes the amortized expansion cost plus the expected weekly production costs. Constraints (4.1b) enforce the requirement that the number of hours per week of usage of each machine to produce each part type must not exceed the assigned weekly machine hours of operation plus the newly acquired hours. Constraint (4.1c) bounds the total number of hours of machine operation before scheduled maintenance. The requirement that weekly number of hours of machine operation must not exceed the limit of total usage hours is given by constraints (4.1d). Constraints (4.1e) are nonnegativity restrictions on the first-stage decision variables.

In the second-stage, the objective function (4.2a) minimizes the weekly production costs, i.e., the cost of producing the parts plus the cost of unmet demand. Constraints (4.2b) enforce the requirement that demand must be satisfied without building any inventories. The requirement that the number of hours per week of usage of each machine to produce each part type must not exceed the number of acquired hours in the first-stage is given by (4.2c). Finally, constraints (4.2d) are nonnegativity restrictions on the second-stage decision variables.

We can also write the DEP or extensive formulation for CEP. To do this, we have to assume that $\tilde{\omega}$ is discretely distributed with $|\Omega| < \infty$ and with each outcome $\omega \in \Omega$ having a probability of occurrence p_ω, where $\sum_{\omega \in \Omega} p_\omega = 1$. We shall now explicitly write the second-stage decision variables for each outcome $\omega \in \Omega$ explicitly as, y_{ij}^ω and s_i^ω, for all $i \in I$ and $j \in J$. We can now write the DEP formulation for CEP as follows:

$$
\begin{aligned}
\text{Min} \ & \sum_{j \in J} c_j x_j + \sum_{\omega \in \Omega} p_\omega \Big(\sum_{i \in I} \sum_{j \in J} g_{ij} y_{ij}^\omega + \sum_{i \in I} \pi_i s_i^\omega\Big) && (4.3) \\
\text{s.t.} \ & -x_j + z_j \leq h_j, && \forall j \in J \\
& \sum_{j \in J} t_j z_j \leq H \\
& z_j \leq u_j, && \forall j \in J \\
& z_j - \sum_{i \in I} y_{ij}^\omega \geq 0, && \forall j \in J,\ \omega \in \Omega \\
& \sum_{j \in J} a_{ij} y_{ij}^\omega + s_i^\omega \geq d_i^\omega, && \forall i \in I,\ \omega \in \Omega \\
& x_j,\ z_j \geq 0, && \forall j \in J \\
& y_{ij}^\omega,\ s_i^\omega \geq 0, && \forall i \in I,\ j \in J,\ \omega \in \Omega.
\end{aligned}
$$

We can also characterize the size of CEP using the number decision variables (columns) and the number constraints (rows). We observe that the number of first-stage decision variables is $n_1 := 2n$, where n is the number of machines. The number of first-stage constraints is $m_1 := 2n + 1$. Similarly, the number of second-stage decision variables is $n_2 := m(n+1)$, while the number of second-stage constraints is $m_2 := m+n$, where m is the number of part types. Letting the number of scenarios $S = |\Omega|$, the DEP has $2n+m(n+1)S$ decision variables and $2n+(m+n)S+1$ constraints. For example, for a CEP instance with $m = 10$, $n = 10$, and $S = 1000$, the DEP has $2n+m(n+1)S = 2(10)+(10)(10+1)(1000) = 110,020$ decision variables and $2n+(m+n)S+1 = 2(10)+(10+10)(1000)+1 = 20{,}021$ constraints. Although LP solvers can be applied directly to solve DEP (4.3), it is computationally challenging to solve instances with a large of scenarios. In fact, for some extremely large-scale instances, even reading the data into the LP solver becomes prohibitive.

4.3 Stochastic Server Location Problem (SSLP)

Another classical application of SMIP is SSLP [25]. This problem involves making optimal strategic decisions regarding where to locate "servers" here-and-now under future uncertainty in resource demand based on whether or not a "client" will be available in the future for service. This problem arises in different applications where "servers" (e.g., facilities) have to be located and built to provide a resource/service to "clients" (e.g., customers) before the availability (and/or demand) of these "clients" is known. Naturally, this problem arises in fields such as telecommunications, supply chain planning, electric power planning, and wildfire response planning.

SSLP involves n potential server locations to serve a set of m potential clients in the future whose availability is unknown but follows a Bernoulli distribution. Typically, $n < m$ and the potential server locations are divided into multiple zones, each with a minimum number of servers that must be located. An illustration of SSLP is given in Fig. 4.2. The notation needed to formulate SSLP is summarized in Tables 4.3 and 4.4.

The two-stage SSLP problem can be written as follows:

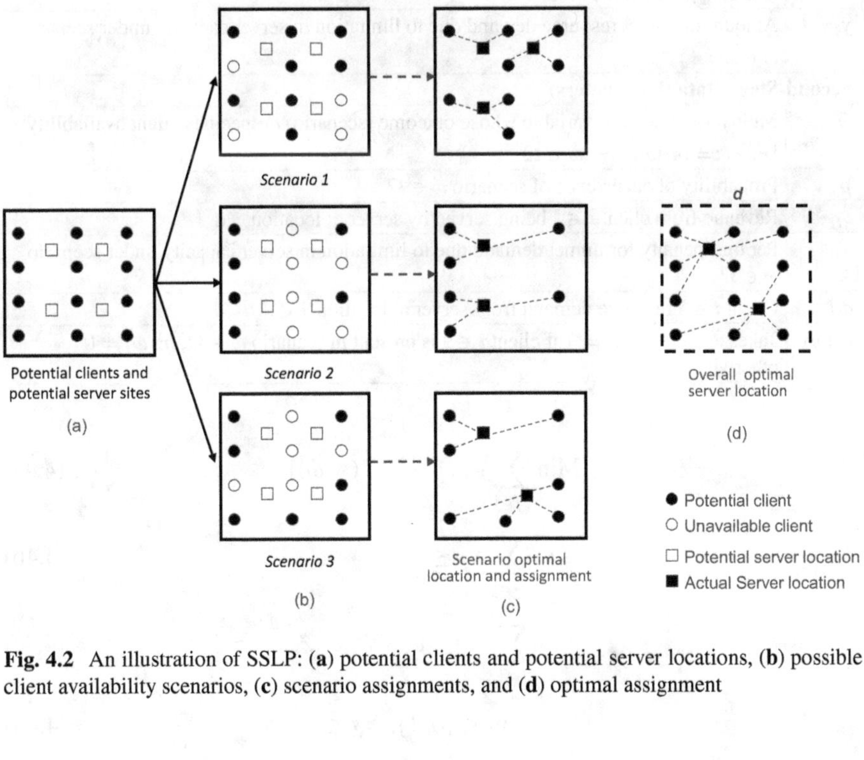

Fig. 4.2 An illustration of SSLP: (**a**) potential clients and potential server locations, (**b**) possible client availability scenarios, (**c**) scenario assignments, and (**d**) optimal assignment

Table 4.3 Notation used to define the first-stage SSLP model

Sets and Indices	
$\mathbb{I}$	Set of clients, indexed $i \in \mathbb{I} := \{i = 1, \cdots, m\}$
$\mathbb{J}$	Set of potential server locations, indexed $j \in \mathbb{J} := \{j = 1, \cdots, n\}$
$\mathscr{Z}$	Set of zones across potential server locations, indexed $z \in \mathscr{Z}$
$\mathbb{J}_z$	Subset of potential server locations in zone $z \in \mathscr{Z}$, indexed $j \in \mathbb{J}_z \subset \mathbb{J}$
First-Stage Decision Variables	
x_j	Decision variable $x_j = 1$ if a server is located at potential site $j \in \mathbb{J}$, $x_j = 0$ otherwise
First-Stage Data (Parameters)	
c_j	Cost of locating a server at potential location $j \in J$
u_j	The capacity of server $j \in J$
v	Upper bound on the total number of servers that can be located
w_z	Minimum number of servers that must located in zone $z \in \mathscr{Z}$

Table 4.4 Notation used to define the second-stage SSLP model

Sets and Indices	
Ω	Set of outcomes (scenarios) for parts demand, indexed $\omega \in \Omega$
Second-Stage Decision Variables	
y_{ij}	Decision variable $y_{ij} = 1$ if client i is served by a server at location j under scenario $\omega \in \Omega$, $y_{ij} = 0$ otherwise
y_{j0}	Amount of unmet resource demand due to limitation in server capacity under scenario $\omega \in \Omega$
Second-Stage Data (Parameters)	
$\tilde{\omega}$	Multivariate random variable whose outcome (scenario) ω specifies client availability, i.e., $\omega := \{h_i(\omega)\}_{i \in \mathbb{I}},\ \omega \in \Omega$
p_ω	Probability of occurrence of scenario $\omega \in \Omega$
q_{ij}	Revenue from client $i \in \mathbb{I}$ being served by server at location $j \in \mathbb{J}$
q_{j0}	Per unit penalty for unmet demand due to limitation in server capacity under scenario $\omega \in \Omega$
d_{ij}	Client $i \in \mathbb{I}$ resource demand from server at location $j \in \mathbb{J}$
$h^i(\omega)$	Takes a value $h^i(\omega) = 1$ if client $i \in \mathbb{I}$ is present in scenario $\omega \in \Omega$, $h^i(\omega) = 0$ otherwise

$$\text{Min} \sum_{j \in \mathbb{J}} c_j x_j - \mathbb{E}[f(x, \tilde{\omega})] \tag{4.4a}$$

$$\text{s.t.} \sum_{j \in \mathbb{J}} x_j \leq v \tag{4.4b}$$

$$\sum_{j \in \mathbb{J}_z} x_j \geq w_z, \ \forall z \in \mathscr{Z} \tag{4.4c}$$

$$x_j \in \{0, 1\}, \ \forall j \in \mathbb{J}, \tag{4.4d}$$

where for any x satisfying the constraints (4.4b)–(4.4d) and for each outcome $\omega \in \Omega$ of $\tilde{\omega}$, the second-stage problem is given by

$$f(x,\omega) = \text{Min} \; -\sum_{i\in\mathbb{I}}\sum_{j\in\mathbb{J}} q_{ij}y_{ij} + \sum_{j\in\mathbb{J}} q_{j0}y_{j0} \tag{4.5a}$$

$$\text{s.t.} \quad \sum_{i\in\mathbb{I}} d_{ij}y_{ij} - y_{j0} \leq u_j x_j, \; \forall j \in \mathbb{J} \tag{4.5b}$$

$$\sum_{j\in\mathbb{J}} y_{ij} = h^i(\omega), \; \forall i \in \mathbb{I} \tag{4.5c}$$

$$y_{ij} \in \{0,1\}, \; \forall i \in \mathbb{I}, \; j \in \mathbb{J} \tag{4.5d}$$

$$y_{j0} \geq 0, \; \forall j \in \mathbb{J}. \tag{4.5e}$$

The objective function (4.4a) maximizes the expected profit for serving the clients. Constraint (4.4b) enforces the requirement that only up to a total of v available servers can be installed. The zonal constraints are given in (4.4c) and dictate that at least w_z servers must be located in zone $z \in \mathscr{Z}$. Constraints (4.4d) are the binary restrictions on the first-stage stage decision variables.

In the second-stage, the objective function (4.5a) computes the revenue for that scenario. Constraints (4.5b) require that a server located at site j serve only up to its capacity u provided it has been located in the first-stage. Observe that the continuous variable y_{j0} accommodates any overflows that are not served due to limitation in server capacity. This results in a loss of revenue at a rate of q_{j0}. The requirement that each available client be served by only one server is given by constraints (4.5c). Finally, constraints (4.5d) and (4.5e) impose binary restrictions on the tactical decision variables and the overflow decision variables, respectively.

Let us also write the DEP for SSLP assuming that $\tilde{\omega}$ is discretely distributed with $|\Omega| < \infty$ and with each outcome $\omega \in \Omega$ having a probability of occurrence p_ω such that $\sum_{\omega\in\Omega} p_\omega = 1$. Let the second-stage decision variables for each outcome $\omega \in \Omega$ be explicitly written as y_{ij}^ω and y_{j0}^ω for all $i \in I$ and $j \in J$. The DEP for CEP can now be written as follows:

$$\text{Min} \sum_{j\in\mathbb{J}} c_j x_j - \sum_{\omega\in\Omega} p_\omega \Big(\sum_{i\in\mathbb{I}}\sum_{j\in\mathbb{J}} q_{ij}y_{ij}^\omega + \sum_{j\in\mathbb{J}} q_{j0}y_{j0}^\omega\Big) \tag{4.6a}$$

$$\text{s.t.} \quad \sum_{j\in\mathbb{J}} x_j \leq v$$

$$\sum_{j\in\mathbb{J}_z} x_j \geq w_z, \; \forall z \in \mathscr{Z}$$

$$\sum_{i\in\mathbb{I}} d_{ij}y_{ij}^\omega - y_{j0}^\omega \leq u_j x_j, \; \forall j \in \mathbb{J}, \; \omega \in \Omega$$

$$\sum_{j \in \mathbb{J}} y_{ij}^{\omega} = h^i(\omega), \ \forall i \in \mathbb{I}, \ \omega \in \Omega$$

$$x_j \in \{0, 1\}, y_{ij}^{\omega} \in \{0, 1\}, y_{j0}^{\omega} \geq 0, \ \forall i \in \mathbb{I}, \ j \in \mathbb{J}, \ \omega \in \Omega$$

$$y_{j0}^{\omega} \geq 0, \ \forall j \in \mathbb{J}, \ \omega \in \Omega.$$

As we did with the SSLP, we can also characterize the size of SSLP in terms of the number of decision variables (columns) and the number constraints (rows). The number of first-stage decision variables (x) is $n_1 := n$, and the number of first-stage constraints is $m_1 := 1 + |\mathscr{Z}|$, where n is the number of potential server locations and $|\mathscr{Z}|$ is the number of zones across the potential server locations. The number of second-stage decision variables is $n_2 := n(m + 1)$, while the number of second-stage constraints is $m_2 := m + n$, where m is the number of clients. Letting the number of scenarios $S = |\Omega|$, the DEP has $n + n(m + 1)S$ decision variables and $1 + |\mathscr{Z}| + (m + n)S$ constraints. For example, a CEP instance with $m = 100, n = 10, |\mathscr{Z}| = 1$, and $S = 100$ has a corresponding DEP with $n + n(m + 1)S = 10 + 10(100 + 1)(100) = 101{,}010$ decision variables and $1 + |\mathscr{Z}| + (m + n)S = 1 + 1 + (100 + 10)(100) = 11{,}002$ constraints. Although MIP solvers can be applied directly to solve DEP (4.6), for relatively large size instances the problem is computationally challenging to solve. In general, MIPs are difficult to solve even for relatively medium-sized instances.

4.4 Stochastic Supply Chain Planning Problem

Another area that has seen the application of SP is supply chain planning under uncertainty. We give an illustration of simplified formulation of the stochastic supply chain (SSCh) planning problem described in [3] to maximize the profit of operating a supply chain that involves getting revenue from supplying products and incurring operation costs in the process. The first-stage decision variables are binary and involve plant sizing, while the second-stage decision variables are mixed-binary and involve product allocation, vendor selection, and operational decisions for running the supply chain. The data uncertainty in this problem appears in the objective function in the form of unknown net profit from supplying products and in the RHS of the constraints in terms of random product demand.

We shall now give an illustration of the two-stage SSCh problem using a simplified generic formulation based on the following notation:

First-Stage Decision Variables (Strategic)

$x \in \{0, 1\}^{n_1}$: binary decisions concerning the following: (a) if product/raw material is selected for processing/supplying, (b) if product/raw material is processed in a particular plant/supplied by a particular vendor, and (c) if a given plant has a particular capacity level at least at some given time period.

First-Stage Problem Data

$c \in \mathbb{R}^{n_1}$: first-stage cost vector.
$A \in \mathbb{R}^{m_1 \times n_1}$: first-stage constraint matrix.
$b \in \mathbb{R}^{m_1}$: first-stage RHS vector.

Second-Stage Decision Variables (Operational)

$y(\omega) \in \{0, 1\}^{n_2}$: binary decisions concerning expansion of facilities, which vendors to supply from, and which products are processes under scenario $\omega \in \Omega$, where Ω is the set of all scenarios.
$z(\omega) \in \mathbb{R}^{n_2}_+$: continuous decision variables for the operation of the supply chain $\omega \in \Omega$.

Second-Stage Problem Data

$\tilde{\omega}$: multivariate random variable whose outcome (scenario) $\omega \in \Omega$ specifies the revenue and product demand, i.e., $\omega := \{q_z(\omega), r(\omega)\},\ \omega \in \Omega$.
$q_y \in \mathbb{R}^{n_2}$: second-stage cost vector associated with facilities expansion, vendor suppliers, and product processing decisions y.
$q_z(\omega) \in \mathbb{R}^{n_2}$: second-stage random cost vector associated with supply chain operation decisions z under scenario $\omega \in \Omega$.
$T \in \mathbb{R}^{m_2 \times n_2}$: technology constraint matrix.
$W_y \in \mathbb{R}^{m_2 \times n_2}$: recourse matrix associated with and product processing decisions y.
$W_z \in \mathbb{R}^{m_2 \times n_2}$: recourse matrix associated with supply chain operation decisions z.
$r(\omega) \in \mathbb{R}^{m_2}$: second-stage RHS vector under scenario $\omega \in \Omega$.

Using the above notation, the two-stage SSCh problem is given as follows:

$$\begin{aligned} &\text{Min } c^\top x + \mathbb{E}[f(x, \tilde{\omega})] \\ &\text{s.t. } Ax = b \\ &\quad x \in \{0, 1\}^{n_1}, \end{aligned} \tag{4.7}$$

where for any x satisfying the constraints in Problem (4.7) and for each outcome $\omega \in \Omega$ of $\tilde{\omega}$, the second-stage problem is given by

$$\begin{aligned} f(x, \omega) = &\text{Min } q_y^\top y(\omega) - q_z(\omega)^\top z(\omega) \\ &\text{s.t. } W_y y(\omega) + W_z z(\omega) = r(\omega) - Tx \\ &\quad y(\omega) \in \{0, 1\}^{n_2},\ z(\omega) \geq 0. \end{aligned} \tag{4.8}$$

The objective value is to minimize the cost of setting up the supply chain minus the expected revenue from supplying the products. The constraints in the first-stage include investment budget and plant capacity constraints. The constraints in the second-stage include operational and capacity expansion requirements. Details of the model formulation and computational results are given in [3]. Notice that the first-stage decision variables are pure binary, while the second-stage decision variables are mixed-binary. Thus, the problem is amenable to SMIP decomposition methods described in Chap. 9. Computational results for SSCh based on using the disjunctive programming approach are reported in [33].

4.5 Fuel Treatment Planning

Wildfires have burn millions of acres and devastate communities every year in many parts of the world. For example, the US National Interagency Fire Center reported that in 2021 there were 58,985 wildfires that burned over seven million acres in the USA [23]. This included over 9,000 wildfires in the state of California that burned more than 2,000 acres. In fact, California was the state with most wildfires that year, followed by Texas with more than 5,000 wildfires. The impact of fire in terms of lives lost as well as economical losses is staggering. The deadliest wildfire in California in 2018 resulted in 85 deaths and cost around $1.5 billion in insured losses.

Fuel accumulation over time contributes to a higher risk of intense and extensive wildfires. Therefore, in wildfire management fire managers must periodically consider *fuel treatment* options to minimize wildfire risk under limited budgets and data uncertainties pertaining to vegetation growth (fuels), weather, and fire behavior. The impact of a fire, such as burned area and losses, is unknown before fire occurrence. Fuel treatment planning (FTP) deals with the removal of all or some of the vegetation from a landscape or altering the vegetation varieties to reduce the potential for fires and their severity. Examples of commonly applied fuel treatment options in the USA include *prescribed burning*, *mechanical thinning*, *mowing*, and *grazing*.

FTP plays an important role in fire management to reduce hazardous fuels to protect life and property. However, selecting the optimal locations for each treatment option under limited budgets is challenging due to the uncertainty in vegetation growth, weather, and fire behavior. Furthermore, for wildfires along the wildland urban interface (WUI), fire managers typically have to be prepared to provide a *standard response* to fight and prevent the fires from becoming large (escaped fires). A *standard response* is a combination of firefighting resources located within a maximum response time (distance) that are required to fully contain a fire within a target final size. Thus, FTP can be integrated with wildfire response planning by optimally deploying firefighting resources to areas of concern to minimize wildfire risk (see Fig. 4.3). Let us consider a generic two-stage stochastic FTP (SFTP) model that combines fuel treatment planning and wildfire response planning for a fire planning unit.

A fire planning unit is a region with a set of fuel treatment areas and a set of *operations bases*, i.e., locations where resources are deployed before fires happen and dispatched from when fires happen. Each operations base needs to have a mix of

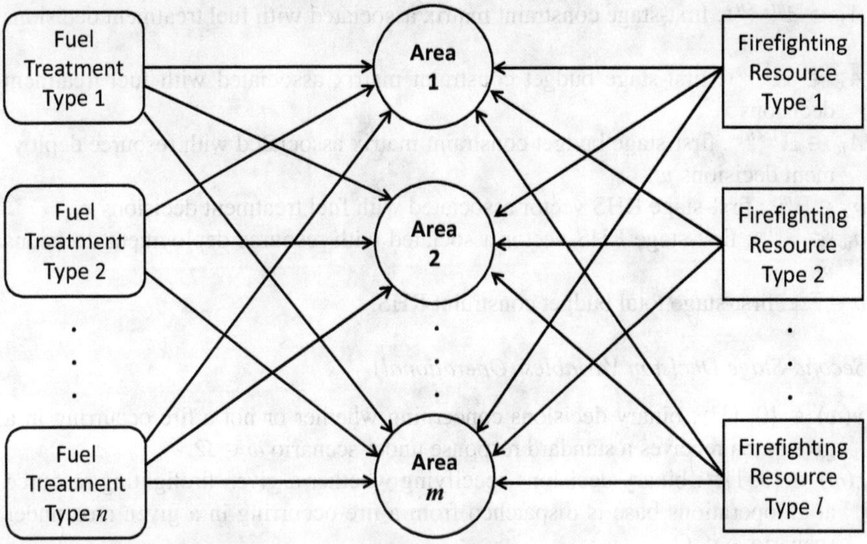

Fig. 4.3 SFTP involves assigning fuel treatment types and firefighting resources to different areas under uncertainty in fire occurrence/behavior

firefighting resources (dozers, plows, fire engines, crews, etc.) required to effectively respond to future fires in the fire planning unit. Each type of firefighting resource has its own line production rate, and a combination of resources is typically required to effectively respond to a reported fire. Fuel treatment decisions and deployment decisions (location/relocation of firefighting resources from one operations base to another) are decided in the first-stage before fires occur, while operational decisions regarding the optimal mix of resources to dispatch to reported fires are made in the second-stage. The goal of the SFTP model is to determine fuel treatment and firefighting resource deployment decisions to minimize the impact of the fires (e.g., area burned and/or fire damage). Since vegetation growth and wildfire impact are uncertain, they are simulated using standard vegetation growth and wildfire behavior software, respectively, to get the data needed for the SFTP model. To mathematically state the SFTP model, we define the following notation:

First-Stage Decision Variables (Strategic)

$x \in \{0, 1\}^{n_x}$: binary decisions specifying if a given area receives a particular type of fuel treatment at a certain level of coverage.

$w \in \{0, 1\}^{n_w}$: binary decisions specifying if a firefighting resource initially located at a given operations base is relocated (deployed) to another operations base.

First-Stage Problem Data

$c_x \in \mathbb{R}^{n_x}$: first-stage cost vector associated with fuel treatment decisions x.

$c_w \in \mathbb{R}^{n_w}$: first-stage cost vector associated with resource deployment decisions w.

$A_x \in \mathbb{R}^{m_x \times n_x}$: first-stage constraint matrix associated with fuel treatment decisions x.
$\bar{A}_x \in \mathbb{R}^{1 \times n_x}$: first-stage budget constraint matrix associated with fuel treatment decisions x.
$\bar{A}_w \in \mathbb{R}^{1 \times n_w}$: first-stage budget constraint matrix associated with resource deployment decisions w.
$b_x \in \mathbb{R}^{m_x}$: first-stage RHS vector associated with fuel treatment decisions x.
$b_w \in \mathbb{R}^{m_w}$: first-stage RHS vector associated with resource deployment decisions w.
$\bar{b} \in \mathbb{R}_+$: first-stage total budget constraint RHS.

Second-Stage Decision Variables (Operational)

$y(\omega) \in \{0, 1\}^{n_y}$: binary decisions concerning whether or not a fire occurring in a given area receives a standard response under scenario $\omega \in \Omega$.
$z(\omega) \in \{0, 1\}^{n_z}$: binary decisions specifying whether a given firefighting resource at an operations base is dispatched from a fire occurring in a given area under scenario $\omega \in \Omega$.

Second-Stage Problem Data

$\tilde{\omega}$: multivariate random variable whose outcome (scenario) ω specifies, for a given feasible x, the impact of the fire and the standard resource line production rate required for a given fire in that scenario, i.e., $\omega := \{q_y(x, \omega), W_y(x, \omega)\}$, $\omega \in \Omega$.
$q_y(x, \omega) \in \mathbb{R}^{n_2}$: second-stage random cost vector associated with fire standard response decisions y for a given x (fuel treatment decisions) under scenario $\omega \in \Omega$.
$q_z(\omega) \in \mathbb{R}^{n_2}$: second-stage random cost vector associated with resource dispatch decisions z under scenario $\omega \in \Omega$.
$T_z \in \mathbb{R}^{m_2 \times n_2}$: technology constraint matrix associated with resource dispatch decisions z.
$W_y(x, \omega) \in \mathbb{R}^{m_2 \times n_2}$: recourse matrix associated with fire standard response decisions y for a given x under scenario $\omega \in \Omega$.
$W_{zb} \in \mathbb{R}^{m_2 \times n_2}$: recourse matrix associated with resource dispatch decisions z and the budget.
$\bar{W}_{zd} \in \mathbb{R}^{m_2 \times n_2}$: recourse matrix associated with resource dispatch decisions z and the dispatching of firefighting resources.
$W_{zp} \in \mathbb{R}^{m_2 \times n_2}$: recourse matrix associated with resource dispatch decisions z and the production rate of firefighting resources.

Using the above notation, a generic two-stage SFTP formulation can be given as follows:

$$\text{Min } c_x^\top x + c_w^\top w + \mathbb{E}[f((x, w), \tilde{\omega})] \tag{4.9a}$$

$$\text{s.t. } A_x x = b_x \tag{4.9b}$$

$$\bar{A}_x x + \bar{A}_w w = \bar{b} \tag{4.9c}$$

$$x \in \{0,1\}^{n_x},\ w \in \{0,1\}^{n_w}, \tag{4.9d}$$

where for any (x, w) satisfying the constraints in Problem (4.9) and for each outcome $\omega \in \Omega$ of $\tilde{\omega}$, the second-stage problem is given by

$$f((x,w),\omega) = \text{Min } q_y(x,\omega)^\top y(\omega) + q_z(\omega)^\top z(\omega) \tag{4.10a}$$

$$\text{s.t. } W_{zb} z(\omega) \le \bar{b} - \bar{A}_x x - \bar{A}_w w \tag{4.10b}$$

$$W_{zd} z(\omega) \le T_z w \tag{4.10c}$$

$$W_y(x,\omega) y(\omega) + W_{zp} z(\omega) \ge 0 \tag{4.10d}$$

$$y(\omega) \in \{0,1\}^{n_y},\ z(\omega) \in \{0,1\}^{n_z}. \tag{4.10e}$$

The objective function (4.9a) minimizes the fuel treatment cost and firefighting resource deployment costs in the first-stage, i.e., before fires happen, plus the expected cost of the impact of the fires in each treatment area (e.g., fire damages) and the cost of dispatching firefighting resources when fires occur. The first-stage constraints are as follows: Constraint (4.9b) enforces the requirement that exactly one treatment option must be selected for each treatment area. Constraint (4.9c) is a budget constraint limiting the sum of the fuel treatment and deployment costs not to exceed the total budget. Constraint (4.9d) specifies the binary restrictions on the decision variables.

The second-stage objective function (4.10a) minimizes the cost per unit area of the impact of the fire plus the cost of dispatching resources to the fires for a given scenario. Constraint (4.10b) is a budget restriction on the operational cost. Constraint (4.10c) enforces the requirement that a firefighting resource can only be dispatched from an operations base to a fire if it was deployed to that base in the first-stage. Constraint (4.10d) computes the production rate of the resources dispatched to a fire location in a given treatment area to meet a standard response requirement. Constraint (4.10e) specifies the binary restrictions on the decision variables.

We observe that the first- and second-stage decision variables in the SFTP model are pure binary, and thus, the problem is amenable to the decomposition methods described in Chap. 9. However, the issue is that this problem has random recourse and the recourse matrix depends on the first-stage decision variable, leading to the endogenous uncertainty. Thus, one has to resort to algorithms for SMIP with endogenous uncertainty. A computational study of an SFTP model that does not incorporate fire response and does not have endogenous uncertainty is reported in [18].

4.6 Appointment Scheduling in Nuclear Medicine

One of the subspecialties of radiology, nuclear medicine, deals with the diagnosis and treatment of patients using radiopharmaceuticals administered using different technologies. Radiopharmaceuticals are radioactive isotopes (e.g., Indium-111, Iodine-131, and Technetium-99m) with a short half-life (minutes) whose radioactivity is used for imaging and treatment of specific parts of the body. Relatively small amounts produce radiation that is considered safe for the body, and using special imaging equipment, pictures are produced. The images are used by radiologists to study how a body organ is functioning and to detect disease or tumors that may be present in the organ. Relatively larger amounts of the radiopharmaceuticals are administered to treat disease. In this case, the radiation is absorbed by the tissue of the body and the type of radiopharmaceutical used and how it is administered depends on the organ.

Nuclear medicine procedures involve *multiple steps* with each step requiring a specific resource and strict time window constraints. Strict time windows for each step are needed because of the short half-life of the radiopharmaceuticals. Therefore, scheduling patients and resources in order to obtain high quality scans or effective treatment is very challenging. A scan can take minutes to hours to complete. Poor scheduling can result in low quality scans requiring the patient to be rescheduled for another day after being exposed to radiation and is a waste of time and valuable resources. Consequently, nuclear medicine departments are concerned with providing high quality service to patients while maximizing their resource utilization.

We consider the nuclear medicine online scheduling problem (NMOSP) model by [28]. This model is based on a real nuclear medicine clinic described in [29]. NMOSP involves scheduling patient requests online (one at a time) and the needed resources under uncertainty in future patient request arrivals. The resources used in performing nuclear medicine procedures include humans, equipment, and radiopharmaceuticals. Human resources include *technologists*, *nurses*, *EKG technicians*, and *radiologists*. Equipment resources involve various types of *gamma cameras*, which can cost up to a million dollars and thus require managing efficiently. Radiopharmaceuticals are expensive and are often produced offsite and must be delivered on time to ensure that the radioactivity will be at its optimum level at the scheduled time of taking a scan or doing treatment. Table 4.5 shows three example nuclear medicine human resources and the tasks they perform. Table 4.6 shows a list of five nuclear medicine procedures, i.e., Current Procedural Terminology (CPT) codes. A detailed sample of one of the procedures, CPT P78315, is given in Table 4.7 with an example schedule shown in Fig. 4.4. The column *station* in the table lists the equipment resources needed at each step of the procedure.

We are now in a position to state a generic two-stage NMOSP formulation using the following notation:

Table 4.5 Example nuclear medicine human resources and tasks they perform

Nurse	Technologist	Radiologist
Hydrate patient	Hydrate patient	Hydrate patient
Radiopharmaceutical administration	Radiopharmaceutical preparation	Radiopharmaceutical administration
Draw doses	Imaging	Draw doses

Table 4.6 Examples of nuclear medicine procedures

CPT Code	Name
78006	ENC-Thyroid Imaging
78315	MSC-Bone Imaging (Three Phase)
78464	CVD-Myocardial Imaging (SP-R ORS)
78465	Cardiovascular Event (CVE) Myocardial Imaging (SP-M)
78815	Positron Emission Tomography (PET)/Computed Tomography (CT) Skull to Thigh

Table 4.7 CPT 78315: MSC-Bone Imaging (Three Phase)

Step	Activity	Time (Minutes)	Station	Human Resource
1	Radiopharmaceutical injection	20	Axis, P2000, Meridian, TRT	Technologist Nurse
2	Imaging	15	Axis, P2000, Meridian	Technologist
3	Patient Wait	150–180	Waiting	
4	Imaging	45	Axis, P3000, Meridian	Technologist

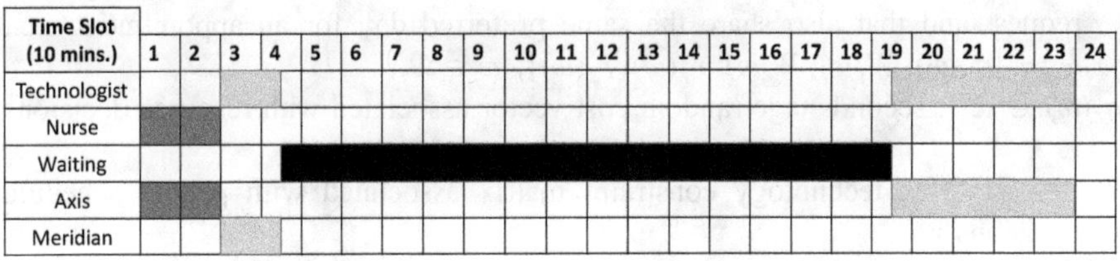

Fig. 4.4 An example schedule for CPT 78315 (MSC-Bone Imaging—Three Phase) given in Table 4.7

First-Stage Decision Variables

$x \in \{0, 1\}^{n_x}$: binary decisions specifying if the patient requesting a given procedure is scheduled to use a specific resource at a given time slot when the procedure is started at some specified time for a step of the procedure.

$w \in \{0, 1\}^{n_w}$: binary decisions specifying if a resource is selected to serve the patient in some step of a procedure when the procedure is started at a given time.

First-Stage Problem Data

$c_w \in \mathbb{R}^{n_w}$: first-stage cost vector associated with resource decisions w.

$A_x \in \mathbb{R}^{m_x \times n_x}$: first-stage constraint matrix associated with patient schedule decisions x.

$e_x \in \mathbb{R}^{m_x}$: first-stage RHS vector of ones associated with patient schedule decisions x.

$A_w \in \mathbb{R}^{m_w \times n_w}$: first-stage constraint matrix associated with resource decisions w.

$e_w \in \mathbb{R}^{m_w}$: first-stage RHS vector of ones associated with resource decisions w.

$\bar{A}_x \in \mathbb{R}^{1 \times n_x}$: first-stage constraint matrix associated with x to ensure that the same resource is scheduled for the duration of each procedure step.

$\bar{A}_w \in \mathbb{R}^{1 \times n_w}$: first-stage constraint matrix associated with w to ensure that the same resource is scheduled for the duration of each procedure step.

Second-Stage Decision Variables

$y(\omega) \in \{0, 1\}^{n_y}$: binary decisions specifying if a given patient requesting a given procedure is scheduled to use a specific resource at a given time slot when the procedure is started at some given time for a specific step of the procedure under scenario $\omega \in \Omega$.

$z(\omega) \in \{0, 1\}^{n_z}$: binary decisions specifying if a resource is selected to serve a given patient in some step when the procedure is started at a given time under scenario $\omega \in \Omega$.

Second-Stage Problem Data

$\tilde{\omega}$: multivariate random variable whose outcome (scenario) ω specifies a set of possible patient requests that could arrive after the current (first-stage) patient request and that also share the same preferred day for an appointment, i.e., $\omega := \{q_z(\omega), W_y\omega), W_z\omega), r_y(\omega), r_z(\omega)\}$, $\omega \in \Omega$.

$q_z(\omega) \in \mathbb{R}^{n_2}$: second-stage random cost vector associated with resource decisions z.

$T_y \in \mathbb{R}^{m_y \times n_x}$: technology constraint matrix associated with patient schedule decisions y.

$W_y(\omega) \in \mathbb{R}^{m_y \times n_y}$: recourse matrix associated with patient schedule decisions y.

$e_y \in \mathbb{R}^{m_y}$: second-stage RHS vector of ones associated with y for the linking constraints.

$\bar{W}_y(\omega) \in \mathbb{R}^{m_2 \times n_y}$: recourse constraint matrix associated with y to ensure that the same resource is scheduled for the duration of each procedure step.

$W_z(\omega) \in \mathbb{R}^{m_z \times n_z}$: recourse matrix associated with resource decisions z.

$\bar{e}_y \in \mathbb{R}^{m_y}$: second-stage RHS vector of ones associated with patient schedule decisions y.

$\bar{W}_z(\omega) \in \mathbb{R}^{m_2 \times n_z}$: recourse constraint matrix associated with resource decisions z to ensure that the same resource is scheduled for the duration of each procedure step.

Using the above notation, a generic two-stage NMOSP formulation can be given as follows:

$$\text{Min } c_w^\top w + \mathbb{E}[f(x, \tilde{\omega})] \tag{4.11a}$$

$$\text{s.t. } A_x x \leq e_x \tag{4.11b}$$

$$A_w w \leq e_w \tag{4.11c}$$

$$A_w w = \mathbf{0} \tag{4.11d}$$

$$\bar{A}_x x - \bar{A}_w w = \mathbf{0} \tag{4.11e}$$

$$x \in \{0, 1\}^{n_x}, w \in \{0, 1\}^{n_w}, \tag{4.11f}$$

where for any x satisfying the constraints in Problem (4.11) and for each outcome $\omega \in \Omega$ of $\tilde{\omega}$, the second-stage problem is given by

$$f(x, \omega) = \text{Max } q_z(\omega)^\top z(\omega) \tag{4.12a}$$

$$\text{s.t. } W_y(\omega) y(\omega) \leq e_y - T_y x \tag{4.12b}$$

$$\bar{W}_y(\omega) y(\omega) \leq \bar{e}_y \tag{4.12c}$$

$$W_z(\omega) z(\omega) = \mathbf{0} \tag{4.12d}$$

$$\bar{W}_y(\omega) y(\omega) - \bar{W}_z(\omega) z(\omega) = \mathbf{0} \tag{4.12e}$$

$$y(\omega) \in \{0, 1\}^{n_y}, \; z(\omega) \in \{0, 1\}^{n_z}. \tag{4.12f}$$

The objective function (4.11a) minimizes waiting (access) time for the current patient plus the expected the number of patients scheduled on a given day. The first-stage constraints are as follows: Constraint (4.11b) makes sure that for the time slot requirements for procedure completion are satisfied, i.e., for each time period for the procedure the patient is assigned to the required resource. Constraint (4.11c) selects the human resource and station, respectively, for each procedure step and determines the appointment start-time. Constraint (4.11d) makes sure that the human resources and stations selected to serve a patient adhere to sequence protocol of the procedure. In addition, the constraint matches the station to the appropriate human resource for each step of the procedure. Constraint (4.11e) ensures that for the duration of each step of the procedure the same resource is scheduled. Binary restrictions on the decision variables are given by constraint (4.11f).

The second-stage objective function (4.12a) maximizes the number of patients scheduled on a given day for scenario ω (throughput). Constraint (4.12b) makes sure that for each patient all the time slot requirements for procedure completion are met. Constraint (4.12c) selects the human resource and station, respectively, for each procedure step and determines the appointment start-time for each patient. Constraint (4.12d) ensures that the human resources and stations chosen to serve a patient follow the procedure sequence protocol and matches the station to the human resource for each procedure step. To make sure that the same resource is scheduled for the duration of each procedure step, constraint (4.12e) is added. Constraint (4.12f) specifies the binary restrictions on the decision variables.

The NMOSP model has pure binary decision variables in both the first- and second-stages and has random recourse. Thus, the problem is amenable to some of the decomposition methods described in Chap. 9 and the algorithm for SMIP with random recourse described in [24]. The paper by [28] reports on a computational study of NMOSP based on an actual nuclear medicine clinic described in [27, 29]. The authors derive and apply their own algorithm tailored for NMOSP.

4.7 Airport Time Slot Allocation Under Uncertainty

In air transportation, airlines and aircraft operators incur huge expenses due to congestion in air traffic every day. In Western and Central Europe, for example, several airports are congested in terms of air traffic despite being considered *fully coordinated*. At a *fully coordinated* airport the number of flights scheduled per unit time (e.g., one hour) cannot exceed the capacity set for the airport. Given that air traffic has continued to grow, planning to increase airport capacity is needed in the long term. However, in the short to medium term, airport operators, airlines, and aircraft operators are interested in optimizing the use of current limited capacity to alleviate congestion and minimize daily operational costs.

Airport time slot (*slot* for short) allocation deals with the allocation of slots among airlines and aircraft operators to reduce congestion. Slot allocation is handled according to the guidelines of the International Air Transport Association (IATA) [15]. A slot is the amount of capacity (in terms of time) allocated to an aircraft to land or take off by a coordinator for a planned operation and use of airport infrastructure required to arrive or depart at a Level 3 airport on a specific date and time [15]. According to IATA, a Level 3 or fully coordinated airport is defined as follows:

Airport where capacity providers have not developed sufficient infrastructure or where governments have imposed conditions that make it impossible to meet demand. A coordinator is appointed to allocate slots to airlines and other aircraft operators using or planning to use the airport as a means of managing available capacity.

Since 1974, IATA has been the authority for providing standards for the management of airport slots, policies, principles, and procedures for slot allocation to the global air transportation community.

We consider the two-stage stochastic time slot allocation problem (STSAP) described by [9] for the optimal allocation of slots at European Union airports under uncertainty in airport nominal capacity (number of available time slots). The model allocates time slots to airlines and aircraft operators based on their requests to operate specific flight movements, i.e., either the departure (from the airport of origin) or the arrival of a flight (at the destination airport). The STSAP model considers the allocation of slots to airlines and aircraft operators with the aim of obtaining *effective* and *reliable* airline schedules. An effective schedule is one that has assigned slots that respect the airline's preferences, while a reliable schedule is one that factors in the uncertainty in airport capacity. The model considers a network of multiple airports and airlines as well as the interdependencies among flights operated by the same airline in determining the slot allocations at all airports simultaneously.

Assuming a fixed time horizon subdivided into equal size time slots, we shall now state a generic formulation of the STSAP model using the following notation:

First-Stage Decision Variables

$x \in \{0, 1\}^{n_1}$: binary decisions specifying if a given movement is allocated to a given time slot.

First-Stage Problem Data

$c \in \mathbb{R}^{n_1}$: first-stage cost vector associated with movement allocation decisions x.

$A_0 \in \mathbb{R}^{m_0 \times n_1}$: first-stage constraint matrix associated with movement allocation decisions x.

$e \in \mathbb{R}^{m_0}$: first-stage RHS vector of ones associated with movement allocation decisions x.

$A_1 \in \mathbb{R}^{m_1 \times n_1}$: first-stage constraint matrix associated with movement allocation decisions x.

$b \in \mathbb{R}^{m_1}$: first-stage RHS vector associated with movement allocation decisions x.

Second-Stage Decision Variables

$y(\omega) \in \mathbb{R}^{n_2}$: number of delayed departures from airport at a given time and day under scenario $\omega \in \Omega$.

$z(\omega) \in \mathbb{R}^{n_2}$: number of delayed arrivals to airport on a given time and day under scenario $\omega \in \Omega$.

Second-Stage Problem Data

$\tilde{\omega}$: multivariate random variable whose outcome (scenario) ω specifies the departure airport capacity, origin airport capacity, and total capacity, i.e., $\omega := \{r_y(\omega), r_z(\omega), r_{yz}(\omega)\}$, $\omega \in \Omega$.

$\varepsilon \in \mathbb{R}_+$: weight associated with the expected recourse function value.

$q \in \mathbb{R}^{n_2}$: second-stage random cost vector associated with delayed arrivals decisions z.

$T_y \in \mathbb{R}^{m_y \times n_1}$: technology constraint matrix associated with delayed departures decisions y.

$T_z \in \mathbb{R}^{m_z \times n_1}$: technology constraint matrix associated with delayed arrivals decisions z.

$T_{yz} \in \mathbb{R}^{m_{yz} \times n_1}$: technology constraint matrix associated with delayed departures decisions y and delayed arrivals decisions z.
$W_y \in \mathbb{R}^{m_y \times n_2}$: recourse matrix associated with delayed departures decisions y.
$W_z \in \mathbb{R}^{m_z \times n_2}$: recourse matrix associated with delayed arrivals decisions z.
$\bar{W}_y \in \mathbb{R}^{m_{yz} \times n_2}$: recourse constraint matrix associated with delayed departures decisions y for the linking constraints.
$\bar{W}_z \in \mathbb{R}^{m_{yz} \times n_2}$: recourse constraint matrix associated with delayed arrivals decisions z for the linking constraints.
$r_y(\omega) \in \mathbb{R}^{m_y}$: second-stage RHS vector associated with y (departure airport capacity) for $\omega \in \Omega$.
$r_z(\omega) \in \mathbb{R}^{m_z}$: second-stage RHS vector associated with z (arrival airport capacity) for $\omega \in \Omega$.
$r_{yz}(\omega) \in \mathbb{R}^{m_{yz}}$: second-stage RHS vector associated with y and z (total airport capacity) for $\omega \in \Omega$.

Using the above notation, a generic two-stage STSAP formulation can be written as follows:

$$\text{Min } c^\top x + \varepsilon \mathbb{E}[f(x, \tilde{\omega})] \tag{4.13a}$$

$$\text{s.t. } A_0 x = e \tag{4.13b}$$

$$A_1 x \leq b \tag{4.13c}$$

$$x \in \{0, 1\}^{n_1}, \tag{4.13d}$$

where for any x satisfying the constraints in Problem (4.13) and for each outcome $\omega \in \Omega$ of $\tilde{\omega}$, the second-stage problem is given by

$$f(x, \omega) = \text{Max } q(\omega)^\top y(\omega) + q(\omega)^\top z(\omega) \tag{4.14a}$$

$$\text{s.t. } W_y y(\omega) \geq r_y(\omega) - T_x x \tag{4.14b}$$

$$W_z z(\omega) \geq r_z(\omega) - T_z x \tag{4.14c}$$

$$\bar{W}_y y(\omega) - \bar{W}_z z(\omega) \geq r_{yz}(\omega) - T_{yz} x \tag{4.14d}$$

$$y(\omega),\ z(\omega) \geq 0. \tag{4.14e}$$

The objective function (4.13a) minimizes the weighted sum of the discrepancies between the schedules and requests and the expected operational delays. The first-stage constraints are as follows: Constraint (4.13b) ensures that a time slot is assigned to every movement requested, while constraint (4.13c) guarantees that departure, arrival, and total capacity requirements are met for each airport all the time. In addition, the constraint ensures that the flight time for all flights is fixed and that the turnaround time for all aircraft is satisfied. Constraint (4.13d) is the binary restrictions on the first-stage decision variables.

Given the first-stage decision, i.e., the schedule of departures and arrivals at each airport at each time period, the second-stage objective function (4.14a) minimizes the delays assigned to aircraft in order to meet the random airport capacity for the scenario. Constraint (4.14b) computes the number of delayed departures while constraint (4.14c) computes the number of delayed arrivals at each airport and time period. Constraint (4.14d) ensures that the airport capacity constraints are respected for the total number of departures and arrivals. Constraint (4.14e) specifies the binary restrictions on the decision variables.

The STSAP model has pure binary decision variables in the first-stage, continuous decision variables in the second-stage and has fixed recourse with randomness appearing only in the RHS. Thus, the model is amenable to Benders decomposing based methods described in Chaps. 6 and 8. For a computational study of STSAP, we refer the reader to the paper by [9]. This study is based on a set of European airports and uses the SAA method described in Chap. 8 to solve the instances.

4.8 Stochastic Air Traffic Flow Management

Air traffic transportation problems have been studied in SP for quite some time now. As pointed out in the introduction of this chapter, air traffic transportation problems include the classical SP test problems *gbd* [11] and *storm* [21]. Recall that *gbd* involves aircraft allocation to maximize profit while allocating different types of aircraft to routes with uncertain demand, while *storm* is an air freight scheduling problem dealing with the allocation of aircraft routes to minimize cargo handling costs under uncertain demand. Some of the problems airlines face in air traffic management are often motivated by the high cost of flight delays due to uncertainty in weather. An example of such a problem is the stochastic air traffic flow management (SATFM) problem described in [10]. This problem deals with the tactical or short-term (typically, less than 24 hours) planning of flight schedules to mitigate unforeseen disruptions due to capacity reductions in air traffic flow caused by bad weather.

To resolve demand and capacity imbalances, the SATFM model includes three tactical control options for aircraft: *ground holding delay* (GHD), *airborne holding delay* (AHD), and *rerouting*. The model considers both airport and en route sector congestion of the airspace system, which is characterized by a set of airports and sectors over a planning time horizon (e.g., 24 hours). Each airport and sector has a planned capacity for each time period, i.e., the maximum number of aircraft allowed. In addition, the following assumptions are made:

- Flight schedules over the planning time horizon are known, i.e., scheduled departures and arrivals at all airports, and possible flight routes.
- Flight time to cross any en route sector is known and fixed for each flight.
- AHD can be assigned to a flight only in the terminal airspace around the destination airport.

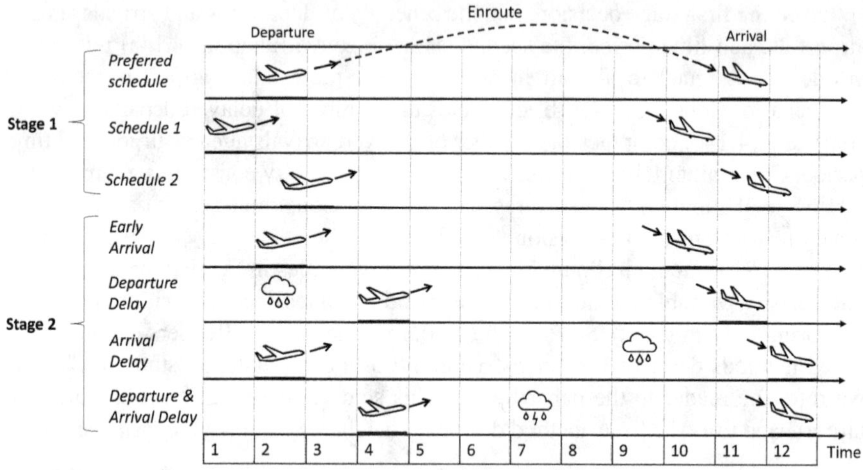

Fig. 4.5 An illustration of SATFM stages and possible schedules (scenarios)

- Airport and sector capacities are uncertain and are modeled as discrete random variables.

Figure 4.5 gives an illustration of the SATFM stages and example schedules for different scenarios. Information regarding the preferred schedule and alternative schedule(s) is provided in the first-stage. Changes to the schedules are decided in the second-stage based on uncertain weather information. Essentially, SATFM makes decisions regarding the release of flights into the air traffic system and the origin–destination route to follow and the amount of AHD to assign to each flight in case of congestion at the destination airport airspace. The goal is to determine flight delays and routes to minimize the cost of total delay while satisfying all origin-destination route capacity constraints of each flight.

Assuming a fixed time horizon subdivided into equal size time periods, we state a generic formulation of the SATFM model using the notation below.

First-Stage Decision Variables

$x_o \in \{0, 1\}^{n_o}$: binary decisions specifying if a flight departs from an origin airport by a given time following a given route.
$x_d \in \{0, 1\}^{n_d}$: binary decisions specifying if a flight arrives at a destination airport by a given time.

First-Stage Problem Data

$c_o \in \mathbb{R}^{n_o}$: first-stage cost vector associated with flight departure decisions x_o.
$c_d \in \mathbb{R}^{n_d}$: first-stage cost vector associated with flight arrival decisions x_d.
$A_o \in \mathbb{R}^{m_o \times n_o}$: first-stage constraint matrix associated with flight departure decisions x_o.

$b_o \in \mathbb{R}^{m_o}$: first-stage RHS vector associated with flight departure decisions x_o (nominal origin airport capacity).

$A_d \in \mathbb{R}^{m_d \times n_d}$: first-stage constraint matrix associated with flight arrival decisions x_d.

$b_d \in \mathbb{R}^{m_d}$: first-stage RHS vector associated with flight arrival decisions x_d (nominal destination airport capacity).

$\bar{A}_o \in \mathbb{R}^{m_{od} \times n_o}$: first-stage constraint matrix associated with flight arrival decisions x_d for the linking constraints.

$\bar{A}_d \in \mathbb{R}^{m_{od} \times n_d}$: first-stage constraint matrix associated with flight arrival decisions x_d for the linking constraints.

$b_{od} \in \mathbb{R}^{m_{od}}$: first-stage RHS vector associated with flight departure decisions x_o and flight arrival decisions x_d (nominal total airport capacity).

$A_s \in \mathbb{R}^{m_s \times n_o}$: first-stage constraint matrix associated with departure decisions x_o.

$b_s \in \mathbb{R}^{m_s}$: first-stage RHS vector associated with departure decisions x_o (nominal airspace sector capacity).

Second-Stage Decision Variables

$z_o(\omega) \in \{0, 1\}^{n'_o}$: binary decisions specifying if a flight departs from an origin airport by a given time in the second-stage following a given route under scenario $\omega \in \Omega$.

$z_d(\omega) \in \{0, 1\}^{n'_d}$: binary decisions specifying if a flight arrives at a destination airport by a given time in the second-stage under scenario $\omega \in \Omega$.

Second-Stage Problem Data

$\tilde{\omega}$: multivariate random variable whose outcome (scenario) ω specifies the departure airport capacity, origin airport capacity, total capacity, and sector capacity, i.e., $\omega := \{r_o(\omega), r_d(\omega), r_{od}(\omega), r_s(\omega)\}$, $\omega \in \Omega$.

$\varepsilon \in \mathbb{R}_+$: weight associated with the expected recourse function value.

$T_o \in \mathbb{R}^{m'_o \times n'_o}$: technology constraint matrix associated with flight departure decisions z_o.

$T_d \in \mathbb{R}^{m'_d \times n'_d}$: technology constraint matrix associated with flight arrival decisions z_d.

$T_{od} \in \mathbb{R}^{m'_{od} \times n'_d}$: technology constraint matrix associated with flight departure decisions z_o for the linking constraints.

$W_o \in \mathbb{R}^{m'_o \times n'_o}$: recourse matrix associated with flight departure decisions z_o.

$W_d \in \mathbb{R}^{m'_d \times n'_d}$: recourse matrix associated with flight arrival decisions z_d.

$\bar{W}_o \in \mathbb{R}^{m'_{od} \times n'_o}$: recourse constraint matrix associated with flight departure decisions z_o for the linking constraints.

$\bar{W}_d \in \mathbb{R}^{m'_{od} \times n'_d}$: recourse constraint matrix associated with flight arrival decisions z_d for the linking constraints.

$W_s \in \mathbb{R}^{m'_s \times n'_s}$: recourse matrix associated with z_s for airspace sector capacity constraints.

$r_o(\omega) \in \mathbb{R}^{m'_o}$: second-stage RHS vector associated with flight departure decisions z_o for $\omega \in \Omega$ (origin airport capacity).

$r_d(\omega) \in \mathbb{R}^{m'_d}$: second-stage RHS vector associated with flight arrival decisions z_d for $\omega \in \Omega$ (destination airport capacity).

$r_{od}(\omega) \in \mathbb{R}^{m'_{od}}$: second-stage RHS vector associated with flight departure decisions z_o and flight arrival decisions z_d for $\omega \in \Omega$ (total airport capacity).

$r_s(\omega) \in \mathbb{R}^{m'_s}$: second-stage RHS vector associated with flight departure decisions z_o for $\omega \in \Omega$ (airspace sector capacity).

Using the above notation, a generic two-stage SATFM formulation can be written as follows:

$$\text{Min } c_o^\top x_o + c_d^\top x_d + \varepsilon\mathbb{E}[f(x_o, x_d, \tilde{\omega})] \tag{4.15a}$$
$$\text{s.t. } A_o x_o \le b_o \tag{4.15b}$$
$$A_d x_d \le b_d \tag{4.15c}$$
$$\bar{A}_o x_o + \bar{A}_d x_d \le b_{od} \tag{4.15d}$$
$$A_s x_o \le b_s \tag{4.15e}$$
$$x_o \in \{0,1\}^{n_o},\ x_d \in \{0,1\}^{n_d}, \tag{4.15f}$$

where for any x satisfying the constraints in Problem (4.15) and for each outcome $\omega \in \Omega$ of $\tilde{\omega}$, the second-stage problem is given by

$$f(x_o, x_d, \omega) = \text{Max } c_o^\top z_o(\omega) + c_d^\top z_d(\omega) \tag{4.16a}$$
$$\text{s.t. } W_o z_o(\omega) \le r_o(\omega) - T_o x_o \tag{4.16b}$$
$$W_d z_d(\omega) \le r_d(\omega) - T_d x_o \tag{4.16c}$$
$$\bar{W}_o z_o(\omega) - \bar{W}_d z_d(\omega) \le r_{od}(\omega) - T_{od} x_o \tag{4.16d}$$
$$W_s z_o(\omega) \le r_s(\omega) \tag{4.16e}$$
$$z_o(\omega) \in \{0,1\}^{n'_o},\ z_d(\omega) \in \{0,1\}^{n'_d}. \tag{4.16f}$$

The objective function (4.15a) minimizes AHD and GHD in the first-stage plus the expected total delay assigned to a flight, i.e., the sum of AHD and GHD incurred in the second-stage. Another term that is typically added to the objective function, which we do not show, is a term to compute the cost reduction from when aircraft delay is on the ground before departure. The first-stage constraints are as follows: Constraints (4.15b), (4.15c), and (4.15d) enforce capacity requirements pertaining to the departure, arrival, and total capacity at the airports, respectively. In addition, these constraints ensure that the flight route and connectivity requirements are satisfied. Constraints (4.15e) are the airspace sector constraints to make sure that the number of flights in a given sector does not exceed the maximum number allowed. Constraint (4.15f) is the binary restrictions on the first-stage decision variables.

Given the first-stage decision, i.e., the flight departures and arrivals at each airport at each time period, the second-stage objective function (4.16a) minimizes the sum of AHD and GHD assigned to aircraft that appear in the second-stage. Similar to the first-stage constraints, (4.16b), (4.16c), and (4.16d) impose the departure, arrival, and total capacity constraints at the airports, respectively, and also ensure that the flight route and connectivity requirements are met. Airspace sector requirements are given by Constraint (4.15e), which ensures that the number of flights in a given sector does not exceed the maximum allowed. Constraint (4.16f) specifies the binary restrictions on the first-stage decision variables.

The SATFM model has pure binary decision variables in both stages and fixed recourse with randomness appearing only in the RHS. Thus, the model is amenable to SMIP decomposing methods described in Chap. 9. The computational study of SATFM in [9], based on a set of European airports, shows that it is hard to solve large-scale instances. Therefore, the authors propose a heuristic method based on the DEP with relaxed second-stage decision variables to solve their instances.

4.9 Satellite Constellation Scheduling Under Uncertainty

We now turn to another interesting application of SP, which is the problem of scheduling observations on a constellation of satellites with remote sensors to maximize the quality of the data collections from the sensors. We consider a version of this problem described by [35] involving two- and three-stage SMIP models for scheduling a constellation of one to n satellites to maximize the expected collection quality for a set of scenarios representing cloud cover uncertainty. We shall refer to this problem as the *stochastic satellite constellation scheduling problem* (SSCSP). In the SSCSP model, complex constellation-target geometries are determined via pre-computations associated with orbital propagators and sensor collection simulators in order to reduce model size and complexity. We consider the three-stage stochastic collection scheduling model where the time horizon is divided into time periods that span stages 1, 2, and 3 (see Fig. 4.6). Stage 1 models the planned schedule for the entire horizon before cloud cover information is revealed. Stage 2 models the schedule adjustments that can be taken when cloud front arrival times are known, while stage 3 allows for computing the schedule quality following the realization of cloud front departure time.

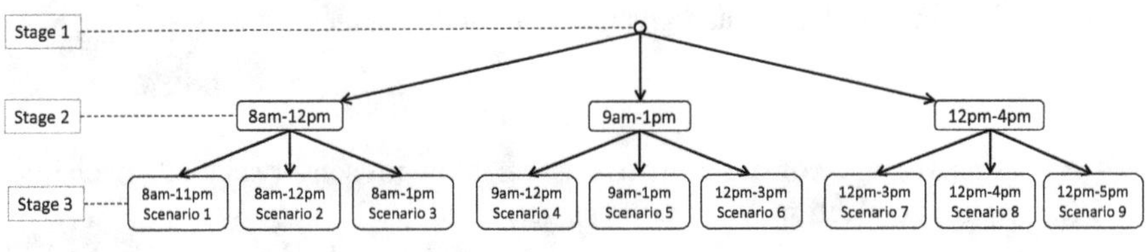

Fig. 4.6 An example of the three stages in SSCSP and cloud cover scenarios

Observation quality is calculated based on cloud cover percentage values, and collection windows have to be scheduled such that this observation quality is maximized. This requirement pertains to several types of remote sensors, including weather satellites in the geosynchronous orbit (GEO). GEO satellites have near fixed field-of-view and allow for many collection windows to be scheduled during the day. For low-earth orbit (LEO) satellites, however, the field-of-view changes with time, and there is a short opportunity to schedule long collection windows. Thus, SSCSP is more suited for sensors such as those on GEO satellites. An example of collection activities includes a collection of images for natural disasters such as wildfires, floods, earthquakes, etc.

Collection windows are divided into two categories: *required* and *optional* collection windows. Required collection windows are those that have to be performed such as forced solar outages and other sensor safety activities. Optional collection windows include Earth observations and customer requests. These constitute the majority of collection windows. It is assumed that the observation period earliest and latest times in which the collection window can be started are known. In addition, the required duration of each collection window is also known. In terms of scheduling, each collection window is assigned a priority relative to the other collection windows. We skip several other model details since our goal is simply to capture the essence of the mathematical formulation and refer the reader to [35].

Let X denote the feasible set for the schedule that covers time periods for stage 1 through stage 2. Similarly, let for a given schedule $x \in X$ and cloud cover scenario $\omega \in \Omega$, $Y(x, \omega)$ denote the feasible set for the schedule that covers time periods in stages 2 and 3. We also define the following additional notation:

Functions

$f_1(x) : X \mapsto \mathbb{R}$: first-stage cost function that computes the observed quality from the sensors based on schedule x.

$f_2(y, \omega) : Y(x, \omega) \mapsto \mathbb{R}$: stage 2 cost function that computes the observed quality from the sensors based on original schedule x and updated schedule y and scenario $\omega \in \Omega$.

$f_3(z, \omega) : Y(x, \omega) \mapsto \mathbb{R}$: stage 3 cost function that computes the observed quality from the sensors based on original schedule x and updated schedule z and scenario $\omega \in \Omega$.

First-Stage Decision Variables

$x \in \{0, 1\}^{n_x}$: binary decisions specifying if a given sensor is scheduled to execute a collection window starting at a given time period in stages 1 and 2.

First-Stage Problem Data

$A \in \mathbb{R}^{m \times n_x}$: first-stage constraint matrix associated with sensor schedule decisions x for required collection windows in stages 1 and 2.

$e \in \mathbb{R}^m$: first-stage RHS vector of ones associated with sensor schedule decisions x for required collection windows in stages 1 and 2.

$A_1 \in \mathbb{R}^{m_1 \times n_x}$: first-stage constraint matrix associated with sensor schedule decisions x for stages 1 and 2.

$e_1 \in \mathbb{R}^{m_1}$: first-stage RHS vector of ones associated with sensor schedule decisions x for stages 1 and 2.

$\bar{A} \in \mathbb{R}^{\bar{m} \times n_x}$: first-stage constraint matrix associated with sensor schedule decisions x to ensure that only collection windows are scheduled in stages 1 and 2.

Stages 2 and 3 Decision Variables

$y \in \{0, 1\}^{n_y}$: binary decisions specifying if a given sensor is scheduled to execute a collection window starting at a given time period in stage 2.

$z \in \{0, 1\}^{n_z}$: binary decisions specifying if a given sensor is scheduled to execute a collection window starting at a given time period in stage 3.

Stages 2 and 3 Problem Data

$\tilde{\omega}$: multivariate random variable whose outcome (scenario) ω specifies a set of cloud cover possibilities with the cloud front arriving in stage 2 and weather updates done in stage 3. A scenario is characterized by the following data:

$$\omega := \{W(\omega), W_2(\omega), W_3(\omega), \bar{W}_2(\omega), \bar{W}_3(\omega), W_z(\omega), \bar{W}_z(\omega)\}, \ \omega \in \Omega.$$

$W(\omega) \in \mathbb{R}^{m_y \times n_y}$: recourse matrix associated with sensor schedule decisions y that links stages 1 and 2.

$T \in \mathbb{R}^{m_y \times n_x}$: technology constraint matrix associated with sensor schedule decisions x stages 1 and 2.

$W_2(\omega) \in \mathbb{R}^{m_2 \times n_y}$: recourse matrix associated with sensor schedule decisions y for stage 2.

$W_3(\omega) \in \mathbb{R}^{m_2 \times n_z}$: recourse matrix associated with sensor schedule decisions z for stage 3.

$e_2 \in \mathbb{R}^{m_2}$: second-stage RHS vector associated with sensor schedule decisions y and z for the linking constraints.

$\bar{W}_2(\omega) \in \mathbb{R}^{m_3 \times n_y}$: recourse matrix associated with sensor schedule decisions y for stage 2.

$\bar{W}_3(\omega) \in \mathbb{R}^{m_3 \times n_z}$: recourse matrix associated with sensor schedule decisions z for stage 3.

$e_3 \in \mathbb{R}^{m_3}$: second-stage RHS vector associated with sensor schedule decisions y and z for the linking constraints.

$W_z(\omega) \in \mathbb{R}^{m_4 \times n_z}$: recourse matrix associated with sensor schedule decisions z.

$e_4 \in \mathbb{R}^{m_4}$: second-stage RHS vector associated with sensor schedule decisions z.

$\bar{W}_z(\omega) \in \mathbb{R}^{\bar{m} \times n_z}$: recourse matrix associated with sensor schedule decisions z.

Using the above notation, a generic three-stage SSCSP formulation can be written as follows:

$$\text{Max } f_1(x) + \mathbb{E}[f(x, \tilde{\omega})] \tag{4.17a}$$

$$\text{s.t. } Ax = e \tag{4.17b}$$

$$A_1 x \leq e_1 \tag{4.17c}$$

$$\bar{A}x = \mathbf{0} \tag{4.17d}$$

$$x \in \{0, 1\}^{n_x}, \tag{4.17e}$$

where for any x satisfying the constraints in Problem (4.17) and for each outcome $\omega \in \Omega$ of $\tilde{\omega}$, stages 2 and 3 problem are given by

$$f(x, \omega) = \text{Max } f_2(y, \omega) + f_3(z, \omega) \tag{4.18a}$$

$$\text{s.t. } W(\omega)y(\omega) = Tx \tag{4.18b}$$

$$W_2(\omega)y(\omega) + W_3(\omega)z(\omega) = e_2 \tag{4.18c}$$

$$\bar{W}_2(\omega)y(\omega) + \bar{W}_3(\omega)z(\omega) \leq e_3 \tag{4.18d}$$

$$W_z(\omega)z(\omega) \leq e_4 \tag{4.18e}$$

$$\bar{W}_z(\omega)z(\omega) = \mathbf{0} \tag{4.18f}$$

$$y(\omega) \in \{0, 1\}^{n_y}, \; z(\omega) \in \{0, 1\}^{n_z}. \tag{4.18g}$$

The objective function (4.17a) maximizes the weighted sum of observed quality (schedule performance of the sensors) in the first-stage plus the expected observed quality in stages 2 and 3. First-stage constraint (4.17b) makes sure that all required collection windows are scheduled, while constraint (4.17c) prevents scheduling a collection window if another collection window was scheduled within the range of time steps in stages 1 and 2. Constraint (4.17d) ensures that only collection windows during stages 1 and 2 are scheduled. Binary restrictions on the decision variables are given by constraint (4.17e).

The objective function (4.18a) maximizes the schedule performance of the sensors in stages 2 and 3, i.e., the observed quality. Constraint (4.18b) makes sure that the collection windows have been scheduled for both stage 1 and stage 2. Constraint (4.18c) ensures that required collection windows are observed either in stage 2 or in stage 3. Constraint (4.18d) imposes the requirement that each collection window is observed at most once. To make sure that multiple collection windows are not scheduled at the same time on the same sensor in stage 3, constraint (4.18e) is added. Constraint (4.18f) ensures that only collection windows during stage 3 can be scheduled. Constraint (4.18g) specifies the binary restrictions on the decision variables.

The SSCSP model has pure binary decision variables in all the stages and has random recourse. Notice that the random recourse in this problem stems from the fact that cloud cover uncertainty affects the time periods, which are linked to the decision variables. However, the problem is still amenable to the decomposition methods described in Chap. 9. In the computational study by [35], the SSCSP model was implemented as DEP using the open-source Pyomo modeling language [13]. The study considers several instances involving one and two satellites which are solved using an MIP solver. The computational results show that the SSCSP model yields better collection quality relative to the deterministic MIP model. Furthermore, the results demonstrate that the stochastic model produces provably optimal or near-optimal schedules within the computational time suitable for sensor operations.

4.10 Wildfire Response Planning

Wildfires cause deaths and millions of dollars in property losses every year as pointed out in Sect. 4.5. Therefore, fire managers are tasked with planning how to respond to wildfires before they happen so that they can provide sufficient resources to contain the fire before they become escaped fires. Wildfire containment is the result of an effectively performed *initial attack*, which is a combination of firefighting resources located within a maximum response time/distance that are first to arrive to contain a fire immediately after it has been reported. To perform an effective initial attack requires strategic and operational planning. The former involves planning the deployment of limited firefighting resources to operations bases before fires happen, while the latter deals with planning the dispatch of the resources from the operations bases to the fires when the fires are reported.

We consider the probabilistically constrained stochastic programming wildfire response (PC-SPWR) model described in [5] for dozer deployment planning for wildfire initial attack. This model incorporates the fire manager's level of risk, denoted as α, into the deployment and dispatch plans. Furthermore, to generate wildfire scenarios, the PC-SPWR model requires a fire behavior simulation as well as a wildfire risk model to estimate the stochastic parameters of the model. The PC-SPWR model was applied to a fire planning unit located in East Texas, USA, to determine deployment and dispatch plans based on historical fire occurrences characterized using representative fire locations (RFLs) (see Fig. 4.7). Such a plan specifies, for a given risk level, the dozers positioned at each operations base at the beginning of the fire season, fires contained and the associated expected cost, and other performance measures. The computational study shows that more resources are deployed to operations bases located in areas with high wildfire risk. Also, the study shows that when the fire manager's level of risk α is decreased, the number of fires contained and the associated expected cost increases.

We shall now illustrate a generic PC-SPWR model using the following notation:

Decision Variables

$x \in \{0, 1\}^{n_x}$: binary decisions specifying if a resource initially located at a given operations base is relocated (deployed) to another operations base.

$y(\omega) \in \{0, 1\}^{n_y}$: binary decisions specifying if a fire in scenario $\omega \in \Omega$ receives a standard response during a given fire season day.

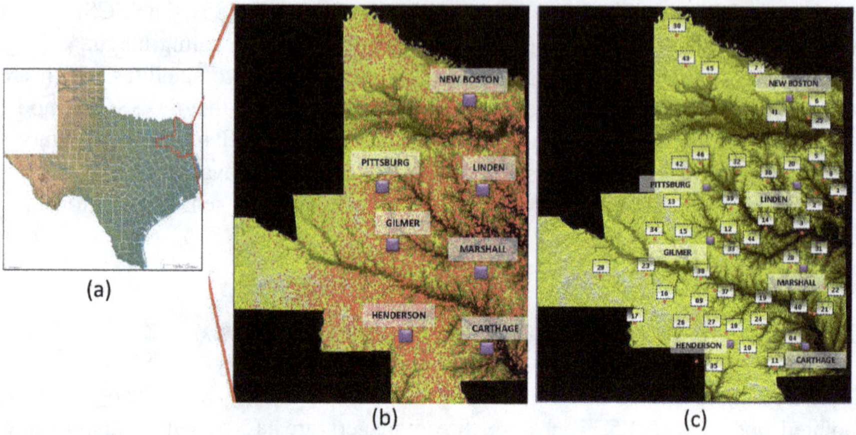

Fig. 4.7 Map of study area in East Texas study showing (**a**) state of Texas map (geology.com), (**b**) historical fire occurrences and operations bases, and (**c**) representative fire locations (RFLs) and operations bases

$z(\omega) \in \{0, 1\}^{n_z}$: binary decisions specifying if a resource from a given operations base which was initially at another operations base is dispatched to a given fire during a specific fire season day in scenario $\omega \in \Omega$.

Problem Data

$c_x \in \mathbb{R}^{n_x}$: cost vector associated with resource deployment decisions x (deployment and relocation cost).

$c_y \in \mathbb{R}^{n_w}$: cost vector associated with fire standard response decisions y (net value change).

$c_z \in \mathbb{R}^{n_z}$: cost vector associated with resource dispatch decisions z (dispatch cost).

$A_x \in \mathbb{R}^{m_x \times n_x}$: constraint matrix associated with resource deployment decisions x.

$e \in \mathbb{R}_+^{m_1}$: vector of ones, $e = (1, \cdots, 1)^\top$.

$\bar{A}_x \in \mathbb{R}^{m_1 \times n_x}$: linking constraint matrix associated with resource deployment decisions x.

$A_z \in \mathbb{R}^{m_1 \times n_z}$: constraint matrix associated with resource dispatch decisions z.

$T_z \in \mathbb{R}^{m_2 \times n_z}$: technology matrix associated with resource dispatch decisions z.

α: fire manager's level of risk, $0 < \alpha < 1$.

$\tilde{\omega}$: multivariate random variable whose outcome (scenario) ω specifies a sequence of fire days during a fire season where each day has a unique pattern of fire occurrences in that scenario, i.e., $\omega := \{T_y(\omega), g(\omega), r_1(\omega), r_2(\omega)\}$, $\omega \in \Omega$.

$T_y(\omega) \in \mathbb{R}^{m_2 \times n_y}$: technology matrix associated with fire standard response decisions y under scenario $\omega \in \Omega$.

$g(\omega) \in \mathbb{R}^{n_y}$: left hand side coefficient vector associated with fire standard response decisions y for normalized wildfire exposure under scenario $\omega \in \Omega$.

$r_1(\omega) \in \mathbb{R}$: second-stage RHS vector associated with fire standard response decisions y for normalized wildfire exposure under scenario $\omega \in \Omega$.

$r_2(\omega) \in \mathbb{R}$: second-stage RHS vector associated with fire standard response decisions y for a number of fires under scenario $\omega \in \Omega$.

Using the above notation, a generic PC-SPWR model can be written as follows:

$$\text{Min } c_x^\top x + c_y^\top y + c_z^\top z \tag{4.19a}$$

$$\text{s.t. } A_x x = e \tag{4.19b}$$

$$A_z z - \bar{A}_x x \leq \mathbf{0} \tag{4.19c}$$

$$\mathbb{P}\{T_z z - T_y(\tilde{\omega})y \geq \mathbf{0},\ g(\tilde{\omega})^\top y \geq r_1(\tilde{\omega}),\ e^\top y \geq r_2(\tilde{\omega})\} \geq 1 - \alpha \tag{4.19d}$$

$$x \in \{0, 1\}^{n_x},\ y \in \{0, 1\}^{n_y},\ z \in \{0, 1\}^{n_z}. \tag{4.19e}$$

In the PC-SPWR model (4.19), the symbol $\mathbf{0}$ denotes an appropriately dimensioned vector of zeros. The objective function (4.19a) minimizes the sum of the deployment cost for the dozers to operations bases and relocation between operations bases, total operational cost, and total fire damage cost. The first-stage constraints are as follows: Constraint (4.19b) enforces the requirement that each dozer can be assigned to exactly one operations base, i.e., it either remains at its current operations base or is relocated to another operations base. Constraint (4.19c) is a linking constraint to ensure that a dozer responds to a single fire in a day and returns to its original operations base.

Constraint (4.19d) is the joint probabilistic constraint with three different constraints that must be satisfied simultaneously with probability $1 - \alpha$. The first constraint guarantees that the mix of dozers dispatched to a fire will construct enough dozer line to contain the fire. The second constraint ensures that the total wildfire risk associated with areas that receive a standard response is above a set NWE threshold. Similarly, the third constraint makes sure that the total number of fires contained exceeds some set threshold. The last constraint (4.19e) specifies the binary restrictions on the decision variables.

The PC-SPWR model has pure binary decision variables and is very challenging to solve in general (see Chap. 2). However, with $|\Omega| < \infty$, one can form the DEP and apply an MIP solver toward solving PC-SPWR. This is what is done in the computational study by [5]. There are few decomposition methods for PC-SPWR in general. We refer the interested reader to [8] for a decomposition method for this class of SP.

4.11 Optimal Vaccine Allocation for Epidemics

Our last example application of SP is from epidemiology. We consider the problem of allocating limited vaccines to prevent epidemics. Specifically, we focus on the relatively recent model described by [12] for determining optimal vaccination strategies (policies) for a multi-community heterogeneous (by age) population for the COVID-19 pandemic. We shall refer to this problem as the stochastic vaccine allocation problem (SVAP). An earlier study for influenza is described in [34]. The first Covid-19 outbreak was reported in Wuhan, China, in December of 2019. In early 2020, the World Health Organization (WHO) declared COVID-19 a global pandemic. The disease is caused by the Severe Acute Respiratory Syndrome Coronavirus (SARS-CoV-2), which has continued to mutate.

In epidemiology, the *basic reproduction number*, R_0, is used to measure the transmission of an infectious disease in a community. Given a primary infection case within a completely susceptible population, R_0 is the average number of secondary infections. To prevent the spread of an infectious disease, health authorities usually implement different mitigation measures. These measures include vaccines and non-pharmaceutical interventions such as contact tracing, mask mandates, travel restrictions, border closures, and quarantining. When mitigation measures are in effect, the change in disease transmissibility over time is measured using the *effective reproduction number*, R_t, which is the average number of secondary infections caused by a primary case at a given time t. If $R_t > 1$, it means that an outbreak will continue. Otherwise, if $R_t \leq 1$, it means that the outbreak will come to an end. Therefore, the primary goal of mitigation measures is to bring the value of R_t to below one.

When a vaccination campaign is in effect, the change in disease transmissibility in a community of households is estimated using the *post-vaccination reproduction number*, R_{HV}, which is the average number of secondary infections caused by a primary case. The assumption in this case is that the vaccine is able to prevent re-infections. A vaccination policy specifies the proportion of individuals to vaccinate in each (age) group in a given household in the community. The primary goal of a vaccination policy is to prevent epidemics by keeping $R_{HV} \leq 1$. Given a vaccination policy, the vaccination coverage is the proportion of individuals in a community who are vaccinated. Thus, in terms of optimization, the goal is to determine the vaccination coverage that results in a level of vaccine induced herd immunity that is high enough to prevent epidemics. In other words, we have to seek an optimal vaccination policy, i.e., a policy that provides the minimum number of vaccinations required to have $R_{HV} \leq 1$. The challenge in doing lies in the underlying epidemiology disease spread model that captures well the complex disease spread process. Disease spread in a dynamical process and models involve several parameters that are uncertain at best.

The SVAP model proposed in [12] involves several COVID-19 spread parameters such as age-related *susceptibility* and *infectivity* to SARS-CoV-2, variation in human interactions, and *vaccine efficacy*. Susceptibility, denoted β, is the ability of getting infected if an uninfected individual has a *close contact* with an infected individual. The close contact rate, denoted m, is the average number of contacts that is adequate for transmitting the disease when the contacted person is susceptible. Infectivity, denoted λ, is the ability of an infected individual to spread the disease to others. We shall denote by ε infectivity, which is the percentage reduction in disease among the vaccinated. The SVAP model also considers distribution of the household composition in a community and is based on the epidemiology model by [7], which has the following basic expression for the post vaccination reproduction number for a homogeneous population:

$$R_{HV} = \sum_{n\in\mathbb{N}} \sum_{v\in\mathbb{V}} a_{nv} x_{nv}, \tag{4.20}$$

where the decision variable x_{nv} is the proportion of n-sized households under vaccination policy v and the disease spread parameter

$$a_{nv} = \frac{mh_n}{\mu} \Big((1-b)(n - f(v)\varepsilon) + b(n - f(v)\varepsilon)^2 + bf(v)\varepsilon(1-\varepsilon) \Big). \tag{4.21}$$

This model assumes that the community is composed of different size households $n \in \mathbb{N}$, where m is the average number contacts an infective makes with individuals outside the household, h_n is the proportion of n-sized households, and μ is the average household size. Disease transmission within the household is captured by $b \in [0, 1]$, where a value of zero means that there is no transmission and a value of one implies the total transmission. The function $f(v)$ gives the number of persons to vaccinate under vaccination policy $v \in \mathbb{V}$ and ε is the vaccine efficacy. Notice that the term mh_n/μ captures outside household transmission rate, while the remainder of the term captures inside household transmission rate.

The model expressed in equation (4.20) becomes more complex when dealing with a heterogeneous population and uncertain disease spread parameters. For an age heterogeneous population, i.e., a population that is affected differently by the disease based on age, the household size is characterized as household type n. For example, the population can be divided into three *age* groups A, B, and C, where $A : age \leq 19$, $B : 20 \leq age \leq 64$, and $C : age \geq 65$. Table 4.8 shows the household type, size, and composition for this case as well as the possible vaccination policies for each household type.

The SVAP model extends the disease spread model (4.20) to the heterogeneous and stochastic setting:

$$R_{HV}(\tilde{\omega}) = \sum_{n\in\mathbb{N}} \sum_{v\in\mathbb{V}} a_{nv}(\tilde{\omega}) x_{nv}, \tag{4.22}$$

where the disease spread parameter $a_{nv}(\tilde{\omega})$ is a random variable with $\tilde{\omega}$ describing the uncertain disease spread parameters. The goal of the SVAP model is to determine optimal vaccination policies for a multi-community heterogeneous population to prevent epidemics, i.e., to have $R_{HV}(\tilde{\omega}) \leq 1$. However, achieving this may not be possible, and thus, a probabilistic constraint is imposed:

$$\mathbb{P}\{R_{HV}(\tilde{\omega}) \leq 1\} \geq \alpha,$$

Table 4.8 Example household types and vaccination policies

HH Type	HH Size	HH Composition	Total Vacc. Policies	Possible Vacc. Policies for Type n HH
n	$p(n)$	$(p_A(n), p_B(n), p_C(n))$	$(p_A(n)+1)*$ $(p_B(n)+1)*$ $(p_C(n)+1)$	$(f_A(n, v), f_B(n, v), f_C(n, v))$
1	1	(1, 0, 0)	2	(0, 0, 0), (1, 0, 0)
2	1	(0, 1, 0)	2	(0, 0, 0), (0, 1, 0)
3	1	(0, 0, 1)	2	(0, 0, 0), (0, 0, 1)
4	2	(2, 0, 0)	3	(0, 0, 0), (1, 0, 0), (2, 0, 0)
5	2	(0, 2, 0)	3	(0, 0, 0), (0, 1, 0), (0, 2, 0)
6	2	(0, 0, 2)	3	(0, 0, 0), (0, 0, 1), (0, 0, 2)
7	2	(1, 1, 0)	4	(0, 0, 0), (0, 1, 0), (1, 0, 0), (1, 1, 0)
8	2	(0, 1, 1)	4	(0, 0, 0), (0, 0, 1), (0, 1, 0), (0, 1, 1)
9	2	(1, 0, 1)	4	(0, 0, 0), (0, 0, 1), (1, 0, 0), (1, 0, 1)
10	3	(1, 2, 0)	6	(0, 0, 0), (0, 1, 0), (0, 2, 0), (1, 0, 0), (1, 1, 0), (1, 2, 0)

HH household, *Vacc.* vaccination

where $\alpha \in (0.5, 1)$ is the *reliability level* set by the decision-maker. This means that the constraint holds at least $\alpha \times 100\%$ of amount of the time and may not hold at most $(1 - \alpha) \times 100\%$ of the time. The constraint is violated when the vaccination policy is ineffective and results in an epidemic. This happens when the disease is highly infectious (e.g., very high infectivity and susceptibility) or when the number of available vaccines is not enough or vaccines are ineffective (e.g., low vaccine efficacy).

Without going into further details, let us describe a simplified SVAP model for a single community for illustrative purposes. We use the following notation to state our simplified model:

Sets

$\mathbb{N}$: set of household types, element $n \in \mathbb{N}$.
$\mathbb{I}$: set of person groups (age), element $i \in \mathbb{I}$.
$\mathbb{V}$: set of vaccination policies, element $v \in \mathbb{V}$.
$\mathbb{V}_n$: set of vaccination policies for household type $n \in \mathbb{N}$, element $v \in \mathbb{V}_n$.
Ω: set of outcomes (scenarios) for the community, element $\omega \in \Omega$.

Decision Variables

$x_{nv} \in \mathbb{R}_+$: proportion to vaccinate in household type $n \in \mathbb{N}$ in the community under vaccination policy $v \in \mathbb{V}$.
$y \in \mathbb{R}_+$: excess amount above reliability level for the community.
$z \in \mathbb{R}_+$: deficit amount below reliability level for the community.

Problem Data

V: total number of available vaccines.
H: number of households in the community.
h_n: proportion of type $n \in \mathbb{N}$ households in the community.
α: user-set model reliability level for the community.
$\bar{\alpha}^e$: excess allowed on model reliability level for the community.
$\bar{\alpha}^s$: deficit allowed on model reliability level for the community.
M_e, M_s: sufficiently large numbers ("big-M").
$\tilde{\omega}$: multivariate random variable whose outcome (scenario) ω specifies the uncertain disease spread parameters, i.e., $\omega := \{m(\omega), b(\omega), \epsilon(\omega), \beta(\omega), \gamma(\omega)\}$, $\omega \in \Omega$.

Functions

$a_{nv}(\omega) : \mathbb{V} \times \Omega \mapsto \mathbb{R}$: uncertain $R_{HV}(\omega)$ parameter that captures the impact of vaccination policy $v \in \mathbb{V}_n$ in a type $n \in \mathbb{N}$ household in the community under scenario $\omega \in \Omega$.
$f(n, v) : \mathbb{V} \mapsto \mathbb{R}$: cost function that computes the number of persons vaccinated in a type $n \in \mathbb{N}$ household under vaccination policy $v \in \mathbb{V}$. Furthermore, $f(n, v) = \sum_{i \in \mathbb{I}} f_i(v)$, where $f_i(n, v)$ is the number of persons in age group $i \in \mathbb{I}$ vaccinated in household n under vaccination policy v.

Using the above notation, a generic SVAP model can be written as follows:

$$\text{Min} \sum_{n \in \mathbb{N}} \sum_{v \in \mathbb{V}_n} f(n, v) h_n x_{nv} - M_e y + M_s z \tag{4.23a}$$

$$\text{s.t.} \sum_{v \in \mathbb{V}_n} x_{nv} = 1, \ \forall n \in \mathbb{N} \tag{4.23b}$$

$$\sum_{n \in \mathbb{N}} \sum_{v \in \mathbb{V}_n} Hf(n, v) h_n x_{nv} \leq V \tag{4.23c}$$

$$y \leq \min\{\alpha^e, 1 - \alpha\} \tag{4.23d}$$

$$z \leq \bar{\alpha}^s \tag{4.23e}$$

$$\mathbb{P}\{\sum_{n \in \mathbb{N}} \sum_{v \in \mathbb{V}_n} a_{nv}(\tilde{\omega}) x_{nv} \leq 1\} - y + z \geq \alpha \tag{4.23f}$$

$$x_{nv}, \ y, \ z \geq 0, \ \forall n \in \mathbb{N}, \ \forall v \in \mathbb{V}_n. \tag{4.23g}$$

The objective function (4.23a) minimizes vaccination coverage and the deviation above and below the specified reliability level for the community. Constraint (4.23b) calculates the proportion to vaccinate in each household type, while constraint (4.23c) imposes the requirement that the total number of vaccinations does not exceed the number of available vaccines. Constraints (4.23d) and (4.23e) bound the deviations above and below the reliability level to be within the specified levels. Constraint (4.23f) is the probabilistic (chance) constraint requiring $R_{HV}(\tilde{\omega}) \leq 1$ to prevent epidemics for the community with probability $\alpha + \bar{\alpha}^e - \bar{\alpha}^s$. The last constraint (4.23e) specifies the nonnegativity restrictions on the decision variables.

Although the SVAP model has continuous decision variables, it is nonconvex in general, which makes it challenging to solve (see Chap. 2). As with the PC-SPWR model in Sect. 4.10, with $|\Omega| < \infty$, one can form the DEP and apply a solver toward solving the problem. This is what is done in the computational study reported in [12]. The study involves a set of neighboring Texas counties and demonstrates that SVAP can aid public health decision-makers in determining the optimal allocation of limited vaccines under uncertainty to control epidemics. Among other findings, the results show that, based on specified reliability level, a certain percentage of the population must be vaccinated in each county to control outbreaks. The code and data used for this study are available on GitHub: https://github.com/gujjulakreddy/COVID-CC. The census data used in the model are available at https://data.census.gov/cedsci/.

Problems

4.1 SP Modeling
Choose a real-life application, and create an optimization model to address an interesting problem from the application using one of the SP approaches (SLP, MR-SLP, SMIP, MR-SMIP, or PC-SP).

(a) Clearly define your decision variables and the model data for each of the stages involved.
(b) Write a detailed SP formulation of your model.
(c) Write a description of the objective function, constraints, and restrictions on the decision variables.

4.2 Capacity Expansion Problem (CEP)
Consider problem CEP described in Sect. 4.2. and extend the formulation assuming that machines can fail and the failures follow a particular distribution.

4.3 Stochastic Server Location Problem (SSLP)
Consider problem SSLP described in Sect. 4.3. Suppose that in addition to customer availability being random, customer demand is also random and follows a particular distribution. Extend the formulation to reflect this new requirement.

4.4 Stochastic Supply Chain (SSCh) Planning Problem
Use the SSCh problem described in Sect. 4.4 as a motivation to derive an SP model (MR-SLP, SMIP, MR-SMIP, or PC-SP) of an SSCh problem of your interest. State the motivation for your choice of the problem, and clearly define your decision variables and the model data for each of the stages involved. Write a detailed mathematical formulation followed by the description of the objective function, constraints, and restrictions on the decision variables.

4.5 Stochastic Fuel Treatment Planning (SFTP)
Study the SFTP problem described in Sect. 4.5, and derive your own SP model (MR-SLP, SMIP, MR-SMIP, or PC-SP) of a different version of the problem. State the motivation for your choice of the problem, and clearly define your decision variables and the model data for each of the stages involved. Write a detailed mathematical formulation followed by the description of the objective function, constraints, and restrictions on the decision variables.

4.6 PC-SP Wildfire Response (PC-SPWR)
Consider the PC-SPWR model described in Sect. 4.10, and derive your own SP model (MR-SLP, SMIP, MR-SMIP, or PC-SP) for a problem from wildfire management. State the motivation for your choice of the problem, and clearly define your decision variables and the model data for each of the stages involved. Write a detailed mathematical formulation followed by the description of the objective function, constraints, and restrictions on the decision variables.

4.7 Nuclear Medicine Online Scheduling Problem (NMOSP)
Review the NMOSP described in Sect. 4.6, and derive an SP model (MR-SLP, SMIP, MR-SMIP, or PC-SP) of a problem from healthcare. State the motivation for your choice of the problem, and clearly define your decision variables and the model data for each of the stages involved. Write a detailed mathematical formulation followed by the description of the objective function, constraints, and restrictions on the decision variables.

4.8 Airport Stochastic Time Slot Allocation Problem (STSAP)
Consider the STSAP described in Sect. 4.7, and derive an SP model (MR-SLP, SMIP, MR-SMIP, or PC-SP) of a problem from airport time slot allocation. State the motivation for your choice of the problem, and clearly define your decision variables and the model data for each of the stages involved. Write a detailed mathematical formulation followed by the description of the objective function, constraints, and restrictions on the decision variables.

4.9 Stochastic Air Traffic Flow Management (SATFM)
Review the SATFM problem described in Sect. 4.8, and derive an SP model (MR-SLP, SMIP, MR-SMIP, or PC-SP) of a problem from air traffic flow management. State the motivation for your choice of the problem, and clearly define your decision variables and the model data for each of the stages involved. Write a detailed mathematical formulation followed by the description of the objective function, constraints, and restrictions on the decision variables.

4.10 Stochastic Satellite Constellation Scheduling problem (SSCP)
Study the SSCP model described in Sect. 4.9 and the references therein, and then create an SP model (MR-SLP, SMIP, MR-SMIP, or PC-SP) for a different version of SSCP. State the motivation for your choice of the problem, and clearly define your decision variables and the model data for each of the stages involved. Write a detailed mathematical formulation followed by the description of the objective function, constraints, and restrictions on the decision variables.

4.11 Stochastic Vaccine Allocation Problem (SVAP)
Extend the SVAP model described in Sect. 4.11 by considering a real setting involving multiple communities each with its own community demographic information and random epidemiology parameters. Create a formulation of SVAP that reflects this new requirement assuming that disease spread randomness in each community follows its own distribution, and the total number of available vaccines has to be shared across all the communities involved.

4.12 SP Modeling: Integrated Chance Constraints
Choose an application area of interest (could be from the example applications described in this chapter), and model a real-life problem from this area using the *integrated chance constraints* approach described in Sect. 2.7 of Chap. 2. Clearly define your decision variables and the model data for each of the stages involved, and then write a detailed mathematical formulation followed by the description of the objective function and each of the constraints.

References

1. S. Ahmed and R. Garcia. Dynamic capacity acquisition and assignment under uncertainty. *Annals of Operations Research*, 124:267–283, 2003.
2. S. Ahmed, M. Tawarmalani, and N. V. Sahinidis. A finite branch and bound algorithm for two-stage stochastic integer programs. *Mathematical Programming*, 100:355–377, 2004.
3. A. Alonso-Ayuso, L. Escudero, A. Garín, M.T. Ortuño, and G. Pérez. An approach for strategic supply chain planning under uncertainty based on stochastic 0-1 programming. *Journal of Global Optimization*, 26:97–124, 2003.
4. G. Angulo, S. Ahmed, and S.S. Dey. Improving the integer L-shaped method. *INFORMS Journal on Computing*, 28(3):483–499, 2016.
5. J.A. Gallego Arrubla, L. Ntaimo, and C. Stripling. Wildfire initial response planning using probabilistically constrained stochastic integer programming. *International Journal of Wildland Fire*, 23:825–838, 2014.
6. F. Barahona, S. Bermon, O. Gunluk, , and S. Hood. Robust capacity planning in semiconductor manufacturing. Technical Report RC22196, IBM Research, 2001.
7. N.G. Becker and D.N. Starczak. Optimal vaccination strategies for a community of households. *Mathematical Biosciences*, 139(2):117–132, 1997.
8. G. Canessa, , J.A. Gallego, L. Ntaimo, and B.K. Pagnoncelli. An algorithm for binary linear chance-constrained problems using IIS. *Computational Optimization & Applications*, 72(3):589 – 608, 2019.
9. L. Corolli, G. Lulli, and L. Ntaimo. The time slot allocation problem under uncertain capacity. *Transportation Research Part C: Emerging Technologies*, 46:16–29, 2014.
10. L. Corolli, G. Lulli, L. Ntaimo, and S. Venkatachalam. A two-stage stochastic integer programming model for air traffic flow management. *IMA Journal of Management Mathematics*, 28(1):19–40, 2017.
11. G.B. Dantzig. *Linear Programming and Extensions*. Princeton University Press, Princeton, New Jersey, 1963.
12. K.R. Gujjula, J. Gong, B. Segundo, and L. Ntaimo. Covid-19 vaccination policies under uncertain transmission characteristics using stochastic programming. *PLoS One*, 17(7):e0270524, 2022.
13. W.E. Hart, C. Laird, J.-P. Watson, and D.L. Woodruff. *Pyomo–optimization modeling in python*, volume 67. Springer Science & Business Media, 2012.
14. J.L. Higle and S. Sen. *Stochastic Decomposition*. Kluwer Academic Publishers, 101 Phillip Drive, Norwell, MA 02061, 1996.
15. IATA. Worldwide Slot Guidelines. https://www.iata.org/en/programs/ops-infra/slots/slot-guidelines/, 2023. Accessed: January 2023.
16. S. Jorjani, C.H. Scott, and D.L. Woodruff. Selection of an optimal subset of sizes. Technical report, University of California, Davis, CA, 1995.
17. S. Jorjani, C.H. Scott, and D.L. Woodruff. Selection of an optimal subset of sizes. *International Journal of Production Research*, 37(16):3697–3710, 1999.
18. M. Kabli, J. Gan, and L. Ntaimo. A Stochastic Programming Model for Fuel Treatment Management. *Forests*, 6(6):2148–2162, 2015.
19. F. Louveaux and Y. Smeers. Optimal investments for electricity generation: A stochastic model and a test problem. In Y Ermoliev and R Wets, editors, *Numerical techniques for stochastic optimization problems*, chapter 24, pages 445–452. Springer-Verlag, Berlin, 1988.
20. W.K. Mak, D.P. Morton, and R.K. Wood. Monte Carlo bounding techniques for determining solution quality in stochastic programs. *Operations Research Letters*, 24:47–56, 1999.
21. J.M. Mulvey and A. Ruszczynski. A new scenario decomposition method for large scale stochastic optimization. *Operations Research*, 43:477–490, 1995.
22. F. H. Murphy, S. Sen, and A. L. Soyster. Electric utility capacity expansion with certain load forecasts. *IIE Transactions*, 14:52–59, 1982.

23. NIFC. National Interagency Fire Center. https://www.nifc.gov/fire-information/statistics, no date. Accessed: 12-04-2022.
24. L. Ntaimo. Disjunctive decomposition for two-stage stochastic mixed-binary programs with random recourse. *Operations Research*, 58(1):229–243, 2010.
25. L. Ntaimo and S. Sen. The million-variable 'march' for stochastic combinatorial optimization. *Journal of Global Optimization*, 32(3):385–400, 2005.
26. G. Perboli, L. Gobbato, and F. Maggioni. A progressive hedging method for the multi-path travelling salesman problem with stochastic travel times. *IMA Journal of Management Mathematics*, 28(1):65–86, 2017.
27. E. Pérez, L. Ntaimo, C. Bailey, and P. McCormack. Modeling and simulation of nuclear medicine patient service management in DEVS. *SIMULATION*, 86(8-9):481–501, 2010.
28. E. Pérez, L. Ntaimo, C.O. Malavé, C. Bailey, and P. McCormack. Stochastic online appointment scheduling of multi-step sequential procedures in nuclear medicine. *Health Care Management Science*, 16(4):281–299, 2013.
29. E. Pérez, L. Ntaimo, W.E. Wilhelm, C. Bailey, and P. McCormack. Patient and resource scheduling of multi-step medical procedures in nuclear medicine. *IIE Transactions on Healthcare Systems Engineering*, 1(3):168–184, 2011.
30. F. Qiu, S. Ahmed, S.S. Dey, and L.A. Wolsey. Covering linear programming with violations. *INFORMS Journal on Computing*, 26(3):531–546, 2014.
31. S. Sen, R.D. Doverspike, and S. Cosares. Network planning with random demand. *Telecommunications Systems*, 3(11-30), 1994.
32. R. Tadei, G. Perboli, and F. Perfetti. The multi-path traveling salesman problem with stochastic travel costs. *EURO Journal on Transportation and Logistics*, 6(1):3–23, 2017.
33. M.W. Tanner and L. Ntaimo. IIS branch-and-cut for joint chance-constrained stochastic programs and application to optimal vaccine allocation. *European Journal of Operational Research*, 207(1):290–296, 2010.
34. M.W. Tanner, L. Sattenspiel, and L. Ntaimo. Finding optimal vaccination strategies under parameter uncertainty using stochastic programming. *Mathematical Biosciences*, 215:144–151, 2008.
35. C.G. Valicka, D.G., A. Staid, J-P. Watson, G. Hackebeil, S. Rathinam, and L. Ntaimo. Mixed-integer programming models for optimal constellation scheduling given cloud cover uncertainty. *European Journal of Operational Research*, 275(2):431–445, 2019.

Part III
Deterministic and Risk-Neutral Decomposition Methods

Chapter 5
Deterministic Large-Scale Decomposition Methods

5.1 Introduction

In this chapter, we study decomposition methods for deterministic large-scale linear programming (LP) to lay the foundation for decomposition methods for stochastic programming (SP) that we present in later chapters. These large-scale deterministic methods were developed in the 1960s before the advent of personal computers and precede decomposition methods for SP. We begin our study with Kelley's cutting-plane method [3], which involves optimizing a convex function over a convex compact set using cutting-planes. These cutting-planes are supporting hyperplanes to approximate the convex objective function. We provide numerical examples and a convergence analysis of Kelley's cutting-plane method. We then move on to study Benders decomposition method [1] for solving large-scale LPs with linking decision variables. Benders method can be thought of as a special case of Kelley's method when the convex function is a value function of an LP. As in Kelley's method, Benders method generates cutting-planes and is often referred to as a *row generation* method.

To potentially reduce the number of iterations in Benders decomposition, we introduce *regularization* whereby we add a quadratic term to the objective function to enable the iterates not to deviate too far from the incumbent solution. Doing this prevents the iterates from "bouncing" around by staying close enough to the incumbent solution. We refer to this method as *regularized* Benders decomposition and illustrate how it works.

Another large-scale decomposition method we study is the Dantzig–Wolfe decomposition method [2], also referred to as *column generation*. This method considers the *dual* to Benders problem, which has linking constraints. So unlike Benders method, Dantzig–Wolfe decomposition generates columns instead of rows. It is thus referred to as *column generation*.

L. Ntaimo, *Computational Stochastic Programming*, Springer Optimization and Its Applications 774, https://doi.org/10.1007/978-3-031-52464-6_5

We end the chapter with a subgradient optimization method. In particular, we consider Lagrangian relaxation of the Benders problem by placing the complicating or "difficult" constraints in the objective and penalizing them. The goal is to make the problem separable so it can be solved using a decomposition method. Therefore, we derive the Lagrangian dual to the problem and establish the relationship between the Lagrangian dual and the Dantzig–Wolfe problem. We apply subgradient optimization to solve the problem. In the next section, we begin our study with Kelley's method.

5.2 Kelley's Cutting-Plane Method

Kelley [3] considered a convex programming problem of minimizing a linear form on a compact convex set. Formally, the problem can be stated as follows:

$$\begin{aligned} &\text{Min } c^\top x \\ &\text{s.t. } f(x) \leq d \\ &\quad x \in X, \end{aligned}$$

where $c \in \mathbb{R}^n$ is the cost vector, $x \in \mathbb{R}^n_+$ is the decision variable vector, $f : \mathbb{R}^n \mapsto \mathbb{R}$ is a *convex* function, $d \in \mathbb{R}$ is the right hand side, and $X \subseteq \mathbb{R}^n$ is *compact* convex set. To lay the foundation for cutting-plane algorithms for SLP that we derive later, we consider the following form of Kelley's problem:

$$\begin{aligned} &\text{Min } f(x) \\ &\text{s.t. } x \in X. \end{aligned} \tag{5.1}$$

The problem is to find an $x \in X$ that minimizes the function $f(x)$.

5.2.1 Algorithm

To solve Problem (5.1), the basic idea of Kelley's cutting-plane method is to sequentially approximate the convex function $f(x)$ by cutting-planes and then optimize the approximated piecewise linear convex function over the set X. These cutting-planes are *tangents* to the function $f(x)$, i.e., *supporting hyperplanes* (see Chap. 1). Since a supporting hyperplane is an affine function, we will denote it by a pair $(\beta_k^0, \beta_k) \in \mathbb{R}^{n+1}$, where at algorithm iteration k, $\beta_k^0 \in \mathbb{R}$ is the intercept and $\beta_k \in \mathbb{R}^n$ is the gradient. Let $x^0 \in X$ be a chosen starting (initial) point. Then a basic Kelley's cutting-plane algorithm can be given as follows:

Algorithm *Basic Kelley's Cutting-Plane*
begin

Step 0. **Initialization.** Let $k \leftarrow 0$ and $x^0 \in X$ be given.
Step 1. **Generate Cutting-Plane.** Find supporting hyperplane (β_k^0, β_k) at x^k so that $f(x) \geq \beta_k^0 - \beta_k^\top x, \quad \forall x \in X$.
Step 2. **Solve Approximation.** Solve

$$\begin{aligned} \ell_k := \text{Min}\ & \eta \\ \text{s.t.}\ & \eta + \beta_t^\top x \geq \beta_t^0,\ t = 0, \cdots, k \\ & x \in X, \eta \in \mathbb{R} \end{aligned}$$

to get x^{k+1}.
Step 3. **Termination.** Stop if a *stopping rule* is satisfied, otherwise set $k \leftarrow k+1$ and return to step 1.

---

end

In step 1, the generated cutting-plane holds with equality at $x = x^k$ since it is a supporting hyperplane. The decision variable η in the LP formulation in step 2 is unrestricted in sign (free) and approximates the value of $f(x)$. The collection of constraints for $t = 0, \cdots, k$ is effectively a constraint set on η that can be written as follows:

$$\max_{x \in X}\{\beta_t^0 - \beta_t^\top x \mid t = 0, \cdots, k\}.$$

This set is the *maximum* of affine functions over X, which approximates $f(x)$ and is therefore *convex*. The LP formulation finds the minimum of the maximum of affine functions. Consequently, as k increases, Kelley's cutting-plane method eventually finds the minimizer of $f(x)$.

We give an illustration of Kelley's method in Fig. 5.1. In the figure, the algorithm is initialized with x^0. At this point, the function value is $f(x^0)$ and a supporting hyperplane (β_0^0, β_0) is generated in step 1 of the algorithm. In step 2, solving the approximation that involves this single supporting hyperplane leads to x^1 with the function value $f(x^1)$. Returning to step 1 of the algorithm, a supporting hyperplane (β_1^0, β_1) is generated. In step 2, the approximation now involves the two supporting hyperplanes generated so far. Solving the approximation LP, we get x^2 with function value $f(x^2)$. We can clearly see that if we continue this process, the iterates $\{x^0, x^1, x^2, \cdots\}$ will tend towards x^*, the minimizer of $f(x)$ over X.

Now the important question at this point is, how can one implement Kelley's cutting-plane algorithm? Well, to implement Kelley's algorithm, we need to perform the following operations: (a) evaluate the function $f(x)$; (b) find supporting hyperplane; and (c) apply some *stopping rule* in step 3. Evaluating the function

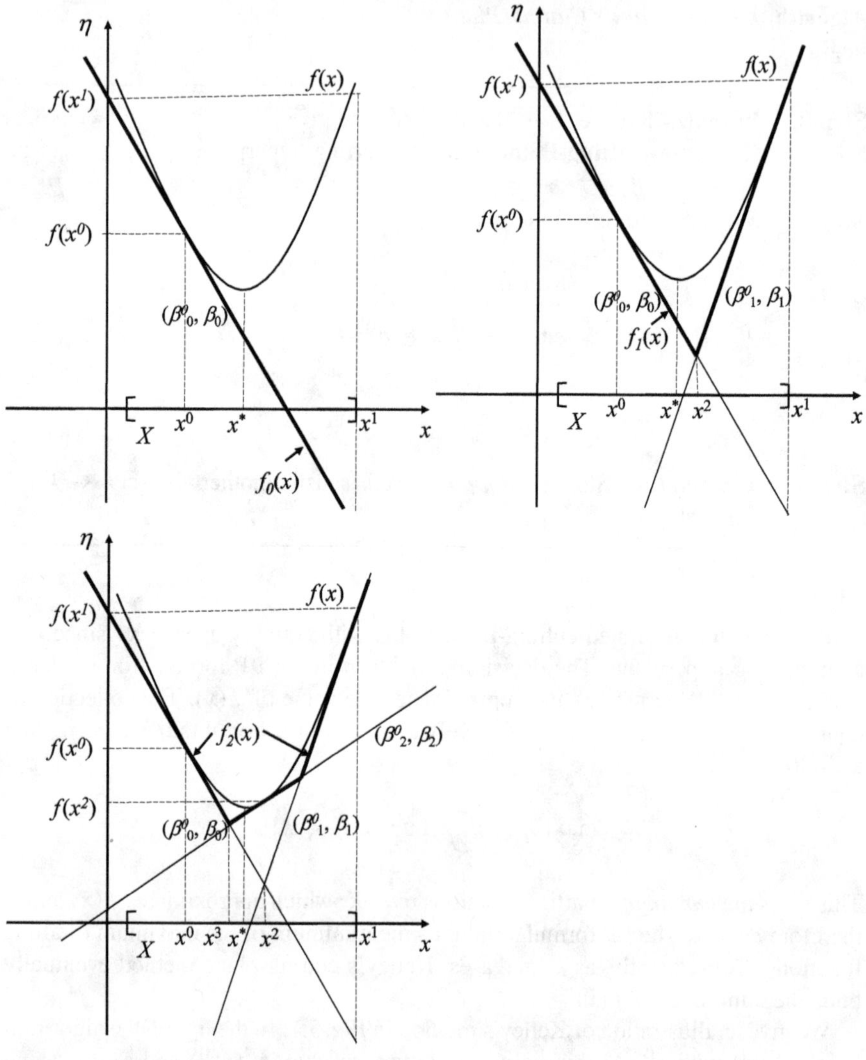

Fig. 5.1 An illustration of Kelley's method

$f(x)$ and finding the supporting hyperplane depend on the nature of the function itself. This includes linearity or nonlinearity of the function, convexity, differentiability, and so on. To establish a stopping criterion, let us consider the details of Kelley's method. Kelley's algorithm involves a sequence of piecewise linear lower bound approximations of f. Therefore, if we have $\{x_t\}_{t=0}^{k}$ and the corresponding supporting hyperplanes $\{\beta_t^0, \beta_t\}_{t=0}^{k}$, then

$$f(x) \geq \beta_t^0 - \beta_t^\top x, \; t = 0, \cdots, k$$

for all $x \in X$, and by convexity of $f(x)$, we have

$$f(x) \geq \max\{\beta_t^0 - \beta_t^\top x \mid t = 0, \cdots, k\}.$$

Now let $f_k(x)$ be the approximation of $f(x)$ at iteration k for all $x \in X$. Then we have the following:

$$\begin{aligned} f_0(x) &= \beta_0^0 - \beta_0^\top x, \\ f_k(x) &= \max\{\beta_t^0 - \beta_t^\top x \mid t = 0, \cdots, k\} \\ &= \max\{f_{k-1}(x),\ \beta_k^0 - \beta_k^\top x\}. \end{aligned}$$

Thus, for all $x \in X$, $f_k(x)$ approximates $f(x)$ from below:

$$f(x) \geq f_k(x) = \underset{x \in X}{\text{Max}}\ \{\beta_t^0 - \beta_t^\top x \mid t = 0, \cdots, k\}.$$

In Fig. 5.1, $f_2(x)$ is represented by the V-shaped function. Let x^* denote the minimizer (optimal point) of $f(x)$. Then we have

$$f(x^*) \geq f_k(x^*) \geq \underset{x \in X}{\text{Min}}\ f_k(x).$$

Thus the right hand side of the above relation provides a *lower bound* on $f(x^*)$. Let us denote this lower bound by ℓ_k. Then

$$\ell_k := \underset{x \in X}{\text{Min}}\ f_k(x).$$

Also, we have

$$f(x^t) \geq f(x^*) \quad \forall t = 0, \cdots, k.$$

Then an upper bound on $f(x)$, denoted u_k, is given by

$$u_k := \min\{f(x^t) \mid x^t \in X, \quad t = 0, \cdots, k\}.$$

Therefore, for a sequence $\{x^t\}_{t=0}^k$ generated by Kelley's algorithm, we have

$$u_k := \text{Min}\{u_{k-1}, f(x^k)\}.$$

Observe that

$$u_k \geq f(x^*) \geq \ell_k.$$

So now that we have established a lower bound ℓ_k and an upper bound u_k on $f(x^*)$, we can suggest some stopping criteria for Kelley's cutting-plane algorithm. To stop the algorithm, we can apply any one of the following stopping rules:

(S1) $u_k - \ell_k \leq \varepsilon,\ \varepsilon > 0.$
(S2) $u_k - \ell_k \leq \varepsilon|u_k|,\ 0 < \varepsilon < 1.$
(S3) $u_k - \ell_k \leq \max\{\varepsilon, \varepsilon|u_k|\}.$

However, we should point out that which of the three stopping rules to apply should be chosen carefully. For example, to use stopping rule (S1), the value of ε has to be chosen to be small enough relative to the range of the values of u_k and ℓ_k. Otherwise, at termination, an optimal solution may not be guaranteed. Unlike rule (S1), stopping rule (S2) uses a fraction of the upper bound (incumbent value), and as such, ε should be selected relative to the u_k values. Finally, stopping rule (S3) is simply a combination of (S1) and (S2). Next we restate Kelley's algorithm to include computing bounds and a stopping rule.

Algorithm *Kelley's Cutting-Plane*
begin

Step 0. **Initialization.** Let $k \leftarrow 0$, $u_{-1} \leftarrow \infty$, $\ell_0 \leftarrow -\infty$, and let $x^0 \in X$ and $\varepsilon > 0$ be given. Select stopping rule (S1), (S2), or (S3).
Step 1. **Generate Cutting-Plane.** Find supporting hyperplane (β_k^0, β_k) at x^k so that $f(x) \geq \beta_k^0 - \beta_k^\top x, \quad \forall x \in X.$
Step 2. **Solve Approximation.** Solve

$$\begin{aligned} \ell_k := \text{Min } & \eta \\ \text{s.t. } & \eta + \beta_t^\top x \geq \beta_t^0,\ t = 0, \cdots, k \\ & x \in X, \eta \in \mathbb{R} \end{aligned}$$

to get x^{k+1}.
Step 3. **Termination.** Set $u_k \leftarrow \min\{u_{k-1}, f(x^k)\}$. If u_k is updated, set incumbent solution $x_\varepsilon^* \leftarrow x^k$. If *stopping rule* is satisfied, stop. Otherwise, set $k \leftarrow k+1$ and return to step 1.

---

end

5.2.2 Numerical Example

Let us now illustrate Kelley's cutting-plane method using a numerical example.

Example 5.1 Apply Kelley's algorithm to $\text{Min}\{f(x) \mid x \in X\}$, where $f(x)$ is a convex function and X is a convex set, specified as follows:

$$f(x) = (x-2)^2 + 1 \text{ and } X = \{-1 \leq x \leq 4.\}$$

Use $\varepsilon = 0.001$ for your stopping rule. For each iteration k of the algorithm, report the following in a table: x^k, $f(x^k)$, $f(x^{k+1})$, $f_k(x^{k+1})$, and $u_k - \ell_k$. Plot $f(x)$ and $f_k(x^k)$ for all k:

(a) Use $x^0 = 0$ as your starting point.
(b) Use $x^0 = -1$ as your starting point.

Part (a): Since $f(x)$ is a differentiable function, the slope β of the cutting-plane at $x \in X$ is $\beta = \nabla f(x^k) = 2(x-2) = 2x - 4$. Given $\nabla f(^k)$, the supporting hyperplane (β^0, β) through the point $x^k \in X$ is given by $f(x) \geq \nabla f(x^k)(x - x^k) + f(x^k)$, where $\beta^0 = -\nabla f(x^k)x^k + f(x^k)$. Now applying Kelley's algorithm, we get the following:

Algorithm *Kelley's Cutting-Plane*
begin

Step 0. **Initialization.** Let $k \leftarrow 0$, $u_{-1} \leftarrow \infty$, $\ell_0 \leftarrow -\infty$, and let $x^0 \leftarrow 0$ and $\varepsilon \leftarrow 0.001$. Select stopping rule (S1).

Step 1. **Generate Cutting-Plane.** Find supporting hyperplane (β_0^0, β_0) at x^0 (tangent) so that $f(x) \geq \beta_0^0 - \beta_0^\top x, \quad \forall x \in X$.
$f(0) = 5$ and $\beta_0 = \nabla f(0) = 2(0) - 4 = -4$. The supporting hyperplane (β_0^0, β_0) is given by the affine function $f(x) = \nabla f(x^0)(x - x^0) + f(x^0)$, i.e., $f(x) = -4(x - (0)) + 5 = -4x + 5$. Thus, $(\beta_0^0, \beta_0) = (5, -4)$.

Remark 5.1 Notice that at x^0 we have $f(x^0) = -4x^0 + 5 = 5$ since $(5, -4)$ is a supporting hyperplane. The piecewise linear approximation $f_0(x)$ of $f(x)$ is given by $f_0(x) = \max_{x \in X}\{-4x + 5\}$, which is written as an LP in the next step.

Step 2. **Solve Approximation.** Solve

$$\begin{aligned} \ell_0 := &\text{Min } \eta \\ &\text{s.t. } \eta + 4x \geq 5, \\ &\quad -1 \leq x \leq 4, \eta \text{ free} \end{aligned}$$

to get $x^1 = 4$ and $\ell_0 = -11$.

Step 3. **Termination.** $u_0 := \text{Min}\{\infty, f(0)\} = \min\{\infty, 5\} = 5$. Set $x_\varepsilon^* \leftarrow 0$ and $u_k - \ell_k \leftarrow 5 - (-11) = 16$. Since $16 > \varepsilon$, set $k \leftarrow 1$ and return to step 1.

Iteration $k = 1$:

Step 1. **Generate Cutting-Plane.** Find supporting hyperplane (β_1^0, β_1) at x^1 (tangent) so that $f(x) \geq \beta_1^0 - \beta_1^\top x, \quad \forall x \in X$.
$f(4) = 5$ and $\nabla f(4) = 2(4) - 4 = 4$. Therefore, the cutting-plane is $f(x) = 4(x - 4) + 5 = 4x - 11$. Thus, $(\beta_1^0, \beta_1) = (-11, 4)$.

Remark 5.2 As in the first iteration, notice that at x^1 we have $f(x^1) = 4 * 4 - 11 = 5$ since $(-11, 4)$ is a supporting hyperplane. Now the piecewise linear approximation $f_1(x)$ of $f(x)$ is given by $f_1(x) = \max_{x \in X}\{-4x + 5, 4x - 11\}$, which is written as an LP in step 2 below.

Step 2. Solve Approximation. Solve

$$
\begin{aligned}
\ell_1 := &\text{Min } \eta \\
&\text{s.t. } \eta + 4x \geq 5, \\
&\quad \eta - 4x \geq -11, \\
&\quad -1 \leq x \leq 4, \eta \text{ free}
\end{aligned}
$$

to get $x^2 \leftarrow 2$ and $\ell_1 = -3$.

Step 3. Termination. $u_1 := \text{Min}\{u_0, f(4)\} = \min\{5, 5\} = 5$. Therefore, set $x^*_\varepsilon \leftarrow 0$ or 4 and $u_k - \ell_k \leftarrow 5 - (-3) = 8$. Since $8 > \varepsilon$, set $k \leftarrow 2$ and return to step 1.

Iteration k $= 2$:

Step 1. Generate cutting-plane. Find supporting hyperplane (β_2^0, β_2) at x^2 so that $f(x) \geq \beta_2^0 - \beta_2^\top x, \quad \forall x \in X$.
$f(2) = 1$ and $\nabla f(2) = 2(2) - 4 = 0$. Therefore, the cutting-plane is $f(x) = 0(x - 4) + 1 = 1$. Thus, $(\beta_1^0, \beta_1) = (1, 0)$.

Remark 5.3 Since $\nabla f(2) = 0$, it means that $x = 2$ is the optimal point. Now the piecewise linear approximation $f_2(x)$ of $f(x)$ is given by $f_2(x) = \max_{x \in X}\{-4x + 5, 4x - 11, 1\}$, which is written as an LP in the next step.

Step 2. Solve Approximation. Solve

$$
\begin{aligned}
\ell_1 := &\text{Min } \eta \\
&\text{s.t. } \eta + 4x \geq 5, \\
&\quad \eta - 4x \geq -11, \\
&\quad \eta \geq 1, \\
&\quad -1 \leq x \leq 4, \eta \text{ free}
\end{aligned}
$$

to get $x^3 \leftarrow 1$ and $\ell_1 = 1$. We should point out here that $x \in [1, 3]$ is optimal (alternative optimal solutions) to this LP.

Step 3. Termination. $u_2 := \text{Min}\{u_1, f(2)\} = \min\{5, 1\} = 1$. Therefore, incumbent solution is not updated. Set $u_k - \ell_k \leftarrow 1 - 1 = 0$. Since $0 < \varepsilon$, stop and declare $x^*_\varepsilon = 2$ optimal with value $f(2) = 1$.

The results of the algorithm are summarized in Table 5.1. Figure 5.2 shows a plot of $f(x)$ and the cutting-planes generated at each iteration.

Table 5.1 Summary of iterations of Kelley's algorithm for part (a)

k	x^k	$f(x^k)$	$f(x^{k+1})$	$f_k(x^{k+1})$	u_k	ℓ_k	$u_k - \ell_k$
0	0	5	5	−11	5	−11	16
1	4	5	1	−3	5	−3	8
2	2	1	2	1	1	1	0

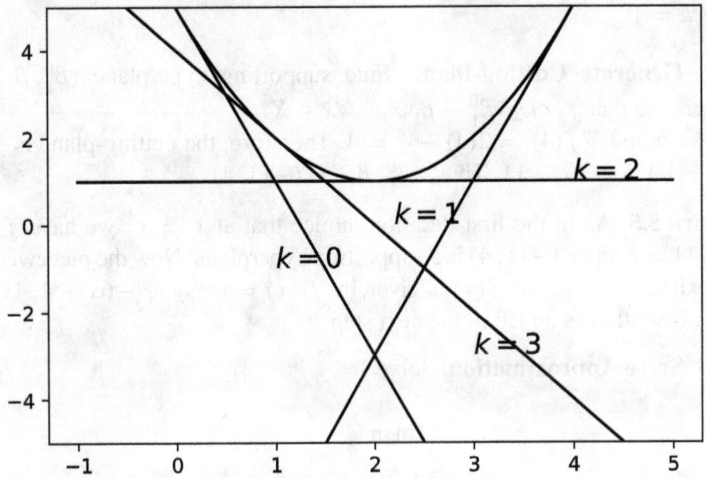

Fig. 5.2 Plot of $f(x)$ and the cutting-planes generated at each iteration

***Part (b)*:** Now applying Kelley's algorithm, we get the following:

Algorithm *Kelley's Cutting-Plane*
begin

Step 0. Initialization. Let $k \leftarrow 0$, $u_{-1} \leftarrow \infty$, $\ell_0 \leftarrow -\infty$, and let $x^0 \leftarrow -1$ and $\varepsilon \leftarrow 0.001$. Select stopping rule (S1).

Step 1. Generate Cutting-Plane. Find supporting hyperplane (β_0^0, β_0) at x^0 (tangent) so that $f(x) \geq \beta_0^0 - \beta_0^\top x, \quad \forall x \in X$.
$f(-1) = 10$ and $\beta_0 = \nabla f(-1) = 2(-1) - 4 = -6$. The supporting hyperplane (β_0^0, β_0) is given by the affine function $f(x) = \nabla f(x^0)(x - x^0) + f(x^0)$, i.e., $f(x) = -6(x - (-1)) + 10 = -6x + 4$. Thus, $(\beta_0^0, \beta_0) = (4, -6)$.

Remark 5.4 Notice that at $x = x^0$ we have $f(x^0) = -6x^0 + 4 = 10$ since $(4, -6)$ is a supporting hyperplane. The piecewise linear approximation $f_0(x)$ of $f(x)$ is given by $f_0(x) = \max_{x \in X}\{-6x + 4\}$, which is written as an LP in the next step.

Step 2. Solve Approximation. Solve

$$\begin{aligned} \ell_0 := & \text{Min } \eta \\ & \text{s.t. } \eta + 6x \geq 4, \\ & -1 \leq x \leq 4, \ \eta \text{ free} \end{aligned}$$

to get $x^1 = 4$ and $\ell_0 = -20$.

Step 3. **Termination.** $u_0 := \text{Min}\{\infty, f(-1)\} = \min\{\infty, 10\} = 10$. Therefore, set $x_\varepsilon^* \leftarrow -1$ and $u_k - \ell_k \leftarrow 10 - (-20) = 30$. Since $30 > \varepsilon$, set $k \leftarrow 1$ and return to step 1.

Iteration $k = 1$:

Step 1. **Generate Cutting-Plane.** Find supporting hyperplane (β_1^0, β_1) at x^1 (tangent) so that $f(x) \geq \beta_1^0 - \beta_1^\top x, \quad \forall x \in X$.
$f(4) = 5$ and $\nabla f(4) = 2(4) - 4 = 4$. Therefore, the cutting-plane is $f(x) = 4(x-4) + 5 = 4x - 11$. Thus, $(\beta_1^0, \beta_1) = (-11, 4)$.

Remark 5.5 As in the first iteration, notice that at $x = x^1$ we have $f(x^1) = 4{*}4 - 11 = 5$ since $(-11, 4)$ is a supporting hyperplane. Now the piecewise linear approximation $f_1(x)$ of $f(x)$ is given by $f_1(x) = \max_{x \in X}\{-6x + 4, 4x - 11\}$, which is written as an LP in the next step.

Step 2. **Solve Approximation.** Solve

$$
\begin{aligned}
\ell_1 := \text{Min}\ & \eta \\
\text{s.t.}\ & \eta + 6x \geq 4, \\
& \eta - 4x \geq -11, \\
& -1 \leq x \leq 4, \eta \text{ free}
\end{aligned}
$$

to get $x^2 \leftarrow 1.5$ and $\ell_1 = -5$.

Step 3. **Termination.** $u_1 := \text{Min}\{u_0, f(4)\} = \min\{10, 5\} = 5$. Therefore, set $x_\varepsilon^* \leftarrow 4$ and $u_k - \ell_k \leftarrow 5 - (-5) = 10$. Since $10 > \varepsilon$, set $k \leftarrow 2$ and return to step 1.

Repeating this process, we get the results summarized in Table 5.2. The optimal solution $x_\varepsilon^* = 2.0078$ is found after eight iterations. A plot of $f(x)$ with the cutting-plane generated at each iteration is given in Fig. 5.3. Observe how the piecewise linear approximation of $f(x)$ in this case requires a lot more iterations that in part (a) by simply changing the starting point.

5.2.3 *Convergence of Kelley's Cutting-Plane Algorithm*

Two basic questions that have to be addressed in algorithm design are: (1) Does the algorithm converge to an optimal solution? (2) If so, what is the convergence rate? In this subsection, we address the first question. We are interested in knowing the answers to the following questions: Does $\{f_k(x^k), f_{k+1}(x^k)\}$ converge as $k \to \infty$?

Table 5.2 Summary of iterations of Kelley's algorithm for part (b)

k	x^k	$f(x^k)$	$f(x^{k+1})$	$f_k(x^{k+1})$	u_k	ℓ_k	$u_k - \ell_k$
0	−1.0000	10.0000	5.0000	−20.0000	10.0000	−20.0000	30.0000
1	4.0000	5.0000	1.2500	−5.0000	5.0000	−5.0000	10.0000
2	1.5000	1.2500	1.5625	0.0000	1.2500	0.0000	1.2500
3	2.7500	1.5625	1.0156	0.6250	1.2500	0.6250	0.6250
4	2.1250	1.0156	1.0352	0.9375	1.0156	0.9375	0.07810
5	1.8125	1.0352	1.0010	0.9766	1.0156	0.9766	0.03908
6	1.9688	1.0010	1.0022	0.9961	1.0010	0.9961	0.0049
7	2.0469	1.0022	1.0001	0.9985	1.0010	0.9985	0.0024
8	2.0078	1.0001	1.0001	0.9998	1.0001	0.9998	0.0003
9	1.9887	1.00013	1.00014	1.00014	1.00013	–	–

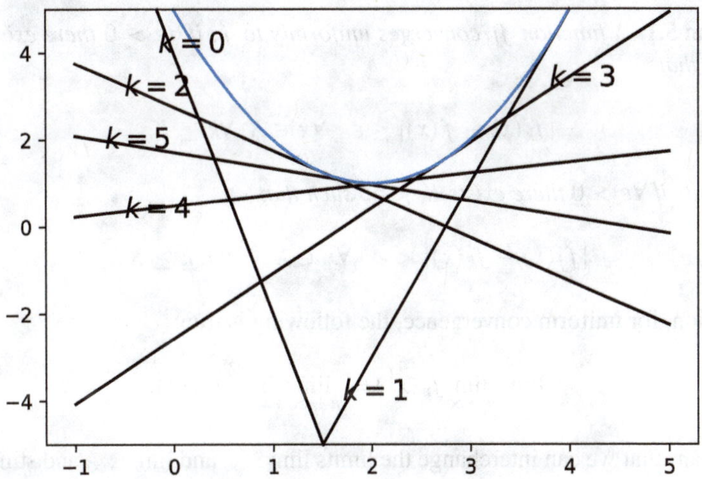

Fig. 5.3 Plot of $f(x)$ and the cutting-planes generated at each iteration

What happens with $\{x^k\}$ as $k \to \infty$? We show that Kelley's cutting-plane algorithm converges to an optimal solution after a finite number of iterations. First, let us restate some basic convergence results before we establish this fact about Kelley's algorithm. Recall from Chap. 1 that the notation $\{\ldots\}$ denotes a *sequence*. For ease of exposition, we denote by f the function $f(x), \forall x \in X$ and by f_k the function approximation $f_k(x), \forall x \in X$. In other words, f refers to the entire function over its domain X, while f_K refers to the function approximation over the same domain.

We established in Chap. 1 that an *infinite sequence* in a compact set has at least one subsequence that converges. For a *limit point* $\bar{x}$, we have $\lim_{n\to\infty} x^n = \bar{x}$. Also,

$$\{x^k\} \to \bar{x} \text{ if } \forall \varepsilon > 0 \text{ there exists } K < \infty \text{ such that } ||\bar{x} - x^k|| < \varepsilon, \quad \forall k \geq K.$$

A function $f(x),\ x \in X$ is *continuous* if

$$x^k \to x \quad \Rightarrow \quad f(x^k) \to f(x), \quad \forall x \in X,$$

while the function approximations $\{f_k(x)\}_{x \in X}$ converge *pointwise* to $f(x)$, i.e.,

$$f_k(x) \to f(x), \quad \forall x \in X.$$

We recall that in *pointwise convergence*, we fix x and then vary the function $f_k(x)$. Therefore, the following condition is sufficient to show pointwise convergence:

$$f(x) \geq f_k(x) \geq f_{k-1}(x), \quad \forall x \in X.$$

Unlike pointwise convergence, *uniform convergence* has stricter requirements. These are summarized in the following result as a theorem:

Theorem 5.1 *A function f_k converges uniformly to f, if $\forall \varepsilon > 0$ there exists $K < \infty$ such that*

$$|f_k(x) - f(x)| \leq \varepsilon \quad \forall x \in X, \forall k \geq K$$

if and only if $\forall \varepsilon > 0$ there exists $K < \infty$ such that

$$|f_j(x) - f_k(x)| < \varepsilon \quad \forall x \in X, \quad \forall j, k \geq K.$$

In addition, for uniform convergence, the following is true:

$$\lim_{k \to \infty} \lim_{n \to \infty} f_n(x^k) = \lim_{n \to \infty} \lim_{k \to \infty} f_n(x^k).$$

This means that we can interchange the limits $\lim_{k \to \infty}$ and $\lim_{n \to \infty}$ and still get the same result. Another useful result is that if X is compact, $\{f_k\}$ are continuous, $\{f_k\}$ converge pointwise to some function $\bar{f}$, and

$$f_k(x) \geq f_{k-1}(x) \quad \forall x \in X, \text{ and } \forall k,$$

then $\{f_k\}$ uniformly converge to $\bar{f}$.

Let us now turn to the convergence of Kelley's cutting-plane algorithm. With Kelley's algorithm, the following is true: X is compact, f is convex, $\{x^k\}$ and $\{f_k\}$ are constructed using Kelley's algorithm, and $f_k(x) \geq f_{k-1}(x) \quad \forall x \in X$, and $\forall k$. Since $x^k \in X \quad \forall k$ and X is compact, every limit point of $\{x^k\}$ is feasible. By construction, we have

$$f_{k-1} \leq f_k \leq f, \quad \forall k.$$

Therefore, this implies that $\{f_k\}$ converge *pointwise* to some function $\bar{f}$, $\{f_k\}$ are continuous as is f, so $\{f_k\}$ converge uniformly. By construction, due to supporting hyperplanes, we have that $f_k(x^k) = f(x^k) \quad \forall k$ and $f_k(x) \le f(x), \quad \forall x \in X, \quad \forall k$.

Now suppose that $\bar{x}$ is a limit point of $\{x^k\}$ with function value f^- so that

$$\{x^k\}_{k \in \mathbb{K}} \to \bar{x}$$

for some index set $\mathbb{K}$. Then $f_k(x^k) = f(x^k) \quad \forall k \in \mathbb{K}$ implies that $\{f_k(x^k)\} \to f(\bar{x})$. Uniform convergence ensures that

$$\{f_k(x^k) - f_{k-1}(x^k)\} \to 0.$$

Let x^* denote the minimizer of f and let q^* be the optimal value corresponding to x^*. In Kelley's algorithm, we have

$$f_{k-1}(x^k) \le f^- \le f(x^k).$$

Taking limits on the left hand side and right hand side, we have

$$\lim_{k \in \mathbb{K}} f_{k-1}(x^k) = f(\bar{x}) \quad \le q^* \le \quad \lim_{k \in \mathbb{K}} f(x^k) = f(\bar{x}).$$

Consequently, $f(\bar{x}) = q^*$, implying that $\bar{x}$ is an optimal solution. Thus, $\bar{x} = x^*$. We have now established that Kelley's cutting-plane algorithm will terminate after a finite number of iterations and will stop at an optimal point. This is indeed amply demonstrated in Example 5.1.

5.3 Benders Decomposition

Benders decomposition [1] deals with large-scale LPs with the following problem structure:

$$\text{Min } c^\top x + q^\top y \tag{5.2}$$

$$\text{s.t. } Ax \qquad \ge b \tag{5.3}$$

$$Tx + Wy \ \ge r \tag{5.4}$$

$$x, \quad y \ \ge 0, \tag{5.5}$$

where $x \in \mathbb{R}_+^{n_1}$ and $y \in \mathbb{R}_+^{n_2}$ are decision variable vectors, $c \in \mathbb{R}^{n_1}$ and $q \in \mathbb{R}^{n_2}$ are cost vectors, $A \in \mathbb{R}^{m_1 \times n_2}$ is a constraint matrix, $b \in \mathbb{R}^{m_1}$ is a right hand side vector, $T \in \mathbb{R}^{m_2 \times n_1}$ and $W \in \mathbb{R}^{m_2 \times n_2}$ are constraint matrices, and $r \in \mathbb{R}^{m_2}$ is a right hand side vector. The decision variable vector x links the two sets of constraints (5.3–5.4), while y only appears in the second set of constraints (5.4). Therefore, x is a linking or "complicating" decision variable, and constraint (5.3) is assumed to be

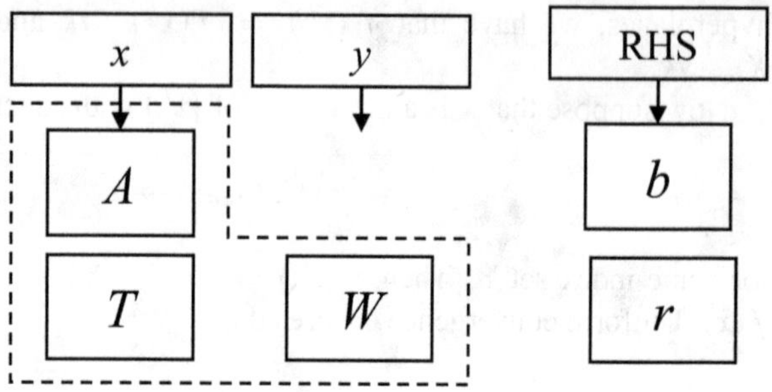

Fig. 5.4 The dual block angular or *L-shape* structure of Benders problem

"easy." The motivation for this model stems from several applications such as those that arise in facility location and logistics that have the Benders problem structure, which is referred to as *dual block angular* or *L-shaped* structure (Fig. 5.4). The L-shaped algorithm , which we study in Chap. 6, gets its name from the structure of the Benders problem.

5.3.1 Decomposition Approach

The basis of Benders decomposition is to project out the x variables and then solve the y variable problem that remains (*subproblem*). Letting $X = \{x \mid Ax \geq b,\ x \geq 0\}$, the Benders problem can be decomposed as follows:

$$\operatorname*{Min}_{x \in X} f(x) = c^\top x + \varphi(x), \tag{5.6}$$

where $\varphi(x)$ is the value function of the following subproblem:

$$\begin{aligned} \varphi(x) := \text{Min}\ & q^\top y \\ \text{s.t.}\ & Wy \geq r - Tx \\ & y \geq 0. \end{aligned} \tag{5.7}$$

We will refer to Problem (5.6) as the *first-stage* problem and to subproblem 5.7 as the *second-stage* problem. We denote by $\pi \in \mathbb{R}^{m_2}$ the vector of dual multipliers corresponding to the constraints of Problem (5.7). Then the dual to the subproblem is given as follows:

$$\begin{aligned} \text{Max}\ & \pi^\top (r - Tx) \\ \text{s.t.}\ & \pi^\top W \leq q \\ & \pi \quad \geq 0. \end{aligned} \tag{5.8}$$

We should point out that the dual feasible region of Problem (5.8) is fixed for all $x \in X$. Let $\Pi = \{\pi^\top W \le q,\ \pi \ge 0\}$. Then for all $x \in X$, the following relation holds:

$$\varphi(x) \ge \underset{\pi \in \Pi}{\text{Max}}\ \pi^\top (r - Tx).$$

Now let π_k be the optimal solution to (5.8). Then we have

$$\varphi(x) \ge \pi_k^\top (r - Tx), \qquad \forall x \in X. \tag{5.9}$$

Observe that relation (5.9) holds with equality at $x = x^k$, i.e., the x that led to π_k. If subproblem (5.7) is *infeasible*, then Problem (5.8) is *unbounded*. In that case, let $\mu_k \in \mathbb{R}^{m_2}$ be the dual *extreme ray*.

The reader should now see that Benders method can be thought of as an application of Kelley's method to approximate $\varphi(x)$ using supporting hyperplanes. Let k continue to denote the algorithm iteration counter. We need to generate supporting hyperplanes of the form (5.9). Thus we will set up a *master program* with $\varphi(x)$ approximated by $\eta \in \mathbb{R}$ as follows:

$$\begin{aligned} \text{Min}\ & c^\top x + \eta \\ \text{s.t.}\ & Ax \ge b \\ & \eta \ge \pi_t^\top (r - Tx),\ t = 0, \cdots, k \\ & x \ge 0, \end{aligned}$$

where $\beta_k^\top := \pi_k^\top T$ and $\beta_k^0 := \pi_k^\top r$. Thus we have

$$\eta \ge \beta_k^0 - \beta_k^\top x,$$

which can be rewritten as follows:

$$\beta_k^\top x + \eta \ge \beta_k^0. \tag{5.10}$$

Inequality (5.10) is called an *optimality cut*, and this inequality represents a supporting hyperplane of the function $\varphi(x)$ at iteration k.

We assumed that subproblem (5.7) is feasible; however, if for some iteration k the subproblem is *infeasible* (the dual is unbounded), then we can get a dual extreme ray $\mu_k \in \mathbb{R}^{m_2}$. In this case, we set $\beta_k^\top := \mu_k^\top T$ and $\beta_k^0 := \mu_k^\top r$. We should point out that $\mu_k^\top (r - Tx) = \infty$. The curious reader should verify this computationally! Therefore, we want $\mu_k^\top (r - Tx) \le 0$. Thus whenever subproblem (5.7) is infeasible, we add the following inequality to the master program:

$$\beta_k^\top x \ge \beta_k^0. \tag{5.11}$$

Inequality (5.11) is called a *feasibility cut* and cuts off the $x \in X$ causing the infeasibility of the subproblem, i.e., the unboundedness of the dual subproblem, at iteration k.

Let Θ_k be the set of iteration indices at which subproblem (5.7) is feasible. Then the Benders *master problem* can be written as follows:

$$
\begin{aligned}
\nu_k := \text{Min } & c^\top x + \eta \\
\text{s.t. } & Ax \geq b \\
& \beta_t^\top x + \eta \geq \beta_t^0, \ t \in \Theta_k \\
& \beta_t^\top x \geq \beta_t^0, \ t \notin \Theta_k \\
& x \geq 0.
\end{aligned}
$$

For a given master problem solution (x^k, η^k), the Benders *subproblem* is given as follows:

$$\varphi(x^k) := \text{Min } q^\top y \tag{5.12}$$

$$\text{s.t. } Wy \geq r - Tx^k \tag{5.13}$$

$$y \geq 0. \tag{5.14}$$

As explained before, if the *subproblem* is feasible, get dual solution π_k and form an optimality cut (5.10). Otherwise, if it is infeasible, get dual extreme ray μ_k and form a feasibility cut (5.11).

5.3.2 *Algorithm*

To solve the Benders Problem (5.2)–(5.5), we first decompose the problem as in (5.6)–(5.7) and then sequentially approximate the subproblem value function $\varphi(x)$ by cutting-planes. Next, the approximated piecewise linear convex function is optimized over the set X. Like in Kelley's method, the cutting-planes are *supporting hyperplanes* of $\varphi(x)$. Because constraints (rows) are added to the master program, a Benders decomposition type algorithm is often referred to as a *row generation* algorithm. A basic Benders decomposition algorithm can be given as follows:

Algorithm *Benders Decomposition*
begin

Step 0. **Initialization.** Let $k \leftarrow 0$, choose $x^0 \in X$, and set $\ell \leftarrow -\infty$ and $u \leftarrow \infty$. Choose a small $\varepsilon > 0$.

Step 1. **Solve Subproblem.** Solve

$$\varphi(x^k) := \text{Min}\{q^\top y \mid Wy = r - Tx^k, y \geq 0\}.$$

If feasible, generate an *optimality* cut:

Get dual solution π_k.
Compute $\beta_k^0 \leftarrow \pi_k^\top r$ and $\beta_k^\top \leftarrow \pi_k^\top T$,

else if infeasible, generate a *feasibility* cut:

Get dual extreme ray μ_k.
Compute $\beta_k^0 \leftarrow \mu_k^\top r$ and $\beta_k^\top \leftarrow \mu_k^\top T$.

Compute *upper bound* if feasible:

Set $v^k \leftarrow c^\top x^k + \varphi(x^k)$.
Set $u \leftarrow \min\{v^k, u\}$.

If u is updated, set *incumbent* solution to $x^* \leftarrow x^k$.

Step 2. Add Cut to Master Program and Solve. If subproblem was feasible:

Add $\beta_k^\top x + \eta \geq \beta_k^0$ to master problem.

Else:

Add $\beta_k^\top x \geq \beta_k^0$ to master problem.

Solve master to get (x^{k+1}, η^{k+1}) and ν^{k+1} as the optimal value. Set $\ell \leftarrow \max\{\nu^{k+1}, \ell\}$.

Step 3. Termination. If $u - \ell \leq \varepsilon|u|$:

Stop, x^* is the ε-optimal solution.

Else:

Set $k \leftarrow k + 1$.

Return to step 1.

---

end

In step 0 of the Benders decomposition algorithm, $x^0 \in X$ should be selected by the modeler based on an understanding of the set X. Otherwise, x^0 can be obtained as follows:

$$x^0 \leftarrow \operatorname{argmin}\{c^\top x \mid Ax \geq b, x \geq 0\}.$$

An illustration of Benders decomposition method is given in Fig. 5.5. In the figure, the algorithm is initialized with x^0. At this point, the function value is $\varphi(x^0)$, and a supporting hyperplane (β_0^0, β_0) is generated in step 1 of the algorithm. In step 2, solving the approximation that involves this single supporting hyperplane leads to x^1. However, at this point, the function value $\varphi(x^1) := \infty$, i.e., the subproblem is *infeasible*. Thus upon returning to step 1 of the algorithm, a feasibility cut (β_1^0, β_1) is generated, and in step 2, the approximation now involves a supporting hyperplane

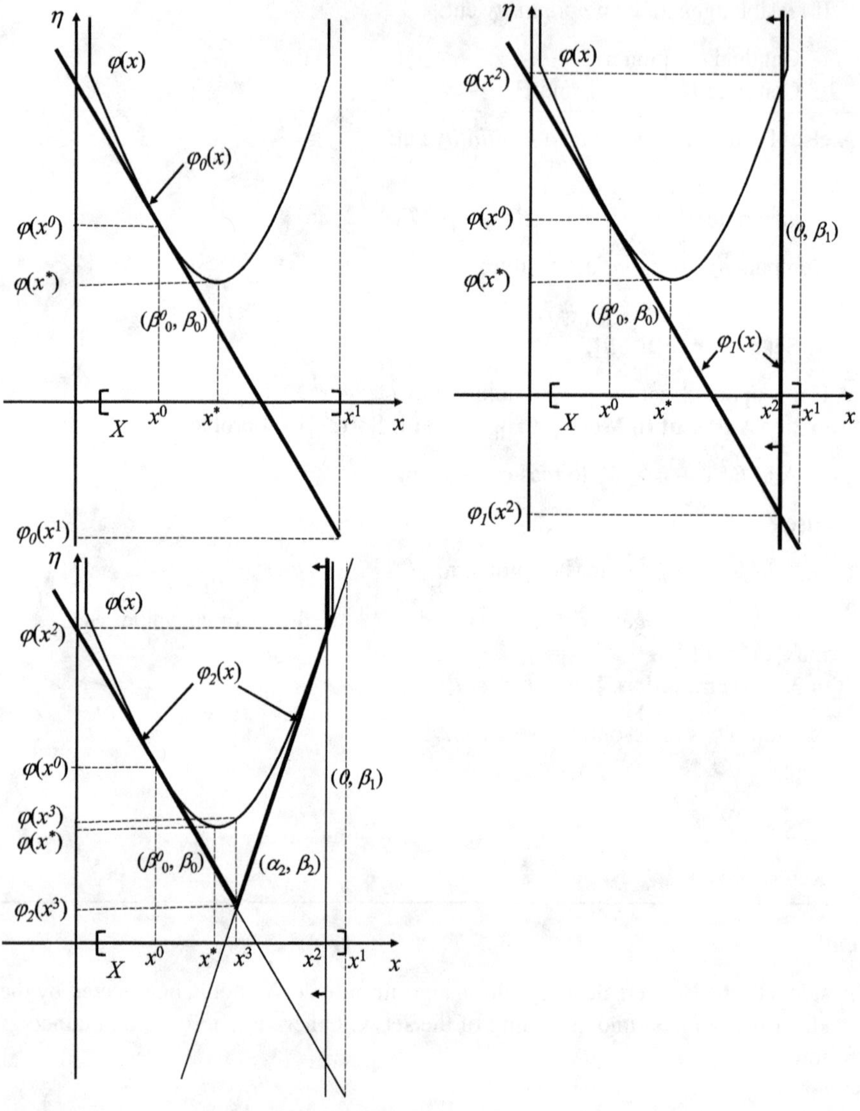

Fig. 5.5 An illustration of Benders decomposition

and a feasibility cut. Solving the master program, we get x^2 with function value $\varphi(x^2)$. The process is repeated to obtain x^3. We can clearly see that continuing with this process, the iterates $\{x^0, x^1, x^2, \cdots\}$ will go towards x^*, the minimizer of $\varphi(x)$ over X.

Even though we assumed a single subproblem in the Benders problem for ease of exposition, Benders decomposition allows for multiple subproblems. We exploit this setting in detail in Chap. 5 when we study the L-shaped method. For now it suffices to say that in step 1 of the algorithm we solve all the subproblems and then compute an optimality cut by aggregating (summing up) the β_k^0's and β_k's, respectively. We illustrate this in Example 5.2. We leave it as an exercise to prove that the Benders decomposition algorithm terminates after a finite number of iterations, and finds an optimal solution if it exists. Next, we illustrate Benders decomposition using numerical examples.

5.3.3 Numerical Examples

Example 5.2 Decompose and solve the following LP using Benders decomposition:

$$\begin{array}{lrrrrl}
\text{Min} & -3x_1 & -8x_2 & -16y_1 & -20y_2 & \\
\text{s.t.} & -x_1 & & & & \geq -1 \\
 & & -x_2 & & & \geq -1 \\
 & -x_1 & & -2y_1 & -3y_2 & \geq -5 \\
 & & -x_2 & -4y_1 & -1y_2 & \geq -5 \\
 & & & -y_1 & & \geq -1 \\
 & & & & -y_2 & \geq -1 \\
 & x_1, & x_2, & y_1, & y_2 & \geq \ 0.
\end{array}$$

Verify your solution by solving the given problem using a direct LP solver.

The first-stage decision variable vector is $x = (x_1, x_2)^\top$ since it appears in all the constraints (linking or complicating decision variable). The second-stage (subproblem) decision variable is $y = (y_1, y_2)^\top$ because it appears only in the last four constraints. We can thus decompose the problem into first- and second-stage subproblems as follows:

First-stage subproblem:

$$\begin{array}{lrrl}
\text{Min} & -3x_1 & -8x_2 & \\
\text{s.t.} & -x_1 & & \geq -1 \\
 & & -x_2 & \geq -1 \\
 & x_1, & x_2 & \geq \ 0.
\end{array}$$

Second-stage subproblem:

$$\begin{array}{llllll}
\text{Min} & -16y_1 & -20y_2 & & & \text{Dual multipliers} \\
\text{s.t.} & -2y_1 & -3y_2 & \geq & -5 + x_1 & \leftarrow \pi_1 \\
& -4y_1 & -y_2 & \geq & -5 + x_2 & \leftarrow \pi_2 \\
& -y_1 & & \geq & -1 & \leftarrow \pi_3 \\
& & -y_2 & \geq & -1 & \leftarrow \pi_4 \\
& y_1, & y_2 & \geq & 0. &
\end{array}$$

Using the dual multipliers, $\pi = (\pi_1, \pi_2, \pi_3, \pi_4)^\top$, the dual to the second-stage subproblem is given as follows:

$$\begin{array}{llllll}
\text{Max} & (-5 + x_1)\pi_1 & +(-5 + x_2)\pi_2 & -\pi_3 & -\pi_4 & \\
\text{s.t.} & -2\pi_1 & -4\pi_2 & -\pi_3 & & \leq -16 \\
& -3\pi_1 & -\pi_2 & & -\pi_4 & \leq -20 \\
& \pi_1, & \pi_2, & \pi_3, & \pi_4 & \geq \ \ 0.
\end{array}$$

The problem data can be summarized as follows:

$$c = (-3, -8)^\top, A = \begin{bmatrix} -1 & 0 \\ 0 & -1 \end{bmatrix}, b = (-1, -1)^\top,$$

$$T = \begin{bmatrix} -1 & 0 \\ 0 & -1 \\ 0 & 0 \\ 0 & 0 \end{bmatrix}, W = \begin{bmatrix} -2 & -3 \\ -4 & -1 \\ -1 & 0 \\ 0 & -1 \end{bmatrix}, \text{ and } r = (-5, -5, -1, -1)^\top.$$

Let set $X = \{-x_1 \geq -1,\ -x_2 \geq -1,\ x_1,\ x_2 \geq 0\}$. We are now ready to apply Benders decomposition algorithm as follows:

Algorithm *Benders Decomposition*
begin

Step 0. Initialization. Let $k \leftarrow 0$, choose $x^0 \in X$, and set $\ell \leftarrow -\infty$ and $u \leftarrow \infty$. Choose $\varepsilon = 10^{-6}$. We can set $x^0 \leftarrow \text{argmin}_{x \in X}\{-3x_1 - 8x_2\}$, i.e., we solve the first-stage problem to get $x^0 = (1, 1)^\top$ as the initial point.

Step 1. Solve Subproblem. Solve

$$\begin{array}{lllll}
\varphi(x^0) := \text{Min} & -16y_1 & -20y_2 & & \\
\text{s.t.} & -2y_1 & -3y_2 & \geq & -4 \\
& -4y_1 & -y_2 & \geq & -4 \\
& -y_1 & & \geq & -1 \\
& & -y_2 & \geq & -1 \\
& y_1, & y_2 & \geq & 0.
\end{array}$$

Problem is feasible: $\varphi(x^0) = -28.8$ and $y^0 = (0.8, 0.8)^\top$. Generate an *optimality* cut:

Get dual solution $\pi_0 = (6.4, 0.8, 0, 0)^\top$.
Compute $\beta_0^0 \leftarrow \pi_0^\top r = (6.4, 0.8, 0, 0)(-5, -5, -1, -1)^\top = -36$

$$\beta_0^\top \leftarrow \pi_0^\top T = (6.4, 0.8, 0, 0)\begin{bmatrix} -1 & 0 \\ 0 & -1 \\ 0 & 0 \\ 0 & 0 \end{bmatrix} = (-6.4, -0.8).$$

Compute *upper bound* if feasible:

Set $v^0 \leftarrow c^\top x^0 + \varphi(x^0) = (-3, -8)(1, 1)^\top + (-28.8) = -39.8$.
Set $u \leftarrow \min\{v^0, u\} = \min\{-39.8, \infty\} = -39.8$.

If u is updated, set *incumbent* solution to $x^* \leftarrow x^0 = (1, 1)^\top$.

Step 2. Add Cut to Master Program and Solve. If subproblem was feasible:

Add $\beta_0^\top x + \eta \geq \beta_0^0$ to master problem:

$$\begin{array}{lrrrl} v^1 := \text{Min} & -3x_1 & -8x_2 & +\eta & \\ \text{s.t.} & -x_1 & & & \geq -1 \\ & & -x_2 & & \geq -1 \\ & -6.4x_1 & -0.8x_2 & +\eta & \geq -36 \\ & x_1, & x_2 & & \geq 0, \ \eta \text{ free.} \end{array}$$

Solve master to get $x^1 = (0, 1)^\top$, $\eta^1 = -35.2$ and $v^1 = -43.2$. Set $\ell \leftarrow \max\{v^1, \ell\} = \max\{-43.2, -\infty\} = -43.2$.

Step 3. Termination. Compute $u - \ell = -39.8 - (-43.2) = 3.4$ and $\varepsilon|u| = 10^{-6}| - 39.8|) = 0.0000398$. Since $u - \ell \geq \varepsilon|u|$:

Set $k \leftarrow 0 + 1 = 1$.

Return to step 1.

Iteration $k = 1$:

Step 1. Solve Subproblem. Solve

$$\begin{array}{lrrl} \varphi(x^1) := \text{Min} & -16y_1 & -20y_2 & \\ \text{s.t.} & -2y_1 & -3y_2 & \geq -5 \\ & -4y_1 & -y_2 & \geq -4 \\ & -y_1 & & \geq -1 \\ & & -y_2 & \geq -1 \\ & y_1, & y_2 & \geq 0. \end{array}$$

Problem is feasible: $\varphi(x^1) = -32$ and $y^0 = (0.75, 1)^\top$. Generate an *optimality* cut:

Get dual solution $\pi_1 = (0, 4, 0, 16)^\top$.
Compute $\beta_1^0 \leftarrow \pi_1^\top r = (0, 4, 0, 16)(-5, -5, -1, -1)^\top = -36$ and $\beta_1^\top \leftarrow$

$$\pi_1^\top T = (0, 4, 0, 16)\begin{bmatrix} -1 & 0 \\ 0 & -1 \\ 0 & 0 \\ 0 & 0 \end{bmatrix} = (0, -4).$$

Compute *upper bound* if feasible:

Set $v^1 \leftarrow c^\top x^1 + \varphi(x^1) = (-3, -8)(0, 1)^\top + (-32) = -40$.
Set $u \leftarrow \min\{v^1, u\} = \min\{-40, -39.8\} = -40$.

If u is updated, set *incumbent* solution to $x^* \leftarrow x^1 = (0, 1)^\top$.

Step 2. Add Cut to Master Program and Solve. If subproblem was feasible:

Add $\beta_1^\top x + \eta \geq \beta_1^0$ to master problem:

$$\begin{array}{llllll} \nu^2 := \text{Min} & -3x_1 & -8x_2 & +\eta & & \\ \text{s.t.} & -x_1 & & & \geq & -1 \\ & & -x_2 & & \geq & -1 \\ & -6.4x_1 & -0.8x_2 & +\eta & \geq & -36 \\ & & -4x_2 & +\eta & \geq & -36 \\ & x_1, & x_2 & & \geq & 0, \ \eta \text{ free.} \end{array}$$

Solve master to get $x^2 = (0.5, 1)^\top$, $\eta^2 = -32$ and $\nu^2 = -41.5$. Set $\ell \leftarrow \max\{\nu^2, \ell\} = \max\{-41.5, -43.2\} = -41.5$.

Step 3. Termination. Compute $u - \ell = -40 - (-41.5) = 1.5$ and $\varepsilon|u| = 10^{-6}|-40|) = 0.00004$. Since $u - \ell \geq \varepsilon|u|$:

Set $k \leftarrow 1 + 1 = 2$.

Return to step 1.

Iteration $k = 2$:

Step 1. Solve Subproblem. Solve

$$\begin{array}{lllll} \varphi(x^2) := \text{Min} & -16y_1 & -20y_2 & & \\ \text{s.t.} & -2y_1 & -3y_2 & \geq & -4.5 \\ & -4y_1 & -y_2 & \geq & -4 \\ & -y_1 & & \geq & -1 \\ & & -y_2 & \geq & -1 \\ & y_1, & y_2 & \geq & 0. \end{array}$$

Problem is feasible: $\varphi(x^2) = -32$ and $y^0 = (0.75, 1)^\top$. Generate an *optimality* cut:

Get dual solution $\pi_2 = (0, 4, 0, 16)^\top$.
Compute $\beta_2^0 \leftarrow \pi_2^\top r = (0, 4, 0, 16)(-5, -5, -1, -1)^\top = -36$ and $\beta_2^\top \leftarrow$

$$\pi_2^\top T = (0, 4, 0, 16)\begin{bmatrix} -1 & 0 \\ 0 & -1 \\ 0 & 0 \\ 0 & 0 \end{bmatrix} = (0, -4).$$

Compute *upper bound* if feasible:

Set $v^2 \leftarrow c^\top x^1 + \varphi(x^1) = (-3, -8)(0.5, 1)^\top + (-32) = -41.5$.
Set $u \leftarrow \min\{v^2, u\} = \min\{-41.5, -40\} = -41.5$.

If u is updated, set *incumbent* solution to $x^* \leftarrow x^1 = (0.5, 1)^\top$.

Step 2. Add Cut to Master Program and Solve. If subproblem was feasible:

Add $\beta_2^\top x + \eta \geq \beta_2^0$ to master problem:

$$\begin{array}{llllll} v^2 := \text{Min} & -3x_1 & -8x_2 & +\eta & & \\ \text{s.t.} & -x_1 & & & \geq & -1 \\ & & -x_2 & & \geq & -1 \\ & -6.4x_1 & -0.8x_2 & +\eta & \geq & -36 \\ & & -4x_2 & +\eta & \geq & -36 \\ & & -4x_2 & +\eta & \geq & -36 \\ & x_1, & x_2 & & \geq & 0, \eta \text{ free}. \end{array}$$

Solve master to get $x^3 = (0.5, 1)^\top$, $\eta^3 = -32$ and $v^3 = -41.5$. Set $\ell \leftarrow \max\{v^2, \ell\} = \max\{-41.5, -41.5\} = -41.5$.

Step 3. Termination. Compute $u - \ell = -41.5 - (-41.5) = 0$ and $\varepsilon|u| = 10^{-6}|-41.5|) = 0.0000415$. Since $u - \ell \leq \varepsilon|u|$:

Stop, $x^* = (0.5, 1)^\top$ is the 10^{-6}-optimal solution, with objective value -41.5.

---

end

Solving the given LP using a direct solver, we get $x^* = (0.5, 1)^\top$ and $y* = (0.75, 1)^\top$ as the optimal solution with objective value -41.5, thus confirming the solution we found using Benders decomposition.

Example 5.3 Decompose the following LP into two second-stage subproblems and then solve using Benders decomposition:

$$
\begin{array}{llllllll}
\text{Min } 3x_1 & +2x_2 & +y_{11} & -0.5y_{21} & +y_{12} & -0.5y_{22} & & \\
\text{s.t.} \quad x_1 & +x_2 & & & & & \geq & 5 \\
x_1 & -2x_2 & +y_{11} & -y_{21} & & & \geq & 3 \\
x_1 & +2x_2 & & +y_{21} & & & \geq & 5 \\
x_1 & -2x_2 & & & +y_{12} & -y_{22} & \geq & 4 \\
x_1 & +2x_2 & & & & +y_{22} & \geq & 8 \\
x_1, & x_2, & y_{11}, & y_{21} & y_{12}, & y_{22} & \geq & 0.
\end{array}
$$

Verify your solution by solving the given problem using a direct LP solver.

The first-stage decision variable vector is $x = (x_1, x_2)^\top$ since it shows up in all the constraints (linking decision variable). The second-stage (subproblem) decision variables are $y_1 = (y_{11}, y_{21})^\top$ that appear only in the second and third constraints and $y_2 = (y_{12}, y_{22})^\top$ that appear only in the last two constraints. We can thus decompose the problem into first- and second-stage subproblems as follows:

First-stage subproblem:

$$
\begin{array}{llll}
\text{Min } 3x_1 & +2x_2 & & \\
\text{s.t.} \quad x_1 & +x_2 & \geq & 5 \\
x_1, & x_2, & \geq & 0.
\end{array}
$$

Second-stage subproblem 1:

$$
\begin{array}{lllll}
\varphi_1(x) := \text{Min } & y_{11} & -0.5y_{21} & & \text{Dual multipliers} \\
\text{s.t.} & y_{11} & -y_{21} & \geq 3 - x_1 + 2x_2 & \leftarrow \pi_{11} \\
& & y_{21} & \geq 5 - x_1 - 2x_2 & \leftarrow \pi_{21} \\
& y_{11}, & y_{21} & \geq \quad 0. &
\end{array}
$$

Second-stage subproblem 2:

$$
\begin{array}{lllll}
\varphi_2(x) := \text{Min } & y_{12} & -0.5y_{22} & & \text{Dual multipliers} \\
\text{s.t.} & y_{12} & -y_{22} & \geq 4 - x_1 + 2x_2 & \leftarrow \pi_{12} \\
& & y_{22} & \geq 8 - x_1 - 2x_2 & \leftarrow \pi_{22} \\
& y_{12}, & y_{22} & \geq \quad 0. &
\end{array}
$$

Using the dual multipliers, $\pi_1 = (\pi_{11}, \pi_{21})^\top$, the dual to *subproblem 1* is given as follows:

$$
\begin{array}{llll}
\text{Max } (4 - x_1 + 2x_2)\pi_{11} & +(8 - x_1 - 2x_2)\pi_{21} & & \\
\text{s.t.} \quad \pi_{11} & & \leq & 1 \\
-\pi_{11} & +\pi_{21} & \leq & -0.5 \\
\pi_{11}, & \pi_{21}, & \geq & 0.
\end{array}
$$

Similarly, using the dual multipliers, $\pi_2 = (\pi_{12}, \pi_{22})^\top$, the dual to *subproblem 2* is given as follows:

$$\begin{array}{llcl} \text{Max } (4 - x_1 + 2x_2)\pi_{12} & +(8 - x_1 - 2x_2)\pi_{22} & & \\ \text{s.t.} \quad \pi_{12} & & \leq & 1 \\ \quad -\pi_{12} & +\pi_{22} & \leq & -0.5 \\ \quad \pi_{12}, & \pi_{22}, & \geq & 0. \end{array}$$

The problem data can be summarized as follows:
$c = (3, 2)^\top$, $A = \begin{bmatrix} 1 & 1 \end{bmatrix}$, $b = 5$,
$T = \begin{bmatrix} 1 & -2 \\ 1 & 2 \end{bmatrix}$, $W = \begin{bmatrix} 1 & -1 \\ 0 & 1 \end{bmatrix}$,
$r_1 = (3, 5)^\top$ for *subproblem 1*, and $r_2 = (4, 8)^\top$ for *subproblem 2*.
Let set $X = \{x_1 + x_2 \geq 5, \ x_1, \ x_2 \geq 0\}$. We can now apply Benders decomposition algorithm as follows:

Algorithm *Benders Decomposition*
begin

Step 0. Initialization. Let $k \leftarrow 0$, choose $x^0 \in X$, and set $\ell \leftarrow -\infty$ and $u \leftarrow \infty$. Choose $\varepsilon = 10^{-6}$. We can set $x^0 \leftarrow \text{argmin}_{x \in X}\{3x_1 + 2x_2\}$, i.e., we solve the first-stage problem to get $x^0 = (0, 5)^\top$ as the initial point.

Step 1. Solve Subproblem.
Solve *subproblem 1*:

$$\begin{array}{rlcl} \varphi_1(x^0) := \text{Min } & y_{11} \ -0.5y_{21} & & \\ \text{s.t.} & y_{11} \quad -y_{21} & \geq & 13 \\ & \quad\quad y_{21} & \geq & -5 \\ & y_{11}, \quad y_{21} & \geq & 0. \end{array}$$

Problem is feasible: $\varphi_1(x^0) :== 13$ and $y^{01} = (13, 0)^\top$. Generate an *optimality* cut:

Get dual solution $\pi_{01} = (1, 0)^\top$.
Compute $\beta_{01}^0 \leftarrow \pi_{01}^\top r = (1, 0)(3, 5)^\top = 3$ and $\beta_{01}^\top \leftarrow \pi_{01}^\top T = (1, 0)\begin{bmatrix} 1 & -2 \\ 1 & 2 \end{bmatrix} = (1, -2)$.

Solve *subproblem 2*:

$$\begin{array}{rlcl} \varphi_2(x^0) :== \text{Min } & y_{11} \ -0.5y_{21} & & \\ \text{s.t.} & y_{11} \quad -y_{21} & \geq & 14 \\ & \quad\quad y_{21} & \geq & -2 \\ & y_{11}, \quad y_{21} & \geq & 0. \end{array}$$

Problem is feasible: $\varphi_2(x^0) :== 14$ and $y^{02} = (1, 0)^\top$. Generate an *optimality* cut:

Get dual solution $\pi_{02} = (1, 0)^\top$.
Compute $\beta_{02}^0 \leftarrow \pi_{02}^\top r = (1, 0)(4, 8)^\top = 4$ and $\beta_{02}^\top \leftarrow \pi_{02}^\top T = (1, 0)\begin{bmatrix} 1 & -2 \\ 1 & 2 \end{bmatrix} = (1, -2)$.

Aggregate the *optimality* cuts:

Compute $\beta_0^0 \leftarrow \beta_{01}^0 + \beta_{02}^0 = 3 + 4 = 7$ and $\beta_0^\top \leftarrow \beta_{01}^\top + \beta_{02}^\top = (1, -2)^\top + (1, -2)^\top = (2, -4)$.

Compute *upper bound* if feasible:

Set $v^0 \leftarrow c^\top x^0 + \varphi_1(x^0) + \varphi_2(x^0) := (3, 2)(0, 5)^\top + 13 + 14 = 37$.
Set $u \leftarrow \min\{v^0, u\} = \min\{37, \infty\} = 37$.

If u is updated, set *incumbent* solution to $x^* \leftarrow x^0 = (0, 5)^\top$.

Step 2. Add Cut to Master Program and Solve. If subproblem was feasible:

Add $\beta_0^\top x + \eta \geq \beta_0^0$ to master problem:

$$
\begin{aligned}
\nu^1 := \text{Min } & 3x_1 + 2x_2 + \eta \\
\text{s.t. } & x_1 + x_2 \geq 5 \\
& 2x_1 - 4x_2 + \eta \geq 7 \\
& x_1, \quad x_2 \geq 0, \ \eta \text{ free.}
\end{aligned}
$$

Solve master to get $x^1 = (5, 0)^\top$, $\eta^1 = -3$ and $\nu^1 = 12$. Set $\ell \leftarrow \max\{\nu^1, \ell\} = \max\{12, -\infty\} = 12$.

Step 3. Termination. Compute $u - \ell = 37 - (12) = 25$ and $\varepsilon|u| = 10^{-6}|37|) = 0.000037$. Since $u - \ell \geq \varepsilon|u|$:

Set $k \leftarrow 0 + 1 = 1$.

Return to step 1.

Iteration $k = 1$:

Step 1. Solve Subproblem.
Solve *subproblem 1*:

$$
\begin{aligned}
\varphi_1(x^1) := \text{Min } & y_{11} - 0.5y_{21} \\
\text{s.t. } & y_{11} - y_{21} \geq -2 \\
& y_{21} \geq 0 \\
& y_{11}, \quad y_{21} \geq 0.
\end{aligned}
$$

Problem is feasible: $\varphi_1(x^1) :== -1$ and $y^{01} = (0, 2)^\top$. Generate an *optimality* cut:

Get dual solution $\pi_{01} = (0.5, 0)^\top$.
Compute $\beta_{01}^0 \leftarrow \pi_{01}^\top r = (0.5, 0)(3, 5)^\top = 1.5$ and $\beta_{01}^\top \leftarrow \pi_{01}^\top T = (0.5, 0)\begin{bmatrix} 1 & -2 \\ 1 & 2 \end{bmatrix} = (0.5, -1)$.

Solve *subproblem 2*:

$$\begin{aligned} \varphi_2(x^1) := \text{Min} \ & y_{11} - 0.5 y_{21} \\ \text{s.t.} \ & y_{11} - y_{21} \geq -1 \\ & y_{21} \geq 3 \\ & y_{11}, \ y_{21} \geq 0. \end{aligned}$$

Problem is feasible: $\varphi_2(x^1) :== 0.5$ and $y^{02} = (2, 3)^\top$. Generate an *optimality* cut:

Get dual solution $\pi_{02} = (1, 0.5)^\top$.
Compute $\beta_{02}^0 \leftarrow \pi_{02}^\top r = (1, 0.5)(4, 8)^\top = 8$ and $\beta_{02}^\top \leftarrow \pi_{02}^\top T = (1, 0.5)\begin{bmatrix} 1 & -2 \\ 1 & 2 \end{bmatrix} = (1.5, -1)$.

Aggregate the *optimality* cuts:

Compute $\beta_0^0 \leftarrow \beta_{01}^0 + \beta_{02}^0 = 1.5 + 8 = 9.5$ and $\beta_0^\top \leftarrow \beta_{01}^\top + \beta_{02}^\top = (0.5, -1)^\top + (1.5, -1)^\top = (2, -2)^\top$.

Compute *upper bound* if feasible:

Set $v^0 \leftarrow c^\top x^0 + \varphi_1(x^1) + \varphi_2(x^1) := (3, 2)(5, 0)^\top - 1 + 0.5 = 14.5$.
Set $u \leftarrow \min\{v^0, u\} = \min\{14.5, 37\} = 14.5$.

If u is updated, set *incumbent* solution to $x^* \leftarrow x^0 = (5, 0)^\top$.

Step 2. Add Cut to Master Program and Solve. If subproblem was feasible:

Add $\beta_0^\top x + \eta \geq \beta_0^0$ to master problem:

$$\begin{aligned} \nu^1 := \text{Min} \ & 3x_1 + 2x_2 + \eta \\ \text{s.t.} \ & x_1 + x_2 \geq 5 \\ & 2x_1 - 4x_2 + \eta \geq 7 \\ & 2x_1 - 2x_2 + \eta \geq 9.5 \\ & x_1, \ x_2 \geq 0, \ \eta \text{ free}. \end{aligned}$$

Solve master to get $x^1 = (5, 0)^\top, \eta^1 = -0.5$ and $\nu^1 = 14.5$. Set $\ell \leftarrow \max\{\nu^1, \ell\} = \max\{14.5, 12\} = 14.5$.

Step 3. **Termination.** Compute $u - \ell = 14.5 - 14.5) = 0$ and $\varepsilon|u| = 10^{-6}|14.5|) = 0.0000145$. Since $u - \ell \leq \varepsilon|u|$:

Stop, $x^* = (5, 0)^\top$ is the 10^{-6}-optimal solution, with objective value 14.5.

---

end

Solving the given LP using a direct solver, we get $x^* = (5, 0)^\top$, $y_1^* = (0, 2)^\top$ and $y_2^* = (2, 3)^\top$ as the optimal solution with objective value 14.5, thus confirming the solution we found using Benders decomposition.

Example 5.4 Let us apply Benders decomposition to the deterministic version of the *abc*-Production Planning problem we formulated in Chap. 3. We restate the formulation below:

$$
\begin{array}{llllllllll}
\text{Min} & 50x_m & +30x_1 & +15x_2 & +10x_3 & -1150y_a & -1525y_b & -1900y_c & & \\
\text{s.t.} & -x_m & & & & & & & \geq & -300 \\
 & & -x_1 & & & & & & \geq & -800 \\
 & & & -x_2 & & & & & \geq & -600 \\
 & & & & -x_3 & & & & \geq & -500 \\
 & & -x_1 & -x_2 & -x_3 & & & & \geq & -1600 \\
 & x_m & & & & -6y_a & -8y_b & -10y_c & \geq & 0 \\
 & & x_1 & & & -20y_a & -25y_b & -28y_c & \geq & 0 \\
 & & & x_2 & & -12y_a & -15y_b & -18y_c & \geq & 0 \\
 & & & & x_3 & -8y_a & -10y_b & -14y_c & \geq & 0 \\
 & & & & & -y_a & & & \geq & -15 \\
 & & & & & & -y_b & & \geq & -10 \\
 & & & & & & & -y_c & \geq & -5 \\
 & x_m, & x_1, & x_2, & x_3, & y_a, & y_b, & y_c & \geq & 0.
\end{array}
$$

The first-stage decision variable vector is $x = (x_m, x_1, x_2, x_3)^\top$ since it is the linking decision variable, and the second-stage (subproblem) decision variable vector is $y = (y_a, y_b, y_c)^\top$. We decompose the problem into the following first- and second-stage subproblems:

First-stage subproblem:

$$
\begin{array}{lllllll}
\text{Min} & 50x_m & +30x_1 & +15x_2 & +10x_3 & & \\
\text{s.t.} & -x_m & & & & \geq & -300 \\
 & & -x_1 & & & \geq & -800 \\
 & & & -x_2 & & \geq & -600 \\
 & & & & -x_3 & \geq & -500 \\
 & & -x_1 & -x_2 & -x_3 & \geq & -1600 \\
 & x_m, & x_1, & x_2, & x_3 & \geq & 0.
\end{array}
$$

Second-stage subproblem:

$$
\begin{array}{lllll}
\varphi(x) := \text{Min} & -1150y_a & -1525y_b & -1900y_c & & \text{Dual multipliers}\\
\text{s.t.} & -6y_a & -8y_b & -10y_c & \geq -x_m & \leftarrow \pi_1\\
& -20y_a & -25y_b & -28y_c & \geq -x_1 & \leftarrow \pi_2\\
& -12y_a & -15y_b & -18y_c & \geq -x_2 & \leftarrow \pi_3\\
& -8y_a & -10y_b & -14y_c & \geq -x_3 & \leftarrow \pi_4\\
& -y_a & & & \geq -15 & \leftarrow \pi_5\\
& & -y_b & & \geq -10 & \leftarrow \pi_6\\
& & & -y_c & \geq -5 & \leftarrow \pi_7\\
& y_a, & y_b, & y_c & \geq 0. &
\end{array}
$$

Using the dual multipliers, $\pi = (\pi_1, \pi_2, \cdots, \pi_7)^\top$, the dual to the second-stage subproblem is given as follows:

$$
\begin{array}{lllllllll}
\text{Max} & -x_m\pi_1 & -x_1\pi_2 & -x_2\pi_3 & -x_3\pi_4 & -15\pi_5 & -10\pi_6 & -5\pi_7 & \\
\text{s.t.} & -6\pi_1 & -20\pi_2 & -12\pi_3 & -8\pi_4 & -\pi_5 & & & \leq -1150\\
& -8\pi_1 & -25\pi_2 & -15\pi_3 & -10\pi_4 & & -\pi_6 & & \leq -1525\\
& -6\pi_1 & -20\pi_2 & -12\pi_3 & -8\pi_4 & & & -\pi_7 & \leq -1900\\
& \pi_1, & \pi_2, & \pi_3, & \pi_4, & \pi_5, & \pi_6, & \pi_7 & \geq 0.
\end{array}
$$

The problem data can be summarized as follows:

$$
c = (50, 30, 15, 10)^\top,\ A = \begin{bmatrix} -1 & 0 & 0 & 0\\ 0 & -1 & 0 & 0\\ 0 & 0 & -1 & 0\\ 0 & 0 & 0 & -1\\ 0 & -1 & -1 & -1 \end{bmatrix},\ b = (-300, -800, -600, -500, -1600)^\top,
$$

$$
T = \begin{bmatrix} 1&0&0&0\\ 0&1&0&0\\ 0&0&1&0\\ 0&0&0&1\\ 0&0&0&0\\ 0&0&0&0\\ 0&0&0&0 \end{bmatrix},\ W = \begin{bmatrix} -6 & -8 & -10\\ -20 & -25 & -28\\ -12 & -15 & -18\\ -8 & -10 & -14\\ -1 & 0 & 0\\ 0 & -1 & 0\\ 0 & 0 & -1 \end{bmatrix},\ \text{and } r = (0, 0, 0, 0, -15, -10, -5)^\top.
$$

Let set $X = \{-x_m \geq -300, -x_1 \geq -800, -x_2 \geq -600, -x_3 \geq -500, -x_1 - x_2 - x_3 \geq -1600, x_m, x_1, x_2, x_3 \geq 0\}$. We are now ready to apply Benders decomposition algorithm as follows:

Algorithm *Benders Decomposition*
begin

Step 0. Initialization. Let $k \leftarrow 0$, choose $x^0 \in X$, and set $\ell \leftarrow -\infty$ and $u \leftarrow \infty$. Choose $\varepsilon = 10^{-6}$. We can set $x^0 \leftarrow \text{argmin}_{x \in X}\{50x_m + 30x_1 + 15x_2 + 10x_3\}$, i.e., we solve the first-stage problem to get $x^0 = (0, 0, 0, 0)^\top$ as the initial point.

Step 1. Solve Subproblem. Solve

$$
\begin{array}{rlllrr}
\varphi(x^0) := \text{Min} & -1150y_a & -1525y_b & -1900y_c & & \text{Dual multipliers} \\
\text{s.t.} & -6y_a & -8y_b & -10y_c \geq & 0 & \leftarrow \pi_1 \\
& -20y_a & -25y_b & -28y_c \geq & 0 & \leftarrow \pi_2 \\
& -12y_a & -15y_b & -18y_c \geq & 0 & \leftarrow \pi_3 \\
& -8y_a & -10y_b & -14y_c \geq & 0 & \leftarrow \pi_4 \\
& -y_a & & \geq & -15 & \leftarrow \pi_5 \\
& & -y_b & \geq & -10 & \leftarrow \pi_6 \\
& & & -y_c \geq & -5 & \leftarrow \pi_7 \\
& y_a, & y_b, & y_c \geq & 0. &
\end{array}
$$

Problem is feasible: $\varphi(x^0) = 0$ and $y^0 = (0, 0, 0)^\top$. Generate an *optimality* cut:

Get dual solution $\pi_0 = (191.667, 0, 0, 0, 0, 0, 0)^\top$.
Compute $\beta_0^0 \leftarrow \pi_0^\top r = (191.667, 0, 0, 0, 0, 0, 0)(0, 0, 0, 0, -15, -10, -5)^\top = 0$

$$
\text{and } \beta_0^\top \leftarrow \pi_0^\top T = (191.667, 0, 0, 0, 0, 0, 0) \begin{bmatrix} 1\,0\,0\,0 \\ 0\,1\,0\,0 \\ 0\,0\,1\,0 \\ 0\,0\,0\,1 \\ 0\,0\,0\,0 \\ 0\,0\,0\,0 \\ 0\,0\,0\,0 \end{bmatrix} = (191.667, 0, 0, 0).
$$

Compute *upper bound* if feasible:

Set $v^0 \leftarrow c^\top x^0 + \varphi(x^0) = 0$.
Set $u \leftarrow \min\{v^0, u\} = \min\{0, \infty\} = 0$.

If u is updated, set *incumbent* solution to $x^* \leftarrow x^0 = (0, 0, 0, 0)^\top$.

Step 2. Add Cut to Master Program and Solve. If subproblem was feasible:

Add $\beta_0^\top x + \eta \geq \beta_0^0$ to master problem:

$$
\begin{array}{llllllll}
\nu^1 := \text{Min} & 50x_m & +30x_1 & +15x_2 & +10x_3 & +\eta & & \\
\text{s.t.} & -x_m & & & & & \geq & -300 \\
 & & -x_1 & & & & \geq & -800 \\
 & & & -x_2 & & & \geq & -600 \\
 & & & & -x_3 & & \geq & -500 \\
 & & -x_1 & -x_2 & -x_3 & & \geq & -1600 \\
 & 191.667x_m & & & & +\eta & \geq & 0 \\
 & x_m, & x_1, & x_2, & x_3 & & \geq & 0 \ \eta \text{ free.}
\end{array}
$$

Solve master to get $x^1 = (300, 0, 0, 0)^\top$, $\eta^1 = -57{,}500$ and $\nu^1 = -42{,}500$. Set $\ell \leftarrow \max\{\nu^1, \ell\} = \max\{-42{,}500, -\infty\} = -42{,}500$.

Step 3. Termination. Compute $u - \ell = 0 - (-42{,}500) = 42{,}500$ and $\varepsilon|u| = 10^{-6}| - 42{,}500|) = 0.0425$. Since $u - \ell \geq \varepsilon|u|$:

Set $k \leftarrow 0 + 1 = 1$.

Return to step 1.

After 12 iterations, the Benders algorithm terminates with optimal solution $x^* = (130, 390, 240, 170)^\top$, $y{*}_2 = (0, 10, 5)^\top$ and objective value -1250. This is the solution we obtained solving the LP with a direct solver. Next, we study an extension of Benders decomposition, regularized Benders decomposition.

5.3.4 Regularized Benders Decomposition

For problems in large dimensions, the master problem solutions x^k can "bounce off" from iteration to iteration during the course of the algorithm. This often leads the algorithm to take too long to converge to the optimal solution. Therefore, in such cases it becomes useful to *regularize* the master program in order to stabilize the *incumbent* solutions during the course of the algorithm. An incumbent solution is simply the best solution (in terms of objective value) found so far. The idea of regularization dates back to deterministic nonsmooth optimization [4–6], where it is common to regularize a nonsmooth function by a Moreau approximation with parameter $\rho > 0$. So to regularize the master program, we add a quadratic term to the objective function to get the following regularized master program at iteration k:

$$
\begin{aligned}
\text{Min } & c^\top x + \eta + \frac{\rho}{2}||x - \bar{x}^k||^2 \\
\text{s.t. } & Ax \qquad \leq b \\
& \beta_t^\top x + \eta \geq \beta_t^0, \ t \in \Theta_k \\
& \beta_t^\top x \qquad \geq \beta_t^0, \ t \notin \Theta_k \\
& x \geq 0,
\end{aligned}
\tag{5.15}
$$

where $\rho > 0$ is a constant and $\bar{x}^k$ is the *incumbent solution* at iteration k. Let x^{k+1} be the solution to master problem (5.15). Regularizing the master program for problems in large dimensions eliminates the "bouncing off" of the x^k solutions from iteration to iteration, thus stabilizing the incumbent solutions during the course of the algorithm.

We need to establish a condition or test for when to update the incumbent solution. We refer to this test as the *incumbent test*. To this end, let $\delta \in (0, 1)$ be given and let subproblem objective value at $\bar{x}^k$ and x^{k+1} be

$$\varphi(\bar{x}^k) := \text{Min}\{q^\top y \mid Wy = r - T\bar{x}^k, y \geq 0\}$$

and

$$\varphi(x^{k+1}) := \text{Min}\{q^\top y \mid Wy = r - Tx^{k+1}, y \geq 0\},$$

respectively. Similarly, let overall objective value at $\bar{x}^k$ and x^{k+1} be given as

$$f(\bar{x}^k) = c^\top \bar{x}^k + \varphi(\bar{x}^k)$$

and

$$f(x^{k+1}) = c^\top x^{k+1} + \varphi(x^{k+1}),$$

respectively. Now let the overall objective value of the approximation at iteration k at the new point x^{k+1} be given as

$$f_k(x^{k+1}) = c^\top x^{k+1} + \eta^{k+1},$$

where

$$\eta^{k+1} = \max\{\beta_t^0 - \beta_t^\top x \mid 0 \leq t \leq k\}.$$

The idea of the incumbent test is to update the incumbent solution $\bar{x}^k$ whenever there is a "significant" improvement in the objective function value from iteration k to $k + 1$. Otherwise, there is no need to update the incumbent solution. This idea can be accomplished by performing the following test:

Incumbent test:

If $|f(x^{k+1}) - f(\bar{x}^k)| \leq \delta |f_k(x^{k+1}) - f(\bar{x}^k)|$
$\quad \bar{x}^{k+1} \leftarrow x^{k+1}$

else
$\quad \bar{x}^{k+1} \leftarrow \bar{x}^k.$

Notice that the left hand side (LHS) of the "if" condition gives the absolute difference between the objective value at x^{k+1} and that at the incumbent solution

$\bar{x}^k$. This is compared to the right hand side (RHS), which is a fraction (defined by δ) of the absolute difference between objective value of the function approximation at x^{k+1} and that at $\bar{x}^k$. Thus if the LHS is less than or equal to the RHS, it means that there is a "significant" improvement in the objective value at x^{k+1}. Therefore, we can update the incumbent solution to be the new point x^{k+1}. Regularizing the master program essentially allows the algorithm at iteration k to choose an x^{k+1} that is not too far from the incumbent solution $\bar{x}^k$ unless there is significant improvement in the objective value at the new point.

Let us now turn to how to implement Benders decomposition algorithm in this case. Observe that $\bar{x}^k$ is a "constant" in the master problem (5.15). Consequently, the quadratic term in the objective can end up being fairly dense, which can slow down the solution routine. To overcome this, let us make a variable substitution $w = x - \bar{x}^k$ and rewrite the master problem as follows:

$$
\begin{aligned}
c^\top \bar{x}^k + \text{Min}\ & c^\top w + \eta + \frac{\rho}{2}||w||^2 \\
\text{s.t.}\ & Aw \qquad = [b - A\bar{x}^k] \\
& \beta_t w + \eta \geq [\beta_t^0 - \beta_t \bar{x}^k],\ t \in \Theta_k \\
& \beta_t w \qquad \geq [\beta_t^0 - \beta_t \bar{x}^k],\ t \notin \Theta_k \\
& w \geq -\bar{x}^k.
\end{aligned} \tag{5.16}
$$

Now this quadratic objective function in terms of w is much easier to work with because the *Hessian* is now an identity matrix. We can formally state the regularized Benders decomposition algorithm as follows:

Algorithm *Regularized Benders Decomposition*

begin

Step 0. **Initialization.** Let $k \leftarrow 0$, choose $x^0 \in X$, and set $\bar{x}^0 = x^0$, $\ell \leftarrow -\infty$ and $u \leftarrow \infty$. Choose a small $\varepsilon > 0$ and $\delta \in (0, 1)$.

Step 1. **Solve Subproblem.** Solve

$$\varphi(x^k) := \text{Min}\{q^\top y \mid Wy = r - Tx^k, y \geq 0\}.$$

If feasible, generate an *optimality* cut:

Get dual solution π_k.
Compute $\beta_k^0 \leftarrow \pi_k^\top r$ and $\beta_k^\top \leftarrow \pi_k^\top T$.

Else if infeasible, generate a *feasibility* cut:

Get dual extreme ray μ_k.
Compute $\beta_k^0 \leftarrow \mu_k^\top r$ and $\beta_k^\top \leftarrow \mu_k^\top T$.

Compute *upper bound* if feasible:

Set $v^k \leftarrow c^\top x^k + \varphi(x^k)$.
Set $u \leftarrow \min\{v^k, u\}$.

If u is updated, set *incumbent* solution to $x^* \leftarrow x^k$.

Step 2. Add Cut to Master Program and Solve. If subproblem was feasible:

Add $\beta_k w + \eta \geq [\beta_k^0 - \beta_k \bar{x}^k]$ to the master problem.

Else:

Add $\beta_k w \geq [\beta_k^0 - \beta_k \bar{x}^k]$ to the master problem.

Solve master to get solution (w^{k+1}, η^{k+1}).
Set $x^{k+1} \leftarrow w^{k+1} + \bar{x}^k$.
Perform *incumbent test* to update $\bar{x}^k$.
Update penalty coefficient ρ.

Step 3. Termination. If $||w^{k+1}||^2 < \varepsilon$:

Stop, x^* is the ε-optimal solution.

Else:

Set $k \leftarrow k + 1$.

Return to step 1.

---

end

In the regularized Benders decomposition algorithm, the master program is a quadratic program and is solved using a quadratic programming solver. The lower bound from solving the master problem is no longer a true lower bound on the objective value due to the regularization term. Therefore, the termination criterion is not based on the lower and upper bounds anymore. The algorithm is stopped when $||w^{k+1}||^2 < \varepsilon$, which means that $x^{k+1} = w^{k+1} + \bar{x}^k \approx \bar{x}^k$.

To guide the algorithm towards the optimal solution, in step 2 of the algorithm, we can update ρ so that the penalization of the deviation from the incumbent is also regulated during the course of the algorithm. To accomplish this, the following rules of thumb can be applied:

Heuristic for Updating ρ

1. ρ is decreased (by some factor, e.g., 0.5) when the incumbent is updated or if there is a relatively small improvement in the overall objective value after a set number of consecutive iterations.
2. ρ is increased if $f(\bar{x}^k + w^k) > f(\bar{x}^k)$.
3. ρ is increased if rejecting $\bar{x}^k + w^k$, i.e., not updating the incumbent for a set number of iterations.

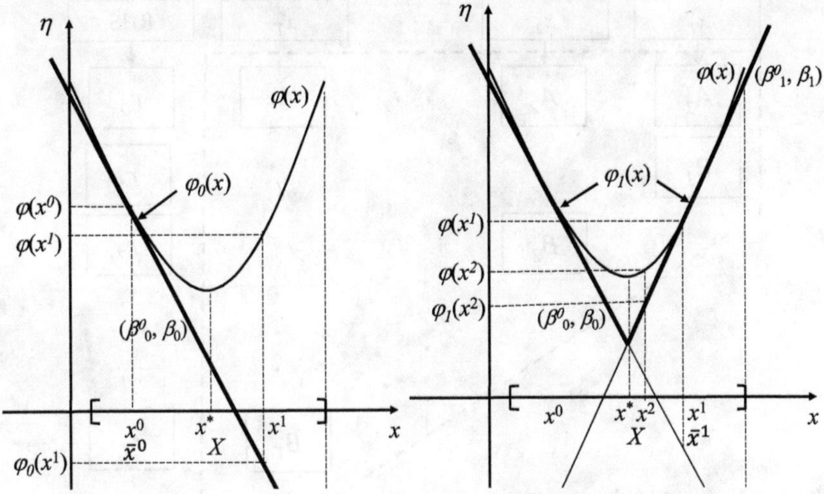

Fig. 5.6 An illustration of regularized Benders decomposition algorithm

We give an illustration of the regularized Benders decomposition algorithm in Fig. 5.6. In the figure, the algorithm is initialized with x^0 with incumbent solution $\bar{x}^0 := x^0$. At this point, the subproblem function value is $\varphi(x^0)$ and a supporting hyperplane (β_0^0, β_0) is generated in step 1 of the algorithm. In step 2, solving the approximation that involves this single supporting hyperplane leads to x^1 with the function value $\varphi(x^1)$. Notice that because of the regularization term x_1 is not at the boundary of the feasible region X. At this point, the incumbent test is applied to see whether or not x_1 provides a better solution based on a chosen δ than x_0. In this case, we assume it does and the incumbent is updated: $\bar{x}^1 := x^1$. Returning to step 1 of the algorithm, a supporting hyperplane (β_1^0, β_1) is generated. In step 2, the approximation now involves the two supporting hyperplanes generated so far. Solving the master program, we get x^2 with function value $\varphi(x^2)$. We can clearly see that if we continue this process, the iterates $\{x^0, x^1, x^2, \cdots\}$ will tend towards the optimal point x^*, the minimizer of $\varphi(x)$ over X, while staying close to the incumbent $\bar{x}^k$ (based on ρ). Next, we study Dantzig–Wolfe decomposition.

5.4 Dantzig–Wolfe Decomposition

Dantzig–Wolfe decomposition [2], also known as *column generation*, deals with large-scale LPs with a large number of decision variables (columns). The problem involves a set of decision variables $x_t \in \mathbb{R}^{n_1}$ for $t = 1, \cdots, T$ sets of constraints of the model that are linked by a set of "complicating" constraints. The Dantzig–Wolfe LP formulation can be given as follows:

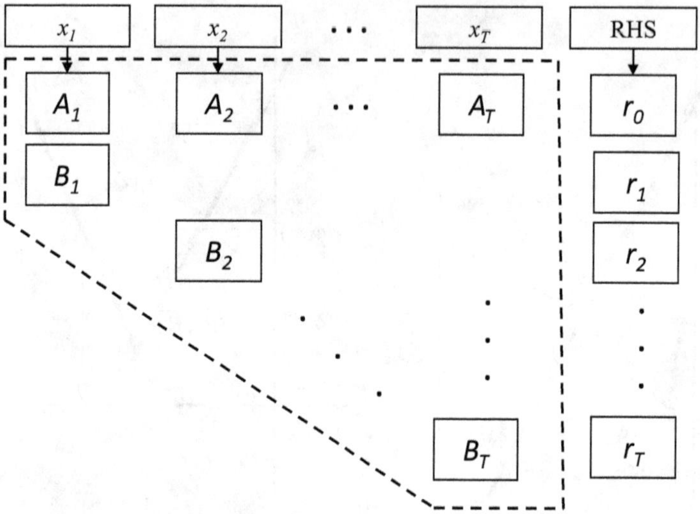

Fig. 5.7 The block angular structure of the Dantzig–Wolfe problem

$$
\begin{aligned}
\text{Min } & c_1^\top x_1 + c_2^\top x_2 + \cdots + c_T^\top x_T \\
\text{s.t. } & A_1^\top x_1 + A_2^\top x_2 + \cdots + A_T^\top x_T && = r_0 \\
& B_1 x_1 && \geq r_1 \\
& \qquad B_2 x_2 && \geq r_2 \\
& \qquad\qquad \ddots && \vdots \\
& \qquad\qquad\qquad B_t x_T && \geq r_T \\
& x_1, \quad x_2, \quad \cdots, \quad x_T \geq 0.
\end{aligned}
\tag{5.17}
$$

In formulation (5.17), $X_t = \{B_t x_t \geq r_t\}$ is considered to be an "easy" set. Notice that the problem is a dual of the Benders problem and has a *diagonal* or *block angular* structure (Fig. 5.7). This means that if we were to ignore the set of complicating constraints, the rest of the problem would become *separable* and the resulting subproblems can be solved separately. This type of formulation arises in many applications including cutting stock and plant location problems.

5.4.1 Decomposition Approach

From here on, we shall assume that Problem (5.17) has a solution. The main idea of Dantzig–Wolfe decomposition is to rewrite the formulation and then decompose the problem into a master problem and subproblems. The master program captures the complicating constraints, while the subproblems capture with the "easy" constraints. To reformulate the problem, we shall apply the *representation (resolution) theorem* from polyhedral theory. However, before we state the representation theorem, let us first rewrite problem (5.17) in a compact way as follows:

$$\begin{aligned} \text{Min} \; & \sum_{t=1}^{T} c_t^\top x_t \\ \text{s.t.} \; & \sum_{t=1}^{T} A_t x_t = r_0 \\ & x_t \in X_t, \quad \forall t = 1, \cdots, T. \end{aligned} \tag{5.18}$$

Theorem 5.2 (Representation Theorem) *Let X_t be a nonempty polyhedron with a least one extreme point. Let $x_{t_i}, i = 1, \cdots, I_t$ be the extreme points and let $w_{t_j}, j = 1, \cdots, J_t$ be the extreme rays of X_t. Then any point $x_t \in X_t$ can be represented in the form*

$$x_t = \sum_{i=1}^{I_t} \lambda_{t_i} x_{t_i} + \sum_{j=1}^{J_t} \theta_{t_j} w_{t_j},$$

where

$$\sum_{i=1}^{I_t} \lambda_{t_i} = 1, \quad \forall t = 1, \cdots, T, \quad \forall i = 1, \cdots, I_t, \forall j = 1, \cdots, J_t,$$

$$\lambda_{t_i}, \theta_{t_j} \geq 0.$$

Substituting the expression for x_t in Problem (5.18), we get the so-called *Dantzig–Wolfe problem* (DWP):

$$\begin{aligned} \text{Min} \; & \sum_{t=1}^{T} \sum_{i=1}^{I_t} \lambda_{t_i} (c_t^\top x_{t_i}) + \sum_{t=1}^{T} \sum_{j=1}^{J_t} \theta_{t_j} (c_t^\top x_{t_j}) \\ \text{s.t.} \; & \sum_{t=1}^{T} \sum_{i=1}^{I_t} \lambda_{t_i} (A_t x_{t_i}) + \sum_{t=1}^{T} \sum_{j=1}^{J_t} \theta_{t_j} (A_t x_{t_j}) = r_0 \\ & \sum_{i=1}^{I_t} \lambda_{t_i} = 1, \quad \forall t = 1, \cdots, T \\ & \lambda_{t_i} \geq 0, \quad \forall t = 1, \cdots, T; \; i = 1, \cdots, I_t \\ & \theta_{t_j} \geq 0, \quad \forall t = 1, \cdots, T; \; j = 1, \cdots, J_t. \end{aligned} \tag{5.19}$$

If X_t is a polytope, then we do not have to deal with the extreme rays. Therefore, we have

$$x_t = \sum_{i=1}^{I_t} \lambda_{t_i} x_{t_i}, \ \sum_{i=1}^{I_t} \lambda_{t_i} = 1, \quad \forall i = 1, \cdots, I_t, \ \lambda_{t_i} \geq 0.$$

For simplicity of exposition, let us assume from here on that X_t is a polytope. Substituting x_t in Problem (5.18), we get the following DWP:

$$\begin{aligned} \text{Min} \ & \sum_{t=1}^{T} \sum_{i=1}^{I_t} \lambda_{t_i} (c_t^\top x_{t_i}) \\ \text{s.t.} \ & \sum_{t=1}^{T} \sum_{i=1}^{I_t} \lambda_{t_i} (A_t x_{t_i}) = r_0 \\ & \sum_{i=1}^{I_t} \lambda_{t_i} = 1, \quad \forall t = 1, \cdots, T \\ & \lambda_{t_i} \geq 0, \quad \forall t = 1, \cdots, T; \ i = 1, \cdots, I_t. \end{aligned} \tag{5.20}$$

We now see that problem DWP does not involve any θ_{t_j} variables, and this makes it relatively easy to explain the Dantzig–Wolfe method.

Let m_0 be the number of complicating constraints in the original problem formulation (5.17) and m_t be the number of constraints in each block t. Then the original problem has $m_0 + \sum_{t=1}^{T} m_t$ constraints. However, the Dantzig–Wolfe Problem (5.20) has $m_0 + T$ constraints, i.e., m_0 complicating constraints and one convexity constraint for each $t = 1, \cdots, T$. This problem has the advantage of having fewer constraints than the original problem. However, the number of decision variables (columns) can be very large since we can have an exponential number of extreme points for each polytope. Observe that a decision variable is created for each extreme point $x_t \in X_t$ for all t. This may not be practical in terms of computer memory for data storage, and hence, a decomposition approach is needed.

The fundamental idea of Dantzig–Wolfe decomposition is that we actually do not need *all* the extreme points to characterize an optimal solution. Therefore, we can form and solve a *restricted* master program (RMP), i.e., a master program with only a subset of the columns in DWP (5.20), and then generate columns "on the fly" as needed. Let $I_t^k < I_t$ for all $t = 1, \cdots, T$ denote the number of such columns. We need to have at least $m_0 + T$ columns in RMP to form a feasible basis. Then RMP takes the following form:

$$
\begin{aligned}
\text{Min} \sum_{t=1}^{T} \sum_{i=1}^{I_t^k} \lambda_{t_i} (c_t^\top x_{t_i}) \\
\text{s.t.} \sum_{t=1}^{T} \sum_{i=1}^{I_t^k} \lambda_{t_i} (A_t x_{t_i}) = r_0 \\
\sum_{i=1}^{I_t^k} \lambda_{t_i} = 1, \quad \forall t = 1, \cdots, T \\
\lambda_{t_i} \geq 0, \quad \forall t = 1, \cdots, T;\ i = 1, \cdots, I_t^k.
\end{aligned}
\tag{5.21}
$$

Let π_0 and π_t, $t = 1, \cdots, T$, be the vector of dual multipliers associated with the complicating constraint and each subproblem constraint, respectively. Then we can rewrite RMP (5.21) more explicitly as follows:

$$
\begin{array}{llll}
\text{Min} & \sum_{i=1}^{I_1^k} \lambda_{1_i} (c_1^\top x_{1_i}) + \sum_{i=1}^{I_2^k} \lambda_{2_i} (c_2^\top x_{2_i}) + \cdots + \sum_{i=1}^{I_T^k} \lambda_{T_i} (c_T^\top x_{T_i}) & & \\
\text{s.t.} & \sum_{i=1}^{I_1^k} \lambda_{1_i} (A_1 x_{1_i}) + \sum_{i=1}^{I_2^k} \lambda_{2_i} (A_2 x_{2_i}) + \cdots + \sum_{i=1}^{I_T^k} \lambda_{T_i} (A_T x_{T_i}) & = r_0 & \leftarrow \pi_0 \\
& \sum_{i=1}^{I_1^k} \lambda_{1_i} & = 1 & \leftarrow \pi_1 \\
& \qquad\qquad \sum_{i=1}^{I_2^k} \lambda_{2_i} & = 1 & \leftarrow \pi_2 \\
& \qquad\qquad\qquad \ddots & \vdots & \\
& \qquad\qquad\qquad\qquad \sum_{i=1}^{I_T^k} \lambda_{T_i} & = 1 & \leftarrow \pi_T \\
& \lambda_{1_i} & \geq 0, & \forall i = 1, \cdots, I_1^k \\
& \qquad\qquad \lambda_{2_i} & \geq 0, & \forall i = 1, \cdots, I_2^k \\
& & \vdots & \\
& \qquad\qquad\qquad\qquad \lambda_{T_i} & \geq 0, & \forall i = 1, \cdots, I_T^k.
\end{array}
\tag{5.22}
$$

We can now see that for a given t and i in RMP (5.22), a column has the form $\begin{bmatrix} c_t^\top x_{t_i} \\ A_t x_{t_i} \\ 1 \end{bmatrix}$. Therefore, to form RMP we need the data $\{c_t, A_T, x_{t_i}\}$, where both c_t and A_t are problem parameters, while x_{t_i} is the ith *extreme point* (vertex) of subproblem t feasible set. Before we discuss how to get x_{t_i}, let us first look at the dual to RMP:

$$\begin{aligned} &\text{Max } \pi_0^\top r_0 + \sum_{t=1}^{T} \pi_t \\ &\text{s.t. } \pi_0^\top (A_t x_{t_i}) + \pi_t \le c_t^\top x_{t_i}, \quad \forall t = 1, \cdots, T; i \in \Phi_t. \end{aligned} \tag{5.23}$$

Notice that in the dual formulation (5.23), the dual decision variables π_0 and $\pi_t, t = 0, \cdots, T$, are free or unrestricted in sign. This is due to the equality constraints in the primal Problem (5.22). From the constraints in (5.23), we see that for any t_i such that $\pi_t > (c_t^\top - \pi_0^\top A_t)x_{t_i}$, it means that the dual constraint is *violated*. Consequently, we have to find the *violated* column in RMP. To do that, we need to minimize $(c_t^\top - \pi_0^\top A_t)x_{t_i} - \pi_t$ subject to subproblem t_i feasible set. Observe that this is simply the *reduced cost* of column (decision variable) λ_{t_i} in RMP. To determine the reduced cost, we have to solve the following problem Dantzig–Wolfe subproblem (DWS):

$$\begin{aligned} &\text{Max } (c_t^\top - \pi_0^\top A_t)x_t \\ &\text{s.t. } B_t x_t \ge r_t \\ &\qquad x_t \ge 0. \end{aligned} \tag{5.24}$$

Let x_{t_i} denote the optimal solution to (5.24); then the reduced cost of variable λ_{t_i} is $(c_t^\top - \pi_0^\top A_t)x_{t_i} - \pi_t$. If $(c_t^\top - \pi_0^\top A_t)x_{t_i} - \pi_t \ge 0$ for all t and i, then the optimal solution to RMP is the optimal solution to DWP.

5.4.2 *Algorithm*

We are now ready to state the Dantzig–Wolfe decomposition algorithm. To begin the algorithm, we need to convert Problem (5.18) into a RMP (5.21) and form an initial basis. The algorithm will then proceed by sequentially generating columns to enter the basis and columns to leave the basis. If at least one column *prices out* negative (minimization problem) for some t and i, this column is added to RMP. Since columns (decision variables) are sequentially added to RMP, the algorithm often takes the name *column generation*. This is in contrast to *row generation* in Benders decomposition.

A basic Dantzig–Wolfe decomposition algorithm can be given as follows:

Algorithm *Dantzig–Wolfe Decomposition Algorithm*
begin

Step 0. **Initialization.** Let $k \leftarrow 0$ and for all $t = 1, \cdots, T$, initialize $I_t^k \geq 1$ and find $x_{t_i}^k \in \operatorname{argmin}\{c_t^\top x_t \mid x_t \in X_t\}$, $i = 1, \cdots, I_t^k$ to create an initial feasible basis.
Using this data form RMP (5.21).

Step 1. **Solve Restricted Master Program.** Solve RMP (5.21) to get dual solution π_0^k and π_t^k, and primal solution $\{\lambda_{t_i}\}_{i=1}^{I_t^k}$, for all $t = 1, \cdots, T$.

Step 2. **Solve Subproblems.** For all $t = 1, \cdots, T$, solve DWS subproblem (5.24) to get x_t^k: $x_t^k \in \operatorname{argmin}\{(c_t^\top - (\pi_0^k)^\top A_t)x_t \mid x_t \in X_t\}$.

Step 3. **Add New Column to RMP.** If $\pi_t^k > (c_t^\top - (\pi_0^k)^\top A_t)x_t^k$ for some t (dual constraint violation!), add x_t^k to RMP and set $I_t^{k+1} \leftarrow I_t^k + 1$.
For the rest of the t's set $I_t^{k+1} \leftarrow I_t^k$.

Step 4. **Termination.** If *no* new column is added in step 3:

Stop, $\{\lambda_{t_i}^k\}_{i=1}^{I_t^k}$ for all $t = 1, \cdots, T$ are optimal.

Report optimal solution $x_t^* = \sum_{i=1}^{I_t^k} \lambda_{t_i}^k x_{t_i}^k$ for all $t = 1, \cdots, T$.

Else:

Set $k \leftarrow k + 1$ and return to step 1.

---

end

In step 0 of the Dantzig–Wolfe decomposition algorithm, we need to create an *initial feasible basis*. This requires finding an appropriate combination of extreme points from each polytope for $t = 1, \cdots, T$. Assuming we have more columns than constraints in RMP (which is typically the case), the *basis* should be of size $m_0 + T$, which is the number of constraints in RMP. This number will determine how many $x_{t_i}^0$'s are needed at initialization, $x_{t_i}^0 \in \{B_t x_t \geq r_t, x_t \geq 0\}$, for each t. This means that each t must contribute at least one column to RMP. For example, if $m_0 = 3$ (three complicating constraints) and $T = 2$ (two subproblems), it means $m_0 + T = 3 + 2 = 5$ columns are needed in the basis. Therefore, we need to form RMP with five columns. For instance, for $t = 1$, we can have λ_{1_i}, $i = 0, 1, 2$, and for $t = 2$, we can have λ_{2_i}, $i = 0, 1$ to form an initial feasible basis. Thus for $t = 1$, we need to get $x_{1_i}^0 \in X_1$ for $i = 0, 1, 2$, while for $t = 2$ we need to get $x_{2_i}^0 \in X_2$ for $i = 0, 1$.

In step 3 of the Dantzig–Wolfe decomposition algorithm, a new column is added for a dual constraint violation by some t. At this point, we can also determine which column should leave the basis as is done in the simplex method in LP. To accomplish that, we can use the *minimum ratio test* and such a column *can* be deleted from RMP. Deleting a column that leaves the basis provides an advantage in terms of keeping the size of RMP fixed and thus not using unnecessary computer memory.

In summary, the Dantzig–Wolfe decomposition algorithm works with a subset of the subproblem feasible regions and a subset of the feasible regions of the original problem.

5.4.3 Numerical Example

Example 5.5 Solve the following LP using Dantzig–Wolfe decomposition algorithm and verify your solution using a direct LP solver.

$$\begin{array}{llllllll}
\text{Min} & -5x_1 & -3x_2 & -5x_3 & -4x_4 & -8x_5 & & \\
\text{s.t.} & -x_1 & -x_2 & -x_3 & -x_4 & -x_5 & \geq & -3 \\
 & -x_1 & +2x_2 & -2x_3 & +2x_4 & -2x_5 & \geq & -2 \\
 & & -x_2 & & & & \geq & -1 \\
 & & x_2 & -x_3 & & & \geq & 0.5 \\
 & & & & -x_4 & & \geq & -1 \\
 & & & & x_4 & -x_5 & \geq & 0.5 \\
 & x_1, & x_2, & x_3, & x_4, & x_5 & \geq & 0.
\end{array}$$

Note: This is the dual to the LP in Example 5.3 written as a minimization problem, which we solved using Benders decomposition.

Notice that the first and second constraints of the problem are complicating/linking constraints since they involve all the five decision variables, x_1, x_2, x_3, x_4 and x_5. The rest of the constraints involve a subset of these decision variables. Thus, the problem has a dual block angular structure and can be solved using Dantzig–Wolfe decomposition. The extreme points and extreme rays are obtained from the subproblems formed after removing the complicating constraints. We can decompose this problem into three subproblems. The first subproblem involves decision variable x_1 only and is given as follows:

$$\begin{array}{lll}
t = 1: & \text{Min} & -5x_1 \\
 & \text{s.t.} & x_1 \geq 0.
\end{array}$$

Subproblem 1: $\text{Min}_{\{x_{1i} \in X_1\}}\{-5x_1\}$, where

$$x_{1i} = x_1, \; X_1 = \{x_1 \geq 0\}.$$

The second subproblem involves the third and fourth constraints with decision variables x_2 and x_3 only:

$$\begin{array}{llllll}
t = 2: & \text{Min} & -3x_2 & -5x_3 & & \\
 & \text{s.t.} & -x_2 & & \geq & -1 \\
 & & x_2 & -x_3 & \geq & 0.5 \\
 & & x_2, & x_3 & \geq & 0.
\end{array}$$

Subproblem 2: $\text{Min}_{\{x_{2i} \in X_2\}}\{-3x_2 - 5x_3\}$, where

$$x_{2i} = (x_2, x_3)^\top, \; X_2 = \{-x_2 \geq -1, \; x_2 - x_3 \geq 0.5, \; x_2, \; x_3 \geq 0\}.$$

The third subproblem involves the fifth and sixth constraints with decision variables x_4 and x_5 only:

$$\begin{array}{rlrl} t = 3: & \text{Min} & -4x_4 & -8x_5 \\ & \text{s.t.} & -x_4 & \quad \geq -1 \\ & & x_4 & -x_5 \geq 0.5 \\ & & x_4, & \quad x_5 \geq \; 0. \end{array}$$

Subproblem 3: $\text{Min}_{\{x_{3i} \in X_3\}}\{-4x_4 - 8x_5\}$, where

$$x_{3i} = (x_4, x_5)^\top, \; X_3 = \{-x_4 \geq -1, \; x_4 - x_5 \geq 0.5, \; x_4, \; x_5 \geq 0\}.$$

The problem data r_0, $\{c_t, \; A_t, \; B_t, \; r_t\}$, for $t = 1, 2, 3$ are as follows:

$$r_0 = \begin{pmatrix} -3 \\ -2 \end{pmatrix};$$

$$c_1 = -5, \; A_1 = \begin{bmatrix} -1 \\ -1 \end{bmatrix}, \; B_1 = 0, r_1 = 0;$$

$$c_2 = (-3, -5)^\top, \; A_2 = \begin{bmatrix} -1 & -1 \\ 2 & -2 \end{bmatrix}, \; B_2 = \begin{bmatrix} -1 & 0 \\ 1 & -1 \end{bmatrix}, \; r_2 = \begin{pmatrix} -1 \\ 0.5 \end{pmatrix}; \text{ and}$$

$$c_3 = (-4, -8)^\top, \; A_3 = \begin{bmatrix} -1 & -1 \\ 2 & -2 \end{bmatrix}, \; B_3 = \begin{bmatrix} -1 & 0 \\ 1 & -1 \end{bmatrix}, \; r_3 = \begin{pmatrix} -1 \\ 0.5 \end{pmatrix}.$$

Given that the original LP has two complicating constraints and three subproblems, we need five decision variables in RMP to form an *initial basis*. Let us choose one extreme point from X_1, two from X_2, and two from X_3 by solving each subproblem:

Subproblem 1: is unbounded with extreme ray $w_{1_1} = 1$ and one extreme point $x_{1_1} = 0$.

$$c_1^\top x_{1_1} = -5(0) = 0 \text{ and } A_1 x_{1_1} = \begin{bmatrix} -1 \\ -1 \end{bmatrix}(0) = 0.$$

$$\textit{Column 1:} \quad \begin{bmatrix} c_1^\top x_{1_1} \\ A_t x_{1_1} \\ 1 \\ 0 \\ 0 \end{bmatrix} = \begin{pmatrix} 0 \\ 0 \\ 0 \\ 1 \end{pmatrix}.$$

Subproblem 2: has three extreme points: $x_{2_1} = (1, 0.5)^\top$, $x_{2_2} = (1, 0)^\top$, $x_{2_3} = (0.5, 0)^\top$. Picking the first two points, we get

$c_2^\top x_{2_1} = (-3, -5)(1, 0.5)^\top = -5.5$ and $A_2 x_{2_1} = \begin{bmatrix} -1 & -1 \\ 2 & -2 \end{bmatrix} \begin{pmatrix} 1 \\ 0.5 \end{pmatrix} = \begin{pmatrix} -1.5 \\ 1 \end{pmatrix}$.

$$\textit{Column 2:} \begin{bmatrix} c_2^\top x_{2_1} \\ A_2 x_{2_1} \\ 0 \\ 1 \\ 0 \end{bmatrix} = \begin{pmatrix} -5.5 \\ -1.5 \\ 1 \\ 0 \\ 1 \end{pmatrix}.$$

$c_2^\top x_{2_2} = (-3, -5)(1, 0)^\top = -3$ and $A_2 x_{2_2} = \begin{bmatrix} -1 & -1 \\ 2 & -2 \end{bmatrix} \begin{pmatrix} 1 \\ 0 \end{pmatrix} = \begin{pmatrix} -1 \\ 2 \end{pmatrix}$.

$$\textit{Column 3:} \begin{bmatrix} c_2^\top x_{2_2} \\ A_2 x_{2_2} \\ 0 \\ 1 \\ 0 \end{bmatrix} = \begin{pmatrix} -3 \\ -1 \\ 2 \\ 0 \\ 1 \end{pmatrix}.$$

Subproblem 3: has three extreme points: $x_{3_1} = (1, 0.5)^\top$, $x_{3_2} = (1, 0)^\top$, $x_{3_3} = (0.5, 0)^\top$. Picking the first two points, we get

$c_3^\top x_{3_1} = (-4, -8)(1, 0.5)^\top = -8$ and $A_3 x_{3_1} = \begin{bmatrix} -1 & -1 \\ 2 & -2 \end{bmatrix} \begin{pmatrix} 1 \\ 0.5 \end{pmatrix} = \begin{pmatrix} -1.5 \\ 1 \end{pmatrix}$.

$$\textit{Column 4:} \begin{bmatrix} c_3^\top x_{3_1} \\ A_3 x_{3_1} \\ 0 \\ 0 \\ 1 \end{bmatrix} = \begin{pmatrix} -8 \\ -1.5 \\ 1 \\ 0 \\ 0 \\ 1 \end{pmatrix}.$$

$c_3^\top x_{3_2} = (-4, -8)(1, 0)^\top = -4$ and $A_3 x_{3_2} = \begin{bmatrix} -1 & -1 \\ 2 & -2 \end{bmatrix} \begin{pmatrix} 1 \\ 0 \end{pmatrix} = \begin{pmatrix} -1 \\ 2 \end{pmatrix}$.

$$\textit{Column 5:} \begin{bmatrix} c_3^\top x_{3_2} \\ A_3 x_{3_2} \\ 0 \\ 0 \\ 1 \end{bmatrix} = \begin{pmatrix} -4 \\ -1 \\ 2 \\ 0 \\ 0 \\ 1 \end{pmatrix}.$$

We can now apply the steps of the Dantzig–Wolfe decomposition algorithm as follows:

Algorithm *Dantzig–Wolfe Decomposition Algorithm*
begin

Step 0. **Initialization.** Let $k \leftarrow 0$, and for all $t = 1, 2, 3$, initialize $I_t^k \geq 1$ and find $x_{t_i}^k \in \text{argmin}\{c_t^\top x_t \mid x_t \in X_t\}$, $i = 1, \cdots, I_t^k$ to create an initial feasible basis. Set $I_1^0 = 1$, $I_2^0 = 2$, and $I_3^0 = 2$. Using this data form RMP (5.21):

$$\begin{array}{lllllll}
\text{Min} & 0\lambda_{1_1} & -5.5\lambda_{2_1} & -3\lambda_{2_2} & -8\lambda_{3_1} & -4\lambda_{3_2} & \\
\text{s.t.} & 0\lambda_{1_1} & -1.5\lambda_{2_1} & -\lambda_{2_2} & -1.5\lambda_{3_1} & -\lambda_{3_2} & \geq -3 \leftarrow \pi_{01}^0 \\
 & 0\lambda_{1_1} & +\lambda_{2_1} & +2\lambda_{2_2} & +\lambda_{3_1} & +2\lambda_{3_2} & \geq -2 \leftarrow \pi_{02}^0 \\
 & \lambda_{1_1} & & & & & = \ \ 1 \leftarrow \pi_1^0 \\
 & & \lambda_{2_1} & +\lambda_{2_2} & & & = \ \ 1 \leftarrow \pi_2^0 \\
 & & & & \lambda_{3_1} & +\lambda_{3_2} & = \ \ 1 \leftarrow \pi_3^0 \\
 & \lambda_{1_1}, & \lambda_{2_1}, & \lambda_{2_2}, & \lambda_{3_1}, & \lambda_{3_2} & \geq \ 0.
\end{array}$$

Step 1. **Solve Restricted Master Program.** Solve RMP (5.21) to get solution:
Dual solution:

$$\pi_0^0 = (\pi_{01}^0, \pi_{02}^0)^\top = (0, 0)^\top, \ \pi_1^0 = 0, \ \pi_2^0 = -5.5, \ \pi_3^0 = -8.$$

Primal solution:

$$\lambda_{1_1}^0 = 1, \ \lambda_{2_1}^0 = 1, \ \lambda_{2_2}^0 = 0, \ \lambda_{3_1}^0 = 1, \ \lambda_{3_2}^0 = 0.$$

Objective value: -13.5.

Step 2. **Solve Subproblems.** For all $t = 1, 2, 3$, solve DWS subproblem (5.24) to get x_t^k: $x_t^k \in \text{argmin}\{(c_t^\top - (\pi_0^k)^\top A_t)x_t \mid x_t \in X_t\}$.

$$t = 1 : x_1^0 \in \text{argmin}\{(c_1^\top - (\pi_0^0)^\top A_1)x_1 \mid x_1 \in X_1\} =$$
$$\{(-5 - (0, 0)\begin{bmatrix} -1 \\ -1 \end{bmatrix})x_1 \mid x_1 \in X_1\} =$$
$$\{-5x_1 \mid x_1 \in X_1\}.$$

Subproblem is unbounded, *extreme ray* $w_{1_1}^0 = 1$.

$$t = 2 : x_2^0 \in \text{argmin}\{(c_2^\top - (\pi_0^0)^\top A_2)(x_2, x_3) \mid (x_2, x_3) \in X_2\} =$$
$$\{((-3, -5) - (0, 0)\begin{bmatrix} -1 & -1 \\ 2 & -2 \end{bmatrix})(x_2, x_3) \mid (x_2, x_3) \in X_2\} =$$
$$\{-3x_2 - 5x_3) \mid (x_2, x_3) \in X_2\}.$$

Subproblem solution: $x_2^0 = (1, 0.5)^\top$.

$$t = 3 : x_3^0 \in \operatorname{argmin}\{(c_3^\top - (\pi_0^0)^\top A_3)(x_4, x_5) \mid (x_4, x_5) \in X_3\} =$$

$$\{((-4, -8) - (0, 0)\begin{bmatrix} -1 & -1 \\ 2 & -2 \end{bmatrix})(x_4, x_5) \mid (x_4, x_5) \in X_3\} =$$

$$\{-4x_2 - 8x_3) \mid (x_4, x_5) \in X_3\}.$$

Subproblem solution: $x_3^0 = (1, 0.5)^\top$.

Step 3. Add New Column to RMP. If $\pi_t^k > (c_t^\top - (\pi_0^k)^\top A_t)x_t^k$ for some t (dual constraint violation!), add x_t^k to RMP and set $I_t^{k+1} \leftarrow I_t^k + 1$.
For the rest of the t's set $I_t^{k+1} \leftarrow I_t^k$.

$$\begin{aligned} t = 1 : (c_1^\top - (\pi_0^0)^\top A_1)x_1^0 =& (-5 - 0)x_1^0 \\ =& -\infty \text{ since } \textit{subproblem 1} \text{ is unbounded.} \\ & \Rightarrow \text{constraint violation since } \pi_1^0 = 0. \end{aligned}$$

Add new column θ_{1_1}: *Column 1:* $\begin{bmatrix} c_1^\top w_{1_1} \\ A_1 w_{1_1} \\ 0 \\ 0 \\ 0 \end{bmatrix} = \begin{pmatrix} -5 \\ -1 \\ -1 \\ 0 \\ 0 \\ 0 \end{pmatrix}$.

Observe that there is no dual constraint violation for *subproblems 2* and *3*:

$$\begin{aligned} t = 2 : (c_2^\top - (\pi_0^0)^\top A_2)x_2^0 =& ((-3, -5) - (0, 0))(1, 0.5)^\top \\ =& -5.5. \\ & \Rightarrow \text{no constraint violation since } \pi_2^0 = -5.5. \end{aligned}$$

$$\begin{aligned} t = 3 : (c_2^\top - (\pi_0^0)^\top A_2)x_3^0 =& ((-4, -8) - (0, 0))(1, 0.5)^\top \\ =& -8. \\ & \Rightarrow \text{no constraint violation since} \pi_3^0 = -8. \end{aligned}$$

Minimum ratio test:

Basis $(\lambda_{2_1}, \lambda_{2_2})$ matrix $B = \begin{bmatrix} -1.5 & -1 \\ 1 & 2 \end{bmatrix}$.

Let

$$
\begin{aligned}
u =& B^{-1}[\text{new column}] \\
=& \begin{bmatrix} -1 & -0.5 \\ 0.5 & 0.75 \end{bmatrix} \begin{pmatrix} -1 \\ -1 \end{pmatrix} \\
=& \begin{pmatrix} 1.5 \\ -1.25 \end{pmatrix}.
\end{aligned}
$$

The ratio is $\frac{\lambda}{u}$ for positive values of u only. Thus, the ratio for λ_{2_1} is $\frac{1}{1.5} = \frac{2}{3}$. Therefore, λ_{2_1} leaves the basis. Set

$$I_1^1 \leftarrow 1,\ J_1^1 \leftarrow 1,\ I_2^1 \leftarrow 1,\ I_3^1 \leftarrow 2.$$

RMP now takes the following form:

$$
\begin{array}{lllllll}
\text{Min } 0\lambda_{1_1} & -5\theta_{1_1} & -3\lambda_{2_2} & -8\lambda_{3_1} & -4\lambda_{3_2} & & \\
\text{s.t. } 0\lambda_{1_1} & -1\theta_{1_1} & -\lambda_{2_2} & -1.5\lambda_{3_1} & -\lambda_{3_2} & \geq -3 & \leftarrow \pi_{01}^1 \\
\quad 0\lambda_{1_1} & -1\theta_{1_1} & +2\lambda_{2_2} & +\lambda_{3_1} & +2\lambda_{3_2} & \geq -2 & \leftarrow \pi_{02}^1 \\
\quad \lambda_{1_1} & & & & & = \ \ 1 & \leftarrow \pi_1^1 \\
 & & +\lambda_{2_2} & & & = \ \ 1 & \leftarrow \pi_2^1 \\
 & & & \lambda_{3_1} & +\lambda_{3_2} & = \ \ 1 & \leftarrow \pi_3^1 \\
\quad \lambda_{1_1}, & \theta_{1_1}, & \lambda_{2_2}, & \lambda_{3_1}, & \lambda_{3_2} & \geq \ 0. &
\end{array}
$$

Step 4. Termination. If *no* new column is added in step 3, stop. Since the termination condition is not satisfied, set
$k \leftarrow 1$ return to step 1:

Step 1. Solve Restricted Master Program. Solve RMP (5.21) to get solution:
Dual solution:

$$\pi_0^1 = (\pi_{01}^1, \pi_{02}^1)^\top = (5, 0)^\top,\ \pi_1^1 = 0,\ \pi_2^1 = 2,\ \pi_3^1 = -0.5.$$

Primal solution:

$$\lambda_{1_1}^1 = 1,\ \theta_{1_1}^1 = 0.5,\ \lambda_{2_2}^1 = 1,\ \lambda_{3_1}^1 = 1,\ \lambda_{3_2}^1 = 0.$$

Objective value: -13.5.

Step 2. Solve Subproblems. For all $t = 1, 2, 3$, solve DWS subproblem (5.24) to get x_t^k: $x_t^k \in \text{argmin}\{(c_t^\top - (\pi_0^k)^\top A_t)x_t \mid x_t \in X_t\}$.

$$
\begin{aligned}
t = 1 : x_1^1 \in & \ \text{argmin}\{(c_1^\top - (\pi_0^1)^\top A_1)x_1 \mid x_1 \in X_1\} = \\
& \{(-5 - (5, 0)\begin{bmatrix} -1 \\ -1 \end{bmatrix})x_1 \mid x_1 \in X_1\} = \\
& \{0x_1 \mid x_1 \in X_1\}.
\end{aligned}
$$

Subproblem solution: $x_1^1 = 0$.

$$t = 2 : x_2^1 \in \operatorname{argmin}\{(c_2^\top - (\pi_0^1)^\top A_2)(x_2, x_3) \mid (x_2, x_3) \in X_2\} =$$
$$\{((-3, -5) - (5, 0)\begin{bmatrix} -1 & -1 \\ 2 & -2 \end{bmatrix})(x_2, x_3) \mid (x_2, x_3) \in X_2\} =$$
$$\{2x_2 + 0x_3) \mid (x_2, x_3) \in X_2\}.$$

Subproblem solution: $x_2^1 = (0.5, 0)^\top$.

$$t = 3 : x_3^1 \in \operatorname{argmin}\{(c_3^\top - (\pi_0^1)^\top A_3)(x_4, x_5) \mid (x_4, x_5) \in X_3\} =$$
$$\{((-4, -8) - (5, 0)\begin{bmatrix} -1 & -1 \\ 2 & -2 \end{bmatrix})(x_4, x_5) \mid (x_4, x_5) \in X_3\} =$$
$$\{x_2 - 3x_3) \mid (x_4, x_5) \in X_3\}.$$

Subproblem solution: $x_3^1 = (1, 0.5)^\top$.

Step 3. Add New Column to RMP. If $\pi_t^k > (c_t^\top - (\pi_0^k)^\top A_t)x_t^k$ for some t (dual constraint violation!), add x_t^k to RMP and set $I_t^{k+1} \leftarrow I_t^k + 1$.
For the rest of the t's set $I_t^{k+1} \leftarrow I_t^k$.

$$\begin{aligned} t = 1 : (c_1^\top - (\pi_0^1)^\top A_1)x_1^1 =& (-5 - (5, 0)\begin{bmatrix} -1 \\ -1 \end{bmatrix})(0) \\ =& \, 0. \\ & \Rightarrow \text{no constraint violation since } \pi_1^1 = 0. \end{aligned}$$

$$\begin{aligned} t = 2 : (c_2^\top - (\pi_0^1)^\top A_2)x_2^1 =& ((-3, -5) - (5, 0)\begin{bmatrix} -1 & -1 \\ 2 & -2 \end{bmatrix})(0.5, 0)^\top \\ =& \, 1. \\ & \Rightarrow \text{constraint violation since } \pi_2^1 = 2. \end{aligned}$$

$$\text{Add new column } \lambda_{23}\text{: } \textit{Column 1:} \begin{bmatrix} c_2^\top x_2^1 \\ A_2 x_2^1 \\ 0 \\ 1 \\ 0 \end{bmatrix} = \begin{pmatrix} -1.5 \\ -0.5 \\ 1 \\ 0 \\ 1 \\ 0 \end{pmatrix}.$$

$$t = 3 : (c_2^\top - (\pi_0^1)^\top A_2)x_3^1 = ((-4, -8) - (5, 0) \begin{bmatrix} -1 & -1 \\ 2 & -2 \end{bmatrix})(1, 0.5)^\top$$
$$= -0.5.$$
$$\Rightarrow \text{no constraint violation since } \pi_3^1 = -0.5.$$

Minimum ratio test:
Basis $(\lambda_{3_1}, \lambda_3)$ matrix $B = \begin{bmatrix} -1.5 & -1 \\ 1 & 2 \end{bmatrix}$.
Let

$$\begin{aligned} u =& B^{-1}[\text{new column}] \\ =& \begin{bmatrix} -1 & -0.5 \\ 0.5 & 0.75 \end{bmatrix} \begin{pmatrix} -0.5 \\ 1 \end{pmatrix} \\ =& \begin{pmatrix} 0 \\ 0.5 \end{pmatrix}. \end{aligned}$$

The ratio is $\frac{\lambda}{u}$ for positive values of u only. Therefore, λ_{3_2} leaves the basis. Set

$$I_1^2 \leftarrow 1,\ J_1^2 \leftarrow 1,\ I_2^2 \leftarrow 2,\ I_3^2 \leftarrow 1.$$

RMP now takes the following form:

$$\begin{array}{llllllll} \text{Min} & 0\lambda_{1_1} & -5\theta_{1_1} & -3\lambda_{2_2} & -8\lambda_{3_1} & -1.5\lambda_{2_3} & & \\ \text{s.t.} & 0\lambda_{1_1} & -1\theta_{1_1} & -\lambda_{2_2} & -1.5\lambda_{3_1} & -\lambda_{2_3} & \geq -3 & \leftarrow \pi_{01}^2 \\ & 0\lambda_{1_1} & -1\theta_{1_1} & +2\lambda_{2_2} & +\lambda_{3_1} & +2\lambda_{2_3} & \geq -2 & \leftarrow \pi_{02}^2 \\ & \lambda_{1_1} & & & & & = \ \ 1 & \leftarrow \pi_1^2 \\ & & & +\lambda_{2_2} & & +\lambda_{2_3} & = \ \ 1 & \leftarrow \pi_2^2 \\ & & & & \lambda_{3_1} & & = \ \ 1 & \leftarrow \pi_3^2 \\ & \lambda_{1_1}, & \theta_{1_1}, & \lambda_{2_2}, & \lambda_{3_1}, & \lambda_{2_3} & \geq \ \ 0. & \end{array}$$

Step 4. Termination. If *no* new column is added in step 3, stop. Since the termination condition is not satisfied, set
$k \leftarrow 2$ return to step 1:

Step 1. Solve Restricted Master Program. Solve RMP (5.21) to get solution:
Dual solution:

$$\pi_0^2 = (\pi_{01}^2, \pi_{02}^2)^\top = (5, 0)^\top,\ \pi_1^2 = 0,\ \pi_2^2 = 1,\ \pi_3^2 = -0.5.$$

Primal solution:

$$\lambda_{1_1}^2 = 1,\ \theta_{1_1}^2 = 1,\ \lambda_{2_2}^2 = 0,\ \lambda_{2_3}^2 = 1,\ \lambda_{3_1}^2 = 1.$$

Objective value: -14.5.

Step 2. Solve Subproblems. For all $t = 1, 2, 3$, solve DWS subproblem (5.24) to get x_t^k: $x_t^k \in \operatorname{argmin}\{(c_t^\top - (\pi_0^k)^\top A_t)x_t \mid x_t \in X_t\}$.

$$\begin{aligned} t = 1 : x_1^2 \in \operatorname{argmin}\{(c_1^\top - (\pi_0^2)^\top A_1)x_1 \mid x_1 \in X_1\} = \\ \{(-5 - (5,0)\begin{bmatrix}-1\\-1\end{bmatrix})x_1 \mid x_1 \in X_1\} = \\ \{0x_1 \mid x_1 \in X_1\}. \end{aligned}$$

Subproblem solution: $x_1^2 = 0$.

$$\begin{aligned} t = 2 : x_2^2 \in \operatorname{argmin}\{(c_2^\top - (\pi_0^2)^\top A_2)(x_2, x_3) \mid (x_2, x_3) \in X_2\} = \\ \{((-3,-5) - (5,0)\begin{bmatrix}-1 & -1\\ 2 & -2\end{bmatrix})(x_2, x_3) \mid (x_2, x_3) \in X_2\} = \\ \{2x_2 + 0x_3) \mid (x_2, x_3) \in X_2\}. \end{aligned}$$

Subproblem solution: $x_2^2 = (0.5, 0)^\top$.

$$\begin{aligned} t = 3 : x_3^2 \in \operatorname{argmin}\{(c_3^\top - (\pi_0^0)^\top A_3)(x_4, x_5) \mid (x_4, x_5) \in X_3\} = \\ \{((-4,-8) - (5,0)\begin{bmatrix}-1 & -1\\ 2 & -2\end{bmatrix})(x_4, x_5) \mid (x_4, x_5) \in X_3\} = \\ \{x_2 - 3x_3) \mid (x_4, x_5) \in X_3\}. \end{aligned}$$

Subproblem solution: $x_3^2 = (1, 0.5)^\top$.

Step 3. Add New Column to RMP. If $\pi_t^k > (c_t^\top - (\pi_0^k)^\top A_t)x_t^k$ for some t (dual constraint violation!), add x_t^k to RMP and set $I_t^{k+1} \leftarrow I_t^k + 1$. For the rest of the t's set $I_t^{k+1} \leftarrow I_t^k$.

$$\begin{aligned} t = 1 : (c_1^\top - (\pi_0^2)^\top A_1)x_1^2 =& (-5 - (5,0)\begin{bmatrix}-1\\-1\end{bmatrix})(0) \\ =& 0. \\ &\Rightarrow \text{no constraint violation since } \pi_1^1 = 0. \end{aligned}$$

$$t = 2 : (c_2^\top - (\pi_0^2)^\top A_2)x_2^2 = ((-3, -5) - (5, 0)\begin{bmatrix} -1 & -1 \\ 2 & -2 \end{bmatrix})(0.5, 0)^\top$$

$$= 1.$$

$$\Rightarrow \text{no constraint violation since } \pi_2^2 = 1.$$

$$t = 3 : (c_2^\top - (\pi_0^2)^\top A_2)x_3^1 = ((-4, -8) - (5, 0)\begin{bmatrix} -1 & -1 \\ 2 & -2 \end{bmatrix})(1, 0.5)^\top$$

$$= -0.5.$$

$$\Rightarrow \text{no constraint violation since } \pi_3^2 = -0.5.$$

Step 4. Termination. If *no* new column is added in step 3, stop. Since the termination condition is satisfied, terminate the algorithm and report the following optimal solution:

$$\text{RMP:} \lambda_{1_1}^2 = 1,\ \theta_{1_1}^2 = 1,\ \lambda_{2_2}^2 = 0,\ \lambda_{2_3}^2 = 1,\ \lambda_{3_1}^2 = 1.$$

$$\text{Original problem: } x_t = \sum_{i=1}^{I_t} \lambda_{t_i} x_{t_i} + \sum_{j=1}^{J_t} \theta_{t_j} w_{t_j} \quad \text{for all } t = 1, 2, 3.$$

$$\begin{aligned} x_1^* = x_1 &= \lambda_{1_1}^2 x_1^2 + \theta_{1_1}^2 w_{1_1}^2 \\ &= 1(0) + 1(1) \\ &= 1. \end{aligned}$$

$$\begin{aligned} x_2^* = (x_2, x_3)^\top &= \lambda_{2_3}^2 x_1^2 \\ &= 1(0.5, 0)^\top \\ &= (0.5, 0)^\top. \end{aligned}$$

$$\begin{aligned} x_3^* = (x_4, x_5)^\top &= \lambda_3^2 x_3^2 \\ &= 1(1, 0.5)^\top \\ &= (1, 0.5)^\top. \end{aligned}$$

Objective value: −14.5.

---

end

Notice that we have arrived at the same optimal value as in Example 5.3, i.e., $-1(-14.5) = 14.5$. The negative sign is due to minimization in objective of Example 5.5.

5.5 Lagrangian Decomposition

Lagrangian decomposition is typically applied to a problem with "hard" or "complicating" constraints as in the Dantzig–Wolfe problem. The idea behind Lagrangian approaches is to "relax" the problem by placing the complicating constraints into the objective with a large enough penalty to enforce feasibility of those constraints. This is referred to as *dualizing the constraints*. For the sake of continuity to the ongoing discourse, we shall apply the Lagrangian approach to the Dantzig–Wolfe problem and see how the two different approaches are related. For convenience, we restate the problem as follows:

$$\begin{aligned} z_P := \text{Min} & \sum_{t=1}^{T} c_t^\top x_t \\ \text{s.t.} & \sum_{t=1}^{T} A_t x_t = r_0 \\ & x_t \in X_t, \quad \forall t = 1, \cdots, T. \end{aligned} \tag{5.25}$$

We continue to assume that Problem (5.25) has a solution and that the sets $X_t = \{B_t x_t \geq r_t\}$, for all $t = 1, \cdots, T$, are *polytopes*. Therefore, we will not be concerned with *extreme rays* in characterizing the points in X_t. The issue remains that the linking constraints make the problem difficult. However, if we were to "ignore" the linking constraints, the rest of the problem would become separable. Thus this will enable decomposing the problem into smaller subproblems that can be solved relatively easy.

5.5.1 *Decomposition Approach*

Let $\pi_0 \in \mathbb{R}^{m_0}$ be the penalty, typically referred to as the *Lagrangian multiplier*. This penalty is simply the unit cost of violating the linking constraint $\sum_{t=1}^{T} A_t x_t = r_0$. Relaxing or *dualizing* this constraint results in the following *Lagrangian relaxation* problem:

$$D(\pi_0) := \underset{x_t \in X_t,\ t=1,\cdots,T}{\text{Min}} \sum_{t=1}^{T} c_t^\top x_t + \pi_0^\top (r_0 - \sum_{t=1}^{T} A_t x_t). \tag{5.26}$$

For a specified value of π_0, Problem (5.26) can be expressed as follows:

$$D(\pi_0) = \pi_0^\top r_0 + \underset{x_t \in X_t,\ t=1,\cdots,T}{\text{Min}} \sum_{t=1}^{T} c_t^\top x_t - \sum_{t=1}^{T} \pi_0^\top (A_t x_t)$$

$$= \sum_{t=1}^{T} (\frac{1}{T}\pi_0^\top r_0 + \underset{x_t \in X_t}{\text{Min}}\ c_t^\top x_t - \pi_0^\top (A_t x_t)).$$

Thus, the problem now becomes separable by t. Since the first term is a constant, we take it out of the minimization and write the subproblem for each $t = 1, \cdots, T$ as follows:

$$D_t(\pi_0) = \frac{1}{T}\pi_0^\top r_0 + \underset{x_t \in X_t}{\text{Min}}\ \{c_t^\top x_t - \pi_0^\top (A_t x_t)\}. \tag{5.27}$$

The remaining task now is to find the "largest" penalty π_0 to enforce the complicating constraint. This problem is the *Lagrangian dual problem* and is given as follows:

$$z_D = \underset{\pi_0}{\text{Max}}\ D(\pi_0) = \sum_{t=1}^{T} D_t(\pi_0). \tag{5.28}$$

The following result relates the optimal value of the Lagrangian dual problem to that of the original problem, and the proof of the result provides a link to Dantzig–Wolfe decomposition:

Theorem 5.3 *Suppose Problem* (5.25) *has a finite optimal value* q_P^*. *Then the value* $z_P \geq z_D$ *for every feasible solution* x_t, $t = 1, \cdots, T$ *and feasible solution* π_0, *and their optimal values coincide,* $q_D^* = q_P^*$.

Proof By definition,

$$D(\pi_0) := \underset{x_t \in X_t,\ t=1,\cdots,T}{\text{Min}} \sum_{t=1}^{T} c_t^\top x_t + \pi_0^\top (r_0 - \sum_{t=1}^{T} A_t x_t).$$

Since the objective function is a linear function of x_t, the optimal value remains the same if we take convex combinations of the elements of X_t, $t = 1, \cdots, T$. Using Theorem 5.2, let x_{t_i}, $i = 1, \cdots, I_t$ be the extreme points of X_t for all $t = 1, \cdots, T$. Then for any fixed π_0, we have

$$D(\pi_0) := \underset{\forall i=1,\cdots,I_t;\ t=1,\cdots,T}{\text{Min}} \sum_{t=1}^{T} c_t^\top x_{t_i} + \pi_0^\top (r_0 - \sum_{t=1}^{T} A_t x_{t_i}).$$

Thus, the Lagrangian dual problem (5.28) is equivalent to the following problem:

$$z_D = \underset{\pi_0}{\text{Max}} \quad \underset{\forall i=1,\cdots,I_t;\ t=1,\cdots,T}{\text{Min}} \quad \sum_{t=1}^{T} c_t^\top x_{t_i} + \pi_0^\top (r_0 - \sum_{t=1}^{T} A_t x_{t_i}).$$

Rewriting the above problem, we get

$$z_D = \text{Max } y$$

$$\text{s.t.} \quad y + \pi_0^\top (-r_0 + \sum_{t=1}^{T} A_t x_{t_i}) \leq \sum_{t=1}^{T} c_t^\top x_{t_i}, \quad \forall i=1,\cdots,I_t;\ t = 1,\cdots,T.$$

Taking the dual to the above problem, we obtain the following problem:

$$\begin{aligned} z_D := \text{Min} \ & \sum_{t=1}^{T}\sum_{i=1}^{I_t} \lambda_{t_i}(c_t^\top x_{t_i}) \\ \text{s.t.} \ & \sum_{i=1}^{I_t} \lambda_{t_i} = 1, \quad \forall t = 1,\cdots,T \\ & \sum_{t=1}^{T}\sum_{i=1}^{I_t} \lambda_{t_i}(A_t x_{t_i}) = r_0 \\ & \lambda_{t_i} \geq 0, \quad \forall t = 1,\cdots,T;\ i = 1,\cdots,I_t. \end{aligned} \tag{5.29}$$

By invoking *weak* and *strong* duality, the result follows. ■

Observe that Problem (5.29) is the same as RMP (5.21). It is interesting to see that the proof of Theorem 5.3 reveals the connection between Lagrangian relaxation and Dantzig–Wolfe decomposition. Essentially, Dantzig–Wolfe decomposition in a way involves solving the Lagrangian dual problem!

Let us now characterize the objective function of the Lagrangian dual problem. For a given π_0, the function $D(\pi_0)$ is *affine*. The Lagrangian dual objective function is the maximum of these affine functions of all possible values of π_0 and is therefore a *piecewise linear concave function* of π_0. Figure 5.8 gives an illustration of the Lagrangian dual objective function. Since the objective function is piecewise linear and concave, we can find the maximizer by applying some type of *steepest ascent algorithm* , similar to Kelley's algorithm for a *convex function*. However, we now have a *nondifferentiable* function due to the "kinks" in the function. Consequently, we need to generate supporting hyperplanes or *subgradients* of the Lagrangian dual function.

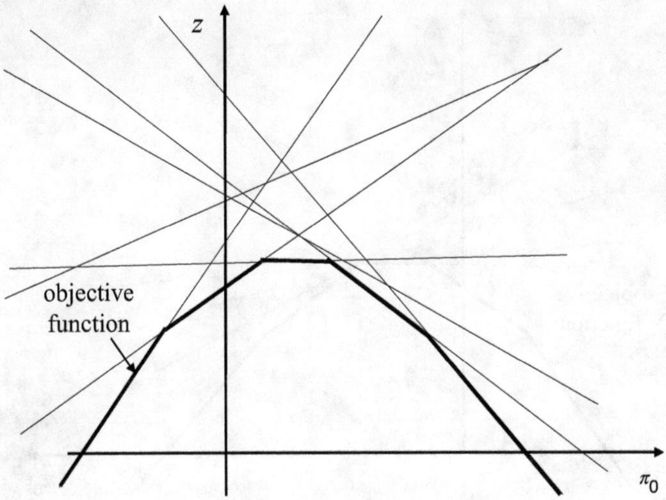

Fig. 5.8 An illustration of the Lagrangian dual objective function

Let $CH(.)$ denote the *convex hull* of a given set of vectors. We can characterize the *subdifferential*, $\partial D(\pi_0)$, of the piecewise linear concave function $D(\pi_0)$ as follows:

Lemma 5.1 *Let*

$$D(\pi_0) := \underset{\ell=1,\cdots,L}{Min} \ h_\ell + \pi_0^\top s_\ell,$$

$$E(\pi_0) = \{\ell \mid D(\pi_0) = h_\ell + \pi_0^\top s_\ell\}.$$

Then:

(i) For every $\ell \in E(\pi_0^*)$*,* s_ℓ *is a subgradient of the function* $D(.)$ *at* π_0^**.*
(ii) A vector is a subgradient of the function $D(.)$ *at* π_0^* *if and only if* $D(\pi_0^*)$ *is a convex combination of the vectors* s_ℓ*,* $\ell \in E(\pi_0^*)$*, i.e.,* $\partial D(\pi_0^*) = CH(\{s_\ell,\ \ell \in E(\pi_0^*)\})$*.*

Figure 5.9 gives an illustration of Lemma 5.1. We can conclude from Lemma 5.1 that the subgradient of the Lagrangian function $D(.)$ (5.26) at π_0 is $s = r_0 - \sum_{t=1}^{T} A_t x_t$, where $x_t, t = 1, \cdots, T$, is the optimal solution to Problem (5.26).

5.5.2 Algorithm

We are now in a position to state a basic subgradient optimization algorithm for solving Problem (5.28) as follows:

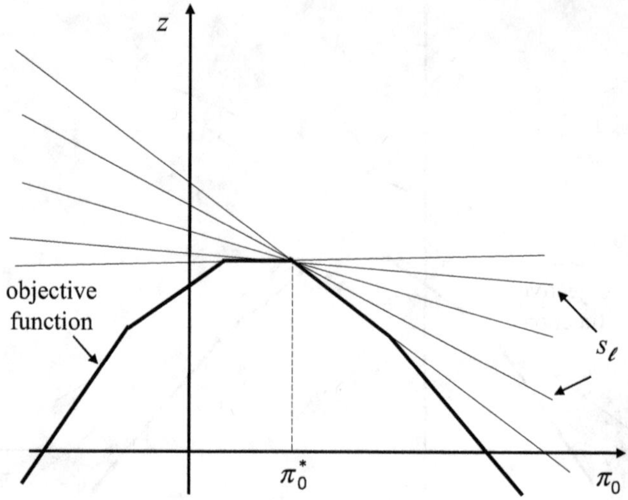

Fig. 5.9 An illustration of Lemma 5.1

Algorithm *Basic Subgradient Optimization*
begin

Step 0. **Initialization.** Let $k \leftarrow 0$, choose a starting point π_0^k and $\varepsilon > 0$.
Step 1. **Compute Subgradient.** Given π_0^k, calculate subgradient s^k of the function $D(.)$ at π_0^k.
Step 2. **Termination.** **If** a termination condition is satisfied,
stop, $\pi_0^* \leftarrow \pi_0^k$ is ε-optimal.
Else continue to step 3.
Step 3. **Compute Stepsize.** Compute stepsize ρ^k and update π_0^{k+1} using s^k, ρ^k, and π_0^k. Set $k \leftarrow k+1$ and return to step 1.

---

end

The subgradient optimization algorithm is simple to state, but the issue in practice is that convergence of the algorithm to an optimal solution is generally slow. It can be shown that $D(\pi_0^k)$ converges to the maximum of $D(.)$ for any stepsize sequence ρ^k such that

$$\sum_{k=1}^{\infty} \rho^k = \infty \text{ and } \lim_{k\to\infty} \rho^k = 0.$$

For example, the sequence $\rho^k = \frac{1}{k}$ satisfies the condition but leads to slow convergence. Another choice for ρ^k is to choose an initial ρ^0 and then set

$$\rho^k = \rho^0 \theta^k, k = 1, 2, \cdots,$$

where θ^k is a scalar satisfying $0 < \theta^k < 1$. If an estimate of the optimal value z_D, denoted $\hat{z}_D$, is known, then the following rule can be used:

$$\rho^k = \frac{\hat{z}_D - D(\pi_0^k)}{||s^k||^2} \theta^k.$$

If the function $D(\pi_0)$ is differentiable, then $s = \nabla D(\pi_0)$ always exists. Therefore, in step 2 the algorithm can be terminated when $||s^k|| < \varepsilon$, while the update in step 3 can be performed using

$$\pi_0^{k+1} \leftarrow \pi_0^k + \rho^k s^k.$$

However, in our case, $D(\pi_0)$ is not differentiable, and thus both the termination criterion and the π_0^k update rule have to be adapted according the given problem. In fact, the termination criterion $\mathbf{0} \in \partial D(\pi_0)$ is often not satisfied in practice. Consequently, the algorithm can be stopped after a fixed number of iterations. Next, we give an illustration of performing a few steps of the basic subgradient optimization algorithm.

5.5.3 *Numerical Example*

Example 5.6 Decompose the following problem into three subproblems using Lagrangian decomposition and then apply three iterations of the basic subgradient optimization algorithm to the decomposed problem:

$$\begin{array}{llllll}
\text{Min} & -5x_1 & -3x_2 & -5x_3 & -4x_4 & -8x_5 \\
\text{s.t.} & -x_1 & -x_2 & -x_3 & -x_4 & -x_5 \geq -3 \\
& -x_1 & +2x_2 & -2x_3 & +2x_4 & -2x_5 \geq -2 \\
& -x_1 & & & & \geq -1 \\
& & -x_2 & & & \geq -1 \\
& & x_2 & -x_3 & & \geq 0.5 \\
& & & & -x_4 & \geq -1 \\
& & & & x_4 & -x_5 \geq 0.5 \\
& x_1, & x_2, & x_3, & x_4, & x_5 \geq 0.
\end{array}$$

Notice that this is the same LP as in Example 5.5 except that we have now added the constraint $-x_1 \geq -1$ to avoid unboundedness of the first subproblem.

Let $\pi_0 \in \mathbb{R}^2_+$ be the *Lagrangian multiplier*, i.e., $\pi_0 = (\pi_{01}, \pi_{02})^\top$. We first need to dualize the first two complicating constraints and rewrite the problem as follows:

$$
\begin{array}{lrrrrrl}
\text{Max}_{\pi_0 \geq 0} \text{ Min} & -5x_1 & -3x_2 & -5x_3 & -4x_4 & -8x_5 & \\
& +\pi_{01}(-3 + x_1 & +x_2 & +x_3 & +x_4 & +x_5) & \\
& +\pi_{02}(-2 + x_1 & -2x_2 & +2x_3 & -2x_4 & +2x_5) & \\
\text{s.t.} & -x_1 & & & & & \geq -1 \\
& & -x_2 & & & & \geq -1 \\
& & x_2 & -x_3 & & & \geq 0.5 \\
& & & & -x_4 & & \geq -1 \\
& & & & x_4 & -x_5 & \geq 0.5 \\
& x_1, & x_2, & x_3, & x_4, & x_5 & \geq 0.
\end{array}
$$

This can be rewritten in the following form:

$$
D(\pi_0) := \underset{x_t \in X_t,\ t=1,\cdots,T}{\text{Min}} \sum_{t=1}^{T} c_t^\top x_t + \pi_0^\top (r_0 - \sum_{t=1}^{T} A_t x_t),
$$

where $T = 3$ and

$$
\begin{aligned}
X_1 &= \{-x_1 \geq -1, x_1 \geq 0\}, \\
X_2 &= \{-x_2 \geq -1,\ x_2 - x_3 \geq 0.5,\ x_2,\ x_3 \geq 0\}, \\
X_3 &= \{-x_4 \geq -1,\ x_4 - x_5 \geq 0.5,\ x_4,\ x_5 \geq 0\},
\end{aligned}
$$

and the problem data r_0, $\{c_t,\ A_t,\ r_t\}$, for $t = 1, 2, 3$ is as follows:

$r_0 = \begin{pmatrix} -3 \\ -2 \end{pmatrix}$,

$c_1 = -5,\ A_1 = \begin{bmatrix} -1 \\ -1 \end{bmatrix}$,

$c_2 = (-3, -5)^\top,\ A_2 = \begin{bmatrix} -1 & -1 \\ 2 & -2 \end{bmatrix}$, and

$c_3 = (-4, -8)^\top,\ A_3 = \begin{bmatrix} -1 & -1 \\ 2 & -2 \end{bmatrix}$.

We can now write the subproblem

$$
D_t(\pi_0) = \frac{1}{T}\pi_0^\top r_0 + \underset{x_t \in X_t}{\text{Min}} \{c_t^\top x_t - \pi_0^\top (A_t x_t)\}
$$

for each t as follows:

$$
D_1(\pi_0) = \frac{1}{3}(\pi_{01}, \pi_{02}) \begin{pmatrix} -3 \\ -2 \end{pmatrix} + \underset{x_1 \in X_1}{\text{Min}} \{-5x_1 - (\pi_{01}, \pi_{02})(\begin{bmatrix} -1 \\ -1 \end{bmatrix} x_1)\}
$$

$$= -\pi_{01} - \frac{2}{3}\pi_{02} + \underset{x_1 \in X_1}{\text{Min}} \{-5x_1 + (\pi_{01} + \pi_{02})x_1\}.$$

$$D_2(\pi_0) = \frac{1}{3}(\pi_{01}, \pi_{02})\begin{pmatrix}-3\\-2\end{pmatrix} + \underset{(x_2,x_3)\in X_2}{\text{Min}} \{(-3,-5)\begin{pmatrix}x_2\\x_3\end{pmatrix} - (\pi_{01}, \pi_{02})$$

$$(\begin{bmatrix}-1 & -1\\ 2 & -2\end{bmatrix}\begin{pmatrix}x_2\\x_3\end{pmatrix})\}$$

$$= -\pi_{01} - \frac{2}{3}\pi_{02} + \underset{(x_2,x_3)\in X_2}{\text{Min}} \{-3x_2 - 5x_3 + (\pi_{01} - 2\pi_{02})x_2$$

$$+ (\pi_{01} + 2\pi_{02})x_3\}.$$

$$D_3(\pi_0) = \frac{1}{3}(\pi_{01}, \pi_{02})\begin{pmatrix}-3\\-2\end{pmatrix} + \underset{(x_4,x_5)\in X_3}{\text{Min}} \{(-4,-8)\begin{pmatrix}x_4\\x_5\end{pmatrix}$$

$$- (\pi_{01}, \pi_{02})(\begin{bmatrix}-1 & -1\\ 2 & -2\end{bmatrix}\begin{pmatrix}x_4\\x_5\end{pmatrix})\}$$

$$= -\pi_{01} - \frac{2}{3}\pi_{02} + \underset{(x_4,x_5)\in X_3}{\text{Min}} \{-4x_4 - 8x_5 + (\pi_{01} - 2\pi_{02})x_4$$

$$+ (\pi_{01} + 2\pi_{02})x_5\}.$$

The *Lagrangian dual problem* is

$$z_D = \underset{\pi_0 \geq 0}{\text{Max}}\ D(\pi_0) = \sum_{t=1}^{T} D_t(\pi_0).$$

Given a solution from each subproblem, the subgradient s of the Lagrangian function $D(.)$ at π_0 is

$$s = r_0 - \sum_{t=1}^{T} A_t x_t$$

$$= \begin{pmatrix} -3 + x_1 + x_2 + x_3 + x_4 + x_5 \\ -2 + x_1 - 2x_2 + 2x_3 - 2x_4 + 2x_5 \end{pmatrix}.$$

We should point out that in this problem $\pi_0 \geq 0$ since we have complicating constraints written as "$\geq$" inequalities. Therefore, we have to enforce the nonnegativity requirement on π_0 in step 3 of the basic subgradient optimization algorithm by computing π_0^{k+1} as follows:

$$\pi_0^{k+1} \leftarrow \max\{\pi_0^k + s^k \rho^k,\ 0\}.$$

Let us now apply three iterations of the basic subgradient optimization algorithm to the decomposed problem:

Algorithm *Basic Subgradient Optimization*
begin

Step 0. **Initialization.** Let $k \leftarrow 0$, choose a starting point $\pi_0^k \geq 0$, say, $\pi_0^0 \leftarrow (0, 0)^\top$ and $\varepsilon \leftarrow 0.001$.
Step 1. **Compute Subgradient.** Given π_0^0, calculate subgradient s^0 of the function $D(.)$ at π_0^0.
Solve subproblems:

$$D_1(\pi_0^0) = -0 - \frac{2}{3}(0) + \underset{x_1 \in X_1}{\text{Min}} \{-5x_1 + (0+0)x_1\}.$$
$$= \underset{x_1 \in X_1}{\text{Min}} \{-5x_1\}.$$

Solution: $x_1 = 1,\ D_1(\pi_0^0) = -5.$

$$D_2(\pi_0^0) = -0 - \frac{2}{3}(0) + \underset{(x_2,x_3) \in X_2}{\text{Min}} \{-3x_2 - 5x_3 + (0 - 2(0))x_2 + (0 + 2(0))x_3\}.$$
$$= \underset{(x_2,x_3) \in X_2}{\text{Min}} \{-3x_2 - 5x_3\}.$$

Solution: $(x_2, x_3)^\top = (1, 0.5),\ D_2(\pi_0^0) = -5.5.$

$$D_3(\pi_0^0) = -0 - \frac{2}{3}(0) + \underset{(x_4,x_5) \in X_3}{\text{Min}} \{-4x_4 - 8x_5 + (0 - 2(0))x_4 + (0 + 2(0))x_5\}.$$
$$= \underset{(x_4,x_5) \in X_3}{\text{Min}} \{-4x_4 - 8x_5\}.$$

Solution: $(x_4, x_5)^\top = (1, 0.5),\ D_3(\pi_0^0) = -8.$

$$\sum_{t=1}^{0} D_t(\pi_0^0) = -5 - 5.5 - 8 = -18.5.$$

Step 2. **Termination.** Compute

$$s^0 = \begin{pmatrix} -3 + x_1 + x_2 + x_3 + x_4 + x_5 \\ -2 + x_1 - 2x_2 + 2x_3 - 2x_4 + 2x_5 \end{pmatrix}.$$

$$= \begin{pmatrix} -3+1+1+0.5+1+0.5 \\ -2+1-2(1)+2(0.5)-2(1)+2(0.5) \end{pmatrix}.$$

$$= \begin{pmatrix} 1 \\ -3 \end{pmatrix}.$$

Continue to step 3.

Step 3. Compute Stepsize. Compute stepsize ρ^0; set $\pi_0^1 \leftarrow \max\{\pi_0^0 + \rho^0 s^0,\ 0\}$;

$$\text{Set } \rho^0 \leftarrow 1.$$

$$\begin{aligned} \pi_{01}^1 &= \max\{\pi_1^0 + \rho^0 s_1^0,\ 0\} \\ &= \max\{0 + 1(1),\ 0\} \\ &= 1. \\ \pi_{02}^1 &= \max\{\pi_2^0 + \rho^0 s_2^0,\ 0\} \\ &= \max\{0 + 1(-3),\ 0\} \\ &= 0. \\ \therefore \pi_0^1 &= \begin{pmatrix} 1 \\ 0 \end{pmatrix}. \end{aligned}$$

Set $k \leftarrow 1$ and return to step 1:

Step 1. Compute Subgradient. Given π_0^1, calculate subgradient s^1 of the function $D(.)$ at π_0^1.

Solve subproblems:

$$\begin{aligned} D_1(\pi_0^1) &= -1 - \frac{2}{3}(0) + \min_{x_1 \in X_1} \{-5x_1 + (1+0)x_1\}. \\ &= -1 + \min_{x_1 \in X_1} \{-4x_1\}. \end{aligned}$$

Solution: $x_1 = 1,\ D_1(\pi_0^1) = -5.$

$$\begin{aligned} D_2(\pi_0^1) &= -1 - \frac{2}{3}(0) + \min_{(x_2,x_3) \in X_2} \{-3x_2 - 5x_3 + (1-2(0))x_2 + (1+2(0))x_3\}. \\ &= -1 + \min_{(x_2,x_3) \in X_2} \{-2x_2 - 4x_3\}. \end{aligned}$$

Solution: $(x_2, x_3)^\top = (1, 0.5),\ D_2(\pi_0^1) = -5.$

$$D_3(\pi_0^1) = -1 - \frac{2}{3}(0) + \underset{(x_4,x_5)\in X_3}{\text{Min}} \{-4x_4 - 8x_5 + (1-2(0))x_4 + (1+2(0))x_5\}.$$
$$= -1 + \underset{(x_4,x_5)\in X_3}{\text{Min}} \{-3x_4 - 7x_5\}.$$

Solution: $(x_4, x_5)^\top = (1, 0.5)$, $D_3(\pi_0^1) = -7.5$.

$$\sum_{t=1}^{3} D_t(\pi_0^1) = -5 - 5 - 7.5 = -17.5.$$

Step 2. **Termination.** Compute

$$s^1 = \begin{pmatrix} -3 + x_1 + x_2 + x_3 + x_4 + x_5 \\ -2 + x_1 - 2x_2 + 2x_3 - 2x_4 + 2x_5 \end{pmatrix}.$$
$$= \begin{pmatrix} -3 + 1 + 1 + 0.5 + 1 + 0.5 \\ -2 + 1 - 2(1) + 2(0.5) - 2(1) + 2(0.5) \end{pmatrix}.$$
$$= \begin{pmatrix} 1 \\ -3 \end{pmatrix}.$$

Continue to step 3.

Step 3. **Compute Stepsize.** Compute stepsize ρ^1; set $\pi_0^2 \leftarrow \max\{\pi_0^1 + \rho^1 s^1, 0\}$;

$$\text{Set } \rho^1 \leftarrow \frac{\rho^0}{2} = \frac{1}{2}.$$
$$\pi_{01}^2 = \max\{\pi_1^1 + \rho^1 s_1^1, 0\}$$
$$= \max\{1 + \frac{1}{2}(1), 0\}$$
$$= 1.5$$
$$\pi_{02}^2 = \max\{\pi_2^1 + \rho^1 s_2^1, 0\}$$
$$= \max\{0 + \frac{1}{2}(-3), 0\}$$
$$= 0.$$
$$\therefore \pi_0^2 = \begin{pmatrix} 1.5 \\ 0 \end{pmatrix}.$$

Set $k \leftarrow 2$ and return to step 1:

Step 1. **Compute Subgradient.** Given π_0^2, calculate subgradient s^2 of the function $D(.)$ at π_0^2.

Solve subproblems:

$$\begin{aligned} D_1(\pi_0^2) &= -1.5 - \frac{2}{3}(0) + \underset{x_1 \in X_1}{\text{Min}} \{-5x_1 + (1.5+0)x_1\}. \\ &= -1.5 + \underset{x_1 \in X_1}{\text{Min}} \{-3.5x_1\}. \end{aligned}$$

Solution: $x_1 = 1,\ D_1(\pi_0^2) = -5.$

$$\begin{aligned} D_2(\pi_0^2) &= -1.5 - \frac{2}{3}(0) + \underset{(x_2,x_3)\in X_2}{\text{Min}} \{-3x_2 - 5x_3 + (1.5 - 2(0))x_2 \\ &\qquad + (1.5 + 2(0))x_3\}. \\ &= -1.5 + \underset{(x_2,x_3)\in X_2}{\text{Min}} \{-1.5x_2 - 3.5x_3\}. \end{aligned}$$

Solution: $(x_2, x_3)^\top = (1, 0.5),\ D_2(\pi_0^2) = -4.75.$

$$\begin{aligned} D_3(\pi_0^2) &= -1.5 - \frac{2}{3}(0) + \underset{(x_4,x_5)\in X_3}{\text{Min}} \{-4x_4 - 8x_5 + (1.5 - 2(0))x_4 \\ &\qquad + (1.5 + 2(0))x_5\}. \\ &= -1.5 + \underset{(x_4,x_5)\in X_3}{\text{Min}} \{2.5x_4 - 6.5x_5\}. \end{aligned}$$

Solution: $(x_4, x_5)^\top = (1, 0.5),\ D_3(\pi_0^2) = -7.25.$

$$\sum_{t=1}^{T} D_t(\pi_0^2) = -5 - 4.75 - 7.25 = -17.$$

Step 2. Termination. Compute

$$\begin{aligned} s^2 &= \begin{pmatrix} -3 + x_1 + x_2 + x_3 + x_4 + x_5 \\ -2 + x_1 - 2x_2 + 2x_3 - 2x_4 + 2x_5 \end{pmatrix}. \\ &= \begin{pmatrix} -3 + 1 + 1 + 0.5 + 1 + 0.5 \\ -2 + 1 - 2(1) + 2(0.5) - 2(1) + 2(0.5) \end{pmatrix}. \\ &= \begin{pmatrix} 1 \\ -3 \end{pmatrix}. \end{aligned}$$

Continue to step 3.

Step 3. Compute Stepsize. Compute stepsize ρ^2; set $\pi_0^3 \leftarrow \max\{\pi_0^2 + \rho^2 s^2,\ 0\}$;

$$\text{Set } \rho^2 \leftarrow \frac{\rho^1}{2} = \frac{1}{4}.$$

$$\begin{aligned}
\pi_{01}^3 &= \max\{\pi_1^2 + \rho^2 s_1^2,\ 0\} \\
&= \max\{1.5 + \frac{1}{4}(1),\ 0\} \\
&= 1.75. \\
\pi_{02}^3 &= \max\{\pi_2^2 + \rho^2 s_2^2,\ 0\} \\
&= \max\{0 + \frac{1}{4}(-3),\ 0\} \\
&= 0. \\
\therefore \pi_0^3 &= \begin{pmatrix} 1.75 \\ 0 \end{pmatrix}.
\end{aligned}$$

Set $k \leftarrow 3$ and return to step 1:

---

end

The algorithm will run for several iterations until $\pi_0^k \approx \begin{pmatrix} 5 \\ 0 \end{pmatrix}$ at which point there will be no further significant change in π_0^k and the algorithm can be terminated. Notice that the summation $\sum_{t=1}^{T} D_t(\pi_0^k)$ will continue to increase as expected until $\sum_{t=1}^{T} D_t(\pi_0^k) = -14.5$, which corresponds to the optimal value with optimal solution $x_1 = 1, x_2 = 0.5, x_3 = 0, x_4 = 1, x_5 = 0.5$.

Problems

5.1. Kelley's Cutting-Plane Method
Use Kelley's algorithm to solve $\varphi(x) = (x - 1.5)^2 + 1$ with $X = \{-1 \leq x \leq 4\}$. Use $x = -1$ as your initial point. At each iteration k of the algorithm, state x^k, $f_k(x^k)$, and $f(x^k)$. Finally, plot $f(x)$ and $f_k(x^k)$ for all k on a single 2D graph.

5.2. Kelley's Cutting-Plane Method
Use Kelley's algorithm to solve $f(x) = (x - 2)^2 + 1$ with $X = \{-1 \leq x \leq 4\}$. Use $x = -1$ as your initial point. At each iteration k of the algorithm, state x^k, $f_k(x^k)$, and $f(x^k)$. Finally, plot $f(x)$ and $f_k(x^k)$ for all k on a single 2D graph.

5.3. Kelley's Cutting-Plane Method
Use Kelley's algorithm to solve $f(x) = f(x_1, x_2) = (x_1 - 1.5)^2 + (x_2 - 1)^2 + 1$ with $X = \{-1 \leq x_1 \leq 4, -1 \leq x_2 \leq 3\}$. Use $x = (x_1, x_2) = (-1, -1)$ as your initial point. At each iteration k of the algorithm, state x^k, $f_{k-1}(x^k)$, and $f(x^k)$. Finally, plot $f(x)$ and $f_k(x^k)$ for all k on a single 3D graph.

5.4. Kelley's Cutting-Plane Method
Use Kelley's algorithm to solve $f(x) = f(x_1, x_2) = (x_1 - 2)^2 + (x_2 - 1)^2 + 1$ with $X = \{-1 \leq x_1 \leq 4, -1 \leq x_2 \leq 4\}$. Use $x = (x_1, x_2) = (-1, -1)$ as your initial point. At each iteration k of the algorithm, state x^k, $f_{k-1}(x^k)$, and $f(x^k)$. Finally, plot $f(x)$ and $f_k(x^k)$ for all k on a single 3D graph.

5.5. Benders Decomposition
Use Benders decomposition to decompose the following problem into a master program and a single subproblem:

$$\begin{aligned}
\text{Min } & 15x + 2y_1 - y_2 + 2y_3 + 2y_4 - y_5 - y_6 \\
\text{s.t. } & x \leq 3 \\
& 3x + y_1 - y_2 \geq 3 \\
& 4x + y_3 \geq 5 \\
& 3x + y_4 - y_5 \geq 4 \\
& 4x - y_6 \geq -1 \\
& x,\ y_1,\ y_2,\ y_3,\ y_4,\ y_5,\ y_6 \geq 0.
\end{aligned}$$

Apply Benders decomposition algorithm to your decomposed problem starting with $x = 0$. At each iteration k of the algorithm, clearly state x^k, the master program with all the cuts, subproblem and its primal and dual solution, lower bound ℓ, and upper bound u.

5.6. Benders Decomposition
Use Benders decomposition to decompose the formulation in Problem 5.5 into a master program and two subproblems, one involving decision variables y_1, y_2 and y_3, and the other with y_4, y_5 and y_6. Now apply Benders decomposition algorithm to solve your decomposed problem starting with $x^0 = 0$. At each iteration k of the algorithm, clearly state x^k, the master program with all the cuts, subproblems and their primal and dual solutions, lower bound ℓ, and upper bound u.

5.7. Dantzig–Wolfe Decomposition
Apply Dantzig–Wolfe decomposition (column generation) to decompose the problem below into a master program and a single subproblem:

$$\begin{aligned}
\text{Min } & -3x_1 - 5x_2 - 4x_3 + 1x_4 \\
\text{s.t. } & 3x_1 + 4x_2 + 3x_3 + 4x_4 = 15
\end{aligned}$$

$$
\begin{aligned}
1 \leq x_1 &\leq 2 \\
0 \leq x_2 &\leq 2 \\
1 \leq x_3 &\leq 2 \\
1 \leq x_4 &\leq 3.
\end{aligned}
$$

Use Dantzig–Wolfe decomposition algorithm to solve your decomposed problem. At each iteration k of the algorithm, clearly state the *restricted* master program (with all the columns generated up to k), current solution, and the subproblem and its solution.

5.8. Dantzig–Wolfe Decomposition
Apply Dantzig–Wolfe decomposition (column generation) to decompose the problem below into a master program and a single subproblem:

$$
\begin{aligned}
\text{Min } & -3x_1 - 5x_2 - 4x_3 + 1x_4 \\
\text{s.t. } & 3x_1 + 4x_2 + 3x_3 + 4x_4 = 15 \\
& 1 \leq x_1 \leq 2 \\
& 0 \leq x_2 \leq 2 \\
& 1 \leq x_3 \leq 2 \\
& x_4 \geq 1.
\end{aligned}
$$

Use Dantzig–Wolfe decomposition algorithm to solve your decomposed problem. At each iteration k of the algorithm, clearly state the *restricted* master program (with all the columns generated up to k), current solution, and the subproblem and its solution.

5.9. Dantzig–Wolfe Decomposition
Apply Dantzig–Wolfe decomposition (column generation) to decompose the problem below into a master program and two subproblems, one with x_1 and x_2, and the other with x_3 and x_4:

$$
\begin{aligned}
\text{Min } & -3x_1 - 5x_2 - 4x_3 + 1x_4 \\
\text{s.t. } & 3x_1 + 4x_2 + 3x_3 + 4x_4 \leq 15 \\
& 1 \leq x_1 \leq 2 \\
& 0 \leq x_2 \leq 2 \\
& 1 \leq x_3 \leq 2 \\
& x_4 \geq 1.
\end{aligned}
$$

Use Dantzig–Wolfe decomposition algorithm to solve your decomposed problem. At each iteration k of the algorithm, clearly state the *restricted master program* (with all the columns generated so far) and its solution, and the subproblems and their solutions.

5.10. Lagrangian (Dual) Decomposition

Apply Lagrangian relaxation (dual decomposition) to dualize the formulation in Problem 5.7 by placing the first constraint (complicating constraint) into the objective. Now use a subgradient optimization algorithm to solve the problem. At each iteration k of the algorithm, clearly state the master program and its solution, subproblems and their solutions, lower bound ℓ, and upper bound u.

5.11. Lagrangian (Dual) Decomposition

Consider the following problem:

$$
\begin{aligned}
\text{Min } & 7.5x_1 + 7.5x_2 + 2y_1 - y_2 + 2y_3 + 2y_4 - y_5 - y_6 \\
\text{s.t. } & x_1 - x_2 = 0 \\
& x_1 \leq 3 \\
& 3x_1 + y_1 - y_2 \geq 3 \\
& 4x_1 + y_3 \geq 5 \\
& x_2 \leq 3 \\
& 3x_2 + y_4 - y_5 \geq 4 \\
& 4x_2 - y_6 \geq -1 \\
& x_1,\ x_2,\ y_1,\ y_2,\ y_3,\ y_4,\ y_5,\ y_6 \geq 0.
\end{aligned}
$$

Apply Lagrangian relaxation (dual decomposition) to solve the problem as follows: Dualize the first constraint and create two subproblems, one involving decision variables x_1, y_1, y_2 and y_3, and the other with x_2, y_4, y_5 and y_6. Use a subgradient optimization algorithm to solve the decomposed problem. At each iteration k of the algorithm, clearly state the master program and its solution, subproblems and their solutions, lower bound ℓ, and upper bound u.

Note: The formulation given in this problem is equivalent to that in Problem 5.5.

5.12. Benders decomposition Algorithm Implementation

Implement (code) Benders decomposition algorithm using your choice of software and LP solver. Test your code on the standard SLP test instances by solving the "CORE" problem of each test instance. Report the computation time, the number of iterations, optimal solution, and optimal value. Solve the core problem using a direct solver to verify your solutions, and compare the computation time with using Benders decomposition. For each test instance, plot a convergence plot showing the objective value (y-axis) versus the iteration number (x-axis).

5.13. Regularized Benders Decomposition Algorithm Implementation
Implement (code) regularized Benders decomposition algorithm using your choice of software and LP and quadratic programming solver. Perform computational experiments as in Problem 5.12. Compare and contrast the Benders and regularized Benders decomposition algorithms.

5.14. Dantzig–Wolfe Decomposition Algorithm Implementation
Implement (code) Dantzig–Wolfe decomposition algorithm using your choice of software and LP solver. Test your code on the standard SLP test instances by solving the dual to the "CORE" problem of each test instance. Report the computation time, the number of iterations, optimal solution, and optimal value. Solve the dual to the core problem using a direct solver to verify your solutions, and compare the computation time with using Benders decomposition. For each test instance, plot a convergence plot showing the objective value (y-axis) versus the iteration number (x-axis).

5.15. Lagrangian Decomposition Algorithm Implementation
Implement (code) the Lagrangian decomposition algorithm using your choice of software and LP solver. Perform computational experiments as in Problem 5.14. Compare and contrast the Dantzig–Wolfe and Lagrangian decomposition algorithms.

References

1. J. F. Benders. Partitioning procedures for solving mixed-variable programming problems. *Numerische Mathematik*, 54:238–252, 1962.
2. G.B. Dantzig and P. Wolfe. The decomposition algorithm for linear programs. *Econometrica*, 29(4):767–778, 1961.
3. J.E. Kelley. The cutting-plane method for solving convex programs. *Journal of the Society of Industrial and Applied Mathematics*, 8(4):703–712, 1960.
4. K.C. Kiwiel. Methods of descent for nondifferentiable optimization. In *Lecture Notes in Mathematics No. 1133*. Springer-Verlag, Berlin, 1985.
5. R.T. Rockafellar. Monotone operators and the proximal point algorithm. *SIAM Journal on Control and Optimization*, 14:877–898, 1976.
6. A. Ruszczyn'ski. A regularized decomposition method for minimizing a sum of polyhedral functions. *Mathematical Programming*, 35:309–333, 1986.

Chapter 6
Risk-Neutral Stochastic Linear Programming Methods

6.1 Introduction

Derivation of algorithms for solving two-stage risk-neutral stochastic linear programming (RN-SLP) problems requires understanding the *structural properties* of the model and the nature of the recourse function. We derive these structural properties (e.g., dual block angular structure, convexity, etc.) in Chaps. 2 and 5. In this chapter we shift our focus to decomposition methods for two-stage RN-SLP, starting with the classical L-shaped method and the multicut method. We then move on to cover the adaptive multicut method. Recall that for a large number of scenarios the deterministic equivalent problem (DEP) can be very large and even impossible to solve using a direct solver. Therefore, we need to exploit *problem structure* and decompose the problem into smaller problems which can be solved. The basic idea is to decompose the DEP into a master program and subproblems based on the scenarios. Then through some coordination and iterative process, we can use the solutions from the master and subproblems to compute the optimal solution to the original problem.

In *stage-wise* decomposition of two-stage RN-SLP, the first stage involves only the here-and-now decision variable vector $x \in \mathbb{R}^{n_1}_+$, while the second stage has the recourse decision variable vector $y(\tilde{\omega}) \in \mathbb{R}^{n_2}_+$. The multivariate random variable $\tilde{\omega}$ is defined on a probability space $(\Omega, \mathscr{A}, \mathbb{P})$. The realization (scenario) of $\tilde{\omega}$ is denoted by ω, $\omega \in \Omega$. The random cost based on $\tilde{\omega}$ is represented by the random cost function $f(x, \tilde{\omega})$. The decision-making process in this setting is illustrated in Fig. 6.1. In the first stage, we make the decision x without full information on the future realization ω of $\tilde{\omega}$. Then in the second stage, we make a "corrective" action $y(\omega)$ based on both the decision x that we made in the first stage and the realization ω, which becomes known in the second stage. The recourse decision $y(\omega)$ is only committed after ω becomes known. This means that the decision $y(\omega)$ adapts to a given scenario ω. Essentially, the two-stage recourse decision-making

L. Ntaimo, *Computational Stochastic Programming*, Springer Optimization and Its Applications 774, https://doi.org/10.1007/978-3-031-52464-6_6

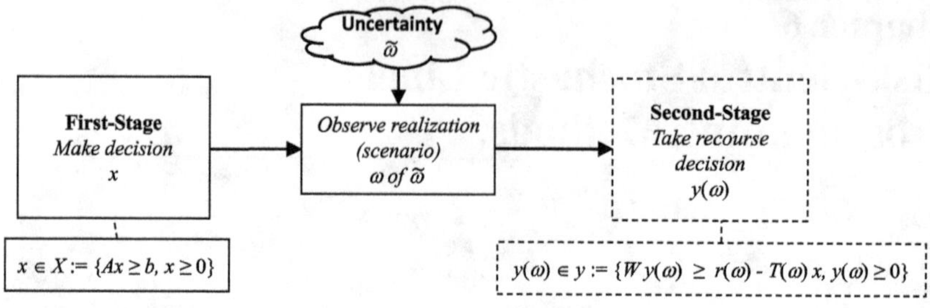

Fig. 6.1 The two-stage recourse decision-making process

process enables the determination of x while taking into account all possible future scenarios, where the future is represented by a probability distribution.

Let us restate the two-stage RN-SLP (recourse model) from Chap. 2 as follows:

$$\underset{x \in X}{\text{Min}}\ \mathbb{E}[f(x, \tilde{\omega})], \tag{6.1}$$

where $\mathbb{E} : \mathscr{F} \mapsto \mathbb{R}$ denotes the expected value. We focus on the risk-neutral case in this chapter and deal with the risk-averse case in Chap. 6. The set $X = \{Ax \geq b,\ x \geq 0\}$ is a nonempty polyhedron that defines the set of first-stage feasible solutions. The matrix $A \in \mathbb{R}^{m_1 \times n_1}$ and the vector $b \in \mathbb{R}^{m_1}$ are the first-stage matrix and right hand side vector, respectively. The family of real random cost variables $\{f(x, \tilde{\omega})\}_{x \in X} \subseteq \mathscr{F}$ is defined on $(\Omega, \mathscr{A}, \mathbb{P})$, where $\mathscr{F}$ is the space of all real random cost variables $f : \Omega \mapsto \mathbb{R}$ satisfying $\mathbb{E}[|f(\tilde{\omega})|] < \infty$. For a given $x \in X$, the real random cost variable $f(x, \tilde{\omega})$ is given by

$$f(x, \tilde{\omega}) := c^\top x + \varphi(x, \tilde{\omega}). \tag{6.2}$$

For a given realization ω of $\tilde{\omega}$, the *recourse function* $\varphi(x, \omega)$ is given by

$$\begin{aligned} \varphi(x, \omega) := &\text{Min}\ q(\omega)^\top y(\omega) \\ &\text{s.t. } Wy(\omega) \geq r(\omega) - T(\omega)x \\ &\qquad y(\omega) \geq 0, \end{aligned} \tag{6.3}$$

where $q(\omega) \in \mathbb{R}^{n_2}$ is the second-stage cost vector and $y(\omega) \in \mathbb{R}^{n_2}_+$ is the recourse decision. The matrix $W \in \mathbb{R}^{m_2 \times n_2}$ is the *recourse matrix*, $T(\omega) \in \mathbb{R}^{m_2 \times n_1}$ is the *technology matrix*, and $r(\omega) \in \mathbb{R}^{m_2}$ is the right hand side vector. By *scenario* ω we mean the realization of the stochastic problem data, i.e., $\omega := (q(\omega), T(\omega), r(\omega))$. Since W does not depend on ω in Problem (6.3), we say that the problem has *fixed recourse*. In the literature on stochastic programming, the term $T(\omega)x$ is often referred to as the *tender variable*. Therefore, if $T(\omega) = T$ for all $\omega \in \Omega$, we say that Problem (6.3) has *fixed* tender variables.

To ensure that Problem (6.1) is well-defined for *computational* purposes, we made the following assumptions in Chap. 2:

(A1) The multivariate random variable $\tilde{\omega}$ is discretely distributed with finitely many scenarios $\omega \in \Omega$, each with the probability of occurrence $p(\omega)$.
(A2) For all $x \in X$, $\{Wy(\omega) \geq r(\omega) - T(\omega)x,\ y(\omega) \geq 0\} \neq \emptyset$.

Assumption (A1) is needed to make the problem tractable. Assumption (A2) is the *relatively complete recourse* assumption (A2), which guarantees the feasibility of the second-stage problem for every $x \in X$. This assumption implies that $\varphi(x, \omega) < \infty$ with probability one for all $x \in X$. Problem (6.1) is said to have *complete recourse* if $\varphi(x, \omega) < \infty$ with probability one for all $x \in \mathbb{R}_+^{n_1}$. It is often not necessary to require complete or relatively complete recourse since the Benders decomposition [1] framework allows for the second stage to be infeasible. So we shall relax assumption (A2) in this chapter and instead consider generating *feasibility cuts* in the algorithms we derive for every $x \in X$ that leads to infeasibility in the second stage. The feasibility cuts will aim to restrict the set X to feasible values of x. So as a convention, we continue to have $\varphi(x, \omega) = +\infty$ if Problem (6.3) is infeasible and $\varphi(x, \omega) = -\infty$ if Problem (6.3) is unbounded. Next, we begin with the derivation of the classical L-shaped algorithm and then give a detailed numerical example to illustrate the algorithm.

6.2 The L-Shaped Method

Historically, SP with recourse (recourse model) can be traced back to Dantzig [5]. We saw in Chap. 5 how the development of algorithms progressed, starting with Kelley's method [8] in 1960. This was followed by the Dantzig–Wolfe decomposition method [6] in 1961 and then Benders decomposition method [1] in 1962. All these three methods deal with deterministic problems. The L-shaped method [14] for RN-SLP appears later in 1969. Let us restate MR-SLP (6.1) for $\lambda := 0$ as follows:

$$\begin{aligned} \text{Min}\ & c^\top x + \mathbb{E}[\varphi(x, \tilde{\omega})] \\ \text{s.t.}\ & Ax \geq b \\ & x \geq 0, \end{aligned} \tag{6.4}$$

where $Q(x) = \mathbb{E}[\varphi(x, \tilde{\omega})]$, and for each realization ω of $\tilde{\omega}$, the recourse function is given by (6.3). Now let $\pi \in \mathbb{R}^{m_2}$ denote the dual multipliers associated with the constraints of subproblem (6.3). Then by LP *weak duality* we have

$$\begin{aligned} \varphi(x,\omega) \geq \ & \text{Max}\ \pi(\omega)^\top(r(\omega) - T(\omega)x) \\ & \text{s.t.}\ W^\top \pi(\omega) \leq q(\omega) \\ & \pi(\omega) \geq 0. \end{aligned} \tag{6.5}$$

Let $\Pi(\omega) = \{\pi(\omega) \mid W^\top \pi(\omega) \leq q(\omega)\}$ denote the subproblem dual feasible set. Notice that if $q(\omega) = q$ for all $\omega \in \Omega$, then $\Pi(\omega) = \Pi$ remains fixed regardless of the choice of x . This property is exploited in the *stochastic decomposition* algorithm of Higle and Sen [7], which we discuss later in Chap. 7 when we deal with statistical methods for SLP.

In Chap. 2, we established that the expected recourse function $Q(x)$ is *convex* over its effective domain. We also continue to consider Problem (6.1) under assumptions (A1) requiring having a discrete probability distribution for $\tilde{\omega}$ to allow for a computationally tractable problem. Nevertheless, if the probability distribution is continuous, then one can create a discrete approximation to achieve this assumption. It has been shown that this can be done with some given level of accuracy [13].

To simplify notation in stating some of the algorithms, let us index the scenarios by $s = 1, \cdots, S,\ S = |\Omega| < \infty$ such that $\Omega := \{\omega^s\}_{s=1}^S$, with each scenario ω^s having probability of occurrence $p(\omega^s)$. Then Problem (6.1) can be rewritten as the following large-scale deterministic equivalent problem (DEP):

$$\begin{aligned} & \text{Min}\ c^\top x + \sum_{s=1}^{S} p(\omega^s) q(\omega^s)^\top y(\omega^s) \\ & \text{s.t.}\ \ Ax \qquad\qquad \geq b \\ & \qquad T(\omega^s)x + Wy(\omega^s) \geq r(\omega^s),\ s = 1, \cdots, S \\ & \qquad\quad x, \qquad y(\omega^s) \geq 0,\ \ s = 1, \cdots, S. \end{aligned}$$

The data for scenario ω^s is given by $(q(\omega^s), T(\omega^s), r(\omega^s))$ for all $s = 1, \cdots, S$. The L-shaped method is named after the "L" shape (dual block angular) structure of the above extensive formulation (see Fig. 6.2). The L-shaped method is in fact an extension of Benders decomposition to the stochastic case.

6.2.1 Decomposition Approach

Now let for a given scenario ω^s the recourse function be denoted by $\varphi(x, \omega^s)$. Then, in terms of *Benders decomposition*, the large-scale LP is

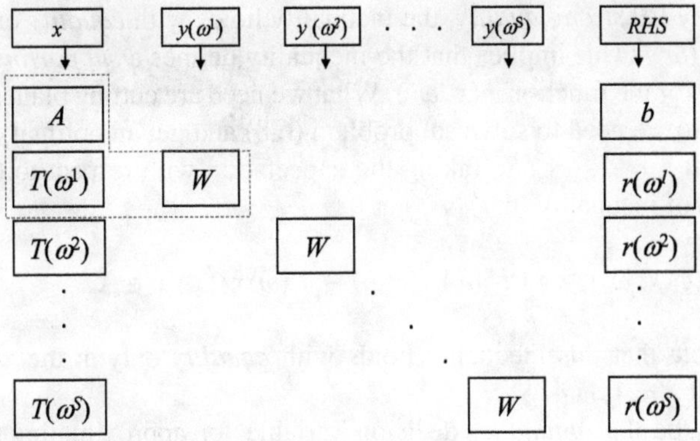

Fig. 6.2 The L-shape or the dual block angular structure

$$
\begin{aligned}
\text{Min } & c^\top x + Q(x) \\
\text{s.t. } & Ax \geq b \\
& x \geq 0,
\end{aligned}
\tag{6.6}
$$

where for $s = 1, \cdots, S$, $Q(x) = \mathbb{E}[\varphi(x, \tilde{\omega})] = \sum_{s=1}^{S} p(\omega^s)\varphi(x, \omega^s)$, and

$$
\begin{aligned}
\varphi(x, \omega^s) = \text{Min } & q(\omega^s)^\top y(\omega^s) \\
\text{s.t. } & Wy(\omega^s) \geq r(\omega^s) - T(\omega^s)x \\
& y(\omega^s) \geq 0.
\end{aligned}
\tag{6.7}
$$

Observe that the recourse function $\varphi(x, \omega^s)$ allows for the ability to adapt to the scenario ω^s. Following Benders decomposition (Chap. 5), we need cutting planes for $Q(x)$. Let $\pi(\omega^s)$ be the dual multipliers associated with constraints of subproblem (6.7). Then the dual to the subproblem is

$$
\begin{aligned}
\varphi(x, \omega^s) \geq \text{ Max } & \pi(\omega^s)^\top (r(\omega^s) - T(\omega^s)x) \\
\text{s.t. } & W^\top \pi(\omega^s) \leq q(\omega^s) \\
& \pi(\omega^s) \geq 0.
\end{aligned}
\tag{6.8}
$$

Let the optimal solution to dual subproblem (6.8) be $\pi^*(\omega^s)$. Keep in mind that actually $\pi^*(\omega^s) := \pi^*(x, \omega^s)$ because the dual solution depends on the first-stage decision x. Nevertheless, we shall continue to use $\pi^*(\omega^s)$ in what follows. Letting $X = \{x \mid Ax \geq b,\ x \geq 0\}$, we have

$$
\varphi(x, \omega^s) \geq \pi^*(\omega^s)^\top (r(\omega^s) - T(\omega^s)x), \ \forall x \in X. \tag{6.9}
$$

Notice that, by LP *strong duality*, the inequality holds with *equality* only at the x's that let to $\pi^*(\omega^s)$. This implies that the inequality defines a *supporting hyperplane* (see Chap. 1) for the function $\varphi(x, \omega^s)$. What we need are cutting planes for $Q(x) = \mathbb{E}[\varphi(x, \tilde{\omega})]$. So we need to solve subproblem (6.8) and get an optimal dual solution $\pi^*(\omega^s)$ for all $s = 1, \cdots, S$. Taking the expectation with respect to the scenarios on both sides of inequality (6.9), we get

$$\mathbb{E}[\varphi(x, \tilde{\omega})] \geq \mathbb{E}[\pi^*(\tilde{\omega})^\top (r(\tilde{\omega}) - T(\tilde{\omega})x)], \ \forall x \in X. \tag{6.10}$$

Again, we note that this inequality holds with *equality* only at the x's that let to $\pi^*(\omega^s)$ for all $s = 1, \cdots, S$.

Now let η be the *optimality* decision variable for approximating the value of $Q(x) := \mathbb{E}[\varphi(x, \tilde{\omega})]$. Furthermore, let

$$\beta^0 = \mathbb{E}[\pi^*(\tilde{\omega})^\top r(\tilde{\omega})] = \sum_{s=1}^{S} p(\omega^s)\pi^*(\omega^s)^\top r(\omega^s)$$

and

$$\beta^\top = \mathbb{E}[\pi^*(\tilde{\omega})^\top T(\tilde{\omega})] = \sum_{s=1}^{S} p(\omega^s)\pi^*(\omega^s)^\top T(\omega^s).$$

Then, using inequality (6.10), we get the following:

$$\eta + \beta^\top x \geq \beta^0, \ \forall x \in X. \tag{6.11}$$

Inequality (6.11) is known as the *optimality cut* and is a supporting hyperplane of the expected recourse function $Q(x)$.

In the case when assumption (A2) does *not* hold, subproblem (6.7) can *infeasible*. If the subproblem is infeasible for some s, then we can get the dual extreme ray μ_s^* associated with the subproblem constraints and set

$$(\mu_s^*)^\top (r(\omega^s) - T(\omega^s)x) \leq 0, \ \forall x \in X.$$

By letting $\beta^0 = (\mu_s^*)^\top r(\omega^s)$ and $\beta^\top = (\mu_s^*)^\top T(\omega^s)$, we obtain the following inequality:

$$\beta^\top x \geq \beta^0, \ \forall x \in X. \tag{6.12}$$

This inequality is called the *feasibility cut* and cuts off a part of the first-stage feasible set X to get rid of the x's leading to the infeasibility of the subproblem for scenario ω^s.

We now have the theoretical results we need in order to state the L-shaped algorithm. In the next subsections, we give a concise statement of the algorithm with a goal of using it as a "blueprint" for implementing (coding) the algorithm on the computer. This is followed by suggestions of how to code certain parts of the algorithm for efficient running codes.

6.2.2 The L-Shaped Algorithm

Let k be the algorithm iteration counter and let ℓ and u denote the lower and upper bounds, respectively, on the optimal value during the course of the algorithm. We can formally state the L-shaped algorithm as follows:

Algorithm *L-shaped*
begin

Step 0. Initialization. $k \leftarrow 0$, x^0 given, $\ell \leftarrow -\infty$, $u \leftarrow \infty$, and $\epsilon > 0$. x^0 can be obtained as follows: $x^0 \leftarrow \text{argmin}\{c^\top x \mid Ax \geq b, x \geq 0\}$.

Step 1. Solve Subproblems and Generate Cut. For $s = 1, \cdots, S$ solve

$$\varphi^k(x^k, \omega^s) := \text{Min}\{q^\top(\omega^s)y | Wy \geq r(\omega^s) - T(\omega^s)x^k, y \geq 0\}.$$

If infeasible for some s, generate a *feasibility* cut:

Get dual extreme ray μ_s^k.
Compute $\beta_k^0 \leftarrow (\mu_s^k)^\top r$ and $\beta_k^\top \leftarrow (\mu_s^k)^\top T(\omega^s)$.
Go to step 2.

else if feasible for all s, generate an *optimality* cut:

Get dual solutions π_s^k.
Compute $\beta_k^0 \leftarrow \sum_{s=1}^S p(\omega^s)(\pi_s^k)^\top r(\omega^s)$ and $\beta_k^\top \leftarrow \sum_{s=1}^S p(\omega^s)(\pi_s^k)^\top T(\omega^s)$.
Compute *upper bound*:

Set $u^k \leftarrow c^\top x^k + \sum_{s=1}^S p(\omega^s)\varphi^k(x^k, \omega^s)$.
Set $u \leftarrow \min\{u^k, u\}$.

If u is updated, set *incumbent* solution to $x^* \leftarrow x^k$.

Step 2. Add Cut to Master Program and Solve. If some subproblem was infeasible

Add $\beta_k^\top x \geq \beta_k^0$ to master problem.

else

Add $\beta_k^\top x + \eta \geq \beta_k^0$ to master problem.

Solve master problem

$$
\begin{aligned}
\ell^{k+1} := \text{Min}\ & c^\top x + \eta \\
\text{s.t.}\ & Ax \geq b \\
& \beta_t^\top x + \eta \geq \beta_t^0,\ t \in \Theta_k \\
& \beta_t^\top x \geq \beta_t^0,\ t \notin \Theta_k \\
& x \geq 0
\end{aligned}
\tag{6.13}
$$

and get solution (x^{k+1}, η^{k+1}).
Set $\ell \leftarrow \max\{\ell^{k+1}, \ell\}$.

Step 3. Termination. If $u - \ell \leq \epsilon|u|$

Stop, x^* is ϵ-optimal.

else

Set $k \leftarrow k + 1$.

Return to Step 1.

---

end

To implement the L-shaped algorithm using a computer software programming package, we must understand how to generate feasibility cuts and how to initialize the optimality cut variable in the master problem. We consider two options for generating a feasibility cut in step 1 of the L-shaped algorithm at a given iteration. The first option, which we take in the stated L-shaped algorithm, is to find the first infeasible subproblem s and generate a feasibility cut by setting $\beta^0 := p(\omega^s)\pi(\omega^s)^\top r(\omega^s)$ and $\beta^\top := p(\omega^s)\pi(\omega^s)^\top T(\omega^s)$. We then add the cut to the master problem. The second option is to find all infeasible subproblems at each iteration, generate a feasibility cut for each one, and then add the cuts to the master program. The first option has the advantage that you can quickly end the current iteration by generating a single feasibility cut which can potentially lead to a feasible x^{k+1}. However, sometimes the problem instance may require generating multiple feasibility cuts at a given iteration to be able to get a feasible first-stage solution. This means that under the first option you would require multiple iterations before generating a feasible solution, and in this case, the second option would be computationally advantageous. In any event, which option is better is not known *a priori* as it is problem instance dependent. One can decide which option to use through experimentation.

In the first iteration of the L-shaped algorithm, the master program can be initialized as follows:

$$
\begin{aligned}
\text{Min}\ & c^\top x \\
\text{s.t.}\ & Ax \geq b \\
& x \geq 0.
\end{aligned}
$$

This problem is solved to get x^0 if it is not given. Otherwise, if x^0 is given, then it is not necessary to solve the given problem in step 0. Often the modeler familiar with the problem instance can know feasible solutions to the problem and can specify x^0. Initializing the master program without the η variable is necessary because in some problem instances at least one of the subproblems is infeasible in the first few iterations and feasibility cuts have to be generated. Thus if η is added at the beginning, it can lead to an unbounded master problem. This requires that η be added to the master program (6.13) *only* after the first optimality cut has been generated. If η is added at the beginning, however, then it should be set to zero in the constraints and only set free (unrestricted in sign) after the first optimality cut is generated and added to the master problem.

To efficiently create a feasibility or optimality cut in the code in step 2, the cuts should be generated "on the fly." This can be accomplished using the following steps:

- ◇ Let $\beta^\top$ be a one-dimensional array (type *double*) and β^0 be a scalar (type *double*).
- ◇ Initialize $\beta^\top \leftarrow \mathbf{0}$ and $\beta^0 \leftarrow 0$.
- ◇ Then for all $s = 1, \cdots, S$, do the following:

Get cut coefficients and update β^0 and β as follows:

$$\begin{aligned} \beta^0 += p(\omega^s)\pi(\omega^s)^\top r(\omega^s) &\quad\equiv\quad \beta^0 \leftarrow \beta^0 + p(\omega^s)\pi(\omega^s)^\top r(\omega^s). \\ \beta^\top += p(\omega^s)\pi(\omega^s)^\top T(\omega^s) &\quad\equiv\quad \beta^\top \leftarrow \beta + p(\omega^s)\pi(\omega^s)^\top T(\omega^s). \end{aligned}$$

The operator $+=$ is the computer programming summing operator and adds to the current values in β^0 and $\beta^\top$, respectively. Therefore, we can update step 1 of the L-shaped algorithm as follows:

Begin

Step 1. Solve Subproblems and Generate Cut.

Initialize $\beta_k^\top \leftarrow \mathbf{0}$, $\beta_k^0 \leftarrow 0$, and $\bar{Q}^k \leftarrow 0$.
For all ω^s, $s = 1, \cdots, S$, solve subproblem (6.7):

If subproblem is *infeasible* for ω^s:

Get dual extreme ray $\mu^k(\omega^s)$.
Compute $\beta_k^\top \leftarrow \mu^k(\omega^s)^\top T(\omega^s)$.
Compute $\beta_k^0 \leftarrow \mu^k(\omega^s)^\top r(\omega^s)$.
Generate *feasibility* cut: $\beta_k^\top x \geq \beta_k^0$.
Go to step 2.

else

Get $\varphi(x^k, \omega^s)$ and dual solution $(\pi^k(\omega^s)$.
Compute $\bar{Q}^k \leftarrow \bar{Q}^k + p(\omega^s)\varphi(x^k, \omega^s)$.
Compute $\beta_k^\top \leftarrow \beta_k^\top + p(\omega^s)\pi^k(\omega^s)^\top T(\omega)$.
compute $\beta_k^0 \leftarrow \beta_k^0 + p(\omega^s)\pi^k(\omega^s)^\top r(\omega^s)$.

If *feasible* for all ω^s:

Generate an *optimality* cut: $\beta_k^\top x + \eta \geq \beta_k^0$.
Compute *upper bound*:

Set $u^k \leftarrow c^\top x^k + \bar{Q}^k$.
Set $u \leftarrow \min\{u^k, u\}$.

If u is updated, set *incumbent* solution $x^* \leftarrow x^k$.

---

end

Next, we give a numerical example to illustrate the L-shaped algorithm in detail.

6.2.3 Numerical Example

Example 6.1 Apply two iterations of the L-shaped algorithm to the instance of the two-stage RN-SLP *abc-Production Planning* problem described in Chap. 3.

Recall the two-stage recourse formulation of the problem:

$$\begin{array}{lllllll}
\text{Min} & 50x_1 & +30x_2 & +15x_3 & +10x_4 & + \sum_{s=1}^{5} p(\omega^s)\varphi(x,\omega^s) & \\
\text{s.t.} & -x_1 & & & & \geq & -300 \\
& & -x_2 & & & \geq & -700 \\
& & & -x_3 & & \geq & -600 \\
& & & & -x_4 & \geq & -500 \\
& & -x_2 & -x_3 & -x_4 & \geq & -1600 \\
& x_1, & x_2, & x_3, & x_4 & \geq & 0,
\end{array}$$

where for a given scenario $s = 1, \cdots, 5$ the second-stage subproblem is given as follows:

$$\begin{array}{llllll}
\varphi(x,\omega^s) := \text{Min} & -1150y_1 & -1525y_2 & -1900y_3 & & \text{Dual multipliers} \\
\text{s.t.} & -6y_1 & -8y_2 & -10y_3 \geq & -x_1 & \leftarrow \pi_1 \\
& -20y_1 & -25y_2 & -28y_3 \geq & -x_2 & \leftarrow \pi_2 \\
& -12y_1 & -15y_2 & -18y_3 \geq & -x_3 & \leftarrow \pi_3 \\
& -8y_1 & -10y_2 & -14y_3 \geq & -x_4 & \leftarrow \pi_4 \\
& -y_1 & & \geq & -d_1(\omega^s) & \leftarrow \pi_5 \\
& & -y_2 & \geq & -d_2(\omega^s) & \leftarrow \pi_6 \\
& & & -y_3 \geq & -d_3(\omega^s) & \leftarrow \pi_7 \\
& y_1, & y_2, & y_3 \geq & 0. &
\end{array}$$

Using vector notation,

$$x = (x_1, x_2, x_3, x_4)^\top,$$

$$y = (y_1, y_2, y_3)^\top$$

and the RHS

$$r(\omega^s) = (0, 0, 0, 0, d_1(\omega^s), d_2(\omega^s), d_3(\omega^s))^\top.$$

The RHS data $d_1(\omega^s)$, $d_2(\omega^s)$, and $d_3(\omega^s)$ are the demand for products a, b, and c, respectively, under scenario ω^s. Using the dual multipliers,

$$\pi = (\pi_1, \pi_2, \cdots, \pi_7)^\top,$$

the dual to the second-stage subproblem is given as follows:

$$\begin{array}{lrrrrrrrl}
\text{Max} & -x_1\pi_1 & -x_2\pi_2 & -x_3\pi_3 & -x_4\pi_4 & -d_1(\omega^s)\pi_5 & -d_2(\omega^s)\pi_6 & -d_3(\omega^s)\pi_7 & \\
\text{s.t.} & -6\pi_1 & -20\pi_2 & -12\pi_3 & -8\pi_4 & -\pi_5 & & & \le -1150 \\
 & -8\pi_1 & -25\pi_2 & -15\pi_3 & -10\pi_4 & & -\pi_6 & & \le -1525 \\
 & -10\pi_1 & -28\pi_2 & -18\pi_3 & -14\pi_4 & & & -\pi_7 & \le -1900 \\
 & \pi_1, & \pi_2, & \pi_3, & \pi_4, & \pi_5, & \pi_6, & \pi_7 & \ge \quad 0.
\end{array}$$

Notice that even though π depends on ω^s, i.e., $\pi := \pi(\omega^s)$, for convenience we do not show this dependency explicitly.

The problem data are summarized as follows:

First-stage:

$$c = (50, 30, 15, 10)^\top, \ A = \begin{bmatrix} -1 & 0 & 0 & 0 \\ 0 & -1 & 0 & 0 \\ 0 & 0 & -1 & 0 \\ 0 & 0 & 0 & -1 \\ 0 & -1 & -1 & -1 \end{bmatrix},$$
$$b = (-300, -700, -600, -500, -1600)^\top.$$

Second-stage:

Number of scenarios $|\Omega| = S = 5$.
Scenario probabilities $p(\omega^s)$, $s = 1, \cdots, 5$:

$$p(\omega^1) = 0.15, \ p(\omega^2) = 0.30, p(\omega^3) = 0.30, p(\omega^4) = 0.20, p(\omega^5) = 0.05.$$

Objective coefficient vector: $q(\omega^s)$, $s = 1, \cdots, 5$,

$$q(\omega^1) = q(\omega^2) = \cdots = q(\omega^5) = (-1150, -1525, -1900)^\top.$$

Recourse matrix:

$$W = \begin{bmatrix} -6 & -8 & -10 \\ -20 & -25 & -28 \\ -12 & -15 & -18 \\ -8 & -10 & -14 \\ -1 & 0 & 0 \\ 0 & -1 & 0 \\ 0 & 0 & -1 \end{bmatrix}.$$

Technology matrix:

$$T(\omega^1) = T(\omega^2) = \cdots = T(\omega^5) = T = \begin{bmatrix} 1 & 0 & 0 & 0 \\ 0 & 1 & 0 & 0 \\ 0 & 0 & 1 & 0 \\ 0 & 0 & 0 & 1 \\ 0 & 0 & 0 & 0 \\ 0 & 0 & 0 & 0 \\ 0 & 0 & 0 & 0 \end{bmatrix}.$$

RHS vectors $r(\omega^s)$, $s = 1, \cdots, 5$:

$$r(\omega^1) = (0, 0, 0, 0, -15, -10, -5)^\top.$$

$$r(\omega^2) = (0, 0, 0, 0, -20, -15, -15)^\top.$$

$$r(\omega^3) = (0, 0, 0, 0, -25, -20, -25)^\top.$$

$$r(\omega^4) = (0, 0, 0, 0, -30, -25, -30)^\top.$$

$$r(\omega^5) = (0, 0, 0, 0, -10, -10, -10)^\top.$$

First-stage feasible set:

$$\begin{array}{rrrrrl} X = \{ & -x_1 & & & & \geq -300 \\ & & -x_2 & & & \geq -700 \\ & & & -x_3 & & \geq -600 \\ & & & & -x_4 & \geq -500 \\ & & -x_2 & -x_3 & -x_4 & \geq -1600 \\ & x_1, & x_2, & x_3, & x_4 & \geq 0\}. \end{array}$$

Next, we now apply two iterations of the L-shaped algorithm.

Algorithm *L-shaped*
begin

Step 0. Initialization. Let $k \leftarrow 0$, choose $x^0 \in X$, and set $\ell \leftarrow -\infty$ and $u \leftarrow \infty$. Choose $\epsilon = 10^{-6}$ and $x^0 \leftarrow \text{argmin}_{x \in X}\{50x_1 + 30x_2 + 15x_3 + 10x_4\} = (0, 0, 0, 0)^\top$ as the initial point.

Step 1. Solve Subproblems and Generate Cut.
Initialize $\beta_0^\top \leftarrow (0, 0, 0, 0)$, $\beta_0^0 \leftarrow 0$, and $\bar{Q}^0 \leftarrow 0$.
For $s = 1, \cdots, 5$ solve

$$
\begin{array}{rrrrl}
\varphi^0(x^0, \omega^s) := \text{Min} & -1150y_1 & -1525y_2 & -1900y_3 & \\
\text{s.t.} & -6y_1 & -8y_2 & -10y_3 & \geq 0 \\
 & -20y_1 & -25y_2 & -28y_3 & \geq 0 \\
 & -12y_1 & -15y_2 & -18y_3 & \geq 0 \\
 & -8y_1 & -10y_2 & -14y_3 & \geq 0 \\
 & -y_1 & & & \geq d_a^s \\
 & & -y_2 & & \geq d_b^s \\
 & & & -y_3 & \geq d_c^s \\
 & y_1, & y_2, & y_3 & \geq 0.
\end{array}
$$

$s = 1$: feasible, $\varphi^0(x^0, \omega^1) = 0$, $y(\omega^1) = (0, 0, 0)^\top$, $\pi(\omega^1) = (191.667, 0, 0, 0, 0, 0, 0)^\top$.
$\bar{Q}^0 \leftarrow \bar{Q}^0 + \varphi^0(x^0, \omega^1) = 0 + 0 = 0.$
$\beta_0^\top \leftarrow \beta_0^\top + p(\omega^1)\pi(\omega^1)^\top T$

$$
= (0, 0, 0, 0) + 0.15(191.667, 0, 0, 0, 0, 0, 0)) \begin{bmatrix} 1\,0\,0\,0 \\ 0\,1\,0\,0 \\ 0\,0\,1\,0 \\ 0\,0\,0\,1 \\ 0\,0\,0\,0 \\ 0\,0\,0\,0 \\ 0\,0\,0\,0 \end{bmatrix} =
$$

$(28.750, 0, 0, 0).$
$\beta_0^0 \leftarrow 0 + p(\omega^1)\pi(\omega^1)^\top r(\omega^1)$
$= 0 + 0.15(191.667, 0, 0, 0, 0, 0, 0)(0, 0, 0, 0, -15, -10, -5)^\top = 0.$
$s = 2$: feasible, $\varphi^0(x^0, \omega^2) = 0$, $y(\omega^2) = (0, 0, 0)^\top$, $\pi(\omega^2) = (191.667, 0, 0, 0, 0, 0, 0)^\top$.
$\bar{Q}^0 \leftarrow \bar{Q}^0 + \varphi^0(x^0, \omega^2) = 0 + 0 = 0.$
$\beta_0^\top \leftarrow \beta_0^\top + p(\omega^2)\pi(\omega^2)^\top T$

$$= (28.750, 0, 0, 0) + 0.30(191.667, 0, 0, 0, 0, 0, 0)) \begin{bmatrix} 1\,0\,0\,0 \\ 0\,1\,0\,0 \\ 0\,0\,1\,0 \\ 0\,0\,0\,1 \\ 0\,0\,0\,0 \\ 0\,0\,0\,0 \\ 0\,0\,0\,0 \end{bmatrix} =$$

(86.250, 0, 0, 0).
$\beta_0^0 \leftarrow 0 + p(\omega^2)\pi(\omega^2)^\top r(\omega^2)$
$= 0 + 0.30(191.667, 0, 0, 0, 0, 0, 0)(0, 0, 0, 0, -15, -10, -5)^\top = 0.$
$s = 3$: feasible, $\varphi^0(x^0, \omega^3) = 0$, $y(\omega^3) = (0, 0, 0)^\top$, $\pi(\omega^3) = (191.667, 0, 0, 0, 0, 0, 0)^\top$.
$\bar{Q}^0 \leftarrow \bar{Q}^0 + \varphi^0(x^0, \omega^3) = 0 + 0 = 0.$
$\beta_0^\top \leftarrow \beta_0^\top + p(\omega^3)\pi(\omega^3)^\top T$

$$= (86.250, 0, 0, 0) + 0.30(191.667, 0, 0, 0, 0, 0, 0)) \begin{bmatrix} 1\,0\,0\,0 \\ 0\,1\,0\,0 \\ 0\,0\,1\,0 \\ 0\,0\,0\,1 \\ 0\,0\,0\,0 \\ 0\,0\,0\,0 \\ 0\,0\,0\,0 \end{bmatrix} =$$

(143.750, 0, 0, 0).
$\beta_0^0 \leftarrow 0 + p(\omega^1)\pi(\omega^3)^\top r(\omega^3)$
$= 0 + 0.30(191.667, 0, 0, 0, 0, 0, 0)(0, 0, 0, 0, -15, -10, -5)^\top = 0.$
$s = 4$: feasible, $\varphi^0(x^0, \omega^4) = 0$, $y(\omega^4) = (0, 0, 0)^\top$, $\pi(\omega^4) = (191.667, 0, 0, 0, 0, 0, 0)^\top$.
$\bar{Q}^0 \leftarrow \bar{Q}^0 + \varphi^0(x^0, \omega^4) = 0 + 0 = 0.$
$\beta_0^\top \leftarrow \beta_0^\top + p(\omega^4)\pi(\omega^4)^\top T$

$$= (143.750, 0, 0, 0) + 0.20(191.667, 0, 0, 0, 0, 0, 0)) \begin{bmatrix} 1\,0\,0\,0 \\ 0\,1\,0\,0 \\ 0\,0\,1\,0 \\ 0\,0\,0\,1 \\ 0\,0\,0\,0 \\ 0\,0\,0\,0 \\ 0\,0\,0\,0 \end{bmatrix} =$$

(182.084, 0, 0, 0).
$\beta_0^0 \leftarrow 0 + p(\omega^4)\pi(\omega^1)^\top r(\omega^4)$
$= 0 + 0.20(191.667, 0, 0, 0, 0, 0, 0)(0, 0, 0, 0, -15, -10, -5)^\top = 0.$
$s = 5$: feasible, $\varphi^0(x^0, \omega^5) = 0$, $y(\omega^5) = (0, 0, 0)^\top$, $\pi(\omega^5) = (191.667, 0, 0, 0, 0, 0, 0)^\top$.
$\bar{Q}^0 \leftarrow \bar{Q}^0 + \varphi^0(x^0, \omega^5) = 0 + 0 = 0.$
$\beta_0^\top \leftarrow \beta_0^\top + p(\omega^5)\pi(\omega^5)^\top T$

$$= (182.084, 0, 0, 0) + 0.05(191.667, 0, 0, 0, 0, 0, 0)) \begin{bmatrix} 1\,0\,0\,0 \\ 0\,1\,0\,0 \\ 0\,0\,1\,0 \\ 0\,0\,0\,1 \\ 0\,0\,0\,0 \\ 0\,0\,0\,0 \\ 0\,0\,0\,0 \end{bmatrix} =$$

$(191.667, 0, 0, 0)$.

$\beta_0^0 \leftarrow 0 + p(\omega^5)\pi(\omega^5)^\top r(\omega^5)$

$\quad = 0 + 0.05(191.667, 0, 0, 0, 0, 0, 0)(0, 0, 0, 0, -15, -10, -5)^\top = 0.$

Compute *upper bound* if feasible:

Set $v^0 \leftarrow c^\top x^0 + \bar{Q}^0 = 0 + 0 = 0.$
Set $u \leftarrow \min\{v^0, u\} = \min\{0, \infty\} = 0.$

If u is updated, set *incumbent* solution to $x^* \leftarrow x^0 = (0, 0, 0, 0)^\top$.

Step 2. Add Cut to Master Program and Solve. If all subproblems were feasible:

Add $\beta_0^\top x + \eta \geq \beta_0^0$ to master problem:

$$\begin{array}{llllllll} \nu^0 := \text{Min} & 50x_1 & +30x_2 & +15x_3 & +10x_4 & +\eta & & \\ \text{s.t.} & -x_1 & & & & & \geq & -300 \\ & & -x_2 & & & & \geq & -700 \\ & & & -x_3 & & & \geq & -600 \\ & & & & -x_4 & & \geq & -500 \\ & & -x_2 & -x_3 & -x_4 & & \geq & -1600 \\ & 191.667x_1 & & & & +\eta & \geq & 0 \\ & x_1, & x_2, & x_3, & x_4 & & \geq & 0. \end{array}$$

Solve master to get $x^1 = (300, 0, 0, 0)^\top$, $\eta^0 = -57{,}500$, and $\nu^0 = -42{,}500$. Set $\ell \leftarrow \max\{\nu^0, \ell\} = \max\{-42{,}500, -\infty\} = -42{,}500$.

Step 3. Termination. Compute $u - \ell = 0 - (-42{,}500) = 42{,}500$ and $\epsilon|u| = 10^{-6}|-42{,}500|) = 0.0425$. Since $u - \ell \geq \epsilon|u|$,

Set $k \leftarrow 0 + 1 = 1$. Return to step 1.

Iteration k=1:

Step 1. Solve Subproblems and Generate Cut.
Initialize $\beta_1^\top \leftarrow (0, 0, 0, 0)$, $\beta_1^0 \leftarrow 0$, and $\bar{Q}^1 \leftarrow 0$.
For $s = 1, \cdots, 5$ solve

$$
\begin{array}{lrrrl}
\varphi^1(x^1,\omega^s) := \text{Min} & -1150y_1 & -1525y_2 & -1900y_3 & \\
\text{s.t.} & -6y_1 & -8y_2 & -10y_3 & \geq 0 \\
 & -20y_1 & -25y_2 & -28y_3 & \geq 0 \\
 & -12y_1 & -15y_2 & -18y_3 & \geq 0 \\
 & -8y_1 & -10y_2 & -14y_3 & \geq 0 \\
 & -y_1 & & & \geq d_a^s \\
 & & -y_2 & & \geq d_b^s \\
 & & & -y_3 & \geq d_c^s \\
 & y_1, & y_2, & y_3 & \geq 0.
\end{array}
$$

$s = 1$: feasible, $\varphi^1(x^1,\omega^1) = 0$, $y(\omega^1) = (0,0,0)^\top$, $\pi(\omega^1) = (0,0,78.3333,35,0,0,0)^\top$.
$\bar{Q}^1 \leftarrow \bar{Q}^1 + \varphi^1(x^1,\omega^1) = 0+0 = 0$.
$\beta_1^\top \leftarrow \beta_1^\top + p(\omega^1)\pi(\omega^1)^\top T$

$$
= (0,0,0,0) + 0.15(0,0,78.3333,35,0,0,0)) \begin{bmatrix} 1\,0\,0\,0 \\ 0\,1\,0\,0 \\ 0\,0\,1\,0 \\ 0\,0\,0\,1 \\ 0\,0\,0\,0 \\ 0\,0\,0\,0 \\ 0\,0\,0\,0 \end{bmatrix} =
$$

$(28.750, 0, 0, 0)$.
$\beta_1^0 \leftarrow 0 + p(\omega^1)\pi(\omega^1)^\top r(\omega^1)$
$= 0 + 0.15(191.667,0,0,0,0,0,0)(0,0,0,0,-15,-10,-5)^\top = 0$.
$s = 2$: feasible, $\varphi^1(x^1,\omega^2) = 0$, $y(\omega^2) = (0,0,0)^\top$, $\pi(\omega^2) = (191.667,0,0,0,0,0,0)^\top$.
$\bar{Q}^1 \leftarrow \bar{Q}^1 + \varphi^1(x^1,\omega^2) = 0+0 = 0$.
$\beta_1^\top \leftarrow \beta_1^\top + p(\omega^2)\pi(\omega^2)^\top T$

$$
= (28.750,0,0,0) + 0.30(191.667,0,0,0,0,0,0)) \begin{bmatrix} 1\,0\,0\,0 \\ 0\,1\,0\,0 \\ 0\,0\,1\,0 \\ 0\,0\,0\,1 \\ 0\,0\,0\,0 \\ 0\,0\,0\,0 \\ 0\,0\,0\,0 \end{bmatrix} =
$$

$(86.250, 0, 0, 0)$.
$\beta_1^0 \leftarrow 0 + p(\omega^2)\pi(\omega^2)^\top r(\omega^2)$
$= 0 + 0.30(191.667,0,0,0,0,0,0)(0,0,0,0,-15,-10,-5)^\top = 0$.
$s = 3$: feasible, $\varphi^1(x^1,\omega^3) = 0$, $y(\omega^3) = (0,0,0)^\top$, $\pi(\omega^3) = (191.667,0,0,0,0,0,0)^\top$.
$\bar{Q}^1 \leftarrow \bar{Q}^1 + \varphi^1(x^1,\omega^3) = 0+0 = 0$.
$\beta_1^\top \leftarrow \beta_1^\top + p(\omega^3)\pi(\omega^3)^\top T$

$$= (86.250, 0, 0, 0) + 0.30(191.667, 0, 0, 0, 0, 0, 0)) \begin{bmatrix} 1\,0\,0\,0 \\ 0\,1\,0\,0 \\ 0\,0\,1\,0 \\ 0\,0\,0\,1 \\ 0\,0\,0\,0 \\ 0\,0\,0\,0 \\ 0\,0\,0\,0 \end{bmatrix} =$$

(143.750, 0, 0, 0).
$\beta_1^0 \leftarrow 0 + p(\omega^1)\pi(\omega^3)^\top r(\omega^3)$
$= 0 + 0.30(191.667, 0, 0, 0, 0, 0, 0)(0, 0, 0, 0, -15, -10, -5)^\top = 0.$
$s = 4$: feasible, $\varphi^1(x^1, \omega^4) = 0$, $y(\omega^4) = (0, 0, 0)^\top$, $\pi(\omega^4) = (191.667, 0, 0, 0, 0, 0, 0)^\top$.
$\bar{Q}^1 \leftarrow \bar{Q}^1 + \varphi^1(x^1, \omega^4) = 0 + 0 = 0.$
$\beta_1^\top \leftarrow \beta_1^\top + p(\omega^4)\pi(\omega^4)^\top T$

$$= (143.750, 0, 0, 0) + 0.20(191.667, 0, 0, 0, 0, 0, 0)) \begin{bmatrix} 1\,0\,0\,0 \\ 0\,1\,0\,0 \\ 0\,0\,1\,0 \\ 0\,0\,0\,1 \\ 0\,0\,0\,0 \\ 0\,0\,0\,0 \\ 0\,0\,0\,0 \end{bmatrix} =$$

(182.084, 0, 0, 0).
$\beta_1^0 \leftarrow 0 + p(\omega^4)\pi(\omega^1)^\top r(\omega^4)$
$= 0 + 0.20(191.667, 0, 0, 0, 0, 0, 0)(0, 0, 0, 0, -15, -10, -5)^\top = 0.$
$s = 5$: feasible, $\varphi^1(x^1, \omega^5) = 0$, $y(\omega^5) = (0, 0, 0)^\top$, $\pi(\omega^5) = (191.667, 0, 0, 0, 0, 0, 0)^\top$.
$\bar{Q}^1 \leftarrow \bar{Q}^1 + \varphi^1(x^1, \omega^5) = 0 + 0 = 0.$
$\beta_1^\top \leftarrow \beta_1^\top + p(\omega^5)\pi(\omega^5)^\top T$

$$= (182.084, 0, 0, 0) + 0.05(191.667, 0, 0, 0, 0, 0, 0)) \begin{bmatrix} 1\,0\,0\,0 \\ 0\,1\,0\,0 \\ 0\,0\,1\,0 \\ 0\,0\,0\,1 \\ 0\,0\,0\,0 \\ 0\,0\,0\,0 \\ 0\,0\,0\,0 \end{bmatrix} =$$

(0, 0, 78.3333, 35).
$\beta_1^0 \leftarrow 0 + p(\omega^5)\pi(\omega^5)^\top r(\omega^5)$
$= 0 + 0.05(191.667, 0, 0, 0, 0, 0, 0)(0, 0, 0, 0, -15, -10, -5)^\top = 0.$
Compute *upper bound* if feasible:

Set $v^1 \leftarrow c^\top x^1 + \bar{Q}^1 = 15{,}000 + 0 = 15{,}000.$
Set $u \leftarrow \min\{v^1, u\} = \min\{15{,}000, 0\} = 0.$

If u is updated, set *incumbent* solution to $x^* \leftarrow x^1 = (300, 0, 0, 0)^\top$.

Step 2. Add Cut to Master Program and Solve. If all subproblems were feasible:

Add $\beta_1^\top x + \eta \geq \beta_1^0$ to master problem:

$$
\begin{array}{rrrrrl}
\nu^1 := \text{Min} & 50x_1 & +30x_2 & +15x_3 & +10x_4 & +\eta \\
\text{s.t.} & -x_1 & & & & \geq -300 \\
& & -x_2 & & & \geq -700 \\
& & & -x_3 & & \geq -600 \\
& & & & -x_4 & \geq -500 \\
& & -x_2 & -x_3 & -x_4 & \geq -1600 \\
& 191.667x_1 & & & & +\eta \geq 0 \\
& & & 78.3333x_3 & +35.0000x_4 & +\eta \geq 0 \\
& x_1, & x_2, & x_3, & x_4 & \geq 0.
\end{array}
$$

Solve master to get $x^2 = (300, 0, 600, 300)^\top$, $\eta^1 = -57{,}500$, and $\nu^1 = -30{,}500$.
Set $\ell \leftarrow \max\{\nu^1, \ell\} = \max\{-30{,}500, -42{,}500\} = -30{,}500$.

Step 3. Termination. Compute $u - \ell = 0 - (-30{,}500) = 30{,}500$ and $\epsilon|u| = 10^{-6}| - 30{,}500|) = 0.0305$. Since $u - \ell \geq \epsilon|u|$,

Set $k \leftarrow 1 + 1 = 2$.

Return to step 1.

After 15 iterations, the L-shaped algorithm terminates with optimal solution $x^* = (236.4, 690, 432, 318)^\top$ and objective value -2268.5. This is the solution we obtained from solving the DEP of the problem instance with a direct solver.

6.3 The Multicut Method

In this subsection, we study two variants of the L-shaped method that involve adding multiple optimality cuts to the master problem at each iteration of the algorithm. Recall that in step 1 of the L-shaped method, we have to solve a subproblem for each scenario to get the dual solution. The dual solutions of all the scenario subproblems are then *aggregated* to generate a single optimality cut to add to the master program in step 2. However, the dual block angular structure of the problem allows to place multiple cuts instead of a single cut at each iteration of the algorithm. Birge and Louveaux [3] proposed placing one optimality cut for each scenario at every iteration of the algorithm. The resulting method is referred to as the *multicut L-shaped algorithm*. Trukhanov et al. [15] proposed choosing and aggregating subsets of scenarios and then adding multiple cuts based on the selected level of aggregation. This method is referred to as the *adaptive multicut L-shaped method*. Next, we derive and state the pure multicut method.

6.3.1 Multicut Decomposition

Let η_s be the optimality cut variable and $t(s)$ be the iteration index at which an optimality cut is generated for scenario ω^s. Let l denote the feasibility cut counter and let $u(s)$ denote the number of optimality cuts generated for scenario ω^s at iteration k of the algorithm. Also, let v be the number of feasibility cuts generated so far. Then the master program for the multicut algorithm at iteration k takes the following form:

$$
\begin{aligned}
\text{Min } & c^\top x + \sum_{s=1}^{S} \eta_s \\
\text{s.t. } & Ax \geq b \\
& \beta_{t(s)}^\top x + \eta_s \geq \beta_{t(s)}^0, \ \forall t(s) = 1, \cdots, u(s), \ s = 1, \cdots, S \qquad (6.14) \\
& \beta_l^\top x \geq \beta_l^0, \ \forall l = 1, \cdots, v \\
& x \geq 0.
\end{aligned}
$$

Let the master program solution at iteration k be $(x^k, \eta_1^k, \cdots, \eta_S^k)$. If no optimality constraint (6.14) is present for some scenario ω^s, do not consider s in the master program formulation for getting x^k. In this case, set $\eta_s^k \leftarrow -\infty$. Let us continue to have $X = \{x \mid Ax \geq b, x \geq 0\}$. Following our derivation of the L-shaped algorithm, we need cutting planes for $\varphi(x, \omega^s)$ given in subproblem (6.7) for each scenario ω^s. Solving the subproblem, we get an optimal dual solution $\pi^*(\omega^s)$. We can now generate an optimality cut for scenario ω^s as follows:

$$
\begin{aligned}
& \eta_s \geq \pi^*(\omega^s)^\top (r(\omega^s) - T(\omega^s)x), \ \forall x \in X \\
\Rightarrow \ & \eta_s + \beta_s^\top x \geq \beta_s^0, \ \forall x \in X,
\end{aligned} \qquad (6.15)
$$

where $\beta_s^0 = p(\omega^s)\pi(\omega^s)^\top r(\omega^s)$ and $\beta_s^\top = p(\omega^s)\pi(\omega^s)^\top T(\omega^s)$. If some subproblem for scenario ω^s is infeasible, get the dual extreme ray μ_s^*, and generate a feasibility cut as follows:

$$
\begin{aligned}
& \mu_s^{*\top} (r(\omega^s) - T(\omega^s)x) \leq 0, \ \forall x \in X \\
\Rightarrow \ & \beta_s^\top x \geq \beta_s^0, \ \forall x \in X,
\end{aligned} \qquad (6.16)
$$

where $\beta_s^0 = \mu_s^{*\top} r(\omega^s)$ and $\beta_s^\top = \mu_s^{*\top} T(\omega^s)$.

6.3.2 Multicut L-Shaped Algorithm

Let us continue to let ℓ and u denote the lower and upper bounds, respectively, on the optimal value during the course of the algorithm. The multicut L-shaped algorithm can be stated as follows:

Algorithm *Multicut L-shaped*
begin

Step 0. Initialization. Set $k \leftarrow 1$, $u(s) \leftarrow 0$, $v \leftarrow 0$, $\ell \leftarrow -\infty$, $u \leftarrow \infty$, $x^1 \leftarrow \text{argmin}\{c^\top x \mid Ax \geq b, x \geq 0\}$, or x^1 be given, $x^* \leftarrow x^1$, and $\eta_s^1 \leftarrow -\infty$, $s = 1, \cdots, S$.

Step 1. Solve Subproblems and Generate Cut. For $s = 1, \cdots, S$ solve

$$\varphi^k(x^k, \omega^s) := \text{Min}\{q^\top(\omega^s)y \mid Wy \geq r(\omega^s) - T(\omega^s)x^k, y \geq 0\}.$$

If subproblem is *infeasible*, generate *feasibility* cut:

Get dual extreme ray μ_s^k.
Set $v \leftarrow v + 1$.
Compute $\beta_v^0 \leftarrow (\mu_s^k)^\top r$ and $\beta_v^\top \leftarrow (\mu_s^k)^\top T(\omega^s)$.

else if *feasible*, generate *optimality* cut:

Get dual solution π_s^k.
Compute $\beta_s^0 \leftarrow p(\omega^s)(\pi_s^k)^\top r(\omega^s)$ and $\beta(\omega^s)^\top \leftarrow p(\omega^s)(\pi_s^k)^\top T(\omega^s)$.
If

$$\eta_s^k < \beta_s^0 - \beta(\omega^s)^\top x^k \tag{6.17}$$

Set $u(s) \leftarrow u(s) + 1$.
Set $\beta_{u(s)}^0 \leftarrow \beta_s^0$ and $\beta_{u(s)}^\top \leftarrow \beta(\omega^s)^\top$.

If *all* subproblems are feasible, compute *upper bound*:

Set $u^k \leftarrow c^\top x^k + \sum_{s=1}^S p(\omega^s)\varphi^k(x^k, \omega^s)$.
Set $u \leftarrow \min\{u^k, u\}$.

If u is updated, set *incumbent* solution $x^* \leftarrow x^k$.

Step 2. Termination. If condition (6.17) does *not* hold for *all* $s = 1, \cdots, S$,

Stop, declare x^* as the optimal solution.

else

Set $k \leftarrow k + 1$.

Step 3. Add Cuts to Master Program and Solve.

For *all* infeasible s subproblems

Add $\beta_l{}^\top x \geq \beta_l^0$ to master problem for all new values of l.

else for *all* feasible s subproblems

Add $\beta_{u(s)}^\top x + \eta_s \geq \beta_{u(s)}^0$.

Solve master program

$$\ell^{k+1} := \text{Min } c^\top x + \sum_{s=1}^{S} \eta_s$$

$$\begin{aligned} \text{s.t.}\; & Ax && \geq b \\ & \beta_{t(s)}^\top x + \eta_s && \geq \beta_{t(s)}^0,\ \forall t(s) = 1, \cdots, u(s),\ s = 1, \cdots, S \\ & \beta_l^\top x && \geq \beta_l^0,\ \forall l = 1, \cdots, v \\ & x && \geq 0 \end{aligned}$$

to get $(x^{k+1}, \eta_1^{k+1}, \cdots, \eta_S^{k+1})$.

Set $\ell \leftarrow \max\{\ell^{k+1}, \ell\}$ and return to step 1.

---

end

Let us now make some observations about the multicut L-shaped algorithm. First, in step 1 of the algorithm feasibility cuts are generated for *all* infeasible subproblems. Second, even though we are computing the upper and lower bounds, u and ℓ, on the optimal value during the course of the algorithm, the termination condition is not based on u and ℓ. This is because we are adding separate optimality cuts and we need to make sure that we construct a good recourse function approximation for each scenario before we can terminate the algorithm. Notice that the termination condition is *after* solving all scenario subproblems. Third, one can compute the gap $u - \ell$ at each iteration of the algorithm to keep track of the convergence of the upper and lower bounds to the optimal value. We should point out that one can define the expected recourse function approximation in the objective function (the second term) as $\sum_{s=1}^{S} p(\omega^s)\eta_s$ instead of $\sum_{s=1}^{S} \eta_s$ as we have done. In that case, the optimality cut parameters should be calculated as follows:

$$\beta_s^0 \leftarrow \pi(\omega^s)^\top r(\omega^s) \text{ and } \beta_s^\top \leftarrow \pi(\omega^s)^\top T(\omega^s).$$

Let us now compare and contrast the multicut and single cut L-shaped approaches. Adding separate cuts, one per scenario at each iteration of the multicut method, results in a larger master problem compared to the single cut L-shaped method. However, by sending disaggregated cuts, more detailed information is passed on to the master program in the multicut method. Consequently, the number of algorithm iterations is expected to be less than in the single cut L-shaped algorithm in general. However, the trade-off between having relatively a smaller number of iterations but a larger master problem is instance dependent. Furthermore, the computation time of solving the master program is also instance dependent. So in certain cases the single cut L-shaped algorithm provides better performance than the multicut and vice versa. One has to determine through computational experimentation which of the two algorithms provides better performance. Finally, it should be pointed out that if the number of scenarios is too large it can even be impossible to create the master program in the multicut method due to computer memory storage issues.

6.3.3 Numerical Example

Example 6.2 Apply two iterations of the multicut algorithm to the instance of the two-stage RN-SLP *abc-Production Planning* problem in Example 6.1.

The initial master problem without the optimality cut variables is given as follows:

$$
\begin{aligned}
\nu^0 := \text{Min } & 50x_1 + 30x_2 + 15x_3 + 10x_4 \\
\text{s.t. } & -x_1 \geq -300 \\
& -x_2 \geq -700 \\
& -x_3 \geq -600, \\
& -x_4 \geq -500 \\
& -x_2 - x_3 - x_4 \geq -1600 \\
& x_1, x_2, x_3, x_4 \geq 0
\end{aligned}
$$

and for a given $x^k = (x_1^k, x_2^k, x_3^k, x_4^k)^\top$, the scenario subproblem for all $s = 1, \cdots, 5$ is given as follows:

$$
\begin{aligned}
\varphi^k(x^k, \omega^s) := \text{Min } & -1150y_1 - 1525y_2 - 1900y_3 \\
\text{s.t. } & -6y_1 - 8y_2 - 10y_3 \geq -x_1 \\
& -20y_1 - 25y_2 - 28y_3 \geq -x_2^k \\
& -12y_1 - 15y_2 - 18y_3 \geq -x_3^k \\
& -8y_1 - 10y_2 - 14y_3 \geq -x_4^k \\
& -y_1 \geq d_a^s \\
& -y_2 \geq d_b^s \\
& -y_3 \geq d_c^s \\
& y_1, y_2, y_3 \geq 0,
\end{aligned}
$$

where the RHS $r(\omega^s) = (0, 0, 0, 0, d_a^s, d_b^s, d_c^s)^\top$. Applying the multicut algorithm, we obtain the following:

Algorithm *Multicut L-shaped*
begin

Step 0. Initialization. Set $k \leftarrow 1$, $u(s) \leftarrow 0$, $v \leftarrow 0$, $\ell \leftarrow -\infty$, $u \leftarrow \infty$, $x^1 \leftarrow \text{argmin}_{x \in X}\{50x_1 + 30x_2 + 15x_3 + 10x_4\} = (0, 0, 0, 0)^\top$, $x^* \leftarrow x^1$, and $\eta_s^1 \leftarrow -\infty$, $s = 1, \cdots, S$.

Step 1. Solve Subproblems and Generate Cut. For $s = 1, \cdots, 5$ solve

$$
\begin{array}{rrrrl}
\varphi^1(x^1, \omega^s) := \text{Min} & -1150y_1 & -1525y_2 & -1900y_3 & \\
\text{s.t.} & -6y_1 & -8y_2 & -10y_3 & \geq 0 \\
 & -20y_1 & -25y_2 & -28y_3 & \geq 0 \\
 & -12y_1 & -15y_2 & -18y_3 & \geq 0 \\
 & -8y_1 & -10y_2 & -14y_3 & \geq 0 \\
 & -y_1 & & & \geq d_a^s \\
 & & -y_2 & & \geq d_b^s \\
 & & & -y_3 & \geq d_c^s \\
 & y_1, & y_2, & y_3 & \geq 0.
\end{array}
$$

$s = 1$: feasible, $\varphi^1(x^1, \omega^1) = 0$, $y_1^1 = (0, 0, 0)^\top$, $\pi_1^1 = (191.667, 0, 0, 0, 0, 0, 0)^\top$.
Generate *optimality* cut:

$$
\begin{aligned}
\beta_1^0 \leftarrow p(\omega^1)(\pi_1^1)^\top r(\omega^1) &= 0.15(191.667, 0, 0, 0, 0, 0, 0) \\
&\qquad (0, 0, 0, 0, -15, -10, -5)^\top \\
&= 0.
\end{aligned}
$$

$$
\beta_1^\top \leftarrow p(\omega^1)(\pi_1^1)^\top T = 0.15(191.667, 0, 0, 0, 0, 0, 0) \begin{bmatrix} 1\,0\,0\,0 \\ 0\,1\,0\,0 \\ 0\,0\,1\,0 \\ 0\,0\,0\,1 \\ 0\,0\,0\,0 \\ 0\,0\,0\,0 \\ 0\,0\,0\,0 \end{bmatrix}
$$

$$
= (28.7501, 0, 0, 0).
$$

Optimality cut, $\beta_1^\top + \eta_1 \geq \beta_1^0$: $\quad 28.7501x_1 + \eta_1 \geq 0$.

$s = 2$: feasible, $\varphi^1(x^1, \omega^2) = 0$, $y_2^1 = (0, 0, 0)^\top$, $\pi_2^1 = (191.667, 0, 0, 0, 0, 0, 0)^\top$.
Generate *optimality* cut:

$$\beta_2^0 \leftarrow p(\omega^1)(\pi_2^1)^\top r(\omega^2) = 0.3(191.667, 0, 0, 0, 0, 0, 0)$$
$$(0, 0, 0, 0, -20, -15, -15)^\top$$
$$= 0.$$

$$\beta_2^\top \leftarrow p(\omega^1)(\pi_2^1)^\top T = 0.3(191.667, 0, 0, 0, 0, 0, 0) \begin{bmatrix} 1\,0\,0\,0 \\ 0\,1\,0\,0 \\ 0\,0\,1\,0 \\ 0\,0\,0\,1 \\ 0\,0\,0\,0 \\ 0\,0\,0\,0 \\ 0\,0\,0\,0 \end{bmatrix}$$
$$= (57.5001, 0, 0, 0).$$

Optimality cut, $\beta_2^\top + \eta_2 \geq \beta_2^0$: $57.5001x_1 + \eta_2 \geq 0$.
$s = 3$: feasible, $\varphi^1(x^1, \omega^3) = 0$, $y_3^1 = (0, 0, 0)^\top$, $\pi_3^1 = (191.667, 0, 0, 0, 0, 0, 0)^\top$.
Generate *optimality* cut:

$$\beta_3^0 \leftarrow p(\omega^1)(\pi_3^1)^\top r(\omega^3) = 0.3(191.667, 0, 0, 0, 0, 0, 0)$$
$$(0, 0, 0, 0, -20, -20, -25)^\top$$
$$= 0.$$

$$\beta_3^\top \leftarrow p(\omega^1)(\pi_3^1)^\top T = 0.3(191.667, 0, 0, 0, 0, 0, 0) \begin{bmatrix} 1\,0\,0\,0 \\ 0\,1\,0\,0 \\ 0\,0\,1\,0 \\ 0\,0\,0\,1 \\ 0\,0\,0\,0 \\ 0\,0\,0\,0 \\ 0\,0\,0\,0 \end{bmatrix}$$
$$= (57.5001, 0, 0, 0).$$

Optimality cut, $\beta_3^\top + \eta_3 \geq \beta_3^0$: $57.5001x_1 + \eta_3 \geq 0$.
$s = 4$: feasible, $\varphi^1(x^1, \omega^4) = 0$, $y_4^1 = (0, 0, 0)^\top$, $\pi_4^1 = (191.667, 0, 0, 0, 0, 0, 0)^\top$.
Generate *optimality* cut:

$$\beta_4^0 \leftarrow p(\omega^1)(\pi_4^1)^\top r(\omega^4) = 0.2(191.667, 0, 0, 0, 0, 0, 0)$$
$$(0, 0, 0, 0, -30, -25, -30)^\top$$
$$= 0.$$

$$\beta_4^\top \leftarrow p(\omega^1)(\pi_4^1)^\top T = 0.2(191.667, 0, 0, 0, 0, 0, 0) \begin{bmatrix} 1\,0\,0\,0 \\ 0\,1\,0\,0 \\ 0\,0\,1\,0 \\ 0\,0\,0\,1 \\ 0\,0\,0\,0 \\ 0\,0\,0\,0 \\ 0\,0\,0\,0 \end{bmatrix}$$

$$= (38.3334, 0, 0, 0).$$

Optimality cut, $\beta_4^\top + \eta_4 \geq \beta_4^0$: $38.3334x_1 + \eta_4 \geq 0$.
$s = 5$: feasible, $\varphi^1(x^1, \omega^5) = 0$, $y_5^1 = (0, 0, 0)^\top$, $\pi_5^1 = (191.667, 0, 0, 0, 0, 0, 0)^\top$.
Generate *optimality* cut:

$$\beta_5^0 \leftarrow p(\omega^1)(\pi_5^1)^\top r(\omega^5) = 0.05(191.667, 0, 0, 0, 0, 0, 0) \\ (0, 0, 0, 0, -10, -10, -10)^\top$$

$$= 0.$$

$$\beta_5^\top \leftarrow p(\omega^1)(\pi_5^1)^\top T = 0.05(191.667, 0, 0, 0, 0, 0, 0) \begin{bmatrix} 1\,0\,0\,0 \\ 0\,1\,0\,0 \\ 0\,0\,1\,0 \\ 0\,0\,0\,1 \\ 0\,0\,0\,0 \\ 0\,0\,0\,0 \\ 0\,0\,0\,0 \end{bmatrix}$$

$$= (9.5834, 0, 0, 0).$$

Optimality cut, $\beta_5^\top + \eta_5 \geq \beta_5^0$: $9.5834x_1 + \eta_5 \geq 0$.
Clearly, $\eta_s^1 < \beta_s^0 - \beta(\omega^s)^\top x^1$, $\forall s = 1, \cdots, 5$.

Set $u(s) \leftarrow u(s) + 1 = 0 + 1 = 1$, $\forall s = 1, \cdots, 5$.
Set $\beta_{u(s)}^0 \leftarrow \beta_s^0$ and $\beta_{u(s)}^\top \leftarrow \beta(\omega^s)^\top$, $\forall s = 1, \cdots, 5$.

If *all* subproblems are feasible, compute *upper bound*:

Set $u^1 \leftarrow c^\top x^1 + \sum_{s=1}^S p(\omega^s)\varphi^1(x^1, \omega^s) = 0$.
Set $u \leftarrow \min\{0, \infty\} = 0$.

If u is updated, set *incumbent* solution $x^* \leftarrow x^1$.

Step 2. Termination. Condition (6.17) *does* hold for *all* $s = 1, \cdots, S$.

Set $k \leftarrow k + 1 = 1 + 1 = 2$.

Step 3. Add Cuts to Master Program and Solve.
Solve master program

$$
\begin{array}{llllllllllll}
\ell^2 := \text{Min} & 50x_1 & +30x_2 & +15x_3 & +10x_4 & +\eta_1 & +\eta_2 & +\eta_3 & +\eta_4 & +\eta_5 & & \\
\text{s.t.} & -x_1 & & & & & & & & & \geq & -300 \\
 & & -x_2 & & & & & & & & \geq & -700 \\
 & & & -x_3 & & & & & & & \geq & -600 \\
 & & & & -x_4 & & & & & & \geq & -500 \\
 & & -x_2 & -x_3 & -x_4 & & & & & & \geq & -1600 \\
 & 28.7501x_1 & & & & +\eta_1 & & & & & \geq & 0 \\
 & 57.5001x_1 & & & & & +\eta_2 & & & & \geq & 0 \\
 & 57.5001x_1 & & & & & & +\eta_3 & & & \geq & 0 \\
 & 38.3334x_1 & & & & & & & +\eta_4 & & \geq & 0 \\
 & 9.5834x_1 & & & & & & & & +\eta_5 & \geq & 0 \\
 & x_1, & x_2, & x_3, & x_4 & & & & & & \geq & 0
\end{array}
$$

to get $x^2 = (300, 0, 0, 0)^\top$, $\eta_1^2 = -8625$, $\eta_2^2 = -17{,}250$, $\eta_3^2 = -17{,}250$, $\eta_4^2 = -11{,}500$, $\eta_5^2 = -2875$, and $\ell^2 = -42{,}500$.
Set $\ell \leftarrow \max\{\ell^2, \ell\} = \max\{-42{,}500, -\infty\} = -42{,}500$ and return to step 1.

Iteration $k = 2$:

Step 1. Solve Subproblems and Generate Cut. For $s = 1, \cdots, 5$ solve

$$
\begin{array}{llllll}
\varphi^2(x^2, \omega^s) := \text{Min} & -1150y_1 & -1525y_2 & -1900y_3 & & \\
\text{s.t.} & -6y_1 & -8y_2 & -10y_3 & \geq & -300 \\
 & -20y_1 & -25y_2 & -28y_3 & \geq & 0 \\
 & -12y_1 & -15y_2 & -18y_3 & \geq & 0 \\
 & -8y_1 & -10y_2 & -14y_3 & \geq & 0 \\
 & -y_1 & & & \geq & d_a^s \\
 & & -y_2 & & \geq & d_b^s \\
 & & & -y_3 & \geq & d_c^s \\
 & y_1, & y_2, & y_3 & \geq & 0.
\end{array}
$$

$s = 1$: feasible, $\varphi^2(x^2, \omega^1) = 0$, $y_1^2 = (0, 0, 0)^\top$, $\pi_1^2 = (0, 0, 78.3333, 35, 0, 0, 0)^\top$.
Generate *optimality* cut:

$$
\begin{aligned}
\beta_1^0 \leftarrow p(\omega^1)(\pi_1^2)^\top r(\omega^1) &= 0.15(0, 0, 78.3333, 35, 0, 0, 0) \\
&\qquad (0, 0, 0, 0, -15, -10, -5)^\top \\
&= 0.
\end{aligned}
$$

$$\beta_1^\top \leftarrow p(\omega^1)(\pi_1^2)^\top T = 0.15(0, 0, 78.3333, 35, 0, 0, 0) \begin{bmatrix} 1 & 0 & 0 & 0 \\ 0 & 1 & 0 & 0 \\ 0 & 0 & 1 & 0 \\ 0 & 0 & 0 & 1 \\ 0 & 0 & 0 & 0 \\ 0 & 0 & 0 & 0 \\ 0 & 0 & 0 & 0 \end{bmatrix}$$

$$= (0, 0, 11.75, 5.25).$$

Optimality cut, $\beta_1^\top + \eta_1 \geq \beta_1^0$: $11.75x_3 + 5.25x_4 + \eta_1 \geq 0$.
$s = 2$: feasible, $\varphi^2(x^2, \omega^2) = 0$, $y_2^2 = (0, 0, 0)^\top$, $\pi_2^2 = (0, 0, 78.3333, 35, 0, 0, 0)^\top$.
Generate *optimality* cut:

$$\beta_2^0 \leftarrow p(\omega^2)(\pi_2^2)^\top r(\omega^2) = 0.3(0, 0, 78.3333, 35, 0, 0, 0) \\ (0, 0, 0, 0, -20, -15, -15)^\top \\ = 0.$$

$$\beta_2^\top \leftarrow p(\omega^2)(\pi_2^2)^\top T = 0.3(0, 0, 78.3333, 35, 0, 0, 0) \begin{bmatrix} 1 & 0 & 0 & 0 \\ 0 & 1 & 0 & 0 \\ 0 & 0 & 1 & 0 \\ 0 & 0 & 0 & 1 \\ 0 & 0 & 0 & 0 \\ 0 & 0 & 0 & 0 \\ 0 & 0 & 0 & 0 \end{bmatrix}$$

$$= (0, 0, 23.5, 10.5).$$

Optimality cut, $\beta_2^\top + \eta_2 \geq \beta_2^0$: $23.5x_3 + 10.5x_4 + \eta_2 \geq 0$.
$s = 3$: feasible, $\varphi^2(x^2, \omega^3) = 0$, $y_3^2 = (0, 0, 0)^\top$, $\pi_3^2 = (0, 0, 78.3333, 35, 0, 0, 0)^\top$.
Generate *optimality* cut:

$$\beta_3^0 \leftarrow p(\omega^3)(\pi_3^2)^\top r(\omega^3) = 0.3(0, 0, 78.3333, 35, 0, 0, 0) \\ (0, 0, 0, 0, -20, -20, -25)^\top \\ = 0.$$

$$\beta_3^\top \leftarrow p(\omega^3)(\pi_3^2)^\top T = 0.3(0, 0, 78.3333, 35, 0, 0, 0)\begin{bmatrix} 1&0&0&0\\ 0&1&0&0\\ 0&0&1&0\\ 0&0&0&1\\ 0&0&0&0\\ 0&0&0&0\\ 0&0&0&0 \end{bmatrix}$$

$$= (0, 0, 23.5, 10.5).$$

Optimality cut, $\beta_3^\top + \eta_3 \geq \beta_3^0$: $23.5x_3 + 10.5x_4 + \eta_2 \geq 0$.
$s = 4$: feasible, $\varphi^2(x^2, \omega^4) = 0$, $y_4^2 = (0, 0, 0)^\top$, $\pi_4^2 = (0, 0, 78.3333, 35, 0, 0, 0)^\top$.
Generate *optimality* cut:

$$\beta_4^0 \leftarrow p(\omega^4)(\pi_4^2)^\top r(\omega^4) = 0.2(0, 0, 78.3333, 35, 0, 0, 0)$$
$$(0, 0, 0, 0, -30, -25, -30)^\top$$
$$= 0.$$

$$\beta_4^\top \leftarrow p(\omega^4)(\pi_4^2)^\top T = 0.2(0, 0, 78.3333, 35, 0, 0, 0)\begin{bmatrix} 1&0&0&0\\ 0&1&0&0\\ 0&0&1&0\\ 0&0&0&1\\ 0&0&0&0\\ 0&0&0&0\\ 0&0&0&0 \end{bmatrix}$$

$$= (0, 0, 15.6667, 7).$$

Optimality cut, $\beta_4^\top + \eta_4 \geq \beta_4^0$: $15.6667x_3 + 7x_4 + \eta_4 \geq 0$.
$s = 5$: feasible, $\varphi^2(x^2, \omega^5) = 0$, $y_5^2 = (0, 0, 0)^\top$, $\pi_5^2 = (0, 0, 78.3333, 35, 0, 0, 0)^\top$.
Generate *optimality* cut:

$$\beta_5^0 \leftarrow p(\omega^5)(\pi_5^2)^\top r(\omega^5) = 0.05(0, 0, 78.3333, 35, 0, 0, 0)$$
$$(0, 0, 0, 0, -10, -10, -10)^\top$$
$$= 0.$$

$$\beta_5^\top \leftarrow p(\omega^5)(\pi_5^2)^\top T = 0.05(0,0,78.3333,35,0,0,0)\begin{bmatrix}1&0&0&0\\0&1&0&0\\0&0&1&0\\0&0&0&1\\0&0&0&0\\0&0&0&0\\0&0&0&0\end{bmatrix}$$

$$= (0,0,3.9167,1.75).$$

Optimality cut, $\beta_5^\top + \eta_5 \geq \beta_5^0$: $\quad 3.9167x_3 + 1.75x_4 + \eta_1 \geq 0.$
Clearly, $\eta_s^2 < \beta_s^0 - \beta(\omega^s)^\top x^2,\ \forall s = 1, \cdots, 5.$

Set $u(s) \leftarrow u(s) + 1 = 1 + 1 = 2,\ \forall s = 1, \cdots, 5.$
Set $\beta_{u(s)}^0 \leftarrow \beta_s^0$ and $\beta_{u(s)}^\top \leftarrow \beta(\omega^s)^\top,\ \forall s = 1, \cdots, 5.$

If *all* subproblems are feasible, compute *upper bound*:

Set $u^2 \leftarrow c^\top x^2 + \sum_{s=1}^S p(\omega^s)\varphi^2(x^2, \omega^s) = 15{,}000.$
Set $u \leftarrow \min\{15{,}000, 0\} = 0.$

u is *not* updated, so *incumbent* solution $x^* \leftarrow x^1$.

Step 2. Termination. Condition (6.17) *does* hold for *all* $s = 1, \cdots, S$.

Set $k \leftarrow k + 1 = 2 + 1 = 3.$

Step 3. Add Cuts to Master Program and Solve.
Solve master program

$$\begin{array}{rlllllllllll}
\ell^3 := \text{Min} & 50x_1 & +30x_2 & +15x_3 & +10x_4 & +\eta_1 & +\eta_2 & +\eta_3 & +\eta_4 & +\eta_5 & & \\
\text{s.t.} & -x_1 & & & & & & & & & \geq & -300 \\
 & & -x_2 & & & & & & & & \geq & -700 \\
 & & & -x_3 & & & & & & & \geq & -600 \\
 & & & & -x_4 & & & & & & \geq & -500 \\
 & & -x_2 & -x_3 & -x_4 & & & & & & \geq & -1600 \\
 & 28.7501x_1 & & & & +\eta_1 & & & & & \geq & 0 \\
 & 57.5001x_1 & & & & & +\eta_2 & & & & \geq & 0 \\
 & 57.5001x_1 & & & & & & +\eta_3 & & & \geq & 0 \\
 & 38.3334x_1 & & & & & & & +\eta_4 & & \geq & 0 \\
 & 9.5834x_1 & & & & & & & & +\eta_5 & \geq & 0 \\
 & & & 11.7500x_3 & +5.2500x_4 & +\eta_1 & & & & & \geq & 0 \\
 & & & 23.5000x_3 & +10.5000x_4 & & +\eta_2 & & & & \geq & 0 \\
 & & & 23.5000x_3 & +10.5000x_4 & & & +\eta_3 & & & \geq & 0 \\
 & & & 15.6667x_3 & +7.0000x_4 & & & & +\eta_4 & & \geq & 0 \\
 & & & 3.9167x_3 & +1.7500x_4 & & & & & +\eta_5 & \geq & 0 \\
 & x_1, & x_2, & x_3, & x_4 & & & & & & \geq & 0
\end{array}$$

to get $x^3 = (300, 0, 734.043, 0)^\top$, $\eta_1^3 = -8265$, $\eta_2^3 = -17{,}250$, $\eta_3^3 = -17{,}250$, $\eta_4^3 = -11{,}500$, $\eta_5^3 = -2875$, and $\ell^3 = -31{,}489.4$.

Set $\ell \leftarrow \max\{\ell^3, \ell\} = \max\{-31{,}489.4, -42{,}500\} = -31{,}489.4$, and return to step 1.

After eight iterations, the multicut algorithm terminates with optimal solution $x^* = (236.4, 690, 432, 318)^\top$ and objective value -2268.5. This is the solution we obtained using the L-shaped algorithm, which took 15 iterations.

6.4 Adaptive Multicut Method

A novel approach one can think of is a *hybrid* method that provides a compromise between the L-shaped and the multicut algorithms. Such a method has the potential to provide better performance than either the single cut or the multicut algorithm for certain problem instances. An example of such a method is the *adaptive multicut aggregation* algorithm [15]. Even though the trade-offs in computation time are problem dependent, in general the L-shaped algorithm tends to have more iterations than the multicut algorithm. This is because there is information loss in the L-shaped algorithm due to aggregation of dual information from all the scenarios into one optimality cut in the master problem. On the contrary, the multicut method uses the dual information explicitly by placing an optimality cut for each scenario in the master problem. However, the master problem grows much faster and is expected to have relatively increased solution time when the number of scenarios $S = |\Omega|$ is very large. Recall that at k of the multicut algorithm the size of the master problem is $(m_1 + kS) \times (n_1 + S)$, as compared to $(m_1 + k) \times (n_1 + 1)$ in the L-shaped method. Thus the motivation for the *adaptive multicut* method is to provide some computational trade-offs by adjusting the level of optimality cut aggregation from single cut to pure multicut, which are two extremes. In general, a cut aggregation level between a single cut and pure multicut can result in better computation time for some problem instances.

6.4.1 Adaptive Multicut Decomposition Approach

At the beginning of each iteration of the adaptive multicut method, we need to decide on the *aggregation level* for partitioning the sample space Ω into subsets of scenarios. We refer to these subsets as *aggregates*. Then scenarios within each aggregate are used in computing an optimality cut for that subset of scenarios. Let J_k denote the *aggregation level* at iteration k, i.e., the number of scenario subsets $\Omega_j \subseteq \Omega, j = 1, \cdots, J_k$, such that $\Omega_1 \cup \Omega_2 \cup \cdots \Omega_{J_k} = \Omega$ and $\Omega_i \cap \Omega_j = \emptyset$ for all $\forall i \neq j$. Then each scenario $\omega \in \Omega$ belongs to only one $\Omega_j \subseteq \Omega$. The probability of each aggregate Ω_j, denoted p_j, is calculated as $p_j = \sum_{\omega \in \Omega_j} p(\omega)$

with $\sum_{j=1}^{J_k} p_j = \sum_{\omega \in \Omega} p(\omega) = 1$. Observe that an aggregation level J_k is such that $1 \leq J_k \leq S$, where $J_k = 1$ implies the L-shaped method, while an aggregation level $J_k = S$ implies the pure multicut method.

Essentially, the adaptive multicut method generalizes the L-shaped method by adjusting the number of aggregates during the course of the algorithm. This allows the algorithm to dynamically vary the scenarios in each aggregate in computing the optimality cuts. Therefore, the number of optimality cut variables in the master problem at each iteration of the algorithm will adapt to the number of aggregates dynamically set for that iteration, thus the name "adaptive" multicut algorithm. The basic idea of this approach is to let the algorithm "learn" more information about the expected recourse function and then settle for a level of aggregation that leads to finding the optimal solution faster for a given problem instance. So it is desirable to initialize the algorithm with a relatively large number of aggregates and then reducing it as more information about the expected recourse function is discovered during the course of the algorithm. For instances where the number of scenarios S is not too large, the algorithm can be initialized as pure multicut. In general, however, the algorithm should be initialized with the number of aggregates between 1 and S. From a pragmatic point of view, computer speed and memory will ultimately dictate the level of cut aggregation since a large number of aggregates typically require more computer memory.

To initialize the adaptive multicut algorithm, choose $J_0 \in [1, S]$ and create the set of initial aggregates $\Omega(0) = \{\Omega_1, \Omega_2, \cdots, \Omega_{J_0}\}$ according to some *aggregation rule*. Then add optimality cut variables $\eta_j, j = 1, \cdots, J_0$, to the master program. Then at a given iteration k of the algorithm, if *all* scenario subproblems for aggregate Ω_j are feasible, generate an *optimality cut* as follows:

$$(\beta_j^k)^\top x + \eta_j \geq \beta_{kj}^0,$$

where

$$(\beta_j^k)^\top = \sum_{\omega \in \Omega_j} p(\omega)(\pi(\omega)^k)^\top T(\omega)$$

and

$$\beta_{kj}^0 = \sum_{\omega \in \Omega_j} p(\omega)(\pi(\omega)^k)^\top r(\omega).$$

Recall that $\pi(\omega)$ is the optimal dual solution to the subproblem for scenario ω. Otherwise, if at least one scenario subproblem is infeasible, get a dual extreme ray $\mu(\omega)$ and calculate a *feasibility cut* as follows:

$$(\beta^k)^\top x \geq \beta_k^0,$$

where

$$(\beta^k)^\top = (\mu(\omega)^k)^\top T(\omega)$$

and

$$\beta_k^0 = (\mu(\omega)^k)^\top r(\omega).$$

Let Θ_k denote the set of iteration numbers up to k where all subproblems are feasible and optimality cuts are generated. Then the master program at iteration k takes the following form:

$$\begin{aligned}
\text{Min } c^\top x + \sum_{j=1}^{J_k} \eta_j, \\
\text{s.t. } Ax \geq b, \\
(\beta_j^t)^\top x + \eta_j \geq \beta_{tj}^0,\ t \in \Theta_k,\ j = 1, \cdots, J_k, \quad (6.18a)\\
(\beta^t)^\top x \geq \beta_t^0,\ t \in \{1, \cdots, k\} \setminus \Theta_k, \quad (6.18b)\\
x \geq 0,
\end{aligned}$$

where constraints (6.18a) and (6.18b) are the optimality and feasibility cuts, respectively. Next, we give a formal statement of the adaptive multicut algorithm.

6.4.2 *Basic Adaptive Multicut Algorithm*

We can formally state a basic adaptive multicut algorithm as follows:

Algorithm *Basic Adaptive Multicut*
begin

Step 0. Initialization. Set $k \leftarrow 0$, iteration index set $\Theta_k \leftarrow \emptyset$, choose $J_k \in [1, S]$, and initialize $\Omega(k)$ using some *aggregation scheme.*

Step 1. Solve Master Problem. Set $k \leftarrow k+1$ and solve master problem (6.18). Let $(x^k, \{\eta_j^k\}_{j=1,\cdots,J_k})$ be an optimal solution to Problem (6.18). If no constraint (6.18a) is present for some $j = 1, \cdots, J_k$, $\eta_j^k \leftarrow -\infty$ and is ignored in the computation.

Step 2. Update Cut Aggregation Level J_k.

Step 2a. Cut Aggregation.

Update $J_k \in [1, S]$ if necessary and generate a set of aggregates $\Omega(k)$ using $\Omega(k-1)$ based on some *aggregation scheme*. Each element $a_j \in$

$\Omega(k)$ is a union of some elements from $\Omega(k-1)$. If, according to the *aggregation scheme*, $a_1, a_2, \cdots, a_\ell \in \Omega(k-1)$ are aggregated into $a_j \in \Omega(k)$, then $a_j = \bigcup_{i=1}^{\ell} a_i$ and $p_j = \sum_{i=1}^{\ell} p_i$. Master problem (6.18) will be modified by removing variables $\eta_1, \cdots, \eta_\ell$ and introducing a new one η_j for the new aggregate.

Step 2b. Update Optimality Cuts.

Update the optimality cut coefficients in the master program (6.18) corresponding to new aggregate a_j as follows: For all iterations $t \in \Theta_k$ replace cuts corresponding to $a_1, \cdots, a_\ell$ with one new cut $(\beta_j^t)^\top x + \eta_j \geq \beta_{tj}^0$, where

$$\beta_{tj}^0 = p_j \sum_{i=1}^{\ell} (1/p_i)\beta_{ti}^0 \ \ \forall a_j \in \Omega(k), t \in \Theta_k,$$

and

$$\beta_j^t = p_j \sum_{i=1}^{\ell} (1/p_i)\beta_i^t \ \ \forall a_j \in \Omega(k), t \in \Theta_k.$$

Step 3. Solve Scenario Subproblems. For all $\omega \in \Omega$ solve

$$\text{Min } \{q(\omega)^\top y \mid Wy \geq r(\omega) - T(\omega)x^k, y \geq 0\}. \tag{6.19}$$

Let $\pi(\omega)^k$ be the dual multipliers associated with an optimal solution of Problem (6.19). If subproblem (6.19) is feasible for all $\omega \in \Omega$, set $\Theta_k \leftarrow \Theta_{k-1} \cup \{k\}$. For each $a_j \in \Omega(k)$ if

$$\eta_j^k < p_j \sum_{\omega \in a_j} \pi(\omega)^k (r(\omega) - T(\omega)x^k), \tag{6.20}$$

compute

$$\beta_{kj}^0 \leftarrow \sum_{\omega \in a_j} p(\omega)(\pi(\omega)^k)^\top r(\omega),$$

and

$$(\beta_j^k)^\top \leftarrow \sum_{\omega \in a_j} p(\omega)(\pi(\omega)^k)^\top T(\omega),$$

and add optimality cut $(\beta_j^k)^\top x + \eta_j \geq \beta_{kj}^0$ to master problem (6.18).

Otherwise, if Problem (6.19) is *infeasible* for some $\omega \in a_j$, let $\mu(\omega)^k$ be the associated dual extreme ray and define

$$\beta_k^0 \leftarrow (\mu(\omega)^k)^\top r(\omega),$$

and

$$(\beta^k)^\top \leftarrow (\mu(\omega)^k)^\top T(\omega).$$

Add feasibility cut $(\beta^k)^\top x \geq \beta_k^0$ to master problem (6.18).

Step 4. Termination. If condition (6.20) does not hold for *all* aggregates $a_j \in \Omega(k)$,

stop, x^k is optimal.

else

Return to Step 1.

---

end

Notice that in step 3 of the algorithm a single feasibility cut is generated for each aggregate $a_j \in \Omega(k)$ if at least one of the scenario subproblems in a_j is infeasible. An alternative approach one can take is to generate feasibility cuts for *all* infeasible subproblems in the aggregate at each iteration.

The flexibility of the adaptive multicut algorithm lies in various options for an aggregation scheme to use for obtaining $\Omega(k)$ from $\Omega(k-1)$. A basic aggregation scheme involves two basic rules [15], *redundancy threshold* and *bound on the number of aggregates.* The redundancy threshold rule is based on aggregating redundant or *inactive* optimality cuts in the master program since they contain "little" information about the optimal solution and can be aggregated without information loss. To complement this rule, a bound on the minimum and on the maximum number of aggregates should be imposed to prevent aggregating all scenarios prematurely and to curtail the proliferation of optimality cuts in the master program. This basic aggregation scheme can be summarized as follows:

Basic Aggregation Scheme:

- ◇ *Redundancy Threshold* τ. Set $0 < \delta < 1$. Consider iteration k after solving the master problem for some aggregate $a_j \in \Omega(k)$. Let k_j be the number of iterations when *all* optimality cuts corresponding to a_j are redundant (the associated dual multipliers are zero). Aggregate all a_j such that $k_j/|\Omega(k)| > \tau$ into one new aggregate.
- ◇ *Bound on the Number of Aggregates.* Set a bound on the minimum number of aggregates $|\Omega(k)|$ and a bound on the maximum number of aggregates at each iteration k. Set these bounds *a priori*, and keep them fixed throughout the algorithm run.

We should point out that redundancy threshold rule works best if the number of iterations is sufficiently large. Therefore, it is desirable to impose a "warm up" period during which no aggregation is made to give the algorithm enough time to "learn" information about the nature of the expected recourse function. Also, even though the redundancy threshold rule requires that *all* optimality cuts for each $a_j \in \Omega(k)$ be redundant, one can require to simply have a given fraction of the optimality cuts be redundant. Alternatively, one can consider aggregating scenarios that are "similar" or yield "similar" optimality cuts, i.e., cuts with similar gradients.

The opposite of cut aggregation is *cut disaggregation*, i.e., partitioning an aggregate into two or more aggregates. This would make available more explicit cut information about the recourse function in the master program. Such a scheme, however, would require all cut information to be stored in memory or written to file in order to partition any aggregate. This would involve a careful implementation in order to avoid slowing down the algorithm due to memory issues associated with bookkeeping.

Next, we give an illustration of the adaptive multicut algorithm for a fixed number of aggregates using a numerical example.

6.4.3 Numerical Example

Example 6.3 Apply two iterations of the adaptive multicut algorithm to the instance of the two-stage RN-SLP *abc-Production Planning* problem in Example 6.2 in Chap. 3. Use a fixed aggregation level $J_k = 3$ with $\Omega(k) = \Omega_1 \cup \Omega_2 \cup \Omega_3$ at each iteration k of the algorithm, where $\Omega_1 = \{\omega^1, \omega^2\}$, $\Omega_2 = \{\omega^3, \omega^4\}$, and $\Omega_3 = \{\omega^5\}$.

For $j = 1, \cdots, 3$ the aggregates are $a_1 \leftarrow \Omega_1$, $a_2 \leftarrow \Omega_2$, and $a_3 \leftarrow \Omega_3$ with corresponding optimality cut variables η_1, η_2, and η_3. The initial master problem without the optimality cut variables is given as follows:

$$
\begin{array}{rllllll}
\nu^0 := \text{Min} & 50x_1 & +30x_2 & +15x_3 & +10x_4 & & \\
\text{s.t.} & -x_1 & & & & \geq & -300 \\
& & -x_2 & & & \geq & -700 \\
& & & -x_3 & & \geq & -600 \\
& & & & -x_4 & \geq & -500 \\
& & -x_2 & -x_3 & -x_4 & \geq & -1600 \\
& x_1, & x_2, & x_3, & x_4 & \geq & 0.
\end{array}
$$

For a given $x^k = (x_1^k, x_2^k, x_3^k, x_4^k)^\top$, the scenario subproblem for each $\omega^s \in \Omega$ for $s = 1, \cdots, 5$ can be written as follows:

$$\begin{array}{rrrrl} \varphi^k(x^k,\omega^s) := \text{Min} & -1150y_1 & -1525y_2 & -1900y_3 & \\ \text{s.t.} & -6y_1 & -8y_2 & -10y_3 & \geq -x_1 \\ & -20y_1 & -25y_2 & -28y_3 & \geq -x_2^k \\ & -12y_1 & -15y_2 & -18y_3 & \geq -x_3^k \\ & -8y_1 & -10y_2 & -14y_3 & \geq -x_4^k \\ & -y_1 & & & \geq d_a^s \\ & & -y_2 & & \geq d_b^s \\ & & & -y_3 & \geq d_c^s \\ & y_1, & y_2, & y_3 & \geq 0, \end{array}$$

where the RHS $r(\omega^s) = (0, 0, 0, 0, d_a^s, d_b^s, d_c^s)^\top$. We can now apply the basic adaptive multicut algorithm as follows:

Step 0. Initialization. Set $k \leftarrow 0$, iteration index set $\Theta_0 \leftarrow \emptyset$, choose $J_0 \leftarrow 5$, and initialize $\Omega(0) = \Omega_1 \cup \Omega_2 \cup \Omega_3$, where $\Omega_1 = \{\omega^1, \omega^2\}, \Omega_2 = \{\omega^3, \omega^4\}$, and $\Omega_3 = \{\omega^5\}$.

Step 1. Solve Master Problem. Set $k \leftarrow 0 + 1 = 1$ and solve initial master problem to get $x^1 \leftarrow (0, 0, 0, 0)^\top$. Set $\eta_j^1 \leftarrow -\infty,\ j = 1, 2, 3$.

Step 2. Update Cut Aggregation Level J_1.

Step 2a. Cut Aggregation.

Aggregation level: $J_1 \leftarrow J_0,\ \Omega(1) \leftarrow \Omega(0)$.

Step 2b. Update Optimality Cuts.

Not necessary to update the optimality cut coefficients in the master program.

Step 3. Solve Scenario Subproblems. For $\omega^s \in \Omega\ s = 1, \cdots, 5$ solve subproblem:

$$\begin{array}{rrrrl} \varphi^1(x^1,\omega^s) := \text{Min} & -1150y_1 & -1525y_2 & -1900y_3 & \\ \text{s.t.} & -6y_1 & -8y_2 & -10y_3 & \geq 0 \\ & -20y_1 & -25y_2 & -28y_3 & \geq 0 \\ & -12y_1 & -15y_2 & -18y_3 & \geq 0 \\ & -8y_1 & -10y_2 & -14y_3 & \geq 0 \\ & -y_1 & & & \geq d_a^s \\ & & -y_2 & & \geq d_b^s \\ & & & -y_3 & \geq d_c^s \\ & y_1, & y_2, & y_3 & \geq 0. \end{array}$$

$s = 1, \omega^1$: subproblem feasible, $\varphi^1(x^1, \omega^1) = 0,\ y_1^1 = (0, 0, 0)^\top,\ \pi_1^1 = (191.667, 0, 0, 0, 0, 0, 0)^\top$.

$$\beta_1^0 \leftarrow p(\omega^1)(\pi_1^1)^\top r(\omega^1) = 0.15(191.667, 0, 0, 0, 0, 0, 0)$$

$$(0, 0, 0, 0, -15, -10, -5)^\top$$

$$= 0.$$

$$\beta_1^\top \leftarrow p(\omega^1)(\pi_1^1)^\top T = 0.15(191.667, 0, 0, 0, 0, 0, 0) \begin{bmatrix} 1\,0\,0\,0 \\ 0\,1\,0\,0 \\ 0\,0\,1\,0 \\ 0\,0\,0\,1 \\ 0\,0\,0\,0 \\ 0\,0\,0\,0 \\ 0\,0\,0\,0 \end{bmatrix}$$

$$= (28.7501, 0, 0, 0).$$

$s = 2, \omega^2$: subproblem feasible, $\varphi^1(x^1, \omega^2) = 0,\ y_2^1 = (0, 0, 0)^\top,\ \pi_2^1 = (191.667, 0, 0, 0, 0, 0, 0)^\top$.

$$\beta_2^0 \leftarrow p(\omega^2)(\pi_2^1)^\top r(\omega^2) = 0.3(191.667, 0, 0, 0, 0, 0, 0)$$

$$(0, 0, 0, 0, -20, -15, -15)^\top$$

$$= 0.$$

$$\beta_2^\top \leftarrow p(\omega^2)(\pi_2^1)^\top T = 0.3(191.667, 0, 0, 0, 0, 0, 0) \begin{bmatrix} 1\,0\,0\,0 \\ 0\,1\,0\,0 \\ 0\,0\,1\,0 \\ 0\,0\,0\,1 \\ 0\,0\,0\,0 \\ 0\,0\,0\,0 \\ 0\,0\,0\,0 \end{bmatrix}$$

$$= (57.5001, 0, 0, 0).$$

Generate *optimality* cut for aggregate a_1: aggregate scenario ω^1 and ω^2 cut information to obtain the following optimality cut: $(\beta_1^\top + \beta_2^\top)x + \eta_1 \geq (\beta_1^0 + \beta_2^0)$: $\quad 86.2502x_1 + \eta_1 \geq 0$.
$s = 3, \omega^3$: subproblem feasible, $\varphi^1(x^1, \omega^3) = 0,\ y_3^1 = (0, 0, 0)^\top,\ \pi_3^1 = (191.667, 0, 0, 0, 0, 0, 0)^\top$.

$$\beta_3^0 \leftarrow p(\omega^3)(\pi_3^1)^\top r(\omega^3) = 0.3(191.667, 0, 0, 0, 0, 0, 0)$$

$$(0, 0, 0, 0, -20, -20, -25)^\top$$

$$= 0.$$

$$\beta_3^\top \leftarrow p(\omega^3)(\pi_3^1)^\top T = 0.3(191.667, 0, 0, 0, 0, 0, 0) \begin{bmatrix} 1 & 0 & 0 & 0 \\ 0 & 1 & 0 & 0 \\ 0 & 0 & 1 & 0 \\ 0 & 0 & 0 & 1 \\ 0 & 0 & 0 & 0 \\ 0 & 0 & 0 & 0 \\ 0 & 0 & 0 & 0 \end{bmatrix}$$

$$= (57.5001, 0, 0, 0).$$

$s = 4, \omega^4$: subproblem feasible, $\varphi^1(x^1, \omega^4) = 0,\ y_4^1 = (0, 0, 0)^\top,\ \pi_4^1 = (191.667, 0, 0, 0, 0, 0, 0)^\top$.

$$\beta_4^0 \leftarrow p(\omega^4)(\pi_4^1)^\top r(\omega^4) = 0.2(191.667, 0, 0, 0, 0, 0, 0)$$
$$(0, 0, 0, 0, -30, -25, -30)^\top$$
$$= 0.$$

$$\beta_4^\top \leftarrow p(\omega^4)(\pi_4^1)^\top T = 0.2(191.667, 0, 0, 0, 0, 0, 0) \begin{bmatrix} 1 & 0 & 0 & 0 \\ 0 & 1 & 0 & 0 \\ 0 & 0 & 1 & 0 \\ 0 & 0 & 0 & 1 \\ 0 & 0 & 0 & 0 \\ 0 & 0 & 0 & 0 \\ 0 & 0 & 0 & 0 \end{bmatrix}$$

$$= (38.3334, 0, 0, 0).$$

Generate *optimality* cut for aggregate a_2: aggregate scenario ω^3 and ω^4 cut information to obtain the following optimality cut: $(\beta_3^\top + \beta_4^\top)x + \eta_2 \geq (\beta_3^0 + \beta_4^0)$: $\quad 95.8335x_1 + \eta_2 \geq 0$.
$s = 5, \omega^5$: subproblem feasible, $\varphi^1(x^1, \omega^5) = 0,\ y_5^1 = (0, 0, 0)^\top,\ \pi_5^1 = (191.667, 0, 0, 0, 0, 0, 0)^\top$.

$$\beta_5^0 \leftarrow p(\omega^5)(\pi_5^1)^\top r(\omega^5) = 0.05(191.667, 0, 0, 0, 0, 0, 0)$$
$$(0, 0, 0, 0, -10, -10, -10)^\top$$
$$= 0.$$

$$\beta_5^\top \leftarrow p(\omega^5)(\pi_5^1)^\top T = 0.05(191.667, 0, 0, 0, 0, 0, 0) \begin{bmatrix} 1\,0\,0\,0 \\ 0\,1\,0\,0 \\ 0\,0\,1\,0 \\ 0\,0\,0\,1 \\ 0\,0\,0\,0 \\ 0\,0\,0\,0 \\ 0\,0\,0\,0 \end{bmatrix}$$

$$= (9.5834, 0, 0, 0).$$

Generate *optimality* cut for aggregate a_3: optimality cut, $\beta_5^\top x + \eta_3 \geq \beta_5^0$: $9.5834x_1 + \eta_3 \geq 0$.
Since the subproblems are feasible for all $\omega^s \in \Omega(1)$, set $\Theta_1 \leftarrow \Theta_0 \cup \{1\}$ and add the optimality cuts to the master problem:

$$\begin{array}{llllllllll}
\ell^2 := \text{Min} & 50x_1 & +30x_2 & +15x_3 & +10x_4 & +\eta_1 & +\eta_2 & +\eta_3 & & \\
\text{s.t.} & -x_1 & & & & & & & \geq & -300 \\
 & & -x_2 & & & & & & \geq & -700 \\
 & & & -x_3 & & & & & \geq & -600 \\
 & & & & -x_4 & & & & \geq & -500 \\
 & & -x_2 & -x_3 & -x_4 & & & & \geq & -1600 \\
 & 86.2502x_1 & & & & +\eta_1 & & & \geq & 0 \\
 & 95.8335x_1 & & & & & +\eta_2 & & \geq & 0 \\
 & 9.5834x_1 & & & & & & +\eta_3 & \geq & 0 \\
 & x_1, & x_2, & x_3, & x_4 & & & & \geq & 0.
\end{array}$$

Step 4. Termination. Condition (6.20) does hold for *all* aggregates $a_j \in \Omega(1)$, $j = 1, 2, 3$.
Return to Step 1.
Iteration $k = 2$:

Step 1. Solve Master Problem. Set $k \leftarrow 1 + 1 = 2$ and solve master problem to get $x^2 = (300, 0, 0, 0)^\top$, $\eta_1^2 = -25{,}875$, $\eta_2^2 = -28{,}750$, $\eta_3^2 = -2956$, and $\ell^2 = -42{,}581$. Set $\ell \leftarrow \max\{\ell^2, \ell\} = \max\{-42{,}581, -\infty\} = -42{,}581$.

Step 2. Update Cut Aggregation Level J_2.

Step 2a. Cut Aggregation.
Aggregation level: $J_2 \leftarrow J_1$, $\Omega(2) \leftarrow \Omega(1)$.

Step 2b. Update Optimality Cuts.
Not necessary to update the optimality cut coefficients in the master program.

Step 3. Solve Scenario Subproblems. For $\omega^s \in \Omega$ $s = 1, \cdots, 5$ solve subproblem:

$$
\begin{array}{rrrrrr}
\varphi^2(x^2, \omega^s) := \text{Min} & -1150y_1 & -1525y_2 & -1900y_3 & & \\
\text{s.t.} & -6y_1 & -8y_2 & -10y_3 & \geq & -300 \\
& -20y_1 & -25y_2 & -28y_3 & \geq & 0 \\
& -12y_1 & -15y_2 & -18y_3 & \geq & 0 \\
& -8y_1 & -10y_2 & -14y_3 & \geq & 0 \\
& -y_1 & & & \geq & d_a^s \\
& & -y_2 & & \geq & d_b^s \\
& & & -y_3 & \geq & d_c^s \\
& y_1, & y_2, & y_3 & \geq & 0.
\end{array}
$$

$s = 1, \omega^1$: subproblem feasible, $\varphi^2(x^2, \omega^1) = 0,\ y_1^2 = (0, 0, 0)^\top,\ \pi_1^2 = (0, 0, 78.3333, 35, 0, 0, 0)^\top$.

$$
\begin{aligned}
\beta_1^0 \leftarrow p(\omega^1)(\pi_1^2)^\top r(\omega^1) &= 0.15(0, 0, 78.3333, 35, 0, 0, 0) \\
&\qquad (0, 0, 0, 0, -15, -10, -5)^\top \\
&= 0.
\end{aligned}
$$

$$
\begin{aligned}
\beta_1^\top \leftarrow p(\omega^1)(\pi_1^2)^\top T &= 0.15(0, 0, 78.3333, 35, 0, 0, 0) \begin{bmatrix} 1\,0\,0\,0 \\ 0\,1\,0\,0 \\ 0\,0\,1\,0 \\ 0\,0\,0\,1 \\ 0\,0\,0\,0 \\ 0\,0\,0\,0 \\ 0\,0\,0\,0 \end{bmatrix} \\
&= (0, 0, 7.83333, 3.5).
\end{aligned}
$$

$s = 2, \omega^2$: subproblem feasible, $\varphi^2(x^2, \omega^2) = 0,\ y_2^2 = (0, 0, 0)^\top,\ \pi_2^2 = (0, 0, 78.3333, 35, 0, 0, 0)^\top$.

$$
\begin{aligned}
\beta_2^0 \leftarrow p(\omega^2)(\pi_2^2)^\top r(\omega^2) &= 0.3(0, 0, 78.3333, 35, 0, 0, 0) \\
&\qquad (0, 0, 0, 0, -20, -15, -15)^\top \\
&= 0.
\end{aligned}
$$

$$\beta_2^\top \leftarrow p(\omega^2)(\pi_2^2)^\top T = 0.3(0,0,78.3333,35,0,0,0) \begin{bmatrix} 1\,0\,0\,0 \\ 0\,1\,0\,0 \\ 0\,0\,1\,0 \\ 0\,0\,0\,1 \\ 0\,0\,0\,0 \\ 0\,0\,0\,0 \\ 0\,0\,0\,0 \end{bmatrix}$$

$$= (0,0,23.5,10.5).$$

Generate *optimality* cut for aggregate a_1: aggregate scenario ω^1 and ω^2 cut information to obtain the following optimality cut: $(\beta_1^\top + \beta_2^\top)x + \eta_1 \geq (\beta_1^0 + \beta_2^0)$: $\quad 34.25x_3 + 15.75x_4 + \eta_1 \geq 0.$
$s = 3, \omega^3$: subproblem feasible, $\varphi^2(x^2, \omega^3) = 0,\ y_3^2 = (0,0,0)^\top,\ \pi_3^2 = (0,0,78.3333,35,0,0,0)^\top$.

$$\beta_3^0 \leftarrow p(\omega^3)(\pi_3^2)^\top r^1 = 0.3(0,0,78.3333,35,0,0,0)$$
$$(0,0,0,0,-20,-20,-25)^\top$$
$$= 0.$$

$$\beta_3^\top \leftarrow p(\omega^3)(\pi_3^2)^\top T = 0.3(0,0,78.3333,35,0,0,0) \begin{bmatrix} 1\,0\,0\,0 \\ 0\,1\,0\,0 \\ 0\,0\,1\,0 \\ 0\,0\,0\,1 \\ 0\,0\,0\,0 \\ 0\,0\,0\,0 \\ 0\,0\,0\,0 \end{bmatrix}$$

$$= (0,0,23.5,10.5).$$

$s = 4, \omega^4$: subproblem feasible, $\varphi^2(x^2, \omega^4) = 0,\ y_4^2 = (0,0,0)^\top,\ \pi_4^2 = (0,0,78.3333,35,0,0,0)^\top$.

$$\beta_4^0 \leftarrow p(\omega^4)(\pi_4^2)^\top r^1 = 0.2(0,0,78.3333,35,0,0,0)$$
$$(0,0,0,0,-30,-25,-30)^\top$$
$$= 0.$$

$$\beta_4^\top \leftarrow p(\omega^4)(\pi_4^2)^\top T = 0.2(0, 0, 78.3333, 35, 0, 0, 0) \begin{bmatrix} 1\,0\,0\,0 \\ 0\,1\,0\,0 \\ 0\,0\,1\,0 \\ 0\,0\,0\,1 \\ 0\,0\,0\,0 \\ 0\,0\,0\,0 \\ 0\,0\,0\,0 \end{bmatrix}$$

$$= (0, 0, 15.6667, 7).$$

Generate *optimality* cut for aggregate a_2: aggregate scenario ω^3 and ω^4 cut information to obtain the following optimality cut: $(\beta_3^\top + \beta_4^\top)x + \eta_2 \geq (\beta_3^0 + \beta_4^0)$: $\quad 39.1667x_3 + 17.5x_4 + \eta_2 \geq 0$.
$s = 5, \omega^5$: subproblem feasible, $\varphi^2(x^2, \omega^5) = 0,\ y_5^2 = (0, 0, 0)^\top,\ \pi_5^2 = (0, 0, 78.3333, 35, 0, 0, 0)^\top$.

$$\beta_5^0 \leftarrow p(\omega^5)(\pi_5^2)^\top r^1 = 0.05(0, 0, 78.3333, 35, 0, 0, 0)$$
$$(0, 0, 0, 0, -10, -10, -10)^\top$$
$$= 0.$$

$$\beta_5^\top \leftarrow p(\omega^5)(\pi_5^2)^\top T = 0.05(0, 0, 78.3333, 35, 0, 0, 0) \begin{bmatrix} 1\,0\,0\,0 \\ 0\,1\,0\,0 \\ 0\,0\,1\,0 \\ 0\,0\,0\,1 \\ 0\,0\,0\,0 \\ 0\,0\,0\,0 \\ 0\,0\,0\,0 \end{bmatrix}$$

$$= (0, 0, 3.9167, 1.75).$$

Generate *optimality* cut for aggregate a_3: Optimality cut, $\beta_5^\top x + \eta_3 \geq \beta_5^0$: $3.9167x_3 + 1.75x_4 + \eta_3 \geq 0$.
Since the subproblems are feasible for all $\omega^s \in \Omega(2)$, set $\Theta_2 \leftarrow \Theta_1 \cup \{2\}$ and add the optimality cuts to the master problem.

$$
\begin{array}{llllllll}
\ell^3 := \text{Min} & 50x_1 & +30x_2 & +15x_3 & +10x_4 & +\eta_1 & +\eta_2 & +\eta_3 & \\
\text{s.t.} & -x_1 & & & & & & & \geq -300 \\
 & & -x_2 & & & & & & \geq -700 \\
 & & & -x_3 & & & & & \geq -600 \\
 & & & & -x_4 & & & & \geq -500 \\
 & & -x_2 & -x_3 & -x_4 & & & & \geq -1600 \\
 & 86.2502x_1 & & & & +\eta_1 & & & \geq 0 \\
 & 95.8335x_1 & & & & & +\eta_2 & & \geq 0 \\
 & 9.5834x_1 & & & & & & +\eta_3 & \geq 0 \\
 & & & 34.2500x_3 & +15.7500x_4 & +\eta_1 & & & \geq 0 \\
 & & & 39.1665x_3 & +17.5000x_4 & & +\eta_2 & & \geq 0 \\
 & & & 3.9167x_3 & +1.7500x_4 & & & +\eta_3 & \geq 0 \\
 & x_1, & x_2, & x_3, & x_4 & & & & \geq 0
\end{array}
$$

Step 4. Termination. Condition (6.20) does hold for *all* aggregates $a_j \in \Omega(k)$, Return to Step 1.

---

end

In step 1, we get $x^3 = (300, 0, 600, 338)^\top$, $\eta_1^3 = -25{,}875$, $\eta_2^3 = -28{,}750$, $\eta_3^3 = -2941$, and $\ell^3 = -30{,}186$. We set $\ell \leftarrow \max\{\ell^3, \ell\} = \max\{-30{,}186, -42{,}500\} = -30{,}186$. After nine iterations, the adaptive multicut algorithm terminates with optimal solution $x^* = (236.4, 690, 432, 318)^\top$ and objective value -2268.5. This is the solution we obtained using the L-shaped and multicut algorithms, which took 15 and 8 iterations, respectively.

6.5 Lagrangian Based Methods

To end this chapter let us introduce an alternative to the L-shaped and multicut algorithms. In particular, there are methods based on the *scenario formulation* of the two-stage RN-SLP (6.1) which we introduced in Chap. 2. These methods are often referred to as *scenario (dual) decomposition* methods. Instead of applying *stage-wise* decomposition of RN-SLP, we apply *scenario-wise* decomposition as follows:

$$
\begin{aligned}
\text{Min} \quad & \sum_{\omega \in \Omega} p_\omega \{c^\top x(\omega) + q(\omega)^\top y(\omega)\} \\
\text{s.t.} \quad & Ax(\omega) \geq b, \ \forall \omega \in \Omega \\
& T(\omega)x(\omega) + Wy(\omega) \geq r(\omega), \ \forall \omega \in \Omega \qquad (6.21) \\
& x(\omega) = \sum_{\varpi \in \Omega} p(\varpi)x(\varpi), \ \forall \omega \in \Omega \\
& x, x(\omega), \ y(\omega) \geq 0, \ \forall \omega \in \Omega.
\end{aligned}
$$

The last set of constraints is the *nonanticipativity* constraints that link all scenarios and are thus considered to be "hard" or "complicating" constraints. The basic idea for making the problem separable by scenarios is to apply *Lagrangian relaxation* by *dualizing* the *nonanticipativity* constraints, i.e., placing them in the objective with a penalty term. The Lagrangian relaxation for scenario $\omega \in \Omega$ can be given as follows:

$$
\begin{aligned}
D(\lambda(\omega)) := \text{Min } & p(\omega)\left(c^\top x(\omega)+q(\omega)^\top y(\omega)\right)+\lambda(\omega)^\top\left(x(\omega)-\sum_{\varpi\in\Omega} p(\varpi)x(\varpi)\right)\\
\text{s.t. } & Ax(\omega) \geq b\\
& T(\omega)x(\omega) + Wy(\omega) \geq r(\omega)\\
& x(\omega),\ y(\omega) \geq 0,
\end{aligned}
$$

where $\lambda(\omega) \in \mathbb{R}^{n_1}$ is the Lagrangian dual multiplier associated with the nonanticipativity constraints. Now define $\lambda = (\lambda(\omega^1), \lambda(\omega^2), \cdots, \lambda(\omega^{|\Omega|}))$ and let $D(\lambda) = \sum_{\omega\in\Omega} D(\lambda_\omega)$. Then the *Lagrangian dual problem* for RN-SLP is given as follows:

$$
Max_{\lambda\in\mathbb{R}^{n_1|\Omega|}}\ D(\lambda). \tag{6.23}
$$

Solving Problem (6.23) is tantamount to finding λ, i.e., finding $\lambda(\omega)$'s for all $\omega \in \Omega$ such that the nonanticipativity constraints are satisfied, at minimal cost. The problem is amenable to dual ascent or subgradient optimization based algorithms such as dual decomposition [4]. An alternative approach, which we consider, is the progressive hedging (PH) method [12, 16]. So next, we decompose Problem (6.21) for the PH approach, state the PH algorithm, and give a numerical illustration of the algorithm.

6.5.1 Progressive Hedging Decomposition Approach

Let us now index the scenario by $s = 1, \cdots, S$, where $S = |\Omega|$. We continue to denote by x the nonanticipative decision that does not depend on the scenario ω^s, i.e., $x(\omega^s) = x, \forall s = 1, \cdots, S$. Let the feasible set for scenario ω^s, denoted $X(\omega^s)$, be defined as follows:

$$
X(\omega^s) := \{Ax \geq b,\ T(\omega^s)x + Wy(\omega^s) \geq r(\omega^s),\ x,\ y(\omega^s) \geq 0\}.
$$

Then the *initial* scenario ω subproblem is given as follows:

$$
\underset{(x,y(\omega^s))\in X(\omega^s)}{\text{Min}} \quad p(\omega)\{c^\top x + q(\omega)^\top y(\omega)\}. \tag{6.24}
$$

The basic idea of progressive hedging is to keep this subproblem structure throughout the course of the algorithm and introduce a nonanticipative solution $\bar{x}^k$ at each iteration with $x(\omega^s)^k$ as the solution to the scenario subproblem. The solution $x(\omega^s)^k$ is not necessarily feasible to the overall problem (Problem (6.21)). The nonanticipative solution is calculated using the *expectation* nonanticipativity constraint (introduced in Chap. 2) as follows:

$$\bar{x}^k := \sum_{s=1}^{S} p(\omega^s) x(\omega^s)^k.$$

Next, a penalty factor $\rho > 0$ (multiplier) is introduced together with a termination threshold δ^k as the input parameters. Then, for a given scenario solution $x(\omega^s)^k$, we calculate w_s^k, the weighted discrepancy between $x(\omega^s)^k$ and $\bar{x}^k$ as follows:

$$w_s^k := \rho(x(\omega^s)^k - \bar{x}^k).$$

This allows for implicitly implementing the nonanticipativity constraints, and the scenario subproblem at iteration k is as follows:

$$\operatorname*{Min}_{(x, y(\omega^s)) \in X(\omega^s)} \quad p(\omega) c^\top x + (w_s^{k-1})^\top x + \frac{\rho}{2} ||x - \bar{x}^{k-1}||^2 + p(\omega^s) q(\omega)^\top y(\omega^s). \tag{6.25}$$

Recall from Chap. 5 that the third term in the objective function of Problem (6.25) is a quadratic *regularization* term and provides the advantage of having iterates $x(\omega^s)^k$ that do not deviate "too far" the nonanticipative solution $\bar{x}^k$ depending on the selected value of ρ). However, there is a "no free lunch" because subproblem (6.25) is a linear quadratic programming (QP) problem and must be solved using QP algorithms such as the barrier method.

6.5.2 *Progressive Hedging Algorithm*

The PH algorithm can be formally stated as follows:

Algorithm *PH*
begin

Step 0. Initialization. Set $k \leftarrow 0$ and choose $\epsilon > 0$ and $\rho > 0$.

(a) For $s = 1, \cdots, S$, compute

$$(x(\omega^s)^k, y(\omega^s)^k) \leftarrow \operatorname{argmin}_{(x, y(\omega^s)) \in X(\omega^s)} \{p(\omega) c^\top x + p(\omega^s) q(\omega^s)^\top y(\omega^s)\}.$$

(b) Compute $\bar{x}^k \leftarrow \sum_{s=1}^{S} p(\omega^s)x(\omega^s)^k$.
(c) For $s = 1, \cdots, S$, compute $w_s^k \leftarrow \rho(x(\omega^s)^k - \bar{x}^k)$.

Step 1. Solve Subproblems and Compute Nonanticipative Solution.

(a) Set $k \leftarrow k + 1$.
(b) For $s = 1, \cdots, S$, compute

$$(x(\omega^s)^k, y(\omega^s)^k) \leftarrow \text{argmin}_{(x,y(\omega^s))\in X(\omega^s)} \{p(\omega)c^\top x + (w_s^{k-1})^\top x + \frac{\rho}{2}||x - \bar{x}^{k-1}||^2 + p(\omega^s)q(\omega)^\top y(\omega^s)\}.$$

(c) Compute $\bar{x}^k \leftarrow \sum_{s=1}^{S} p(\omega^s)x(\omega^s)^k$.

Step 2. Update the Weights.

(a) For $s = 1, \cdots, S$, compute

$$w_s^k \leftarrow w(\omega^s)^{k-1} + \rho\left(x(\omega^s)^k - \bar{x}^k\right).$$

(b) Compute $\delta^k \leftarrow \sum_{s=1}^{S} p(\omega^s)||x_s^k - \bar{x}^k||$.

Step 3. Termination.

If $\delta^k < \epsilon$

Stop, $\bar{x}^k$ is ϵ-optimal.

else

Return to step 1.

---

end

The PH algorithm is relatively easy to state and its convergence is based on the proximal point method [11]. When the nonanticipative decision variable x is continuous, the PH algorithm converges linearly to a common solution $\bar{x}$. However, the algorithm is sensitive to the selected values for the penalty ρ. The value of ρ has to be carefully chosen for each instance in proportion to the unit cost of the associated decision variable.

Bibliographic Notes

The algorithms presented in this chapter focus on a two-stage SP. Several algorithms have also been developed for multistage SP (MSP), a framework for modeling problems involving sequential decision-making over time. Decomposition approaches for MSP were first proposed by [2] and are based on the L-shaped method [14]. Two other classical methods for MSP are the progressive hedging method [12] and the stochastic dual dynamic programming (SDDP) algorithm [9], which was developed for solving hydrothermal scheduling problems. A tutorial by [10] provides a single coherent modeling framework for MSP that encompasses different approaches, including dynamic programming, SP, and simulation.

Problems

6.1 L-Shaped Algorithm
Perform two more iterations of the L-shaped algorithm toward solving the problem in Example 6.1. Use an LP solver of your choice to solve the master program and the subproblems.

6.2 Decomposition and L-Shaped Algorithm
Consider an instance of the two-stage RN-SLP with recourse

$$\begin{aligned} \text{Min } & c^\top x + \mathbb{E}[f(x, \tilde{\omega}] \\ \text{s.t. } & Ax \leq b \\ & x \geq 0, \end{aligned}$$

where for a first-stage decision x and realization (scenario) $\omega \in \Omega$ of $\tilde{\omega}$,

$$\begin{aligned} f(x, \omega) := \text{Min } & q(\omega)^\top y \\ \text{s.t. } & Wy \geq r(\omega) - Tx \\ & y \geq 0 \end{aligned}$$

with data given as follows:

First-stage:
Objective coefficient vector:

$$c = (6, 4)^\top$$

$$A = \begin{bmatrix} 1 & 1 \\ 0 & 1 \end{bmatrix}$$

$$b = (5, 1)^\top.$$

Second-stage:
Number of scenarios: $|\Omega| = S = 2$.

Scenario probabilities: $p(\omega^s)$, $s = 1, 2$,

$$p(\omega^1) = 0.5,\ p(\omega^2) = 0.5.$$

Objective coefficient vector: $q(\omega^s)$, $s = 1, 2$,

$$q(\omega^1) = q(\omega^2) = (4, 2)^\top.$$

Recourse matrix:

$$W = \begin{bmatrix} 1 & -1 \\ 0 & 1 \end{bmatrix}.$$

Technology matrix:

$$T(\omega^1) = T(\omega^2) = T = \begin{bmatrix} 1 & -2 \\ 1 & 2 \end{bmatrix}.$$

RHS vector $r(\omega^s)$, $s = 1, 2$:

$$r(\omega^1) = (3, 5)^\top,$$

$$r(\omega^2) = (4, 8)^\top.$$

First-stage feasible set:

$$X = \{\ x_1 \quad x_2 \geq \ \ 5 \\ x_2 \geq -1\,. \\ x_1,\ x_2,\ \geq\ 0\}$$

(a) Write the deterministic equivalent formulation (DEP), and solve the instance using a *direct solver* and report your solution.

(b) Now decompose the DEP in part (a), and write the master program and two scenario subproblems.

(c) Solve the decomposed problem using the *L-shaped algorithm*. Initialize the algorithm with $x^0 \leftarrow \text{argmin}_X\{c^\top x\}$. At each iteration k of the algorithm, clearly state x^k, the master program with all the cuts, subproblem and its primal and dual solutions, lower bound ℓ, and the upper bound u.

6.3 L-Shaped Convergence

Prove that L-shaped algorithm terminates after a finite number of iterations and finds an optimal solution if it exists.

6.4 L-Shaped Algorithm Implementation

(a) Implement (code) the L-shaped algorithm using software and an LP solver of your choice.
(b) Use your L-shaped code to solve the instance in Example 6.1
(c) Use your L-shaped code to solve the instance in Problem 6.2.
(d) Use your L-shaped code to solve standard test instances of your choice.

6.5 Multicut L-Shaped Algorithm
Perform two more iterations of the multicut L-shaped algorithm toward solving the problem in Example 6.2. Use an LP solver of your choice to solve the master program and the subproblems.

6.6 Decomposition and Multicut L-Shaped Algorithm
Consider the instance given in Problem 6.2:

(a) Write the master program and two scenario subproblems.
(b) Perform two iterations of the *Multicut L-shaped algorithm*. Initialize the algorithm with $x^0 \leftarrow \text{argmin}_X\{c^\top x\}$. At each iteration k of the algorithm, clearly state x^k, the master program with all the cuts, subproblem and its primal and dual solutions, lower bound ℓ, and the upper bound u.

6.7 Multicut L-Shaped Convergence
Prove that Multicut L-shaped algorithm terminates after a finite number of iterations and finds an optimal solution if it exists.

6.8 Multicut L-Shaped Algorithm Implementation

(a) Implement (code) the Multicut L-shaped algorithm using software and LP solver of your choice.
(b) Use your Multicut L-shaped code to solve the instance in Example 6.1
(c) Use your Multicut L-shaped code to solve the instance in Problem 6.2.
(d) Use your Multicut L-shaped code to solve standard test instances of your choice.

6.9 Comparison of L-Shaped and Multicut L-Shaped Methods

(a) Show that the Multicut L-shaped algorithm will have at most as many iterations as the single cut L-shaped algorithm, i.e., can have fewer iterations.
(b) Compare and contrast the L-shaped and multicut L-shaped algorithms in terms of generating feasibility cuts, termination condition, number of iterations, and computation time.

6.10 Adapive Multicut L-Shaped Algorithm
Perform two more iterations of the adaptive multicut L-shaped algorithm toward solving the problem in Example 6.3. Use an LP solver of your choice to solve the master program and the subproblems.

6.11 Decomposition and Adaptive Multicut L-Shaped Algorithm
Consider the instance given in Problem 6.3:

(a) Partition Ω into three (3) aggregates of your choice and then write the master program and scenario subproblems under each aggregate.
(b) Perform two iterations of the *adaptive multicut L-shaped algorithm.* Choose your own aggregation scheme and initialize the algorithm with $x^0 \text{argmin}_X\{c^\top x\}$. At each iteration k of the algorithm, clearly state x^k, the master program with all the cuts, subproblem and its primal and dual solutions, lower bound ℓ, and the upper bound u.

6.12 Adaptive Multicut L-Shaped Convergence
Prove that adaptive multicut L-shaped algorithm terminates after a finite number of iterations and finds an optimal solution if it exists.

6.13 Adaptive Multicut L-Shaped Algorithm Implementation

(a) Implement (code) the adaptive multicut L-shaped algorithm using software and LP solver of your choice.
(b) Use your adaptive multicut L-shaped algorithm code to solve the instance in Example 6.3.
(c) Use your adaptive multicut L-shaped algorithm code to solve standard test instances of your choice.

6.14 Comparison of L-Shaped, Multicut, and Adaptive Multicut Methods

(a) Show that the adaptive multicut L-shaped algorithm will have a number of iterations ranging from that of the single cut to the pure multicut method.
(b) Compare and contrast the L-shaped, multicut L-shaped, and adaptive multicut L-shaped algorithms in terms of generating feasibility cuts, termination condition, number of iterations, and computation time.

6.15 Scenario Decomposition and the PH Algorithm
Consider the problem instance in Problem 6.2.

(a) Write the scenario formulation of the problem, and solve the instance using a *direct solver* and report your solution.
(b) Now perform a scenario decomposing of the problem in part (a) for the PH algorithm and write the two scenario subproblems.
(c) Solve the decomposed problem using the *PH algorithm.* Choose an appropriate value for ρ. At each iteration k of the algorithm, clearly state the subproblem solutions $x(\omega^s)^k$, nonanticipative solution $\bar{x}^k$, lower bound ℓ, the weight w_s^k, and the value of δ^k.

6.16 PH Algorithm Implementation

(a) Implement (code) the PH algorithm using software and QP solver of your choice.
(b) Use your PH code to solve the instance in Example 6.1.
(c) Use your PH code to solve the instance in Problem 6.2.
(d) Use your PH code to solve standard test instances of your choice.

6.17 Comparison of Benders Based Methods and the PH Method

Compare and contrast the PH method and Benders based methods (L-shaped, multicut L-shaped, and adaptive multicut L-shaped) in terms of problem decomposition, solution approach, termination condition, a number of iterations, and computation time.

References

1. J.F. Benders. Partitioning procedures for solving mixed-variable programming problems. *Numerische Mathematik*, 54:238–252, 1962.
2. J.R. Birge. Decomposition and partitioning methods for multistage stochastic linear programs. *Operations Research*, 33.5:989–1007, 1985.
3. J.R. Birge and F.V. Louveaux. A multicut algorithm for two-stage stochastic linear programs. *European Journal of Operational Research*, 34:384–392, 1988.
4. C.C. Carøe. *Decomposition in Stochastic Integer Programming*. Ph.D. thesis, Dept. of Operations Research, University of Copenhagen, Denmark, 1998.
5. G.B. Dantzig. Linear programming under uncertainty. *Management Science*, 1(3-4):197–206, 1955. Republished in the 50th anniversary issue of Management Science 50(12):1764–1769, 2004.
6. G.B. Dantzig and P. Wolfe. The decomposition algorithm for linear programs. *Econometrica*, 29(4):767–778, 1961.
7. J.L. Higle and S. Sen. *Stochastic Decomposition*. Kluwer Academic Publishers, 101 Phillip Drive, Norwell, MA 02061, 1996.
8. J.E. Kelley. The cutting-plane method for solving convex programs. *Journal of the Society of Industrial and Applied Mathematics*, 8(4):703–712, 1960.
9. M.V. Pereira and L.M. Pinto. Multi-stage stochastic optimization applied to energy planning. *Mathematical Programming*, 52.1-3:359–375, 1991.
10. W.B. Powell. Clearing the jungle of stochastic optimization. In *Tutorials in Operations Research*, pages 109–137. INFORMS, 2014. https://doi.org/10.1287/educ.2014.0128.
11. R.T. Rockafellar. Monotone operators and the proximal point algorithm. *SIAM Journal on Control and Optimization*, 14:877–898, 1976.
12. R.T. Rockafellar and R.J-B. Wets. Scenario and policy aggregation in optimization under uncertainty. *Mathematics of Operations Research*, 16:119–147, 1991.
13. R. Schultz. Continuity properties of expectation functions in stochastic integer programming. *Mathematics of Operations Research*, 18:578–589, 1993.
14. R. Van Slyke and R.J-B. Wets. L-shaped linear programs with application to optimal control and stochastic programming. *SIAM Journal on Applied Mathematics*, 17:638–663, 1969.
15. S. Trukhanov, L. Ntaimo, and A. Schaefer. Adaptive multicut aggregation for two-stage stochastic linear programs with recourse. *European Journal of Operational Research*, 206:395–406, 2010.
16. J-P. Watson and D.L. Woodruff. Progressive hedging innovations for a class of stochastic mixed-integer resource allocation problems. *Computational Management Science*, 8:355–370, 2011.

Part IV
Risk-Averse, Statistical, and Discrete Decomposition Methods

Chapter 7
Mean-Risk Stochastic Linear Programming Methods

7.1 Introduction

In this chapter, we turn to decomposition methods for two-stage mean-risk stochastic linear programming (MR-SLP), for both quantile and deviation risk measures. We pointed out in Chap. 2 that the deterministic equivalent problem (DEP) can be challenging or even impossible to solve using a direct solver for a relatively large number of scenarios. Therefore, we need to exploit *problem structure* and decompose the DEP into a master program and smaller subproblems that can be solved via a coordination and iterative approach. We also saw from Chap. 6 that derivation of algorithms for solving two-stage risk-neutral SLP problems requires a thorough understanding of the problem *structural properties*. Well, this holds true for MR-SLP as well, and in this case, the algorithm will also depend on the nature of the risk measure under consideration.

To begin, let us now restate the two-stage MR-SLP from Chap. 2 as follows:

$$\underset{x \in X}{\text{Min}}\ \mathbb{E}[f(x, \tilde{\omega})] + \lambda \mathbb{D}[f(x, \tilde{\omega})], \tag{7.1}$$

where $\mathbb{E} : \mathscr{F} \mapsto \mathbb{R}$ denotes the expected value, $\mathbb{D} : \mathscr{F} \mapsto \mathbb{R}$ is the *risk measure*, and $\lambda \geq 0$ is a suitable weight factor that quantifies the trade-off between expected cost and risk. The problem is risk-neutral if $\lambda := 0$. Risk measure $\mathbb{D}$ is chosen so that the problem remains a *convex optimization problem*, allowing it to be solved using convex optimization methods. The set $X = \{Ax \geq b,\ x \geq 0\}$ is a nonempty polyhedron that defines the set of first-stage feasible solutions. The matrix $A \in \mathbb{R}^{m_1 \times n_1}$ and vector $b \in \mathbb{R}^{m_1}$ are the first-stage matrix and right hand side (RHS) vector, respectively. The family of real random cost variables $\{f(x, \tilde{\omega})\}_{x \in X} \subseteq \mathscr{F}$ are defined on $(\Omega, \mathscr{A}, \mathbb{P})$, where $\mathscr{F}$ is the space of all real random cost variables $f : \Omega \mapsto \mathbb{R}$ satisfying $\mathbb{E}[|f(\tilde{\omega})|] < \infty$. For a given $x \in X$, the real random cost variable $f(x, \tilde{\omega})$ is given by

L. Ntaimo, *Computational Stochastic Programming*, Springer Optimization and Its Applications 774, https://doi.org/10.1007/978-3-031-52464-6_7

$$f(x, \tilde{\omega}) := c^\top x + \varphi(x, \tilde{\omega}). \tag{7.2}$$

For a given realization ω of $\tilde{\omega}$, the *recourse function* $\varphi(x, \omega)$ is given by

$$\begin{aligned} \varphi(x, \omega) := \text{Min } & q(\omega)^\top y(\omega) \\ \text{s.t. } & Wy(\omega) \geq r(\omega) - T(\omega)x \\ & y(\omega) \geq 0, \end{aligned} \tag{7.3}$$

where $q(\omega) \in \mathbb{R}^{n_2}$ is the second-stage cost vector and $y(\omega) \in \mathbb{R}^{n_2}_+$ is the recourse decision. The matrix $W \in \mathbb{R}^{m_2 \times n_2}$ is the *recourse matrix*, $T(\omega) \in \mathbb{R}^{m_2 \times n_1}$ is the *technology matrix*, and $r(\omega) \in \mathbb{R}^{m_2}$ is the RHS vector. By *scenario* ω, we mean the realization of the stochastic problem data, i.e., $\omega := (q(\omega), T(\omega), r(\omega))$.

To ensure that Problem (7.1) is well-defined for *computational* purposes, we continue to make the following assumption:

(A1) The multivariate random variable $\tilde{\omega}$ is discretely distributed with finitely many scenarios $\omega \in \Omega$, each with probability of occurrence $p(\omega)$.

Assumption (A1) is needed to make the problem tractable. We relax the following *relatively complete recourse* assumption that we made in Chap. 2:

(A2) For all $x \in X$, $\{Wy(\omega) \geq r(\omega) - T(\omega)x,\ y(\omega) \geq 0\} \neq \emptyset$.

Assumption (A2) guarantees the feasibility of the second-stage problem for every $x \in X$. This assumption implies that $\varphi(x, \omega) < \infty$ with probability one for all $x \in X$. Because we relax this assumption, it means that in the algorithms we derive we shall generate *feasibility cuts* for every $x \in X$ that leads to infeasibility of the second-stage subproblem. The feasibility cuts restrict the set X to feasible values of x. So as a convention, we continue to have $\varphi(x, \omega) = +\infty$ if Problem (7.3) is infeasible and $\varphi(x, \omega) = -\infty$ if Problem (7.3) is unbounded. Also, it is desirable for the risk measure $\mathbb{D}$ to be convexity preserving. So for the risk-averse case, i.e., when $\lambda > 0$, we consider convex risk measures only.

For completeness sake, we first give the mathematical definition of each risk measure from Chap. 2, restate its MR-SLP DEP formulation, and then derive the decomposition approach. We recommend reviewing Chap. 2 for the reader who is not familiar with the definitions of the risk measures we consider in this chapter. A thorough analysis of risk-averse models, optimality theory, and duality is given in [27], and decomposition approaches are studied in [1] and a computational study is given in [3]. In this chapter, we consider two fundamental decomposition approaches. The first involves a single optimality cut in the master program to approximate both the expectation and deviation terms together. We call this the *aggregated cut* approach. The second approach involves two separate cuts, one for the expectation term and the other for the deviation term. We refer to this as the *separate cut* approach. We derive solution algorithms for the aggregated and separate cut approaches, respectively, and give a detailed numerical illustration for each.

7.2 Aggregated Cut Decomposition for QDEV, CVaR, and EE

In this section, we derive the aggregated cut method involving a single cut to approximate the expectation and risk terms of the objective function. We refer to this approach as *aggregated cut* and derive the $\mathbb{D}$-AGG algorithm for $\mathbb{D} \in$ {QDEV, CVaR, EE}. Next, we perform a decomposition of the DEP for risk measure QDEV and then move on to CVaR and EE, in that order. We then give a formal statement of the $\mathbb{D}$-AGG algorithm before ending the section with a numerical example to illustrate the $\mathbb{D}$-AGG algorithm.

7.2.1 Quantile Deviation

In Chap. 2, we defined the risk measure QDEV and derived a suitable mathematical expression of the risk measure for optimization purposes. QDEV is a two-sided risk measure and reflects the expectation of the deviation above and below the α-quantile of the cumulative distribution of $f(x, \tilde{\omega})$. Given $x \in X$ and $\alpha \in (0, 1)$, QDEV is mathematically defined as follows:

$$\phi_{QDEV_\alpha}(x) := \underset{\eta \in \mathbb{R}}{\text{Min}}\ \mathbb{E}[\varepsilon_1(\eta - f(x, \tilde{\omega}))_+ + \varepsilon_2(f(x, \tilde{\omega}) - \eta)_+],$$

where $\varepsilon_1 > 0$ and $\varepsilon_2 > 0$ are such that $\alpha = \varepsilon_2/(\varepsilon_1 + \varepsilon_2)$. The minimum of the given optimization problem is attained at some η, which is the α-quantile of the cumulative distribution of $f(x, \tilde{\omega})$. An MR-SLP with $\mathbb{D} := \phi_{QDEV_\alpha}$ and $\lambda \geq 0$ is given as follows:

$$\underset{x \in X}{\text{Min}}\ \mathbb{E}[f(x, \tilde{\omega})] + \lambda \phi_{QDEV_\alpha}(x).$$

Given $\lambda \in [0, 1/\varepsilon_1]$ under assumption (A1), the DEP for the above optimization problem is given as follows:

$$\begin{aligned}
\text{Min}\quad & (1 - \lambda\varepsilon_1)c^\top x + \lambda\varepsilon_1\eta + (1 - \lambda\varepsilon_1) \sum_{\omega \in \Omega} p(\omega)q(\omega)^\top y(\omega) \\
& + \lambda(\varepsilon_1 + \varepsilon_2) \sum_{\omega \in \Omega} p(\omega)v(\omega) \\
\text{s.t.}\quad & T(\omega)x + Wy(\omega) \geq r(\omega),\ \forall \omega \in \Omega \\
& - c^\top x - q(\omega)^\top y(\omega) + \eta + v(\omega) \geq 0,\ \forall \omega \in \Omega \\
& x \in X,\ \eta \in \mathbb{R},\ y(\omega) \in \mathbb{R}_+^{n_2},\ v(\omega) \in \mathbb{R}_+, \forall \omega \in \Omega.
\end{aligned}$$

Since this problem is a large-scale LP with dual block angular structure, we can employ Benders decomposition and use the L-shaped algorithm to solve it. To decompose the DEP, let (x, η) be first-stage decision variables and $(y(\omega), \upsilon(\omega))$ be second-stage decision variables. Then the master program at iteration k of the L-shaped algorithm will take the following form:

$$\begin{aligned}
\ell^k := \text{Min } & (1 - \lambda\varepsilon_1)c^\top x + \lambda\varepsilon_1\eta + \gamma \\
\text{s.t. } & Ax \geq b \\
& \beta_t^\top x + \beta_t^1\eta + \gamma \geq \beta_t^0, \quad t \in \Theta_k \\
& \beta_t^\top x + \beta_t^1\eta \qquad \geq \beta_t^0, \quad t \notin \Theta_k \\
& x \geq 0, \delta_l \leq \eta \leq \delta_h.
\end{aligned} \tag{7.4}$$

Notice that the variable η takes the value of the α-quantile of the cumulative distribution of $f(x, \tilde{\omega})$ and is therefore unrestricted in sign (free). However, we bound it using the carefully selected parameters δ_l and δ_h to speed up computation. These parameters can be determined for a given instance via experimentation. The free variable γ is the optimality cut variable, and it represents a lower bound approximation for $(1-\lambda\varepsilon_1)\sum_{\omega\in\Omega} p(\omega)q(\omega)^\top y(\omega) + \lambda(\varepsilon_1+\varepsilon_2)\sum_{\omega\in\Omega} p(\omega)\upsilon(\omega)$. The second and third sets of constraints represent the optimality and feasibility cuts, respectively, with left hand side coefficients at iteration k denoted by β_k and β_k^1 and the corresponding RHS by β_k^0. The index set Θ_k is the set of iteration numbers up to k where all subproblems are feasible and an optimality cut is generated.

Given the master problem solution (x^k, η^k) at iteration k, the subproblem for scenario ω is given as follows:

$$\begin{aligned}
\varphi(x^k, \eta^k, \omega) := \text{Min } & (1 - \lambda\varepsilon_1)q(\omega)^\top y + \lambda(\varepsilon_1 + \varepsilon_2)\nu \\
\text{s.t. } & Wy \geq r(\omega) - T(\omega)x^k \\
& -q(\omega)^\top y + \nu \geq -\eta^k + c^\top x^k \\
& y, \nu \geq 0.
\end{aligned} \tag{7.5}$$

Let the dual solution to subproblem (7.5) corresponding to the two constraints be given by $\pi^k(\omega)$ and $\pi_0^k(\omega)$, respectively. Then we can compute the optimality cut coefficients and RHS as follows:

$$\beta_k = \sum_{\omega\in\Omega} p(\omega)\left(T(\omega)^\top\pi^k(\omega) - \pi_0^k(\omega)c\right),$$

$$\beta_k^1 = \sum_{\omega\in\Omega} p(\omega)\pi_0^k(\omega)$$

and

$$\beta_k^0 = \sum_{\omega \in \Omega} p(\omega) \pi^k(\omega)^\top r(\omega).$$

If the subproblem is *infeasible*, then a dual extreme ray $(\mu^k(\omega), \mu_0^k(\omega))$ is obtained and a feasibility cut generated as follows:

$$\beta_k = T(\omega)^\top \mu^k(\omega) - \mu_0^k(\omega)c,$$

$$\beta_k^1 = \mu_0^k(\omega)$$

and

$$\beta_k^0 = \mu^k(\omega)^\top r(\omega).$$

7.2.2 Conditional Value-at-Risk

In Chap. 2, we studied risk measure CVaR that reflects the expectation of the $(1 - \alpha).100\%$ worst outcomes for a given probability level $\alpha \in (0, 1)$. Given $x \in X$ and $\alpha \in (0, 1)$, CVaR can be expressed as follows:

$$\phi_{CVaR_\alpha}(x) := \operatorname*{Min}_{\eta \in \mathbb{R}} \{\eta + \frac{1}{1-\alpha} \mathbb{E}[(f(x, \tilde{\omega}) - \eta)_+]\}.$$

CVaR can also be expressed in terms of $\phi_{QDEV_\alpha}(x)$ as follows:

$$\phi_{CVaR_\alpha}(x) := \mathbb{E}[f(x, \tilde{\omega})] + \frac{1}{\varepsilon_1} \phi_{QDEV_\alpha}(x).$$

For any $\lambda \geq 0$, an MR-SLP with $\mathbb{D} := \phi_{CVaR_\alpha}$ can be written as follows:

$$\operatorname*{Min}_{x \in X} \mathbb{E}[f(x, \tilde{\omega})] + \lambda \phi_{CVaR_\alpha}(x).$$

An alternative formulation, for $0 \leq \lambda \leq 1$, is given as

$$\operatorname*{Min}_{x \in X} (1 - \lambda)\mathbb{E}[f(x, \tilde{\omega})] + \lambda \phi_{CVaR_\alpha}(x).$$

This model has a *convex combination* of the mean and the CVaR dispersion statistic and is coherent. The DEP for this model can be given as follows:

$$
\begin{aligned}
\text{Min} \quad & (1-\lambda)c^\top x + (1-\lambda)\sum_{\omega\in\Omega} p(\omega)q(\omega)^\top y(\omega) + \lambda\eta + \frac{\lambda}{1-\alpha}\sum_{\omega\in\Omega} p(\omega)\upsilon(\omega) \\
\text{s.t.} \quad & T(\omega)x + Wy(\omega) \geq r(\omega),\ \forall\omega\in\Omega \\
& -c^\top x - q(\omega)^\top y(\omega) + \eta + \upsilon(\omega) \geq 0,\ \forall\omega\in\Omega \\
& x\in X,\ \eta\in\mathbb{R},\ y(\omega)\in\mathbb{R}_+^{n_2},\ \upsilon(\omega)\in\mathbb{R}_+, \forall\omega\in\Omega.
\end{aligned}
$$

This DEP is a large-scale LP with dual block angular structure, and we can employ the L-shaped algorithm to solve it. The first-stage decision variables are (x,η) and the second-stage decision variables are $(y(\omega),\upsilon(\omega))$. The master program at iteration k of the L-shaped algorithm will take the following form:

$$
\begin{aligned}
\ell^k := &\text{Min } (1-\lambda)c^\top x + \lambda\eta + \gamma \\
&\text{s.t. } Ax \geq b \\
&\quad \beta_t^\top x + \beta_t^1\eta + \gamma \ \geq \beta_t^0, \quad t\in\Theta_k \\
&\quad \beta_t^\top x + \beta_t^1\eta \qquad \geq \beta_t^0, \quad t\notin\Theta_k \\
&\qquad\quad x\geq 0,\ \delta_l \leq \eta \leq \delta_h.
\end{aligned} \tag{7.6}
$$

In this case, the optimality cut variable γ represents a lower bound approximation for $(1-\lambda)\sum_{\omega\in\Omega} p(\omega)q(\omega)^\top y(\omega) + \lambda\eta + \frac{\lambda}{1-\alpha}\sum_{\omega\in\Omega} p(\omega)\upsilon(\omega)$. Given the master problem solution (x^k,η^k) at iteration k, the subproblem for scenario ω is given as follows:

$$
\begin{aligned}
\varphi(x^k,\eta^k,\omega) := &\text{Min } (1-\lambda)q(\omega)^\top y + \frac{\lambda}{1-\alpha}\upsilon \\
&\text{s.t. } Wy \ \geq r(\omega) - T(\omega)x^k \\
&\quad -q(\omega)^\top y + \upsilon \geq -\eta^k + c^\top x^k \\
&\qquad\quad y,\ \upsilon \geq 0.
\end{aligned} \tag{7.7}
$$

Using the dual solution to the subproblem, $\pi^k(\omega)$ and $\pi_0^k(\omega)$, we can compute the optimality cut coefficients and RHS as follows:

$$
\beta_k = \sum_{\omega\in\Omega} p(\omega)\left(T(\omega)^\top\pi^k(\omega) - \pi_0^k(\omega)c\right),
$$

$$
\beta_k^1 = \sum_{\omega\in\Omega} p(\omega)\pi_0^k(\omega),
$$

and

$$\beta_k^0 = \sum_{\omega \in \Omega} p(\omega)\pi^k(\omega)^\top r(\omega).$$

If the subproblem is *infeasible*, then a dual extreme ray $(\mu^k(\omega), \mu_0^k(\omega))$ is obtained and a feasibility cut generated as follows:

$$\beta_k = T(\omega)^\top \mu^k(\omega) - \mu_0^k(\omega)c,$$

$$\beta_k^1 = \mu_0^k(\omega)$$

and

$$\beta_k^0 = \mu^k(\omega)^\top r(\omega).$$

7.2.3 Expected Excess

In Chap. 2, we define risk measure EE as the expected value of the excess over a given target η. Given $x \in X$, $\eta \in \mathbb{R}$, and $\lambda \geq 0$, EE is mathematically defined as follows:

$$\phi_{EE_\eta}(x) := \mathbb{E}[f(x, \tilde{\omega}) - \eta]_+.$$

An MR-SLP with $\mathbb{D} := \phi_{EE_\eta}$ is given as follows:

$$\operatorname*{Min}_{x \in X} \ \mathbb{E}[f(x, \tilde{\omega})] + \lambda \phi_{EE_\eta}(x).$$

The DEP to this problem can be given as follows:

$$\begin{aligned}
\text{Min} \quad & c^\top x + \sum_{\omega \in \Omega} p(\omega) q(\omega)^\top y(\omega) + \lambda \sum_{\omega \in \Omega} p(\omega) v(\omega) \\
\text{s.t.} \quad & T(\omega)x + Wy(\omega) \geq r(\omega), \ \forall \omega \in \Omega \\
& -c^\top x - q(\omega)^\top y(\omega) + v(\omega) \geq -\eta, \ \forall \omega \in \Omega \\
& x \in X, \ \ y(\omega) \in \mathbb{R}_+^{n_2}, \ v(\omega) \in \mathbb{R}_+, \ \forall \omega \in \Omega.
\end{aligned}$$

As with QDEV and CVaR, this DEP is a large-scale LP with dual block angular structure, and we can use the L-shaped algorithm to solve it. In terms of decomposing the problem, the first-stage decision variable is x and the second-stage decision variables are $(y(\omega), v(\omega))$. Therefore, at iteration k of the L-shaped algorithm, the master program will take the following form:

$$
\begin{aligned}
\ell^k := \text{Min } & c^\top x + \gamma \\
\text{s.t. } & Ax \geq b \\
& \beta_t^\top x + \gamma \geq \beta_t^0, \quad t \in \Theta_k \\
& \beta_t^\top x \qquad \geq \beta_t^0, \quad t \notin \Theta_k \\
& x \geq 0.
\end{aligned} \tag{7.8}
$$

The optimality cut variable γ represents a lower bound approximation for $\sum_{\omega \in \Omega} p(\omega) q(\omega)^\top y(\omega) + \lambda \sum_{\omega \in \Omega} p(\omega) v(\omega))$. Given the master problem solution (x^k, γ^k) at iteration k, the subproblem for scenario ω is given as follows:

$$
\begin{aligned}
\varphi(x^k, \omega) := \text{Min } & q(\omega)^\top y + \lambda v \\
\text{s.t. } & Wy \geq r(\omega) - T(\omega) x^k \\
& - q(\omega)^\top y + v \geq -\eta + c^\top x^k \\
& y, \ v \geq 0.
\end{aligned} \tag{7.9}
$$

Using the dual solution to the subproblem, $\pi^k(\omega)$ and $\pi_0^k(\omega)$, we can compute the optimality cut coefficients and RHS as follows:

$$
\beta_k = \sum_{\omega \in \Omega} p(\omega) \left(T(\omega)^\top \pi^k(\omega) - \pi_0^k(\omega) c \right),
$$

and

$$
\beta_k^0 = \sum_{\omega \in \Omega} p(\omega) \left(\pi^k(\omega)^\top r(\omega) - \pi_0^k(\omega)^\top \eta \right).
$$

If the subproblem is *infeasible*, then a dual extreme ray $(\mu^k(\omega), \mu_0^k(\omega))$ is obtained and a feasibility cut generated as follows:

$$
\beta_k = T(\omega)^\top \mu^k(\omega) - \mu_0^k(\omega) c
$$

and

$$
\beta_k^0 = \mu^k(\omega)^\top r(\omega) - \mu_0^k(\omega)^\top \eta.
$$

Next, we devise and formalize the aggregated cut algorithm for MR-SLP with QDEV, CVaR, or EE.

7.2.4 $\mathbb{D}$-AGG Algorithm

We are now in a position to state the aggregated cut algorithm for the three risk measures, QDEV, CVaR, and EE. To simplify the exposition, we state one generic algorithm with conditions (cases) for each of the risk measures. We refer to the algorithm as $\mathbb{D}$-AGG algorithm, where $\mathbb{D}$:= QDEV, $\mathbb{D}$:= CVaR or $\mathbb{D}$:= EE, and "AGG" stands for "aggregated." Let k continue to denote the algorithm iteration counter and let ℓ and u denote the lower and upper bounds, respectively, on the optimal value during the course of the algorithm. Then we can formally state the $\mathbb{D}$-AGG algorithm as follows:

Algorithm $\mathbb{D}$-*AGG*
begin

Step 0. Initialization. $k \leftarrow 0$, x^0 given or $x^0 \leftarrow \operatorname{argmin}\{c^\top x \mid Ax \geq b, x \geq 0\}$, $\ell \leftarrow -\infty$, $u \leftarrow \infty$, and $\epsilon > 0$.

Case 1: $\mathbb{D}$:= QDEV

Select appropriate values for δ_l and δ_h.
Choose $\lambda \in [0, \frac{1}{\varepsilon_1}]$, $\alpha \in (0, 1)$, and $\varepsilon_1, \varepsilon_2 > 0$ such that $\alpha = \varepsilon_2/(\varepsilon_1 + \varepsilon_2)$ and $\eta^0 \in [\delta_l, \delta_h]$.

Case 2: $\mathbb{D}$:=CVaR

Select appropriate values for δ_l and δ_h.
Choose $\lambda \in [0, 1]$, $\alpha \in (0, 1)$, and $\eta^0 \in [\delta_l, \delta_h]$.

Case 3: $\mathbb{D}$:=EE

Choose target value $\eta \in \mathbb{R}$ and $\lambda \geq 0$.

Step 1. Solve Subproblems.

Case 1: $\mathbb{D}$:= QDEV

Set $\beta_k \leftarrow \mathbf{0}$, $\beta_k^1 \leftarrow 0$, $\beta_k^0 \leftarrow 0$, and $\bar{Q}^k \leftarrow 0$.
For each $\omega \in \Omega$, solve subproblem (7.5):

Get $\varphi(x^k, \eta^k, \omega)$ and dual solution $(\pi^k(\omega), \pi_0^k(\omega))$.
Compute $\bar{Q}^k \leftarrow \bar{Q}^k + p(\omega)\varphi(x^k, \eta^k, \omega)$.
Compute $\beta_k \leftarrow \beta_k + p(\omega)\big(T(\omega)^\top \pi^k(\omega) - \pi_0^k(\omega)c\big)$.
Compute $\beta_k^1 \leftarrow \beta_k^1 + p(\omega)\pi_0^k(\omega)$.
Compute $\beta_k^0 \leftarrow \beta_k^0 + p(\omega)\pi^k(\omega)^\top r(\omega)$.

If subproblem is *infeasible* for some ω:

Get dual extreme ray $(\mu^k(\omega), \mu_0^k(\omega))$.
Compute $\beta_k \leftarrow \beta_k + \big(T(\omega)^\top \mu^k(\omega) - \mu_0^k(\omega)c\big)$.
Compute $\beta_k^1 \leftarrow \beta_k^1 + \mu_0^k(\omega)$.

Compute $\beta_k^0 \leftarrow \beta_k^0 + \mu^k(\omega)^\top r(\omega)$.
Generate *feasibility* cut: $\beta_k^\top x + \beta_k^1 \eta \geq \beta_k^0$.
Go to step 2.

Else if *feasible* for all ω:

Generate an *optimality* cut: $\beta_k^\top x + \beta_k^1 \eta + \gamma \geq \beta_k^0$.
Compute *upper bound*:

Set $u^k \leftarrow (1 - \lambda\varepsilon_1)c^\top x^k + \lambda\varepsilon_1 \eta^k + \bar{Q}^k$.
Set $u \leftarrow \min\{u^k, u\}$.

If u is updated, set *incumbent* solution $(x^*, \eta^*) \leftarrow (x^k, \eta^k)$.

Case 2: $\mathbb{D}$:=CVaR

Set $\beta_k \leftarrow \mathbf{0}$, $\beta_k^1 \leftarrow 0$, $\beta_k^0 \leftarrow 0$, and $\bar{Q}^k \leftarrow 0$.
For each $\omega \in \Omega$ solve subproblem (7.7):

Get $\varphi(x^k, \eta^k, \omega)$ and dual solution $(\pi^k(\omega), \pi_0^k(\omega))$.
Compute $\bar{Q}^k \leftarrow \bar{Q}^k + p(\omega)\varphi(x^k, \eta^k, \omega)$.
Compute $\beta_k \leftarrow \beta_k + p(\omega)\left(T(\omega)^\top \pi^k(\omega) - \pi_0^k(\omega)c\right)$.
Compute $\beta_k^1 \leftarrow \beta_k^1 + p(\omega)\pi_0^k(\omega)$.
Compute $\beta_k^0 \leftarrow \beta_k^0 + p(\omega)\pi^k(\omega)^\top r(\omega)$.

If subproblem is *infeasible* for some ω:

Get dual extreme ray $(\mu^k(\omega), \mu_0^k(\omega))$.
Compute $\beta_k \leftarrow \beta_k + \left(T(\omega)^\top \mu^k(\omega) - \mu_0^k(\omega)c\right)$.
Compute $\beta_k^1 \leftarrow \beta_k^1 + \mu_0^k(\omega)$.
Compute $\beta_k^0 \leftarrow \beta_k^0 + \mu^k(\omega)^\top r(\omega)$.
Generate *feasibility* cut: $\beta_k^\top x + \beta_k^1 \eta \geq \beta_k^0$.
Go to step 2.

Else if *feasible* for all ω:

Generate an *optimality* cut: $\beta_k^\top x + \beta_k^1 \eta + \gamma \geq \beta_k^0$.
Compute *upper bound*:

Set $u^k \leftarrow (1 - \lambda)c^\top x^k + \lambda \eta^k + \bar{Q}^k$.
Set $u \leftarrow \min\{u^k, u\}$.

If u is updated, set *incumbent* solution $(x^*, \eta^*) \leftarrow (x^k, \eta^k)$.

Case 3: $\mathbb{D}$:=EE

Set $\beta_k \leftarrow \mathbf{0}$, $\beta_k^1 \leftarrow 0$, $\beta_k^0 \leftarrow 0$, and $\bar{Q}^k \leftarrow 0$.
For each $\omega \in \Omega$ solve subproblem (7.9):

Get $\varphi(x^k, \omega)$ and dual solution $(\pi^k(\omega), \pi_0^k(\omega))$.
Compute $\bar{Q}^k \leftarrow \bar{Q}^k + p(\omega)\varphi(x^k, \omega)$.
Compute $\beta_k \leftarrow \beta_k + p(\omega)\left(T(\omega)^\top \pi^k(\omega) - \pi_0^k(\omega)c\right)$.
Compute $\beta_k^0 \leftarrow \beta_k^0 + p(\omega)\left(\pi^k(\omega)^\top r(\omega) - \pi_0^k(\omega)^\top \eta\right)$.

If subproblem is *infeasible* for some ω:

Get dual extreme ray $(\mu^k(\omega), \mu_0^k(\omega))$.
Compute $\beta_k \leftarrow \big(T(\omega)^\top \mu^k(\omega) - \mu_0^k(\omega)c\big)$.
Compute $\beta_k^0 \leftarrow \big(\mu^k(\omega)^\top r(\omega) - \mu_0^k(\omega)^\top \eta\big)$.
Generate *feasibility* cut: $\beta_k^\top x \geq \beta_k^0$.
Go to step 2.

Else if *feasible* for all ω:

Generate an *optimality* cut: $\beta_k^\top x + \gamma \geq \beta_k^0$.
Compute *upper bound*:

Set $u^k \leftarrow c^\top x^k + \bar{Q}^k$.
Set $u \leftarrow \min\{u^k, u\}$.

If u is updated, set *incumbent* solution $x^* \leftarrow x^k$.

Step 2. Add Cut to Master Program and Solve.
Case 1: $\mathbb{D}$:= QDEV

If some subproblem was *infeasible*:

Add feasibility cut to master problem (7.4).

Else:

Add optimality cut to master problem (7.4).

Solve master problem (7.4), and get optimal value ℓ_{k+1} and solution (x^{k+1}, η^{k+1}).

Case 2: $\mathbb{D}$:=CVaR

If some subproblem was *infeasible*:

Add feasibility cut to master problem (7.6).

Else:

Add optimality cut to master problem (7.6).

Solve master problem (7.6), and get optimal value ℓ_{k+1} and solution (x^{k+1}, η^{k+1}).

Case 3: $\mathbb{D}$:=EE

If some subproblem was *infeasible*:

Add feasibility cut to master problem (7.8):

Else:

Add optimality cut to master problem (7.8).

Solve master problem (7.8), and get optimal value ℓ_{k+1} and solution x^{k+1}.

Set $\ell \leftarrow \max\{\ell^{k+1}, \ell\}$.

Step 3. **Termination.** If $u - \ell \leq \epsilon|u|$:

Cases 1 and 2: $\mathbb{D}$:= QDEV or CVaR

Stop, solution (x^*, η^*) is ϵ-optimal.

Case 3: $\mathbb{D}$:=EE

Stop, solution x^* is ϵ-optimal.

Else:

Set $k \leftarrow k + 1$.

Return to step 1.

End

The $\mathbb{D}$-AGG algorithm is simply a version of the L-shaped algorithm for the risk-neutral case, i.e., when the risk factor $\lambda = 0$. However, to solve MR-SLP we need to trace the efficient frontier by considering all possible values of λ. Thus we can view the problem as a *multi-objective* optimization problem if λ is not fixed. We propose doing a parametric optimization approach. Notice that the $\mathbb{D}$-AGG algorithm is stated for a specified value of λ at initialization in step 0. To perform parametric optimization over λ values, ignore the initialization of λ and instead perform the following steps for determining the value of λ to use for each $\mathbb{D}$-*AGG* algorithm run:

Algorithm *Parametric* $\mathbb{D}$-AGG
begin

Step a. **Initialization.** Set $\kappa \leftarrow 0$, $\lambda^0 \leftarrow 0$ and choose $\delta > 0$ (very small number) and $\bar{\lambda} > 0$ (very large number).

Step b. **Solve MR-SLP for $\lambda \leftarrow \lambda^\kappa$.** Apply $\mathbb{D}$-AGG algorithm.

Step c. **Apply Parametric Analysis to Master Problem.** Find the range $[\lambda, \lambda^*]$ for which the current master problem basis remains optimal:

Case 1: $\mathbb{D}$:= QDEV

If $\lambda^* < \frac{1}{\varepsilon_1}$:

Set $\lambda^{\kappa+1} \leftarrow \min\{\frac{1}{\varepsilon_1}, \lambda^* + \delta\}$.

Else:

Stop.

Case 2: $\mathbb{D}$:=CVaR

If $\lambda^* < 1$:

Set $\lambda^{\kappa+1} \leftarrow \min\{1, \lambda^* + \delta\}$.

Else:

Stop.

Case 3: $\mathbb{D}$:=EE

If $\lambda^* < \bar{\lambda}$:

Set $\lambda^{\kappa+1} \leftarrow \min\{\bar{\lambda}, \lambda^* + \delta\}$.

Else:

Stop.

$\kappa \leftarrow \kappa + 1$.

Return to step (b).

End

We should point out that the values of δ and $\bar{\lambda}$ are problem instance dependent and should be chosen carefully through experimentation. Next, we give a detailed illustration of the $\mathbb{D}$-AGG algorithm.

7.2.5 Numerical Example

Let us use the *abc-Production Planning* problem from Chap. 3 to illustrate how to apply the $\mathbb{D}$-AGG algorithm.

Example 7.1 Apply two iterations of the $\mathbb{D}$-AGG algorithm for $\mathbb{D}$:= QDEV to the two-stage SLP *abc-Production Planning* problem from Chaps. 3 and 6. Use $\alpha := 0.95$, with $\varepsilon_1 := 1$ and $\varepsilon_2 := 19$ so that $\alpha = \varepsilon_2/(\varepsilon_1 + \varepsilon_2) := 0.95$ and $\lambda \in [0, 1/\varepsilon_1] = [0, 1]$. Set $\lambda := 0.5$ for this example.

Recall that the *abc-Production Planning* problem involves maximizing the profit of producing three different products a, b, and c given a limited supply of raw materials-abc and the number of processing hours for three different processes, respectively. The first-stage decision variables are as follows:

x_1 : Number of units of Material-abc purchased.
x_2 : Number of hours of Process-1 purchased.
x_3: Number of hours of Process-2 purchased.
x_4: Number of hours of Process-3 purchased.

The second-stage decision variables are as follows:

$y_1(\omega)$: Number of units of Product-a produced under scenario ω.
$y_2(\omega)$: Number of units of Product-b produced under scenario ω.
$y_3(\omega)$: Number of units of Product-c produced under scenario ω.

The two-stage recourse formulation of the problem is given as follows:

$$
\begin{array}{llllll}
\text{Min} & 50x_1 & +30x_2 & +15x_3 & +10x_4 + \sum_{s=1}^{5} p(\omega^s) q(x, \omega^s) & \\
\text{s.t.} & -x_1 & & & \geq & -300 \\
& & -x_2 & & \geq & -700 \\
& & & -x_3 & \geq & -600 \\
& & & & -x_4 \geq & -500 \\
& & -x_2 & -x_3 & -x_4 \geq & -1600 \\
& x_1, & x_2, & x_3, & x_4 \geq & 0,
\end{array}
$$

where for a given scenario $s = 1, \cdots, 5$, the second-stage subproblem is given as follows:

$$
\begin{array}{lllllll}
q(x, \omega^s) := \text{Min} & -1150y_1 & -1525y_2 & -1900y_3 & & & \text{Dual multipliers} \\
\text{s.t.} & -6y_1 & -8y_2 & -10y_3 \geq & & -x_1 & \leftarrow \pi_1 \\
& -20y_1 & -25y_2 & -28y_3 \geq & & -x_2 & \leftarrow \pi_2 \\
& -12y_1 & -15y_2 & -18y_3 \geq & & -x_3 & \leftarrow \pi_3 \\
& -8y_1 & -10y_2 & -14y_3 \geq & & -x_4 & \leftarrow \pi_4 \\
& -y_1 & & & \geq & -d_1(\omega^s) & \leftarrow \pi_5 \\
& & -y_2 & & \geq & -d_2(\omega^s) & \leftarrow \pi_6 \\
& & & -y_3 & \geq & -d_3(\omega^s) & \leftarrow \pi_7 \\
& y_1, & y_2, & y_3 & \geq & 0. &
\end{array}
$$

Using vector notation, $x = (x_1, x_2, x_3, x_4)^\top$, $y = (y_1, y_2, y_3)^\top$ and RHS $r(\omega^s) = (0, 0, 0, 0, d_1(\omega^s), d_2(\omega^s), d_3(\omega^s))^\top$. The RHS data $d_1(\omega^s)$, $d_2(\omega^s)$, and $d_3(\omega^s)$ is the demand for products a, b, and c, respectively, under scenario ω^s. Using the dual multipliers, $\pi = (\pi_1, \pi_2, \cdots, \pi_7)^\top$, the dual to the second-stage subproblem for scenario ω^s is given as follows:

$$
\begin{array}{lllllllll}
\text{Max} & -x_1\pi_1 & -x_2\pi_2 & -x_3\pi_3 & -x_4\pi_4 & -d_1(\omega^s)\pi_5 & -d_2(\omega^s)\pi_6 & -d_3(\omega^s)\pi_7 & \\
\text{s.t.} & -6\pi_1 & -20\pi_2 & -12\pi_3 & -8\pi_4 & -\pi_5 & & & \leq -1150 \\
& -8\pi_1 & -25\pi_2 & -15\pi_3 & -10\pi_4 & & -\pi_6 & & \leq -1525 \\
& -10\pi_1 & -28\pi_2 & -18\pi_3 & -14\pi_4 & & & -\pi_7 & \leq -1900 \\
& \pi_1, & \pi_2, & \pi_3, & \pi_4, & \pi_5, & \pi_6, & \pi_7 & \geq 0.
\end{array}
$$

The problem data are summarized as follows:
First-stage:

$$
c := (50, 30, 15, 10)^\top, \; A := \begin{bmatrix} -1 & 0 & 0 & 0 \\ 0 & -1 & 0 & 0 \\ 0 & 0 & -1 & 0 \\ 0 & 0 & 0 & -1 \\ 0 & -1 & -1 & -1 \end{bmatrix},
$$

$$
b := (-300, -700, -600, -500, -1600)^\top.
$$

Second-stage:
Number of scenarios $S = 5$.

Recourse Matrix $W := \begin{bmatrix} -6 & -8 & -10 \\ -20 & -25 & -28 \\ -12 & -15 & -18 \\ -8 & -10 & -14 \\ -1 & 0 & 0 \\ 0 & -1 & 0 \\ 0 & 0 & -1 \end{bmatrix}$.

Scenario probabilities $p(\omega^s), s = 1, \cdots, 5$:
$p(\omega^1) := 0.15,\ p(\omega^2) := 0.30, p(\omega^3) := 0.30, p(\omega^4) := 0.20$, and $p(\omega^5) := 0.05$.

Technology matrix $T(\omega^1) = T(\omega^2) = \cdots = T(\omega^5) = T := \begin{bmatrix} 1\,0\,0\,0 \\ 0\,1\,0\,0 \\ 0\,0\,1\,0 \\ 0\,0\,0\,1 \\ 0\,0\,0\,0 \\ 0\,0\,0\,0 \\ 0\,0\,0\,0 \end{bmatrix}$.

RHS vectors $r(\omega^s), s = 1, \cdots, 5$:
$r(\omega^1) := (0, 0, 0, 0, -15, -10, -5)^\top$,
$r(\omega^2) := (0, 0, 0, 0, -20, -15, -15)^\top$,
$r(\omega^3) := (0, 0, 0, 0, -25, -20, -25)^\top$,
$r(\omega^4) := (0, 0, 0, 0, -30, -25, -30)^\top$, and
$r(\omega^5) := (0, 0, 0, 0, -10, -10, -10)^\top$.
First-stage feasible set:

$$X = \{ \begin{array}{rrrrl} -x_1 & & & & \geq -300 \\ & -x_2 & & & \geq -700 \\ & & -x_3 & & \geq -600 \\ & & & -x_4 & \geq -500 \\ & -x_2 & -x_3 & -x_4 & \geq -1600 \\ x_1, & x_2, & x_3, & x_4 & \geq 0\} \end{array}.$$

We can apply the $\mathbb{D}$-AGG algorithm for $\mathbb{D}$:= QDEV as follows:

Algorithm $\mathbb{D}$-*AGG*
begin

Step 0. **Initialization.** Let $k \leftarrow 0$, choose $x^0 \in X$, and set $\ell \leftarrow -\infty$ and $u \leftarrow \infty$. Choose $\epsilon \leftarrow 10^{-6}$ and $x^0 \leftarrow \text{argmin}_{x \in X}\{50x_1+30x_2+15x_3+10x_4\} = (0, 0, 0, 0)^\top$ as the initial point , and $\lambda \leftarrow 0.5$. *Case 1:* $\mathbb{D}$:= QDEV

Select appropriate values for $\delta_l = -5000$ and $\delta_h = 0$.

Choose $\lambda \in [0, \frac{1}{\varepsilon_1}] = 0.5$, $\alpha \in (0, 1)$, and $\varepsilon_1 = 1, \varepsilon_2 = 19$ such that $\alpha = \varepsilon_2/(\varepsilon_1 + \varepsilon_2) = 0.95$ and $\eta^0 = 0$.

Step 1. **Solve Subproblems.** *Case 1:* $\mathbb{D} :=$ QDEV

Set $\beta_k \leftarrow \mathbf{0}$, $\beta_0^1 \leftarrow 0$, $\beta_0^0 \leftarrow 0$, and $\bar{Q}^k \leftarrow 0$.
$(1-\lambda\epsilon_1)q(\omega^1)=(1-0.5)(-1150, -1525, -1900)^\top=(-575, -762.5, -950)^\top$ and $\lambda(\epsilon_1 + \epsilon_2) = 0.5(20) = 10$
For $s = 1, \cdots, 5$, solve

$$
\begin{array}{rrrrrl}
\varphi^0(x^0, \omega^s) := \text{Min} & -575y_1 & -762.5y_2 & -950y_3 & +10\nu & \\
\text{s.t.} & -6y_1 & -8y_2 & -10y_3 & & \geq 0 \\
 & -20y_1 & -25y_2 & -28y_3 & & \geq 0 \\
 & -12y_1 & -15y_2 & -18y_3 & & \geq 0 \\
 & -8y_1 & -10y_2 & -14y_3 & & \geq 0 \\
 & -y_1 & & & & \geq d_a^s \\
 & & -y_2 & & & \geq d_b^s \\
 & & & -y_3 & & \geq d_c^s \\
 & -1150y_1 & -1525y_2 & -1900y_3 & +\nu & \geq 0 \\
 & y_1, & y_2, & y_3 & & \geq 0.
\end{array}
$$

$s = 1$: feasible, $\varphi^0(x^0, \omega^1) = 0$, $y_1^0 = (0, 0, 0)^\top$, $\nu = 0$, $\pi^0(\omega^1) = (95.8333, 0, 0, 0, 0, 0, 0, 0)^\top$ and $\pi_0^0(\omega^1) = 0$.

Compute $\bar{Q}^0 \leftarrow \bar{Q}^0 + p(\omega^1)\varphi^0(x^0, \omega^1) = 0 + 0.15(0) = 0$.
Compute $\beta_0 \leftarrow \beta_0 + p(\omega^1)\big(T(\omega^1)^\top\pi^0(\omega^1) - \pi_0^0(\omega^1)c\big)$

$$
= \quad (0,0,0,0)^\top \; + \; 0.15(\begin{bmatrix} 1\,0\,0\,0 \\ 0\,1\,0\,0 \\ 0\,0\,1\,0 \\ 0\,0\,0\,1 \\ 0\,0\,0\,0 \\ 0\,0\,0\,0 \\ 0\,0\,0\,0 \end{bmatrix}^\top (95.8333, 0, 0, 0, 0, 0, 0, 0) \; -
$$

$0(50, 30, 15, 10)^\top) = (14.3750, 0, 0, 0)^\top$.
Compute $\beta_0^1 \leftarrow \beta_0^1 + p(\omega^1)\pi_0^0(\omega^1) = 0 + 0.15(0) = 0$.
Compute $\beta_0^0 \leftarrow \beta_0^0 + p(\omega^1)\pi^0(\omega^1)^\top r(\omega^1)$
$= 0 + 0.15(95.8333, 0, 0, 0, 0, 0, 0, 0)(0, 0, 0, 0, -15, -10, -5)^\top = 0$.

$s = 2$: feasible, $\varphi^0(x^0, \omega^2) = 0$, $y_2^0 = (0, 0, 0)^\top$, $\pi^0(\omega^2) = (95.8333, 0, 0, 0, 0, 0, 0, 0)^\top$, and $\pi_0^0(\omega^2) = 0$.

Compute $\bar{Q}^0 \leftarrow \bar{Q}^0 + p(\omega^2)\varphi^0(x^0, \omega^2) = 0 + 0.3(0) = 0$.

Compute $\beta_0 \leftarrow \beta_0 + p(\omega^2)\left(T(\omega^2)^\top \pi^0(\omega^2) - \pi_0^0(\omega^2)c\right)$

$$= \quad (14.3750, 0, 0)^\top \; + \; 0.3(\begin{bmatrix} 1\,0\,0\,0 \\ 0\,1\,0\,0 \\ 0\,0\,1\,0 \\ 0\,0\,0\,1 \\ 0\,0\,0\,0 \\ 0\,0\,0\,0 \\ 0\,0\,0\,0 \end{bmatrix}^\top (95.8333, 0, 0, 0, 0, 0, 0, 0) \; -$$

$0(50, 30, 15, 10)^\top) = (43.150, 0, 0, 0)^\top$.
Compute $\beta_0^1 \leftarrow \beta_0^1 + p(\omega^2)\pi_0^0(\omega^2) = 0 + 0.3(0) = 0$.
Compute $\beta_0^0 \leftarrow \beta_0^0 + p(\omega^2)\pi^0(\omega^2)^\top r(\omega^2)$
$= 0 + 0.3((95.8333, 0, 0, 0, 0, 0, 0, 0)(0, 0, 0, 0, -20, -15, -15)^\top) = 0$.

$s = 3$: feasible, $\varphi^0(x^0, \omega^3) = 0$, $y_3^0 = (0, 0, 0)^\top$, $\pi^0(\omega^3) = (95.8333, 0, 0, 0, 0, 0, 0, 0)^\top$, and $\pi_0^0(\omega^3) = 0$.

Compute $\bar{Q}^0 \leftarrow \bar{Q}^0 + p(\omega^3)\varphi^0(x^0, \omega^3) = 0 + 0.3(0) = 0$.
Compute $\beta_0 \leftarrow \beta_0 + p(\omega^3)\left(T(\omega^3)^\top \pi^0(\omega^3) - \pi_0^0(\omega^3)c\right)$

$$= \; (43.1250, 0, 0, 0)^\top \; + 0.3(\begin{bmatrix} 1\,0\,0\,0 \\ 0\,1\,0\,0 \\ 0\,0\,1\,0 \\ 0\,0\,0\,1 \\ 0\,0\,0\,0 \\ 0\,0\,0\,0 \\ 0\,0\,0\,0 \end{bmatrix}^\top (95.8333, 0, 0, 0, 0, 0, 0, 0) \; -$$

$0(50, 30, 15, 10)^\top) = (71.850, 0, 0, 0)^\top$.
Compute $\beta_0^1 \leftarrow \beta_0^1 + p(\omega^3)\pi_0^0(\omega^3) = 0 + 0.15(0) = 0$.
Compute $\beta_0^0 \leftarrow \beta_0^0 + p(\omega^3)\pi^0(\omega^3)^\top r(\omega^3)$
$= 0 + 0.3(95.8333, 0, 0, 0, 0, 0, 0, 0)(0, 0, 0, 0, -25, -20, -25)^\top = 0$.

$s = 4$: feasible, $\varphi^0(x^0, \omega^4) = 0$, $y_4^0 = (0, 0, 0)^\top$, $\pi^0(\omega^4) = (95.8333, 0, 0, 0, 0, 0, 0, 0)^\top$, and $\pi_0^0(\omega^4) = 0$.

Compute $\bar{Q}^0 \leftarrow \bar{Q}^0 + p(\omega^4)\varphi^0(x^0, \omega^4) = 0 + 0.2(0) = 0$.
Compute $\beta_0 \leftarrow \beta_0 + p(\omega^4)\left(T(\omega^4)^\top \pi^0(\omega^4) - \pi_0^0(\omega^4)c\right)$

$$= \quad (71.850, 0, 0, 0)^\top \; + \; 0.2(\begin{bmatrix} 1\,0\,0\,0 \\ 0\,1\,0\,0 \\ 0\,0\,1\,0 \\ 0\,0\,0\,1 \\ 0\,0\,0\,0 \\ 0\,0\,0\,0 \\ 0\,0\,0\,0 \end{bmatrix}^\top (95.8333, 0, 0, 0, 0, 0, 0, 0) \; -$$

$0(50, 30, 15, 10)^\top) = (91.1667, 0, 0, 0)^\top$.

Compute $\beta_0^1 \leftarrow \beta_0^1 + p(\omega^4)\pi_0^0(\omega^4) = 0 + 0.15(0) = 0$.
Compute $\beta_0^0 \leftarrow \beta_0^0 + p(\omega^4)\pi^0(\omega^4)^\top r(\omega^4)$
$= 0 + 0.2(95.8333, 0, 0, 0, 0, 0, 0, 0)(0, 0, 0, 0, -30, -25, -30)^\top = 0$.

$s = 5$: feasible, $\varphi^0(x^0, \omega^5) = 0$, $y_5^0 = (0, 0, 0)^\top$, $\pi^0(\omega^5) = (95.8333, 0, 0, 0, 0, 0, 0, 0)^\top$, and $\pi_0^0(\omega^5) = 0$.

Compute $\bar{Q}^0 \leftarrow \bar{Q}^0 + p(\omega^5)\varphi^0(x^0, \omega^5) = 0 + 0.05(0) = 0$.
Compute $\beta_0 \leftarrow \beta_0 + p(\omega^5)\left(T(\omega^5)^\top \pi^0(\omega^5) - \pi_0^0(\omega^5)c\right)$

$$= (91.1667, 0, 0, 0)^\top + 0.05(\begin{bmatrix} 1\,0\,0\,0 \\ 0\,1\,0\,0 \\ 0\,0\,1\,0 \\ 0\,0\,0\,1 \\ 0\,0\,0\,0 \\ 0\,0\,0\,0 \\ 0\,0\,0\,0 \end{bmatrix}^\top (95.8333, 0, 0, 0, 0, 0, 0, 0) -$$
$0(50, 30, 15, 10)^\top) = (95.8333, 0, 0, 0)^\top$.
Compute $\beta_0^1 \leftarrow \beta_0^1 + p(\omega^5)\pi_0^0(\omega^5) = 0 + 0.05(0) = 0$.
Compute $\beta_0^0 \leftarrow \beta_0^0 + p(\omega^5)\pi^0(\omega^5)^\top r(\omega^5)$
$= 0 + 0.05(95.8333, 0, 0, 0, 0, 0, 0, 0)(0, 0, 0, 0, -30, -25, -30)^\top = 0$.

Generate an *optimality* cut: $95.8333x_1 + 0\eta + \gamma \geq 0$.
Compute *upper bound*:

Set $u^0 \leftarrow (1-\lambda\varepsilon_1)c^\top x^0 + \lambda\varepsilon_1\eta^0 + \bar{Q}^0 = (1-0.5)0 + 0\eta^0 + 0 = 0$.
Set $u \leftarrow \min\{0, \infty\} = 0$.

u is updated, and set *incumbent* solution $(x^*, \eta^*) \leftarrow (x^0, \eta^0)$, where $x^0 = (0, 0, 0, 0)^\top$ and $\eta^0 = 0$.

Step 2. Add Cut to Master Program and Solve.

Add optimality cut to master problem (7.4).
Add $95.8333x + 0\eta + \gamma \geq 0$ to master problem:

$$\begin{array}{llllllll}
\ell^1 := \text{Min} & 25x_1 & +15x_2 & +7.5x_3 & +5x_4) & +0.5\eta & +\gamma & \\
\text{s.t.} & -x_1 & & & & & & \geq -300 \\
& & -x_2 & & & & & \geq -700 \\
& & & -x_3 & & & & \geq -600 \\
& & & & -x_4 & & & \geq -500 \\
& & -x_2 & -x_3 & -x_4 & & & \geq -1600 \\
& 95.8333x_1 & & & & & +\gamma & \geq 0 \\
& x_1, & x_2, & x_3, & x_4 & & & \geq 0.
\end{array}$$

Solve master to get $x^1 = (300, 0, 0, 0)^\top$, $\eta = 0$, and $\gamma = -28{,}750$ with objective value $\ell^0 = -21{,}250$. Set $\ell \leftarrow \max\{\ell^0, \ell\} = \max\{-\infty, -21{,}250\} = -21{,}250$.
Since $u - \ell = 0 - (-21{,}250) > 10^{-6}|0|$, set $k \leftarrow 1$.

Iteration k=1:

Step 1. Solve Subproblems.

Set $\beta_1 \leftarrow \mathbf{0}$, $\beta_2^1 \leftarrow 0$, $\beta_1^0 \leftarrow 0$, and $\bar{Q}^1 \leftarrow 0$.
For each $\omega \in \Omega$, solve subproblem (7.5). For $s = 1, \cdots, 5$, solve

$$
\begin{array}{lrrrrcr}
\varphi^1(x^1, \omega^s) := \text{Min} & -575y_1 & -762.5y_2 & -950y_3 & +10\nu & & \\
\text{s.t.} & -6y_1 & -8y_2 & -10y_3 & & \geq & 0 \\
& -20y_1 & -25y_2 & -28y_3 & & \geq & -300 \\
& -12y_1 & -15y_2 & -18y_3 & & \geq & 0 \\
& -8y_1 & -10y_2 & -14y_3 & & \geq & 0 \\
& -y_1 & & & & \geq & d_a^s \\
& & -y_2 & & & \geq & d_b^s \\
& & & -y_3 & & \geq & d_c^s \\
& -1150y_1 & -1525y_2 & -1900y_3 & +\nu \geq 0 & & \\
& y_1, & y_2, & y_3 & & \geq & 15{,}000.
\end{array}
$$

$s = 1$: feasible, $\varphi^1(x^1, \omega^1) = 150{,}000$, $y_1^1 = (0, 0, 0)^\top$, $\nu = 15{,}000$, $\pi^1(\omega^1) = (0, 0, 1108.33, 0, 0, 0, 0, 0)^\top$, and $\pi_0^1(\omega^1) = 10$.

Compute $\bar{Q}^1 \leftarrow \bar{Q}^1 + p(\omega^1)\varphi^1(x^2, \omega^1) = 0 + 0.15(150{,}000) = 22{,}500$.
Compute $\beta_1 \leftarrow \beta_1 + p(\omega^1)\big(T(\omega^1)^\top \pi^1(\omega^1) - \pi_0^1(\omega^1)c\big)$

$$
= \quad (0,0,0,0)^\top \quad + \quad 0.15(0,0,1108.33,0,0,0,0,0) \begin{bmatrix} 1\,0\,0\,0 \\ 0\,1\,0\,0 \\ 0\,0\,1\,0 \\ 0\,0\,0\,1 \\ 0\,0\,0\,0 \\ 0\,0\,0\,0 \\ 0\,0\,0\,0 \end{bmatrix}^\top \quad -
$$

$10(50, 30, 15, 10)^\top$
$= (-75, -45, 143.750, -15)^\top$.
Compute $\beta_1^1 \leftarrow \beta_1^1 + p(\omega^1)\pi_0^1(\omega^1) = 0 + 0.15(10) = 1.5$.
Compute $\beta_1^0 \leftarrow \beta_1^0 + p(\omega^1)\pi^1(\omega^1)^\top r(\omega^1)$
$= 0 + 0.15((0, 0, 1108.33, 0, 0, 0, 0, 0)(0, 0, 0, 0, -15, -10, -5)^\top) = 0$.

$s = 2$: feasible, $\varphi^1(x^1, \omega^2) = 150{,}000$, $y_2^1 = (0, 0, 0)^\top$, $\pi^1(\omega^2) = (0, 0, 1108.33, 0, 0, 0, 0, 0)^\top$, and $\pi_0^1(\omega^2) = 10$.

Compute $\bar{Q}^1 \leftarrow \bar{Q}^1 + p(\omega^2)\varphi^1(x^2, \omega^2) = 22{,}500 + 0.3(150{,}000) = 67{,}500$.
Compute $\beta_1 \leftarrow \beta_1 + p(\omega^2)\left(T(\omega^2)^\top \pi^1(\omega^2) - \pi_0^1(\omega^2)c\right)$

$$= (-75, -45, 143.750, -15)^\top + 0.3(\begin{bmatrix} 1\,0\,0\,0 \\ 0\,1\,0\,0 \\ 0\,0\,1\,0 \\ 0\,0\,0\,1 \\ 0\,0\,0\,0 \\ 0\,0\,0\,0 \\ 0\,0\,0\,0 \end{bmatrix}^\top$$

$(0, 0, 1108.33, 0, 0, 0, 0, 0) - 10(50, 30, 15, 10)^\top)$
$= (-225, -135, 431.2485, -45)^\top$.
Compute $\beta_2^1 \leftarrow \beta_2^1 + p(\omega^2)\pi_0^1(\omega^2) = 1.5 + 0.3(10) = 4.5$.
Compute $\beta_1^0 \leftarrow \beta_1^0 + p(\omega^2)\pi^1(\omega^2)^\top r(\omega^2)$
$= 0 + 0.3(0, 0, 1108.33, 0, 0, 0, 0, 0)(0, 0, 0, 0, -20, -15, -15)^\top = 0$.

$s = 3$: feasible, $\varphi^1(x^1, \omega^3) = 150{,}000$, $y_3^1 = (0, 0, 0)^\top$, $\pi^1(\omega^3) = (0, 0, 1108.33, 0, 0, 0, 0, 0)^\top$, and$\pi_0^1(\omega^3) = 10$.

Compute $\bar{Q}^1 \leftarrow \bar{Q}^1 + p(\omega^3)\varphi^1(x^2, \omega^3) = 67{,}500 + 0.3(150{,}000) = 112{,}500$.
Compute $\beta_1 \leftarrow \beta_1 + p(\omega^3)\left(T(\omega^3)^\top \pi^1(\omega^3) - \pi_0^1(\omega^3)c\right)$

$$= (-225, -135, 431.2485, -45)^\top + 0.3(\begin{bmatrix} 1\,0\,0\,0 \\ 0\,1\,0\,0 \\ 0\,0\,1\,0 \\ 0\,0\,0\,1 \\ 0\,0\,0\,0 \\ 0\,0\,0\,0 \\ 0\,0\,0\,0 \end{bmatrix}^\top$$

$(0, 0, 1108.33, 0, 0, 0, 0, 0) - 10(50, 30, 15, 10)^\top)$
$= (-375, -225, 718.7475, -75)^\top$.
Compute $\beta_2^1 \leftarrow \beta_2^1 + p(\omega^3)\pi_0^1(\omega^3) = 4.5 + 0.3(10) = 7.5$.
Compute $\beta_1^0 \leftarrow \beta_1^0 + p(\omega^3)\pi^1(\omega^3)^\top r(\omega^3)$
$= 0 + 0.3(0, 0, 1108.33, 0, 0, 0, 0, 0)(0, 0, 0, 0, -25, -20, -25)^\top = 0$.

$s = 4$: feasible, $\varphi^1(x^1, \omega^4) = 150{,}000$, $y_4^1 = (0, 0, 0)^\top$, $\pi^1(\omega^4) = (0, 0, 1108.33, 0, 0, 0, 0, 0)^\top$, and $\pi_0^1(\omega^4) = 10$.

Compute $\bar{Q}^1 \leftarrow \bar{Q}^1 + p(\omega^4)\varphi^1(x^2, \omega^4) = 112{,}500 + 0.2(150{,}000) = 142{,}500$.

Compute $\beta_1 \leftarrow \beta_1 + p(\omega^4)\big(T(\omega^4)^\top \pi^1(\omega^4) - \pi_0^1(\omega^4)c\big)$

$$= (-375, -225, 718.7475, -75)^\top + 0.2(\begin{bmatrix} 1\,0\,0\,0 \\ 0\,1\,0\,0 \\ 0\,0\,1\,0 \\ 0\,0\,0\,1 \\ 0\,0\,0\,0 \\ 0\,0\,0\,0 \\ 0\,0\,0\,0 \end{bmatrix}^\top$$

$(0, 0, 1108.33, 0, 0, 0, 0, 0) - 10(50, 30, 15, 10)^\top)$
$= (-475, -285, 910.4135, -95)^\top$.
Compute $\beta_2^1 \leftarrow \beta_2^1 + p(\omega^4)\pi_0^1(\omega^4) = 7.5 + 0.2(10) = 9.5$.
Compute $\beta_1^0 \leftarrow \beta_1^0 + p(\omega^4)\pi^1(\omega^4)^\top r(\omega^4)$
$= 0 + 0.2(0, 0, 1108.33, 0, 0, 0, 0, 0)(0, 0, 0, 0, -30, -25, -30)^\top = 0$.

$s = 5$: feasible, $\varphi^1(x^1, \omega^5) = 150{,}000$, $y_5^1 = (0, 0, 0)^\top$, $\pi^1(\omega^5) = (0, 0, 1108.33, 0, 0, 0, 0, 0)^\top$, and $\pi_0^1(\omega^5) = 10$.

Compute $\bar{Q}^1 \leftarrow \bar{Q}^1 + p(\omega^5)\varphi^1(x^2, \omega^5) = 142{,}500 + 0.05(150{,}000) = 150{,}000$.
Compute $\beta_1 \leftarrow \beta_1 + p(\omega^5)\big(T(\omega^5)^\top \pi^1(\omega^5) - \pi_0^1(\omega^5)c\big)$

$$= (-475, -285, 910.4135, -95)^\top + 0.05(\begin{bmatrix} 1\,0\,0\,0 \\ 0\,1\,0\,0 \\ 0\,0\,1\,0 \\ 0\,0\,0\,1 \\ 0\,0\,0\,0 \\ 0\,0\,0\,0 \\ 0\,0\,0\,0 \end{bmatrix}^\top$$

$(0, 0, 1108.33, 0, 0, 0, 0, 0) - 10(50, 30, 15, 10)^\top)$
$= (-500, -300, 958.33, -100)^\top$.
Compute $\beta_2^1 \leftarrow \beta_2^1 + p(\omega^5)\pi_0^1(\omega^5) = 9.5 + 0.05(10) = 10$.
Compute $\beta_1^0 \leftarrow \beta_1^0 + p(\omega^5)\pi^1(\omega^5)^\top r(\omega^5)$
$= 0 + 0.05(0, 0, 1108.33, 0, 0, 0, 0, 0)(0, 0, 0, 0, -30, -25, -30)^\top = 0$.

Generate an *optimality* cut: $-500x_1 - 300x_2 + 958.3333x_3 - 100x_4 + 10\eta + \gamma \geq 0$.
Compute *upper bound*:

Set $u^1 \leftarrow (1 - \lambda\varepsilon_1)c^\top x^1 + \lambda\varepsilon_1\eta^1 + \bar{Q}^1 = (1 - 0.5)(50, 30, 15, 10)$ $(300, 0, 0, 0)^\top$
$+ 0(0) + 0 = 7500$.
Set $u \leftarrow \min\{0, 7500\} = 0$.

u is not updated, the *incumbent* solution $(x^*, \eta^*) \leftarrow (x^0, \eta^0)$, where $x^0 = (0, 0, 0, 0)^\top$ and $\eta^0 = 0$.

Step 2. Add Cut to Master Program and Solve.

Add optimality cut to master problem (7.4).
Add $-500x_1-300x_2+958.3333x_3-100x_4+10\eta+\gamma \geq 0$ to master problem:

$$
\begin{array}{llllllll}
\ell^2 := \text{Min} & 25x_1 & +15x_2 & +7.5x_3 & +5x_4) & +0.5\eta & +\gamma & \\
\text{s.t.} & -x_1 & & & & & & \geq -300 \\
& & -x_2 & & & & & \geq -700 \\
& & & -x_3 & & & & \geq -600 \\
& & & & -x_4 & & & \geq -500 \\
& & -x_2 & -x_3 & -x_4 & & & \geq -1600 \\
& 95.8333x_1 & & & & & +\gamma & \geq 0 \\
& -500x_1 & -300x_2 & +958.3333x_3 & -100x_4 & +10\eta & +\gamma & \geq 0 \\
& x_1, & x_2, & x_3, & x_4 & & & \geq 0.
\end{array}
$$

Solve master to get $x^2 = (300, 0, 186.522, 0)^\top$, $\eta = 0$, and $\gamma = -28{,}750$ with objective value $\ell^2 = -19{,}851.1$. Set $\ell \leftarrow \max\{\ell^1, \ell^2\} = \max\{-21{,}250, -19{,}851.1\} = -19{,}851.1$.
Since $u - \ell = 0 - (-19{,}851.1) > 10^{-6}|0|$, set $k \leftarrow 2$.

After 21 iterations, the $\mathbb{D}$-AGG algorithm terminates with optimal solution $x^* = (234.49, 690, 429.796, 312.857)^\top$, and objective value -1083.19. Recall from Chap. 6 that the risk-neutral (i.e., when $\lambda = 0$) optimal solution is $x^* = (236.4, 690, 432, 318)^\top$ with objective value -2268.5. Compared to the risk-neutral case, the QDEV risk measure gives a solution that invests slightly less in the material and processing time to account for the risk of a less likely scenario ω^5.

Next, we study the separate cut decomposition approach for QDEV, CVaR, and EE.

7.3 Separate Cut Decomposition for QDEV, CVaR, and EE

Let us now consider an alternative approach to the aggregated method involving two separate cuts, one for the expectation term and the other for the deviation term. We refer to this approach as *separate cut* and derive the $\mathbb{D}$-SEP algorithm . Next, we decompose the DEP for risk measure QDEV and then move on to CVaR and EE, in that order. We then give a formal statement of the $\mathbb{D}$-SEP algorithm and end the section with a numerical example to illustrate the algorithm.

7.3.1 Quantile Deviation

Instead of placing a single or *aggregatedd* optimality cut in the master program at a given iteration of the algorithm, an alternative approach is to place *two separate* cuts: one for the expectation term approximated by γ, and the other for the deviation term approximated by ζ. In this case, the master program takes the following form:

$$\begin{aligned}
\ell^k := \text{Min } & (1-\lambda\varepsilon_1)c^\top x + \lambda\varepsilon_1\eta + (1-\lambda\varepsilon_1)\gamma + \lambda(\varepsilon_1+\varepsilon_2)\zeta \\
\text{s.t. } & Ax \geq b \\
& \beta_t^\top x + \gamma \geq \beta_t^0, \qquad t \in \Theta_k \\
& \sigma_t^\top x + \sigma_t^1\eta + \zeta \geq \sigma_t^0, \quad t \in \Theta_k \\
& \beta_t^\top x \qquad\qquad \geq \beta_t^0, \quad t \notin \Theta_k \\
& x \geq 0, \delta_l \leq \eta \leq \delta_h.
\end{aligned} \tag{7.10}$$

Now γ represents a lower bound approximation for the expectation term $\sum_{\omega\in\Omega} p(\omega)q(\omega)^\top y(\omega)$, while ζ represents that for the deviation term $\sum_{\omega\in\Omega} p(\omega)v(\omega)$. Given a solution (x^k, η^k) to the master problem at iteration k, the subproblem for scenario ω is given as follows:

$$\begin{aligned}
\varphi(x^k,\omega) := \text{Min } & q(\omega)^\top y \\
\text{s.t. } & Wy \geq r(\omega) - T(\omega)x^k \\
& y \geq 0,
\end{aligned} \tag{7.11}$$

with its dual solution given by $\pi^k(\omega)$. Then the optimality cut coefficients on the expectation term are calculated as follows:

$$\beta_k = \sum_{\omega\in\Omega} p(\omega)T(\omega)^\top \pi^k(\omega)$$

and

$$\beta_k^0 = \sum_{\omega\in\Omega} p(\omega)\pi^k(\omega)^\top r(\omega).$$

If the subproblem is *infeasible*, then a dual extreme ray $\mu^k(\omega)$ is obtained and a feasibility cut generated as follows:

$$\beta_k = T(\omega)^\top \mu^k(\omega)$$

and

$$\beta_k^0 = \mu^k(\omega)^\top r(\omega).$$

Given a solution (x^k, η^k) to master problem (7.10) and solution $y^k(\omega)$ to subproblem (7.11), the quantile deviation term assumes the value

$$\nu^k(\omega) = \max\left\{c^\top x^k + \varphi(x^k, \omega) - \eta^k, 0\right\}.$$

Thus the optimality cut coefficients for the quantile deviation term are calculated as follows: Let indicator parameter $\mathbb{I}(\omega) = 1$ if $c^\top x^k + \varphi(x^k, \omega) > \eta^k$, and $\mathbb{I}(\omega) = 0$ otherwise. Then

$$\sigma_k = \sum_{\omega \in \Omega} \mathbb{I}(\omega) p(\omega)(T(\omega)^\top \pi^k(\omega) - c),$$

$$\sigma_k^1 = \sum_{\omega \in \Omega} \mathbb{I}(\omega) p(\omega),$$

and

$$\sigma_k^0 = \sum_{\omega \in \Omega} \mathbb{I}(\omega) p(\omega) \pi^k(\omega)^\top r(\omega).$$

7.3.2 Conditional Value-at-Risk

The *separate cut* master program for CVaR takes the following form:

$$\begin{aligned}
\ell^k := \text{Min } & (1-\lambda)c^\top x + \lambda\eta + (1-\lambda)\gamma + \frac{\lambda}{1-\alpha}\zeta \\
\text{s.t. } & Ax \geq b \\
& \beta_t^\top x + \gamma \geq \beta_t^0, \quad t \in \Theta_k \\
& \sigma_t^\top x + \sigma_t^1 \eta + \zeta \geq \sigma_t^0, \quad t \in \Theta_k \\
& \beta_t^\top x \geq \beta_t^0, \quad t \notin \Theta_k \\
& x \geq 0, \delta_l \leq \eta \leq \delta_h.
\end{aligned} \tag{7.12}$$

As in the QDEV case, γ represents a lower bound approximation for the expectation term $\sum_{\omega \in \Omega} p(\omega) q(\omega)^\top y(\omega)$ and ζ represents that for the deviation term $\sum_{\omega \in \Omega} p(\omega) \nu(\omega)$. Given a solution (x^k, η^k) to the master problem at iteration k, the subproblem for scenario ω is as given formulation (7.11), with its dual solution given by $\pi^k(\omega)$. Then the optimality cut coefficients on the expectation term are calculated as follows:

$$\beta_k = \sum_{\omega \in \Omega} p(\omega) T(\omega)^\top \pi^k(\omega)$$

and

$$\beta_k^0 = \sum_{\omega \in \Omega} p(\omega) \pi^k(\omega)^\top r(\omega).$$

If the subproblem is *infeasible*, then a dual extreme ray $\mu^k(\omega)$ is obtained and a feasibility cut generated as follows:

$$\beta_k = T(\omega)^\top \mu^k(\omega)$$

and

$$\beta_k^0 = \mu^k(\omega)^\top r(\omega).$$

Given a solution (x^k, η^k) to master problem (7.12) and solution $y^k(\omega)$ to subproblem (7.11), the CVaR deviation term assumes the value

$$\nu^k(\omega) = \max\left\{c^\top x^k + \varphi(x^k, \omega) - \eta^k, 0\right\}.$$

Thus the optimality cut coefficients for the deviation term are calculated as follows: Let indicator parameter $\mathbb{I}(\omega) = 1$ if $c^\top x^k + \varphi(x^k, \omega) > \eta^k$, and $\mathbb{I}(\omega) = 0$ otherwise. Then

$$\sigma_k = \sum_{\omega \in \Omega} \mathbb{I}(\omega) p(\omega) (T(\omega)^\top \pi^k(\omega) - c),$$

$$\sigma_k^1 = \sum_{\omega \in \Omega} \mathbb{I}(\omega) p(\omega),$$

and

$$\sigma_k^0 = \sum_{\omega \in \Omega} \mathbb{I}(\omega) p(\omega) \pi^k(\omega)^\top r(\omega).$$

7.3.3 Expected Excess

The *separate cut* master program for EE takes the following form:

$$
\begin{aligned}
\ell^k := \text{Min } & c^\top x + \gamma + \lambda\zeta \\
\text{s.t. } & Ax \geq b \\
& \beta_t^\top x + \gamma \geq \beta_t^0, \quad t \in \Theta_k \\
& \sigma_t^\top x + \zeta \geq \sigma_t^0, \quad t \in \Theta_k \\
& \beta_t^\top x \quad\quad \geq \beta_t^0, \quad t \notin \Theta_k \\
& x \geq 0.
\end{aligned} \tag{7.13}
$$

As in the QDEV and CVaR cases, γ represents a lower bound approximation for the expectation term $\sum_{\omega\in\Omega} p(\omega)q(\omega)^\top y(\omega)$ and ζ represents that for the deviation term $\sum_{\omega\in\Omega} p(\omega)\nu(\omega)$. Given a solution x^k to the master problem at iteration k, the subproblem for scenario ω is the same as formulation (7.11) for the QDEV and CVaR with its dual solution given by $\pi^k(\omega)$. The optimality cut coefficients on the expectation term are calculated as follows:

$$\beta_k = \sum_{\omega\in\Omega} p(\omega)T(\omega)^\top \pi^k(\omega)$$

and

$$\beta_k^0 = \sum_{\omega\in\Omega} p(\omega)\pi^k(\omega)^\top r(\omega).$$

If the subproblem is *infeasible*, then a dual extreme ray $\mu^k(\omega)$ is obtained and a feasibility cut generated as follows:

$$\beta_k = T(\omega)^\top \mu^k(\omega)$$

and

$$\beta_k^0 = \mu^k(\omega)^\top r(\omega).$$

Given a solution x^k to master problem (7.10) and subproblem solution $y^k(\omega)$, the EE deviation term assumes the value

$$\nu^k(\omega) = \max\left\{c^\top x^k + \varphi(x^k, \omega) - \eta, 0\right\}.$$

Thus the optimality cut coefficients for the deviation term are calculated as follows: Let indicator parameter $\mathbb{I}(\omega) = 1$ if $c^\top x^k + \varphi(x^k, \omega) > \eta$, and $\mathbb{I}(\omega) = 0$ otherwise. Then

$$\sigma_k = \sum_{\omega\in\Omega} \mathbb{I}(\omega)p(\omega)(T(\omega)^\top \pi^k(\omega) - c)$$

and

$$\sigma_k^0 = \sum_{\omega \in \Omega} \mathbb{I}(\omega) p(\omega)(\pi^k(\omega)^\top r(\omega) - \eta).$$

Next, we devise the separate cut algorithm.

7.3.4 $\mathbb{D}$-SEP Algorithm

We can now formally state a separate cut algorithm for QDEV, CVaR, and EE as follows:

Algorithm $\mathbb{D}$-SEP
begin

Step 0. **Initialization.** $k \leftarrow 0$, x^0 given, $\ell \leftarrow -\infty$, $u \leftarrow \infty$, and $\epsilon > 0$. x^0 can be obtained as follows: $x^0 \leftarrow \text{argmin}\{c^\top x \mid Ax \geq b, x \geq 0\}$.

Case 1: $\mathbb{D}$:= QDEV

Select appropriate values for δ_l and δ_h.
Choose $\lambda \in [0, \frac{1}{\varepsilon_1}]$, $\alpha \in (0, 1)$, and $\varepsilon_1, \varepsilon_2 > 0$ such that $\alpha = \varepsilon_2/(\varepsilon_1 + \varepsilon_2)$ and $\eta^0 \in [\delta_l, \delta_h]$.

Case 2: $\mathbb{D}$:=CVaR

Select appropriate values for δ_l and δ_h.
Choose $\lambda \in [0, 1]$, $\alpha \in (0, 1)$, and $\eta^0 \in [\delta_l, \delta_h]$.

Case 3: $\mathbb{D}$:=EE

Choose target value $\eta \in \mathbb{R}$ and $\lambda \geq 0$.

Step 1. **Solve Subproblems.**

Case 1: $\mathbb{D}$:= QDEV

Set $\beta_k \leftarrow \mathbf{0}$, $\beta_k^1 \leftarrow 0$, $\beta_k^0 \leftarrow 0$, $\sigma_k \leftarrow \mathbf{0}$, $\sigma_k^1 \leftarrow 0$, $\sigma_k^0 \leftarrow 0$, $\bar{Q}_k \leftarrow 0$, and $\bar{v}_k \leftarrow 0$.
For each $\omega \in \Omega$, solve subproblem (7.11):

Get $\varphi(x^k, \omega)$ and dual solution $\pi^k(\omega)$.
If $c^\top x^k + \varphi(x^k, \omega) > \eta^k$:

Set $\mathbb{I}(\omega) \leftarrow 1$.

Else:

Set $\mathbb{I}(\omega) \leftarrow 0$.

Compute $\bar{Q}_k \leftarrow \bar{Q}_k + p(\omega)\varphi(x^k, \omega)$.

Compute $\bar{\nu}_k \leftarrow \bar{\nu}_k + \mathbb{I}(\omega)p(\omega)\left(c^\top x^k + \varphi(x^k, \omega) - \eta^k\right)$.
Compute $\beta_k \leftarrow \beta_k + p(\omega)T(\omega)^\top \pi^k(\omega)$.
Compute $\beta_k^0 \leftarrow \beta_k^0 + p(\omega)\pi^k(\omega)^\top r(\omega)$.
Compute $\sigma_k \leftarrow \sigma_k + \mathbb{I}(\omega)p(\omega)(T(\omega)^\top \pi^k(\omega) - c)$.
Compute $\sigma_k^1 \leftarrow \sigma_k^1 + \mathbb{I}(\omega)p(\omega)$.
Compute $\sigma_k^0 \leftarrow \sigma_k^0 + \mathbb{I}(\omega)p(\omega)\pi^k(\omega)^\top r(\omega)$.

If subproblem is *infeasible* for some ω:

Get dual extreme ray $\mu^k(\omega)$.
Compute $\beta_k \leftarrow T(\omega)^\top \mu^k(\omega)$.
Compute $\beta_k^0 \leftarrow \mu^k(\omega)^\top r(\omega)$.
Generate *feasibility* cut: $\beta_k^\top x \geq \beta_k^0$.
Go to step 2.

Else if *feasible* for all ω:

Generate *optimality* cuts: $\beta_k^\top x + \gamma \geq \beta_k^0$ and $\sigma_k^\top x + \sigma_k^1 \eta + \zeta \geq \sigma_k^0$.
Compute *upper bound*:

Set $u^k \leftarrow (1 - \lambda\varepsilon_1)(c^\top x^k + \bar{Q}_k) + \lambda\varepsilon_1\eta^k + \lambda(\varepsilon_1 + \varepsilon_2)\bar{\nu}_k$.
Set $u \leftarrow \min\{u^k, u\}$.

If u is updated, set *incumbent* solution $(x^*, \eta^*) \leftarrow (x^k, \eta^k)$.

Case 2: $\mathbb{D}$:=CVaR

Set $\beta_k \leftarrow \mathbf{0}$, $\beta_k^1 \leftarrow 0$, $\beta_k^0 \leftarrow 0$, $\sigma_k \leftarrow \mathbf{0}$, $\sigma_k^1 \leftarrow 0$, $\sigma_k^0 \leftarrow 0$, $\bar{Q}_k \leftarrow 0$, and $\bar{\nu}_k \leftarrow 0$.
For each $\omega \in \Omega$, solve subproblem (7.11):

Get $\varphi(x^k, \omega)$ and dual solution $\pi^k(\omega)$.
If $c^\top x^k + \varphi(x^k, \omega) > \eta^k$:

Set $\mathbb{I}(\omega) \leftarrow 1$.

Else:

Set $\mathbb{I}(\omega) \leftarrow 0$.

Compute $\bar{Q}_k \leftarrow \bar{Q}_k + p(\omega)\varphi(x^k, \omega)$.
Compute $\bar{\nu}_k \leftarrow \bar{\nu}_k + \mathbb{I}(\omega)p(\omega)\left(c^\top x^k + \varphi(x^k, \omega) - \eta^k\right)$.
Compute $\beta_k \leftarrow \beta_k + p(\omega)T(\omega)^\top \pi^k(\omega)$.
Compute $\beta_k^0 \leftarrow \beta_k^0 + p(\omega)\pi^k(\omega)^\top r(\omega)$.
Compute $\sigma_k \leftarrow \sigma_k + \mathbb{I}(\omega)p(\omega)(T(\omega)^\top \pi^k(\omega) - c)$.
Compute $\sigma_k^1 \leftarrow \sigma_k^1 + \mathbb{I}(\omega)p(\omega)$.
Compute $\sigma_k^0 \leftarrow \sigma_k^0 + \mathbb{I}(\omega)p(\omega)\pi^k(\omega)^\top r(\omega)$.

If subproblem is *infeasible* for some ω:

Get dual extreme ray $\mu^k(\omega)$.

Compute $\beta_k \leftarrow T(\omega)^\top \mu^k(\omega)$.
Compute $\beta_k^0 \leftarrow \mu^k(\omega)^\top r(\omega)$.
Generate *feasibility* cut: $\beta_k^\top x \geq \beta_k^0$.
Go to step 2.

Else if *feasible* for all ω:

Generate *optimality* cuts: $\beta_k^\top x + \gamma \geq \beta_k^0$ and $\sigma_k^\top x + \sigma_k^1 \eta + \zeta \geq \sigma_k^0$.
Compute *upper bound*:

Set $u^k \leftarrow (1-\lambda)(c^\top x^k + \bar{Q}_k) + \lambda \eta^k + \frac{\lambda}{1-\alpha} \bar{\nu}_k$.
Set $u \leftarrow \min\{u^k, u\}$.

If u is updated, set *incumbent* solution $(x^*, \eta^*) \leftarrow (x^k, \eta^k)$.

Case 3: $\mathbb{D}$:=EE

Set $\beta_k \leftarrow \mathbf{0}$, $\beta_k^0 \leftarrow 0$, $\sigma_k \leftarrow \mathbf{0}$, $\sigma_k^1 \leftarrow 0$, $\sigma_k^0 \leftarrow 0$, $\bar{Q}_k \leftarrow 0$, and $\bar{\nu}_k \leftarrow 0$.
For each $\omega \in \Omega$, solve subproblem (7.11):

Get $\varphi(x^k, \omega)$ and dual solution $\pi^k(\omega)$.
If $c^\top x^k + \varphi(x^k, \omega) > \eta$:

Set $\mathbb{I}(\omega) \leftarrow 1$.

Else:

Set $\mathbb{I}(\omega) \leftarrow 0$.

Compute $\bar{Q}_k \leftarrow \bar{Q}_k + p(\omega)\varphi(x^k, \omega)$.
Compute $\bar{\nu}_k \leftarrow \bar{\nu}_k + \mathbb{I}(\omega)p(\omega)\left(c^\top x^k + \varphi(x^k, \omega) - \eta\right)$.
Compute $\beta_k \leftarrow \beta_k + p(\omega)T(\omega)^\top \pi^k(\omega)$.
Compute $\beta_k^0 \leftarrow \beta_k^0 + p(\omega)\pi^k(\omega)^\top r(\omega)$.
Compute $\sigma_k \leftarrow \sigma_k + \mathbb{I}(\omega)p(\omega)(T(\omega)^\top \pi^k(\omega) - c)$.
Compute $\sigma_k^0 \leftarrow \sigma_k^0 + \mathbb{I}(\omega)p(\omega)(\pi^k(\omega)^\top r(\omega) - \eta)$.

If subproblem is *infeasible* for some ω:

Get dual extreme ray $\mu^k(\omega)$.
Compute $\beta_k \leftarrow T(\omega)^\top \mu^k(\omega)$.
Compute $\beta_k^0 \leftarrow \mu^k(\omega)^\top r(\omega)$.
Generate *feasibility* cut: $\beta_k^\top x \geq \beta_k^0$.
Go to step 2.

Else if *feasible* for all ω:

Generate *optimality* cuts: $\beta_k^\top x + \gamma \geq \beta_k^0$, $\sigma_k^\top x + \zeta \geq \sigma_k^0$.
Compute *upper bound*:

Set $u^k \leftarrow c^\top x^k + \bar{Q}_k + \lambda \bar{\nu}_k$.
Set $u \leftarrow \min\{u^k, u\}$.

If u is updated, set *incumbent* solution $x^* \leftarrow x^k$.

Step 2. **Add Cut to Master Program and Solve.**

Case 1: $\mathbb{D} :=$ QDEV

If some subproblem was *infeasible*:

Add feasibility cut to master problem (7.10).

Else:

Add optimality cut to master problem (7.10).

Solve master problem (7.10), and get optimal value ℓ_{k+1} and solution (x^{k+1}, η^{k+1}).

Case 2: $\mathbb{D} :=$CVaR

If some subproblem was *infeasible*:

Add feasibility cut to master problem (7.12).

Else:

Add optimality cut to master problem (7.12).

Solve master problem (7.12), and get optimal value ℓ_{k+1} and solution (x^{k+1}, η^{k+1}).

Case 3: $\mathbb{D} :=$EE

If some subproblem was *infeasible*:

Add feasibility cut to master problem (7.13).

Else:

Add optimality cut to master problem (7.13).

Solve master problem (7.13), and get optimal value ℓ_{k+1} and solution x^{k+1}.

Set $\ell \leftarrow \max\{\ell^{k+1}, \ell\}$.

Step 3. **Termination.** If $u - \ell \leq \epsilon|u|$:

Cases 1 and 2: $\mathbb{D} :=$ QDEV or CVaR

Stop, solution (x^*, η^*) is ϵ-optimal.

Case 3: $\mathbb{D} :=$EE

Stop, solution x^* is ϵ-optimal.

Else:

Set $k \leftarrow k + 1$.

Return to Step 1.

End

To perform parametric optimization over λ values, ignore the initialization of λ in step 0 and instead perform the parametric $\mathbb{D}$-*SEP* following the same step as outlined for $\mathbb{D}$-AGG in the previous section. Next, we give a numerical illustration of the $\mathbb{D}$-SEP algorithm.

7.3.5 Numerical Example

Example 7.2 Apply two iterations of the $\mathbb{D}$-SEP algorithm for $\mathbb{D} :=$ QDEV to the problem instance in Example 7.1.

Algorithm $\mathbb{D}$-*SEP*
begin

Step 0. **Initialization.** Let $k \leftarrow 0$, choose $x^0 \in X$, and set $\ell \leftarrow -\infty$ and $u \leftarrow \infty$. Choose $\epsilon \leftarrow 10^{-6}$ and $x^0 \leftarrow \text{argmin}_{x\in X}\{50x_1+30x_2+15x_3+10x_4\} = (0, 0, 0, 0)^\top$ as the initial point , and $\lambda \leftarrow 0.5$.
Case 1: $\mathbb{D} :=$ QDEV

Select appropriate values for $\delta_l = -5000$ and $\delta_h = 0$.
Choose $\lambda \in [0, \frac{1}{\varepsilon_1}] = 0.5$, $\alpha \in (0, 1)$, and $\varepsilon_1 = 1$, $\varepsilon_2 = 19$ such that $\alpha = \varepsilon_2/(\varepsilon_1 + \varepsilon_2) = 0.95$ and $\eta^0 = 0$.

Step 1. **Solve Subproblems.** *Case 1:* $\mathbb{D} :=$ QDEV
For $s = 1, \cdots, 5$, solve

$$
\begin{array}{rrrrl}
\varphi^0(x^0, \omega^s) := \text{Min} & -1150y_1 & -1525y_2 & -1900y_3 & \\
\text{s.t.} & -6y_1 & -8y_2 & -10y_3 & \geq 0 \\
 & -20y_1 & -25y_2 & -28y_3 & \geq 0 \\
 & -12y_1 & -15y_2 & -18y_3 & \geq 0 \\
 & -8y_1 & -10y_2 & -14y_3 & \geq 0 \\
 & -y_1 & & & \geq d_a^s \\
 & & -y_2 & & \geq d_b^s \\
 & & & -y_3 & \geq d_c^s \\
 & y_1, & y_2, & y_3 & \geq 0.
\end{array}
$$

$s = 1$: feasible, $\varphi^0(x^0, \omega^1) = 0$, $y_1^0 = (0, 0, 0)^\top$, $\pi^0(\omega^1) = (191.667, 0, 0, 0, 0, 0, 0)^\top$.

Since $c^\top x^0 + \varphi(x^0, \omega^1) = (50, 30, 15, 10)(0, 0, 0, 0)^\top + 0 = 0$, we set $\mathbb{I}(\omega^1) \leftarrow 0$.
Compute $\bar{Q}^0 \leftarrow \bar{Q}^0 + p(\omega^1)\varphi^0(x^0, \omega^1) = 0 + 0.15(0) = 0$.
Compute $\bar{\nu}_0 \leftarrow \bar{\nu}_0 + \mathbb{I}(\omega)p(\omega)\left(c^\top x^0 + \varphi(x^0, \omega) - \eta^0\right)$
$= 0 + 0(0.15)((50, 30, 15, 10)(0, 0, 0, 0)^\top + 0 - 0) = 0.$

Compute $\beta_0 \leftarrow \beta_0 + p(\omega)T(\omega)^\top \pi^0(\omega)$

$$= (0,0,0,0)^\top + 0.15 \begin{bmatrix} 1\,0\,0\,0 \\ 0\,1\,0\,0 \\ 0\,0\,1\,0 \\ 0\,0\,0\,1 \\ 0\,0\,0\,0 \\ 0\,0\,0\,0 \\ 0\,0\,0\,0 \end{bmatrix}^\top (191.667, 0, 0, 0, 0, 0, 0)^\top$$

$= (28.75, 0, 0, 0)^\top$.
$\beta_1^0 \leftarrow \beta_1^0 + p(\omega^1)\pi^1(\omega^1)^\top r(\omega^1)$
$= (0,0,0,0) + 0.15(191.667, 0, 0, 0, 0, 0, 0)(0, 0, 0, 0, -15, -10, -5)^\top =$
0.
Compute $\sigma_0 \leftarrow \sigma_0 + \mathbb{I}(\omega)p(\omega)(T(\omega)^\top \pi^0(\omega) - c)$

$$= (0,0,0,0)^\top + 0(0.15)(\begin{bmatrix} 1\,0\,0\,0 \\ 0\,1\,0\,0 \\ 0\,0\,1\,0 \\ 0\,0\,0\,1 \\ 0\,0\,0\,0 \\ 0\,0\,0\,0 \\ 0\,0\,0\,0 \end{bmatrix}^\top (191.667, 0, 0, 0, 0, 0, 0)^\top -$$

$(50, 30, 15, 10))$
$= 0$.
Compute $\sigma_0^1 \leftarrow \sigma_0^1 + \mathbb{I}(\omega)p(\omega) = 0 + 0(0.15) = 0$.
Compute $\sigma_0^0 \leftarrow \sigma_0^0 + \mathbb{I}(\omega)p(\omega)\pi^0(\omega)^\top r(\omega)$
$= 0 + 0(0.15)(191.667, 0, 0, 0, 0, 0, 0)(0, 0, 0, 0, -15, -10, -5)^\top = 0$.

$s = 2$: feasible, $\varphi^0(x^0, \omega^2) = 0$, $y_1^0 = (0,0,0)^\top$, $\pi^0(\omega^2) = (191.667, 0, 0, 0, 0, 0, 0)^\top$.

Since $c^\top x^0 + \varphi^0(x^0, \omega^2) = (50, 30, 15, 10)(0,0,0,0)^\top + 0 = 0$, we set $\mathbb{I}(\omega^2) \leftarrow 0$.
Compute $\bar{Q}^0 \leftarrow \bar{Q}^0 + p(\omega^2)\varphi^0(x^0, \omega^2) = 0 + 0.3(0) = 0$.
Compute $\bar{\nu}_0 \leftarrow \bar{\nu}_0 + \mathbb{I}(\omega)p(\omega)\left(c^\top x^0 + \varphi(x^0, \omega) - \eta^0\right)$
$= 0 + 0(0.3)((50, 30, 15, 10)(0,0,0,0)^\top + 0 - 0) = 0$.
Compute $\beta_0 \leftarrow \beta_0 + p(\omega)T(\omega)^\top \pi^0(\omega)$

$$= (28.75, 0, 0, 0)^\top + 0.3 \begin{bmatrix} 1\,0\,0\,0 \\ 0\,1\,0\,0 \\ 0\,0\,1\,0 \\ 0\,0\,0\,1 \\ 0\,0\,0\,0 \\ 0\,0\,0\,0 \\ 0\,0\,0\,0 \end{bmatrix}^\top (191.667, 0, 0, 0, 0, 0, 0)^\top$$

$= (86.25, 0, 0, 0)^\top$.

$\beta_1^0 \leftarrow \beta_1^0 + p(\omega^2)\pi^1(\omega^2)^\top r(\omega^2)$
$= 0 + 0.3(191.667, 0, 0, 0, 0, 0, 0)(0, 0, 0, 0, -20, -15, -15)^\top = 0.$
Compute $\sigma_0 \leftarrow \sigma_0 + \mathbb{I}(\omega)p(\omega)(T(\omega)^\top \pi^0(\omega) - c)$

$$= (0, 0, 0, 0)^\top + 0(0.3)(\begin{bmatrix} 1\,0\,0\,0 \\ 0\,1\,0\,0 \\ 0\,0\,1\,0 \\ 0\,0\,0\,1 \\ 0\,0\,0\,0 \\ 0\,0\,0\,0 \\ 0\,0\,0\,0 \end{bmatrix}^\top (191.667, 0, 0, 0, 0, 0, 0)^\top -$$

$(50, 30, 15, 10)) = 0.$
Compute $\sigma_0^1 \leftarrow \sigma_0^1 + \mathbb{I}(\omega)p(\omega) = 0 + 0(0.3) = 0.$
Compute $\sigma_0^0 \leftarrow \sigma_0^0 + \mathbb{I}(\omega)p(\omega)\pi^0(\omega)^\top r(\omega)$
$= 0 + 0(0.3)(191.667, 0, 0, 0, 0, 0, 0)(0, 0, 0, 0, -20, -15, -15)^\top = 0.$

$s = 3$: feasible, $\varphi^0(x^0, \omega^3) = 0$, $y_1^0 = (0, 0, 0)^\top$, $\pi^0(\omega^3) = (191.667, 0, 0, 0, 0, 0, 0)^\top$.

Since $c^\top x^0 + \varphi(x^0, \omega^3) = (50, 30, 15, 10)(0, 0, 0, 0)^\top + 0 = 0$, we set $\mathbb{I}(\omega^3) \leftarrow 0$.
Compute $\bar{Q}^0 \leftarrow \bar{Q}^0 + p(\omega^3)\varphi^0(x^0, \omega^3) = 0 + 0.3(0) = 0.$
Compute $\bar{\nu}_0 \leftarrow \bar{\nu}_0 + \mathbb{I}(\omega)p(\omega)\left(c^\top x^0 + \varphi(x^0, \omega) - \eta^0\right)$
$= 0 + 0(0.3)((50, 30, 15, 10)(0, 0, 0, 0)^\top + 0 - 0) = 0.$
Compute $\beta_0 \leftarrow \beta_0 + p(\omega)T(\omega)^\top \pi^0(\omega)$

$$= (86.25, 0, 0, 0)^\top + 0.3 \begin{bmatrix} 1\,0\,0\,0 \\ 0\,1\,0\,0 \\ 0\,0\,1\,0 \\ 0\,0\,0\,1 \\ 0\,0\,0\,0 \\ 0\,0\,0\,0 \\ 0\,0\,0\,0 \end{bmatrix}^\top (191.667, 0, 0, 0, 0, 0, 0)^\top$$

$= (143.75, 0, 0, 0)^\top$.
$\beta_1^0 \leftarrow \beta_1^0 + p(\omega^3)\pi^1(\omega^3)^\top r(\omega^3)$
$= 0 + 0.3(191.667, 0, 0, 0, 0, 0, 0)(0, 0, 0, 0, -25, -20, -25)^\top = 0.$
Compute $\sigma_0 \leftarrow \sigma_0 + \mathbb{I}(\omega)p(\omega)(T(\omega)^\top \pi^0(\omega) - c)$

$$= (0, 0, 0, 0)^\top + 0(0.3)(\begin{bmatrix} 1\,0\,0\,0 \\ 0\,1\,0\,0 \\ 0\,0\,1\,0 \\ 0\,0\,0\,1 \\ 0\,0\,0\,0 \\ 0\,0\,0\,0 \\ 0\,0\,0\,0 \end{bmatrix}^\top (191.667, 0, 0, 0, 0, 0, 0)^\top -$$

(50, 30, 15, 10))
$= 0.$
Compute $\sigma_0^1 \leftarrow \sigma_0^1 + \mathbb{I}(\omega)p(\omega) = 0 + 0(0.3) = 0.$
Compute $\sigma_0^0 \leftarrow \sigma_0^0 + \mathbb{I}(\omega)p(\omega)\pi^0(\omega)^\top r(\omega)$
$= 0 + 0(0.3)(191.667, 0, 0, 0, 0, 0, 0)(0, 0, 0, 0, -25, -20, -25)^\top = 0.$

$s = 4$: feasible, $\varphi^0(x^0, \omega^4) = 0$, $y_1^0 = (0, 0, 0)^\top$, $\pi^0(\omega^4) = (191.667, 0, 0, 0, 0, 0, 0)^\top$.

Since $c^\top x^0 + \varphi(x^0, \omega^4) = (50, 30, 15, 10)(0, 0, 0, 0)^\top + 0 = 0$, we set $\mathbb{I}(\omega^4) \leftarrow 0$.
Compute $\bar{Q}^0 \leftarrow \bar{Q}^0 + p(\omega^4)\varphi^0(x^0, \omega^4) = 0 + 0.2(0) = 0.$
Compute $\bar{\nu}_0 \leftarrow \bar{\nu}_0 + \mathbb{I}(\omega^4)p(\omega^4)\left(c^\top x^0 + \varphi(x^0, \omega^4) - \eta^0\right)$
$= 0 + 0(0.2)((50, 30, 15, 10)(0, 0, 0, 0)^\top + 0 - 0) = 0.$
Compute $\beta_0 \leftarrow \beta_0 + p(\omega^4)T(\omega^4)^\top \pi^0(\omega^4)$

$$= (143.75, 0, 0, 0)^\top + 0.2 \begin{bmatrix} 1\,0\,0\,0 \\ 0\,1\,0\,0 \\ 0\,0\,1\,0 \\ 0\,0\,0\,1 \\ 0\,0\,0\,0 \\ 0\,0\,0\,0 \\ 0\,0\,0\,0 \end{bmatrix}^\top (191.667, 0, 0, 0, 0, 0, 0)^\top$$

$= (182.083, 0, 0, 0)^\top.$
$\beta_1^0 \leftarrow \beta_1^0 + p(\omega^4)\pi^1(\omega^4)^\top r(\omega^4)$
$= 0 + 0.2(191.667, 0, 0, 0, 0, 0, 0)(0, 0, 0, 0, -30, -25, -30)^\top = 0.$
Compute $\sigma_0 \leftarrow \sigma_0 + \mathbb{I}(\omega^4)p(\omega^4)(T(\omega^4)^\top \pi^0(\omega^4) - c)$

$$= (0, 0, 0, 0)^\top + 0(0.2)(\begin{bmatrix} 1\,0\,0\,0 \\ 0\,1\,0\,0 \\ 0\,0\,1\,0 \\ 0\,0\,0\,1 \\ 0\,0\,0\,0 \\ 0\,0\,0\,0 \\ 0\,0\,0\,0 \end{bmatrix}^\top (191.667, 0, 0, 0, 0, 0, 0)^\top -$$

(50, 30, 15, 10))
$= 0.$
Compute $\sigma_0^1 \leftarrow \sigma_0^1 + \mathbb{I}(\omega^4)p(\omega^4) = 0 + 0(0.2) = 0.$
Compute $\sigma_0^0 \leftarrow \sigma_0^0 + \mathbb{I}(\omega^4)p(\omega^4)\pi^0(\omega^4)^\top r(\omega^4)$
$= 0 + 0(0.2)(191.667, 0, 0, 0, 0, 0, 0)(0, 0, 0, 0, -30, -25, -30)^\top = 0.$

$s = 5$: feasible, $\varphi^0(x^0, \omega^5) = 0$, $y_1^0 = (0, 0, 0)^\top$, $\pi^0(\omega^5) = (191.667, 0, 0, 0, 0, 0, 0)^\top$.

Since $c^\top x^0 + \varphi(x^0, \omega^5) = (50, 30, 15, 10)(0, 0, 0, 0)^\top + 0 = 0$, we set $\mathbb{I}(\omega^5) \leftarrow 0$.
Compute $\bar{Q}^0 \leftarrow \bar{Q}^0 + p(\omega^5)\varphi^0(x^0, \omega^5) = 0 + 0.05(0) = 0.$

Compute $\bar{\nu}_0 \leftarrow \bar{\nu}_0 + \mathbb{I}(\omega^5)p(\omega^5)\left(c^\top x^0 + \varphi(x^0, \omega^5) - \eta^0\right)$
$= 0 + 0(0.05)((50, 30, 15, 10)(0, 0, 0, 0)^\top + 0 - 0) = 0.$
Compute $\beta_0 \leftarrow \beta_0 + p(\omega^5)T(\omega^5)^\top \pi^0(\omega^5)$

$$= (182.083, 0, 0, 0)^\top + 0.05 \begin{bmatrix} 1\,0\,0\,0 \\ 0\,1\,0\,0 \\ 0\,0\,1\,0 \\ 0\,0\,0\,1 \\ 0\,0\,0\,0 \\ 0\,0\,0\,0 \\ 0\,0\,0\,0 \end{bmatrix}^\top (191.667, 0, 0, 0, 0, 0, 0)^\top$$

$= (191.6667, 0, 0, 0)^\top.$
$\beta_1^0 \leftarrow \beta_1^0 + p(\omega^5)\pi^1(\omega^5)^\top r(\omega^5)$
$= 0 + 0.05(191.667, 0, 0, 0, 0, 0, 0)(0, 0, 0, 0, -10, -10, -10)^\top = 0.$
Compute $\sigma_0 \leftarrow \sigma_0 + \mathbb{I}(\omega^5)p(\omega^5)(T(\omega^5)^\top \pi^0(\omega^5) - c)$

$$= \quad (0, 0, 0, 0)^\top \; + \; 0(0.05)(\begin{bmatrix} 1\,0\,0\,0 \\ 0\,1\,0\,0 \\ 0\,0\,1\,0 \\ 0\,0\,0\,1 \\ 0\,0\,0\,0 \\ 0\,0\,0\,0 \\ 0\,0\,0\,0 \end{bmatrix}^\top (191.667, 0, 0, 0, 0, 0, 0)^\top \quad -$$

$(50, 30, 15, 10))$
$= 0.$
Compute $\sigma_0^1 \leftarrow \sigma_0^1 + \mathbb{I}(\omega^5)p(\omega^5) = 0 + 0(0.05) = 0.$
Compute $\sigma_0^0 \leftarrow \sigma_0^0 + \mathbb{I}(\omega^5)p(\omega^5)\pi^0(\omega^5)^\top r(\omega^5)$
$= 0 + 0(0.05)(191.667, 0, 0, 0, 0, 0, 0)(0, 0, 0, 0, -10, -10, -10)^\top = 0.$

Generate *optimality* cuts: $\beta_0^\top x + \gamma \geq \beta_0^0 \leftarrow 191.6667x_1 + \gamma \geq 0$ and $\sigma_0^\top x + \sigma_0^1 \eta + \zeta \geq \sigma_0^0 \leftarrow \eta \geq 0.$
Compute *upper bound*:

Set $u^0 \leftarrow (1 - 0.5)((50, 30, 15, 10)(0, 0, 0, 0)^\top + 0 + 0.5(0) + 0.5(20)(0) = 0.$
Set $u \leftarrow \min\{0, \infty\} = 0.$

Since u is updated, set *incumbent* solution $(x^*, \eta^*) \leftarrow ((0, 0, 0, 0)^\top, 0).$

Step 2. Add Cut to Master Program and Solve.

Add optimality cuts $191.6667x_1 + \gamma \geq 0$ and $\zeta \geq 0$ to master problem (7.10).

Add $\beta_0^\top x + \gamma \geq \beta_0^0$ and $\sigma_0^\top x + \zeta \geq \sigma_0^0$ to master problem:

$$
\begin{array}{llrrrrrrrl}
\nu^0 := \text{Min} & & 25x_1 & +15x_2 & +7.5x_3 & +5x_4 & +0.5\zeta & +\gamma & & \\
\text{s.t.} & & -x_1 & & & & & & \geq & -300 \\
& & & -x_2 & & & & & \geq & -700 \\
& & & & -x_3 & & & & \geq & -600 \\
& & & & & -x_4 & & & \geq & -500 \\
& & & -x_2 & -x_3 & -x_4 & & & \geq & -1600 \\
& & 191.6667x_1 & & & & & +\gamma & \geq & 0 \\
& & & & & & \zeta & & \geq & 0 \\
& & x_1, & x_2, & x_3, & x_4 & & & \geq & 0.
\end{array}
$$

Solve master problem (7.10), and get optimal value $\ell_{k+1} = -21{,}250$ and solution $x^1 = (300, 0, 0, 0)^\top$, $\gamma = -57{,}500$, $\zeta = 0$.

Set $\ell \leftarrow \max\{-21{,}250, -\infty\} = -21{,}250$. $u - \ell \geq 10^{-6}|0|$: Set $k \leftarrow 1$. Return to Step 1.

For $k = 1$:

Step 1. **Solve Subproblems.** For $s = 1, \cdots, 5$, solve

$$
\begin{array}{lrrrrl}
\varphi^1(x^1, \omega^s) := \text{Min} & -1150y_1 & -1525y_2 & -1900y_3 & & \\
\text{s.t.} & -6y_1 & -8y_2 & -10y_3 & \geq & 0 \\
& -20y_1 & -25y_2 & -28y_3 & \geq & -300 \\
& -12y_1 & -15y_2 & -18y_3 & \geq & 0 \\
& -8y_1 & -10y_2 & -14y_3 & \geq & 0 \\
& -y_1 & & & \geq & d_a^s \\
& & -y_2 & & \geq & d_b^s \\
& & & -y_3 & \geq & d_c^s \\
& y_1, & y_2, & y_3 & \geq & 0.
\end{array}
$$

$s = 1$: feasible, $\varphi^1(x^1, \omega^1) = 0$, $y_1^0 = (0, 0, 0)^\top$, $\pi^0(\omega^1) = (0, 0, 105.5556, 0, 0, 0, 0)^\top$.

Since $c^\top x^1 + \varphi(x^1, \omega^1) = (50, 30, 15, 10)(300, 0, 0, 0)^\top + 0 > 0$, we set $\mathbb{I}(\omega^1) \leftarrow 1$.

Compute $\bar{Q}^1 \leftarrow \bar{Q}^1 + p(\omega^1)\varphi^1(x^1, \omega^1) = 0 + 0.15(0) = 0$.

Compute $\bar{\nu}_1 \leftarrow \bar{\nu}_1 + \mathbb{I}(\omega)p(\omega)\left(c^\top x^1 + \varphi(x^1, \omega) - \eta^1\right)$
$= 0 + 1(0.15)((50, 30, 15, 10)(300, 0, 0, 0)^\top + 0 - 0) = 2250.$

Compute $\beta_1 \leftarrow \beta_1 + p(\omega)T(\omega)^\top \pi^1(\omega)$

$$= (0,0,0,0)^\top + 0.15 \begin{bmatrix} 1\,0\,0\,0 \\ 0\,1\,0\,0 \\ 0\,0\,1\,0 \\ 0\,0\,0\,1 \\ 0\,0\,0\,0 \\ 0\,0\,0\,0 \\ 0\,0\,0\,0 \end{bmatrix}^\top (0,0,105.5556,0,0,0,0)^\top$$

$= (0, 0, 15.8333, 0)^\top$.

$\beta_1^0 \leftarrow \beta_1^0 + p(\omega^1)\pi^1(\omega^1)^\top r(\omega^1)$

$= 0 + 0.15(0, 0, 105.5556, 0, 0, 0, 0)(0, 0, 0, 0, -15, -10, -5)^\top = 0.$

Compute $\sigma_1 \leftarrow \sigma_1 + \mathbb{I}(\omega)p(\omega)(T(\omega)^\top \pi^1(\omega) - c)$

$$= (0,0,0,0)^\top + 1(0.15)(\begin{bmatrix} 1\,0\,0\,0 \\ 0\,1\,0\,0 \\ 0\,0\,1\,0 \\ 0\,0\,0\,1 \\ 0\,0\,0\,0 \\ 0\,0\,0\,0 \\ 0\,0\,0\,0 \end{bmatrix}^\top (0,0,105.5556,0,0,0,0)^\top -$$

$(50, 30, 15, 10))$

$= (-7.5, -4.5, 13.5833, -1.5)^\top$.

Compute $\sigma_1^1 \leftarrow \sigma_1^1 + \mathbb{I}(\omega)p(\omega) = 0 + 1(0.15) = 0.15.$

Compute $\sigma_1^0 \leftarrow \sigma_1^0 + \mathbb{I}(\omega)p(\omega)\pi^1(\omega)^\top r(\omega)$

$= 0 + 1(0.15)(0, 0, 105.5556, 0, 0, 0, 0)(0, 0, 0, 0, -15, -10, -5)^\top = 0.$

$s = 2$: feasible, $\varphi^1(x^1, \omega^2) = 0$, $y_1^0 = (0,0,0)^\top$, $\pi^0(\omega^2) = (0, 0, 105.5556, 0, 0, 0, 0)^\top$.

Since $c^\top x^1 + \varphi(x^1, \omega^2) = (50, 30, 15, 10)(300, 0, 0, 0)^\top + 0 > 0$, we set $\mathbb{I}(\omega^2) \leftarrow 1$.

Compute $\bar{Q}^1 \leftarrow \bar{Q}^1 + p(\omega^2)\varphi^1(x^1, \omega^2) = 0 + 0.3(0) = 0.$

Compute $\bar{\nu}_1 \leftarrow \bar{\nu}_1 + \mathbb{I}(\omega)p(\omega)\left(c^\top x^1 + \varphi(x^1, \omega) - \eta^0\right)$

$= 2250 + 1(0.3)((50, 30, 15, 10)(0, 0, 0, 0)^\top + 0 - 0) = 6750.$

Compute $\beta_1 \leftarrow \beta_1 + p(\omega)T(\omega)^\top \pi^0(\omega)$

$$= (0,0,15.8333,0)^\top + 0.3 \begin{bmatrix} 1\,0\,0\,0 \\ 0\,1\,0\,0 \\ 0\,0\,1\,0 \\ 0\,0\,0\,1 \\ 0\,0\,0\,0 \\ 0\,0\,0\,0 \\ 0\,0\,0\,0 \end{bmatrix}^\top (0,0,105.5556,0,0,0,0)^\top =$$

$(0, 0, 47.5, 0)^\top$.

$\beta_1^0 \leftarrow \beta_1^0 + p(\omega^2)\pi^1(\omega^2)^\top r(\omega^2)$
$= 0 + 0.3(0, 0, 105.5556, 0, 0, 0, 0)(0, 0, 0, 0, -20, -15, -15)^\top = 0.$
Compute $\sigma_1 \leftarrow \sigma_1 + \mathbb{I}(\omega)p(\omega)(T(\omega)^\top \pi^0(\omega) - c)$

$$= (-7.5, -4.5, 13.5833, -1.5)^\top + 1(0.3)(\begin{bmatrix} 1\,0\,0\,0 \\ 0\,1\,0\,0 \\ 0\,0\,1\,0 \\ 0\,0\,0\,1 \\ 0\,0\,0\,0 \\ 0\,0\,0\,0 \\ 0\,0\,0\,0 \end{bmatrix}^\top$$

$(0, 0, 105.5556, 0, 0, 0, 0)^\top - (50, 30, 15, 10)) = (-22.5, -13.5, 40.75, -4.5)^\top.$
Compute $\sigma_1^1 \leftarrow \sigma_1^1 + \mathbb{I}(\omega)p(\omega) = 0.15 + 1(0.3) = 0.45.$
Compute $\sigma_1^0 \leftarrow \sigma_1^0 + \mathbb{I}(\omega)p(\omega)\pi^0(\omega)^\top r(\omega)$
$= 0 + 1(0.3)(0, 0, 105.5556, 0, 0, 0, 0)(0, 0, 0, 0, -20, -15, -15)^\top = 0.$

$s = 3$: feasible, $\varphi^1(x^1, \omega^3) = 0$, $y_1^0 = (0, 0, 0)^\top$, $\pi^0(\omega^3) = (0, 0, 105.5556, 0, 0, 0, 0)^\top$.

Since $c^\top x^1 + \varphi(x^1, \omega^3) = (50, 30, 15, 10)(300, 0, 0, 0)^\top + 0 > 0$, we set $\mathbb{I}(\omega^3) \leftarrow 1$.
Compute $\bar{Q}^1 \leftarrow \bar{Q}^1 + p(\omega^3)\varphi^1(x^1, \omega^3) = 0 + 0.3(0) = 0.$
Compute $\bar{\nu}_1 \leftarrow \bar{\nu}_1 + \mathbb{I}(\omega)p(\omega)\left(c^\top x^1 + \varphi(x^1, \omega) - \eta^0\right)$
$= 6750 + 1(0.3)((50, 30, 15, 10)(300, 0, 0, 0)^\top + 0 - 0) = 11{,}250.$
Compute $\beta_1 \leftarrow \beta_1 + p(\omega)T(\omega)^\top \pi^0(\omega)$

$$= (0, 0, 31.6667, 0)^\top + 0.3 \begin{bmatrix} 1\,0\,0\,0 \\ 0\,1\,0\,0 \\ 0\,0\,1\,0 \\ 0\,0\,0\,1 \\ 0\,0\,0\,0 \\ 0\,0\,0\,0 \\ 0\,0\,0\,0 \end{bmatrix}^\top (0, 0, 105.5556,$$

$0, 0, 0, 0)^\top = (0, 0, 79.1667, 0)^\top.$
$\beta_1^0 \leftarrow \beta_1^0 + p(\omega^3)\pi^1(\omega^3)^\top r(\omega^3)$
$= 0 + 0.3(0, 0, 105.5556, 0, 0, 0, 0)(0, 0, 0, 0, -25, -20, -25)^\top = 0.$
Compute $\sigma_1 \leftarrow \sigma_1 + \mathbb{I}(\omega)p(\omega)(T(\omega)^\top \pi^0(\omega) - c)$

$$= (-22.5, -13.5, 40.75, -4.5)^\top + 1(0.3)(\begin{bmatrix} 1\,0\,0\,0 \\ 0\,1\,0\,0 \\ 0\,0\,1\,0 \\ 0\,0\,0\,1 \\ 0\,0\,0\,0 \\ 0\,0\,0\,0 \\ 0\,0\,0\,0 \end{bmatrix}^\top (0, 0, 105.5556,$$

$0, 0, 0, 0)^\top - (50, 30, 15, 10)) = (-37.5, -22.5, 67.9167, -7.5)^\top.$

Compute $\sigma_1^1 \leftarrow \sigma_1^1 + \mathbb{I}(\omega)p(\omega) = 0.45 + 1(0.3) = 0.75.$
Compute $\sigma_1^0 \leftarrow \sigma_1^0 + \mathbb{I}(\omega)p(\omega)\pi^0(\omega)^\top r(\omega)$
$= 0 + 1(0.3)(0, 0, 105.5556, 0, 0, 0, 0)(0, 0, 0, 0, -25, -20, -25)^\top = 0.$

$s = 4$: feasible, $\varphi^1(x^1, \omega^4) = 0$, $y_1^0 = (0, 0, 0)^\top$, $\pi^0(\omega^4) = (0, 0, 105.5556, 0, 0, 0, 0)^\top$.

Since $c^\top x^1 + \varphi(x^1, \omega^4) = (50, 30, 15, 10)(300, 0, 0, 0)^\top + 0 > 0$, we set $\mathbb{I}(\omega^4) \leftarrow 1$.
Compute $\bar{Q}^1 \leftarrow \bar{Q}^1 + p(\omega^4)\varphi^1(x^1, \omega^4) = 0 + 0.2(0) = 0.$
Compute $\bar{\nu}_1 \leftarrow \bar{\nu}_1 + \mathbb{I}(\omega^4)p(\omega^4)\left(c^\top x^1 + \varphi(x^1, \omega^4) - \eta^0\right)$
$= 11{,}250 + 1(0.2)((50, 30, 15, 10)(0, 0, 0, 0)^\top + 0 - 0) = 14{,}250.$
Compute $\beta_1 \leftarrow \beta_1 + p(\omega^4)T(\omega^4)^\top \pi^0(\omega^4)$

$$= (0, 0, 79.1667, 0)^\top + 0.2 \begin{bmatrix} 1\,0\,0\,0 \\ 0\,1\,0\,0 \\ 0\,0\,1\,0 \\ 0\,0\,0\,1 \\ 0\,0\,0\,0 \\ 0\,0\,0\,0 \\ 0\,0\,0\,0 \end{bmatrix}^\top (0, 0, 105.5556, 0, 0, 0, 0)^\top = (0, 0, 100.278, 0)^\top.$$

$\beta_1^0 \leftarrow \beta_1^0 + p(\omega^4)\pi^1(\omega^4)^\top r(\omega^4)$
$= 0 + 0.2(0, 0, 105.5556, 0, 0, 0, 0)(0, 0, 0, 0, -30, -25, -30)^\top = 0.$
Compute $\sigma_1 \leftarrow \sigma_1 + \mathbb{I}(\omega^4)p(\omega^4)(T(\omega^4)^\top \pi^0(\omega^4) - c)$

$$= (-37.5, -22.5, 67.9167, -7.5)^\top + 1(0.2)\left(\begin{bmatrix} 1\,0\,0\,0 \\ 0\,1\,0\,0 \\ 0\,0\,1\,0 \\ 0\,0\,0\,1 \\ 0\,0\,0\,0 \\ 0\,0\,0\,0 \\ 0\,0\,0\,0 \end{bmatrix}^\top (0, 0, 105.5556, 0, 0, 0, 0)^\top - (50, 30, 15, 10)\right) = (-47.5, -28.5, 86.0278, -9.5)^\top.$$

Compute $\sigma_1^1 \leftarrow \sigma_1^1 + \mathbb{I}(\omega^4)p(\omega^4) = 0.75 + 1(0.2) = 0.95.$
Compute $\sigma_1^0 \leftarrow \sigma_1^0 + \mathbb{I}(\omega^4)p(\omega^4)\pi^0(\omega^4)^\top r(\omega^4)$
$= 0 + 1(0.2)(0, 0, 105.5556, 0, 0, 0, 0)(0, 0, 0, 0, -30, -25, -30)^\top = 0.$

$s = 5$: feasible, $\varphi^1(x^1, \omega^5) = 0$, $y_1^0 = (0, 0, 0)^\top$, $\pi^0(\omega^5) = (0, 0, 105.5556, 0, 0, 0, 0)^\top$.

Since $c^\top x^1 + \varphi(x^1, \omega^5) = (50, 30, 15, 10)(300, 0, 0, 0)^\top + 0 > 0$, we set $\mathbb{I}(\omega^5) \leftarrow 1$.
Compute $\bar{Q}^1 \leftarrow \bar{Q}^1 + p(\omega^5)\varphi^1(x^1, \omega^5) = 0 + 0.05(0) = 0.$
Compute $\bar{\nu}_1 \leftarrow \bar{\nu}_1 + \mathbb{I}(\omega^5)p(\omega^5)\left(c^\top x^1 + \varphi(x^1, \omega^5) - \eta^0\right)$
$= 14{,}250 + 1(0.05)((50, 30, 15, 10)(0, 0, 0, 0)^\top + 0 - 0) = 15{,}000.$

Compute $\beta_1 \leftarrow \beta_1 + p(\omega^5)T(\omega^5)^\top \pi^0(\omega^5)$

$$= (0, 0, 100.278, 0)^\top + 0.05 \begin{bmatrix} 1\,0\,0\,0 \\ 0\,1\,0\,0 \\ 0\,0\,1\,0 \\ 0\,0\,0\,1 \\ 0\,0\,0\,0 \\ 0\,0\,0\,0 \\ 0\,0\,0\,0 \end{bmatrix}^\top (0, 0, 105.5556, 0, 0, 0, 0)^\top =$$

$(0, 0, 105.5556, 0)^\top$.

$\beta_1^0 \leftarrow \beta_1^0 + p(\omega^5)\pi^1(\omega^5)^\top r(\omega^5)$

$= 0 + 0.05(0, 0, 105.5556, 0, 0, 0, 0)(0, 0, 0, 0, -10, -10, -10)^\top = 0.$

Compute $\sigma_1 \leftarrow \sigma_1 + \mathbb{I}(\omega^5)p(\omega^5)(T(\omega^5)^\top \pi^0(\omega^5) - c)$

$$= (-47.5, -28.5, 86.0278, -9.5)^\top + 1(0.05)(\begin{bmatrix} 1\,0\,0\,0 \\ 0\,1\,0\,0 \\ 0\,0\,1\,0 \\ 0\,0\,0\,1 \\ 0\,0\,0\,0 \\ 0\,0\,0\,0 \\ 0\,0\,0\,0 \end{bmatrix}^\top$$

$(0, 0, 105.5556, 0, 0, 0, 0)^\top - (50, 30, 15, 10)) = (-50, -30, 90.5556, -10)^\top$.

Compute $\sigma_1^1 \leftarrow \sigma_1^1 + \mathbb{I}(\omega^5)p(\omega^5) = 0.95 + 1(0.05) = 1$.

Compute $\sigma_1^0 \leftarrow \sigma_1^0 + \mathbb{I}(\omega^5)p(\omega^5)\pi^0(\omega^5)^\top r(\omega^5)$

$= 0 + 1(0.05)(0, 0, 105.5556, 0, 0, 0, 0)(0, 0, 0, 0, -10, -10, -10)^\top = 0.$

Generate *optimality* cuts: $\beta_1^\top x + \gamma \geq \beta_0^0 \leftarrow 105.5556x_3 + \gamma \geq 0$ and $\sigma_1^\top x + \sigma_1^1 \eta + \zeta \geq \sigma_1^0 \leftarrow -50x_1 - 30x_2 + 90.5556x_3 - 10x_4 + \eta + \zeta \geq 0$.
Compute *upper bound*:

Set $u^1 \leftarrow (1 - 0.5)((50, 30, 15, 10)(300, 0, 0, 0)^\top + 0 + 0.5(0 + 0 + 0 + 0 + 0) + 0.5(20)15{,}000 = 22{,}500$.
Set $u \leftarrow \min\{0, 22{,}500\} = 0$.

Since u is updated, set *incumbent* solution $(x^*, \eta^*) \leftarrow ((0, 0, 0, 0)^\top, 0)$.

Step 2. Add Cut to Master Program and Solve.
Add optimality cuts $191.6667x_1 + \gamma \geq 0$ and $\zeta \geq 0$ to master problem (7.10).
Add $\beta_0^\top x + \gamma \geq \beta_0^0$ and $\sigma_0^\top x + \zeta \geq \sigma_0^0$ to master problem:

$$
\begin{array}{llllllllll}
\nu^0 := \text{Min} & 25x_1 & +15x_2 & +7.5x_3 & +5x_4 & & +0.5\zeta & +\gamma & & \\
\text{s.t.} & -x_1 & & & & & & & \geq & -300 \\
 & & -x_2 & & & & & & \geq & -700 \\
 & & & -x_3 & & & & & \geq & -600 \\
 & & & & -x_4 & & & & \geq & -500 \\
 & & -x_2 & -x_3 & -x_4 & & & & \geq & -1600 \\
 & 191.6667x_1 & & & & & & +\gamma & \geq & 0 \\
 & & & & & & \zeta & & \geq & 0 \\
 & & & 105.5556x_3 & & & & +\gamma & \geq & 0 \\
 & -50x_1 & -30x_2 & 90.5556x_3 & -10x_4 & +\eta & +\zeta & +\gamma & \geq & 0 \\
 & x_1, & x_2, & x_3, & x_4 \geq 0 & & & & & .
\end{array}
$$

Solve master problem (7.10), and get optimal value $\ell_2 = -17{,}164.5$ and solution $x^1 = (300, 0, 544.737, 0)^\top$, $\gamma = -57{,}500$, $\eta = 0$, $\zeta = 0$.

Set $\ell \leftarrow \max\{-21{,}250, -17{,}164.5\} = -17{,}164.5$. $u - \ell \geq 10^{-6}|0|$: Set $k \leftarrow 2$.

Return to Step 1.

After 12 iterations, the $\mathbb{D}$-SEP algorithm terminates with optimal solution $x^* = (234.49, 690, 429.796, 312.857)^\top$ and objective value -1083.19. Recall that we obtained the same solution with the $\mathbb{D}$-AGG algorithm that took 21 iterations and longer in terms of computation time. Therefore, the $\mathbb{D}$-SEP algorithm is a preferred choice in terms of both the number of iterations and computation time in this case.

Next, we study decomposition for MR-SLP with ASD.

7.4 Aggregated Cut Decomposition for ASD

Another risk measure of interest we studied in Chap. 2 is absolute semideviation (ASD). This risk measure is defined in a similar manner as EE, but with the target η replaced by the mean value $\mathbb{E}[f(x, \tilde{\omega})]$ and risk factor $0 \leq \lambda \leq 1$. Mathematically, ASD is defined as follows:

$$\phi_{ASD}(x) := \mathbb{E}[[f(x, \tilde{\omega}) - \mathbb{E}[f(x, \tilde{\omega})]]_+].$$

It gives the expected value of the excess over the mean value, which depends on the decision variable x. An MR-SLP with ASD is given as follows:

$$\min_{x \in X} \ \mathbb{E}[f(x, \tilde{\omega})] + \lambda \phi_{ASD}(x).$$

Given $\lambda \in [0, 1]$, the DEP formulation for this problem is given as follows:

$$
\begin{aligned}
\text{Min} \quad & (1-\lambda)c^\top x + (1-\lambda)\sum_{\omega \in \Omega} p(\omega)q(\omega)^\top y(\omega) \; + \lambda \sum_{\omega \in \Omega} p(\omega)v(\omega) \\
\text{s.t.} \quad & T(\omega)x + Wy(\omega) \geq r(\omega), \; \forall \omega \in \Omega \\
& -c^\top x - q(\omega)^\top y(\omega) + v(\omega) \geq 0, \; \forall \omega \in \Omega \\
& -c^\top x - \sum_{\varpi \in \Omega} p(\varpi)q(\varpi)^\top y(\varpi) + v(\omega) \geq 0, \; \forall \omega \in \Omega \\
& x \in X, \;\; y(\omega) \in \mathbb{R}_+^{n_2}, \; v(\omega) \in \mathbb{R}, \; \forall \omega \in \Omega.
\end{aligned}
$$

Unlike the DEP formulations for QDEV, CVaR, and EE that have a dual block angular structure and thus are amenable to Benders decomposition, ASD has a block angular structure due to the third set of constraints (linking constraints). Unfortunately, these constraints link all the scenarios, and therefore, standard SLP methods such as the L-shaped method cannot be applied directly to solve MR-SLP with ASD. Nevertheless, a *subgradient optimization* or *column generation* approach can be applied. We take the *subgradient optimization* approach and decompose the DEP into a master problem and scenario subproblems. We then derive subgradients to allow us to compute optimality cuts. Let us define the master problem at iteration k as follows:

$$
\begin{aligned}
\ell_k := \text{Min} \; & (1-\lambda)c^\top x + \gamma \\
\text{s.t.} \; & Ax \geq b \\
& \beta_t^\top x + \gamma \geq \beta_t^0, \quad t = 1, \cdots, k \\
& x \geq 0.
\end{aligned} \tag{7.14}
$$

As in the previous section, the free variable γ is the optimality cut variable. In this case, it gives a lower bound approximation for $(1-\lambda)\sum_{\omega \in \Omega} p(\omega)q(\omega)^\top y(\omega) + \lambda \sum_{\omega \in \Omega} p(\omega)v(\omega)$. The constraints represent the optimality cuts with left hand side coefficients at iteration k denoted by β_k and the corresponding RHS by β_k^0. Given a master problem solution x^k at iteration k, the subproblem for scenario ω is given by

$$
\begin{aligned}
\varphi(x^k, \omega) = \text{Min} \; & q(\omega)^\top y(\omega) \\
\text{s.t.} \; & Wy(\omega) \geq r(\omega) - T(\omega)x^k \\
& y(\omega) \geq 0.
\end{aligned} \tag{7.15}
$$

7.4.1 Subgradient Optimization Approach

We shall now devise a subgradient optimization approach to tackle the decomposed problem. To begin, let the set of dual feasible multipliers to subproblem (7.15) be denoted $\Pi(\omega) = \{\pi(\omega) \in \mathbb{R}_+^{m_2} \mid W^\top \pi(\omega) \le q(\omega)\}$. Then we have the following:

$$\begin{aligned} f(x,\omega) &= c^\top x + \varphi(x,\omega),\ \forall x \in X \\ &= c^\top x + \underset{\pi(\omega)\in\Pi(\omega)}{\text{Max}} \left\{\pi(\omega)^\top (r(\omega) - T(\omega)x)\right\},\ \forall x \in X \\ &\ge c^\top x + \pi^k(\omega)^\top (r(\omega) - T(\omega)x),\ \forall x \in X \end{aligned}$$

The last inequality follows from LP *weak duality*. Applying LP *strong duality*, we get the following:

$$\begin{aligned} f(x,\omega) = c^\top x + \pi^k(\omega)^\top (r(\omega) - T(\omega)x) + f(x^k,\omega) - c^\top x^k \\ - \pi^k(\omega)^\top (r(\omega) - T(\omega)x^k),\ \forall x \in X. \end{aligned}$$

Therefore, we can conclude that

$$f(x,\omega) \ge f(x^k,\omega) + (c - T(\omega)^\top \pi^k(\omega))^\top (x - x^k),\ \forall x \in X.$$

This relation reveals that $c - T(\omega)^\top \pi^k(\omega)$ is a *subgradient* of $f(.,\omega)$ at x^k. Given a solution x^k and optimal value $\varphi(x^k,\omega)$ to subproblem (7.15) at iteration k for all $\omega \in \Omega$, the deviation term $\nu^k(\omega)$ is calculated as follows:

$$\nu^k(\omega) = c^\top x^k + \max\{\varphi(x^k,\omega), \sum_{\varpi\in\Omega} p(\varpi)\varphi(x^k,\varpi)\}. \tag{7.16}$$

Expression (7.16) means that we have two cases to consider because of the "max" operator:

$$\nu^k(\omega) = c^\top x^k + \varphi(x^k,\omega) \text{ or } \nu^k(\omega) = c^\top x^k + \sum_{\varpi\in\Omega} p(\varpi)\varphi(x^k,\varpi).$$

Case 1: $\nu^k(\omega) = c^\top x^k + \varphi(x^k,\omega)$.

For this case, we shall use the subgradient $c - T(\omega)^\top \pi^k(\omega)$ to define $\beta_k(\omega)$ and $\beta_k^0(\omega)$ as follows:

$$\beta_k(\omega) = (1-\lambda)T(\omega)^\top \pi^k(\omega) + \lambda(T(\omega)^\top \pi^k(\omega) - c)$$

and

$$\beta_k^0(\omega) = (1-\lambda)\pi^k(\omega)^\top r(\omega) + \lambda\pi^k(\omega)^\top r(\omega).$$

Case 2: $\nu^k(\omega) = c^\top x^k + \sum_{\varpi\in\Omega} p(\varpi)\varphi(x^k, \varpi)$.
For this case, we shall use the expected subgradient $c - \sum_{\varpi\in\Omega} p(\varpi)T(\varpi)^\top\pi^k(\varpi)$ to calculate β_k and $\beta_k^0(\omega)$ as follows:

$$\beta_k(\omega) = (1-\lambda)T(\omega)^\top\pi^k(\omega) + \lambda(\sum_{\varpi\in\Omega} p(\varpi)T(\varpi)^\top\pi^k(\varpi) - c)$$

and

$$\beta_k^0(\omega) = (1-\lambda)\pi^k(\omega)^\top r(\omega) + \lambda\sum_{\varpi\in\Omega} p(\varpi)\pi^k(\varpi)^\top r(\varpi).$$

Since we have the above two cases to consider for a given scenario ω, it means that at iteration k of our subgradient optimization algorithm the optimality cut coefficients β_k and RHS β_k^0 in the master problem have to be calculated by selecting the $\beta_k(\omega)$ and $\beta_k^0(\omega)$ for each case. The vector β_k and scalar β_k^0 are computed as follows:

$$\beta_k = \sum_{\omega\in\Omega} p(\omega)\beta_k(\omega) \text{ and } \beta_k^0 = \sum_{\omega\in\Omega} p(\omega)\beta_k^0(\omega).$$

The optimality cut to add to the master program is $\beta_k^\top x + \gamma \geq \beta_k^0$. If a subproblem is *infeasible* for some scenario ω, then a dual extreme ray $\mu^k(\omega)$ is obtained and a feasibility cut generated as follows:

$$\beta_k = T(\omega)^\top\mu^k(\omega)$$

and

$$\beta_k^0 = \mu^k(\omega)^\top r(\omega).$$

Next, we give a formal statement of the AGG algorithm for ASD.

7.4.2 ASD-AGG Algorithm

Let ℓ and u continue to denote the lower and upper bounds, respectively, on the optimal value during the course of the subgradient optimization algorithm. We can formally state the ASD-AGG algorithm as follows:

Algorithm ASD-AGG
begin

Step 0. **Initialization.** $k \leftarrow 0$, x^0 given or $x^0 \leftarrow \text{argmin}\{c^\top x \mid Ax \geq b, x \geq 0\}$,
$\ell \leftarrow -\infty$, $u \leftarrow \infty$, $\epsilon > 0$, and $\lambda \in [0, 1]$.

Step 1. **Solve Subproblems.**

Set $\beta_k \leftarrow \mathbf{0}$, $\beta_k^0 \leftarrow 0$, $\bar{Q}^k \leftarrow 0$, and $\bar{\nu}^k \leftarrow 0$.
For each $\omega \in \Omega$, solve subproblem (7.15):

Get and store optimal value $\varphi(x^k, \omega)$.
Get dual solution $\pi^k(\omega)$.
Compute $\bar{Q}^k \leftarrow \bar{Q}^k + p(\omega)\varphi(x^k, \omega)$.

If subproblem is *infeasible* for some ω:

Get dual extreme ray $\mu^k(\omega)$.
Compute $\beta_k \leftarrow T(\omega)^\top \mu^k(\omega)$.
Compute $\beta_k^0 \leftarrow \mu^k(\omega)^\top r(\omega)$.
Generate *feasibility* cut: $\beta_k^\top x \geq \beta_k^0$.
Go to step 2.

Else if *feasible* for all ω, for each $\omega \in \Omega$:

If $\varphi(x^k, \omega) \geq \bar{Q}^k$:

Compute $\bar{\nu}^k \leftarrow \bar{\nu}^k + p(\omega)\varphi(x^k, \omega)$.
Compute $\beta_k \leftarrow \beta_k + p(\omega)\big((1-\lambda)T(\omega)^\top \pi^k(\omega)$
$+\lambda(T(\omega)^\top \pi^k(\omega) - c)\big)$.
Compute $\beta_k^0 \leftarrow \beta_k^0 + p(\omega)\big((1-\lambda)\pi^k(\omega)^\top r(\omega) + \lambda\pi^k(\omega)^\top r(\omega)\big)$.

Else:

Compute $\bar{\nu}^k \leftarrow \bar{\nu}^k + p(\omega)\bar{Q}^k$.
Compute $\beta_k \leftarrow \beta_k + p(\omega)\big((1-\lambda)T(\omega)^\top \pi^k(\omega)$
$+\lambda(\sum_{\varpi \in \Omega} p(\varpi)T(\varpi)^\top \pi^k(\varpi) - c)\big)$.
Compute $\beta_k^0 \leftarrow \beta_k^0 + p(\omega)\big((1-\lambda)\pi^k(\omega)^\top r(\omega)$
$+\lambda\sum_{\varpi \in \Omega} p(\varpi)\pi^k(\varpi)^\top r(\varpi)\big)$.

Generate an *optimality* cut: $\beta_k^\top x + \gamma \geq \beta_k^0$.
Compute *upper bound*:

Set $u^k \leftarrow c^\top x^k + (1-\lambda)\bar{Q}^k + \lambda\bar{\nu}^k$.
Set $u \leftarrow \min\{u^k, u\}$.

If u is updated, set *incumbent* solution $x^* \leftarrow x^k$.

Step 2. **Add Cut to Master Program and Solve.** If some subproblem was infeasible:

Add feasibility cut to master problem (7.14).

Else:

Add optimality cut to master problem (7.14).

Solve master problem (7.14), and get optimal value ℓ_{k+1} and solution x^{k+1}.

Set $\ell \leftarrow \max\{\ell^{k+1}, \ell\}$.

Step 3. **Termination.** If $u - \ell \leq \epsilon|u|$:

Stop, solution x^* is ϵ-optimal.

Else:

Set $k \leftarrow k + 1$.

Return to Step 1.

End

To perform parametric optimization over λ values, simply ignore the initialization of λ in step 0 of the ASD-AGG algorithm and instead perform the following steps to determine the value of λ to use for each algorithm run:

Algorithm *Parametric* ASD-AGG
begin

Step a. **Initialization.** Set $\kappa \leftarrow 0$, $\lambda^0 \leftarrow 0$ and choose $\delta > 0$ (very small number).

Step b. **Solve MR-SLP for $\lambda \leftarrow \lambda^\kappa$.** Apply ASD-AGG algorithm.

Step c. **Apply Parametric Analysis to Master Problem.** Find the range $[\lambda, \lambda^*]$ for which the current master problem basis remains optimal:

If $\lambda^* < 1$:

Set $\lambda^{\kappa+1} \leftarrow \min\{1, \lambda^* + \delta\}$.

Else:

Stop.

$\kappa \leftarrow \kappa + 1$.

Return to step (b).

End

7.4.3 Numerical Example

Example 7.3 Apply two iterations of the ASD-AGG algorithm to the problem instance in Example 7.1.

Algorithm ASD-AGG
begin

Step 0. **Initialization.** Let $k \leftarrow 0$, choose $x^0 \in X$, and set $\ell \leftarrow -\infty$ and $u \leftarrow \infty$. Choose $\epsilon \leftarrow 10^{-6}$ and $x^0 \leftarrow \text{argmin}_{x\in X}\{50x_1+30x_2+15x_3+10x_4\} = (0,0,0,0)^\top$ as the initial point , and $\lambda \leftarrow 0.5$.

Step 1. **Solve Subproblems.** For $s = 1, \cdots, 5$, solve

$$\begin{array}{rrrrl}
\varphi^0(x^0,\omega^s) := \text{Min} & -1150y_1 & -1525y_2 & -1900y_3 & \\
\text{s.t.} & -6y_1 & -8y_2 & -10y_3 & \geq 0 \\
& -20y_1 & -25y_2 & -28y_3 & \geq 0 \\
& -12y_1 & -15y_2 & -18y_3 & \geq 0 \\
& -8y_1 & -10y_2 & -14y_3 & \geq 0 \\
& -y_1 & & & \geq d_a^s \\
& & -y_2 & & \geq d_b^s \\
& & & -y_3 & \geq d_c^s \\
& y_1, & y_2, & y_3 & \geq 0.
\end{array}$$

$s = 1$: feasible, $\varphi^0(x^0,\omega^1)$, $y_1^0 = (0,0,0)^\top$, $\pi^0(\omega^1) = (191.667, 0, 0, 0, 0, 0, 0)^\top$. Compute $\bar{Q}^0 \leftarrow \bar{Q}^0 + p(\omega^1)\varphi(x^0,\omega^1) = 0 + 0.15(0) = 0$.

$s = 2$: feasible, $\varphi^0(x^0,\omega^2) = 0$, $y_2^0 = (0,0,0)^\top$, $\pi^0(\omega^2) = (191.667, 0, 0, 0, 0, 0, 0)^\top$. Compute $\bar{Q}^0 \leftarrow \bar{Q}^0 + p(\omega^2)\varphi(x^0,\omega^2) = 0 + 0.3(0) = 0$.

$s = 3$: feasible, $\varphi^0(x^0,\omega^3) = 0$, $y_3^0 = (0,0,0)^\top$, $\pi^0(\omega^3) = (191.667, 0, 0, 0, 0, 0, 0)^\top$. Compute $\bar{Q}^0 \leftarrow \bar{Q}^0 + p(\omega^3)\varphi(x^0,\omega^3) = 0 + 0.3(0) = 0$.

$s = 4$: feasible, $\varphi^0(x^0,\omega^4) = 0$, $y_4^0 = (0,0,0)^\top$, $\pi^0(\omega^4) = (191.667, 0, 0, 0, 0, 0, 0)^\top$. Compute $\bar{Q}^0 \leftarrow \bar{Q}^0 + p(\omega^4)\varphi(x^0,\omega^4) = 0 + 0.2(0) = 0$.

$s = 5$: feasible, $\varphi^0(x^0,\omega^5) = 0$, $y_5^0 = (0,0,0)^\top$, $\pi^0(\omega^5) = (191.667, 0, 0, 0, 0, 0, 0)^\top$. Compute $\bar{Q}^0 \leftarrow \bar{Q}^0 + p(\omega^5)\varphi(x^0,\omega^5) = 0 + 0.05(0) = 0$.

For $s = 1$, $\varphi^0(x^0,\omega^1) = \bar{Q}^0$:
Compute $\bar{\nu}^0 = \bar{\nu}^0 + p(\omega^1)\varphi^0(x^0,\omega^1) = 0 + 0.15(0) = 0$.
$\beta_0^0 \leftarrow \beta_0^0 + p(\omega^1)\big((1-\lambda)\pi^0(\omega^1)^\top r(\omega^1) + \lambda\pi^0(\omega^1)^\top r(\omega^1)\big)$
$= 0 + 0.15((1-0.5) \times (191.667, 0, 0, 0, 0, 0, 0)(0, 0, 0, 0, -15, -10, -5)^\top$
$+ 0.5 \times (191.667, 0, 0, 0, 0, 0, 0)(0, 0, 0, 0, -15, -10, -5)^\top) = 0.$

Compute $\beta_0 \leftarrow \beta_0 + p(\omega^1)\big((1-\lambda)T(\omega^1)^\top \pi^0(\omega^1) + \lambda(T(\omega^1)^\top \pi^0(\omega^1) - c^\top)\big)$
$= (0, 0, 0, 0)^\top$

$$+ 0.15((1-0.5)\begin{bmatrix} 1&0&0&0\\ 0&1&0&0\\ 0&0&1&0\\ 0&0&0&1\\ 0&0&0&0\\ 0&0&0&0\\ 0&0&0&0 \end{bmatrix}^\top (191.667, 0, 0, 0, 0, 0, 0)^\top$$

$$+ 0.5\begin{bmatrix} 1&0&0&0\\ 0&1&0&0\\ 0&0&1&0\\ 0&0&0&1\\ 0&0&0&0\\ 0&0&0&0\\ 0&0&0&0 \end{bmatrix}^\top (191.667, 0, 0, 0, 0, 0, 0)^\top - (50, 30, 15, 10))$$

$= (21.25, -4.5, -2.25, -1.5)^\top$.
For $s = 2$, $\varphi^0(x^0, \omega^2) = \bar{Q}^0$:
Compute $\bar{\nu}^0 = \bar{\nu}^0 + p(\omega^2)\varphi^0(x^0, \omega^2) = 0 + 0.3(0) = 0$.
$\beta_0^0 \leftarrow \beta_0^0 + p(\omega^2)\big((1-\lambda)\pi^0(\omega^2)^\top r(\omega^2) + \lambda \pi^0(\omega^2)^\top r(\omega^2)\big)$
$= 0 + 0.3((1-0.5)(191.667, 0, 0, 0, 0, 0, 0)^\top (0, 0, 0, 0, -20, -15, -15)$
$+ 0.5(191.667, 0, 0, 0, 0, 0, 0)^\top (0, 0, 0, 0, -20, -15, -15)) = 0$.
Compute $\beta_0 \leftarrow \beta_0 + p(\omega^2)\big((1-\lambda)T(\omega^2)^\top \pi^1(\omega^2) + \lambda(T(\omega^2)^\top \pi^1(\omega^2) - c^\top)\big)$
$= (21.25, -4.5, -2.25, -1.5)$

$$+ 0.3((1-0.5)\begin{bmatrix} 1&0&0&0\\ 0&1&0&0\\ 0&0&1&0\\ 0&0&0&1\\ 0&0&0&0\\ 0&0&0&0\\ 0&0&0&0 \end{bmatrix}^\top (191.667, 0, 0, 0, 0, 0, 0)^\top$$

$$+ 0.5\begin{bmatrix} 1&0&0&0\\ 0&1&0&0\\ 0&0&1&0\\ 0&0&0&1\\ 0&0&0&0\\ 0&0&0&0\\ 0&0&0&0 \end{bmatrix}^\top (191.667, 0, 0, 0, 0, 0, 0)^\top - (50, 30, 15, 10))$$

$= (63.75, -13.5, -6.75, -4.5)^\top$.

For $s = 3$, $\varphi^0(x^0, \omega^3) = \bar{Q}^0$:
Compute $\bar{\nu}^0 = \bar{\nu}^0 + p(\omega^3)\varphi^0(x^0, \omega^3) = 0 + 0.3(0) = 0$.
$\beta_0^0 \leftarrow \beta_0^0 + p(\omega^3)\left((1-\lambda)\pi^0(\omega^3)^\top r(\omega^3) + \lambda\pi^0(\omega^3)^\top r(\omega^3)\right)$
$= 0 + 0.3((1-0.5)(191.667, 0, 0, 0, 0, 0, 0)^\top(0, 0, 0, 0, -25, -20, -25)$
$+ 0.5(191.667, 0, 0, 0, 0, 0, 0)^\top(0, 0, 0, 0, -25, -20, -25)) = 0$.
Compute $\beta_0 \leftarrow \beta_0 + p(\omega^3)\left((1-\lambda)T(\omega^3)^\top\pi^1(\omega^3) + \lambda(T(\omega^3)^\top\pi^1(\omega^3) - c^\top)\right)$
$= (63.75, -13.5, -6.75, -4.5)^\top$

$$+0.3((1-0.5)\begin{bmatrix}1&0&0&0\\0&1&0&0\\0&0&1&0\\0&0&0&1\\0&0&0&0\\0&0&0&0\\0&0&0&0\end{bmatrix}^\top(191.667, 0, 0, 0, 0, 0, 0)^\top$$

$$+0.5\begin{bmatrix}1&0&0&0\\0&1&0&0\\0&0&1&0\\0&0&0&1\\0&0&0&0\\0&0&0&0\\0&0&0&0\end{bmatrix}^\top(191.667, 0, 0, 0, 0, 0, 0)^\top - (50, 30, 15, 10))$$

$= (106.25, -22.5, -11.25, -7.5)^\top$.
For $s = 4$, $\varphi^0(x^0, \omega^4) = \bar{Q}^0$:
Compute $\bar{\nu}^0 = \bar{\nu}^0 + p(\omega^4)\varphi^0(x^0, \omega^4) = 0 + 0.2(0) = 0$.
$\beta_0^0 \leftarrow \beta_0^0 + p(\omega^4)\left((1-\lambda)\pi^0(\omega^4)^\top r(\omega^4) + \lambda\pi^0(\omega^4)^\top r(\omega^4)\right)$
$= 0 + 0.2((1-0.5)(191.667, 0, 0, 0, 0, 0, 0)^\top(0, 0, 0, 0, -30, -25, -30)$
$+ 0.5(191.667, 0, 0, 0, 0, 0, 0)^\top(0, 0, 0, 0, -30, -25, -30)) = 0$.
Compute $\beta_0 \leftarrow \beta_0 + p(\omega^4)\left((1-\lambda)T(\omega^4)^\top\pi^1(\omega^4) + \lambda(T(\omega^4)^\top\pi^1(\omega^4) - c^\top)\right)$
$= (106.25, -22.5, -11.25, -7.5)^\top$

$$+0.3((1-0.5)\begin{bmatrix}1&0&0&0\\0&1&0&0\\0&0&1&0\\0&0&0&1\\0&0&0&0\\0&0&0&0\\0&0&0&0\end{bmatrix}^\top(191.667, 0, 0, 0, 0, 0, 0)^\top$$

$$+\,0.5\begin{bmatrix}1\,0\,0\,0\\0\,1\,0\,0\\0\,0\,1\,0\\0\,0\,0\,1\\0\,0\,0\,0\\0\,0\,0\,0\\0\,0\,0\,0\end{bmatrix}^\top (191.667, 0, 0, 0, 0, 0, 0)^\top - (50, 30, 15, 10))$$

$= (134.583, -28.5, -14.25, -9.5)^\top$.

For $s = 5$, $\varphi^0(x^0, \omega^5) = \bar{Q}^0$:

Compute $\bar{\nu}^0 = \bar{\nu}^0 + p(\omega^5)\varphi^0(x^0, \omega^5) = 0 + 0.2(0) = 0$.

$\beta_0^0 \leftarrow \beta_0^0 + p(\omega^5)\left((1-\lambda)\pi^0(\omega^5)^\top r(\omega^5) + \lambda\pi^0(\omega^5)^\top r(\omega^5)\right)$

$= 0 + 0.2((1-0.5)(191.667, 0, 0, 0, 0, 0, 0)^\top(0, 0, 0, 0, -10, -10, -10)$

$+ 0.5(191.667, 0, 0, 0, 0, 0, 0)^\top(0, 0, 0, 0, -10, -10, -10)) = 0$.

Compute $\beta_0 \leftarrow \beta_0 + p(\omega^5)\left((1-\lambda)T(\omega^5)^\top\pi^1(\omega^5) + \lambda(T(\omega^5)^\top\pi^1(\omega^5) - c^\top)\right)$

$$= (134.583, -28.5, -14.25, -9.5)^\top + 0.3((1-0.5)\begin{bmatrix}1\,0\,0\,0\\0\,1\,0\,0\\0\,0\,1\,0\\0\,0\,0\,1\\0\,0\,0\,0\\0\,0\,0\,0\\0\,0\,0\,0\end{bmatrix}^\top (191.667, 0, 0, 0, 0, 0, 0)^\top$$

$$+\,0.5\begin{bmatrix}1\,0\,0\,0\\0\,1\,0\,0\\0\,0\,1\,0\\0\,0\,0\,1\\0\,0\,0\,0\\0\,0\,0\,0\\0\,0\,0\,0\end{bmatrix}^\top (191.667, 0, 0, 0, 0, 0, 0)^\top - (50, 30, 15, 10))$$

$= (141.667, -30, -15, -10)^\top$.

Generate an *optimality* cut: $\beta_0^\top x + \gamma \geq \beta_k^0 = (141.667, -30, -15, -10)^\top x + \gamma \geq 0$.

Set $u^0 \leftarrow c^\top x^0 + (1-\lambda)\bar{Q}^0 + \lambda\bar{\nu}^k = (50, 30, 15, 10)(0, 0, 0, 0)^\top + (1-0.5)0 + 0.50 = 0$.

Set $u \leftarrow \min\{u^0, u\} = 0$.

Step 2. Add Cut to Master Program and Solve.

Add $\beta_0^\top x + \gamma \geq \beta_0^0$ to master problem:

$$
\begin{array}{llllllll}
\nu^1 := \text{Min} & \gamma & & & & & & \\
\text{s.t.} & -x_1 & & & & & \geq & -300 \\
& & -x_2 & & & & \geq & -700 \\
& & & -x_3 & & & \geq & -600 \\
& & & & -x_4 & & \geq & -500 \\
& & -x_2 & -x_3 & -x_4 & & \geq & -1600 \\
& 141.667x_1 & -30x_2 & -15x_3 & -10x_4 & +\gamma & \geq & 0 \\
& x_1, & x_2, & x_3, & x_4 & & \geq & 0.
\end{array}
$$

Solve master to get $x^1 = (300, 0, 0, 0)^\top, \gamma = -42{,}500$ and $\nu^1 = -42{,}500$. Set $\ell \leftarrow \max\{\nu^0, \ell\} = \max\{-42{,}500, -\infty\} = -42{,}500$.

Step 3. **Termination.** Compute $u - \ell = 0 - (-42{,}500) = 42{,}500$ and $\epsilon|u| = 10^{-6}| - 42{,}500|) = 0.0425$. Since $u - \ell \geq \epsilon|u|$:

Set $k \leftarrow 0 + 1 = 1$. Return to step 1.

Return to Step 1.

For $k = 1$:

Step 1. **Solve Subproblems.** For $s = 1, \cdots, 5$, solve

$$
\begin{array}{lllll}
\varphi^1(x^1, \omega^s) := \text{Min} & -1150y_1 & -1525y_2 & -1900y_3 & \\
\text{s.t.} & -6y_1 & -8y_2 & -10y_3 & \geq 0 \\
& -20y_1 & -25y_2 & -28y_3 & \geq -300 \\
& -12y_1 & -15y_2 & -18y_3 & \geq 0 \\
& -8y_1 & -10y_2 & -14y_3 & \geq 0 \\
& -y_1 & & & \geq d_a^s \\
& & -y_2 & & \geq d_b^s \\
& & & -y_3 & \geq d_c^s \\
& y_1, & y_2, & y_3 & \geq 0.
\end{array}
$$

$s = 1$: feasible, $\varphi^1(x^1, \omega^1) = 0$, $y_1^1 = (0, 0, 0)^\top$, $\pi^1(\omega^1) = (0, 0, 105.556, 0, 0, 0, 0)^\top$. Compute $\bar{Q}^1 \leftarrow \bar{Q}^1 + p(\omega^1)\varphi(x^1, \omega^1) = 0 + 0.15(0) = 0$

$s = 2$: feasible, $\varphi^1(x^1, \omega^2) = 0$, $y_2^1 = (0, 0, 0)^\top$, $\pi^1(\omega^2) = (0, 0, 105.556, 0, 0, 0, 0)^\top$. Compute $\bar{Q}^1 \leftarrow \bar{Q}^1 + p(\omega^2)\varphi(x^1, \omega^2) = 0 + 0.15(0) = 0$.

$s = 3$: feasible, $\varphi^1(x^1, \omega^3) = 0$, $y_3^1 = (0, 0, 0)^\top$, $\pi^1(\omega^3) = (0, 0, 105.556, 0, 0, 0, 0)^\top$. Compute $\bar{Q}^1 \leftarrow \bar{Q}^1 + p(\omega^3)\varphi(x^1, \omega^3) = 0 + 0.15(0) = 0$.

$s = 4$: feasible, $\varphi^1(x^1, \omega^4) = 0$, $y_4^1 = (0, 0, 0)^\top$, $\pi^1(\omega^4) = (0, 0, 105.556, 0, 0, 0, 0)^\top$. Compute $\bar{Q}^1 \leftarrow \bar{Q}^1 + p(\omega^4)\varphi(x^1, \omega^4) = 0 + 0.15(0) = 0$.

$s = 5$: feasible, $\varphi^1(x^1, \omega^5) = 0$, $y_5^1 = (0, 0, 0)^\top$, $\pi^1(\omega^5) = (0, 0, 105.556, 0, 0, 0, 0)^\top$. Compute $\bar{Q}^1 \leftarrow \bar{Q}^1 + p(\omega^5)\varphi(x^1, \omega^5) = 0 + 0.15(0) = 0$.
For $s = 1$, $\varphi^1(x^1, \omega^1) = \bar{Q}^1$:
Compute $\bar{\nu}^1 = \bar{\nu}^1 + p(\omega^1)\varphi^1(x^1, \omega^1) = 0 + 0.15(0) = 0$.
$\beta_1^0 \leftarrow \beta_1^0 + p(\omega^1)\left((1-\lambda)\pi^1(\omega^1)^\top r(\omega^1) + \lambda\pi^1(\omega^1)^\top r(\omega^1)\right)$
$= 0 + 0.15((1-0.5)(0, 0, 105.556, 0, 0, 0, 0)^\top(0, 0, 0, 0, -15, -10, -5)$
$+ 0.5(0, 0, 105.556, 0, 0, 0, 0)^\top(0, 0, 0, 0, -15, -10, -5)) = 0$.
Compute $\beta_1 \leftarrow \beta_1 + p(\omega^1)\left((1-\lambda)T(\omega^1)^\top\pi^1(\omega^1) + \lambda(T(\omega^1)^\top\pi^1(\omega^1) - c^\top)\right) = (0, 0, 0, 0)^\top$

$$+0.15((1-0.5)(\begin{bmatrix} 1\,0\,0\,0 \\ 0\,1\,0\,0 \\ 0\,0\,1\,0 \\ 0\,0\,0\,1 \\ 0\,0\,0\,0 \\ 0\,0\,0\,0 \\ 0\,0\,0\,0 \end{bmatrix}^\top \quad 0, 0, 105.556, 0, 0, 0, 0)^\top$$

$$+0.5\begin{bmatrix} 1\,0\,0\,0 \\ 0\,1\,0\,0 \\ 0\,0\,1\,0 \\ 0\,0\,0\,1 \\ 0\,0\,0\,0 \\ 0\,0\,0\,0 \\ 0\,0\,0\,0 \end{bmatrix}^\top \quad 0, 0, 105.556, 0, 0, 0, 0)^\top - (50, 30, 15, 10))$$

$= (-7.5, -4.5, 13.5833, -1.5)^\top$.
For $s = 2$, $\varphi^1(x^1, \omega^2) = \bar{Q}^1$:
Compute $\bar{\nu}^1 = \bar{\nu}^1 + p(\omega^2)\varphi^1(x^1, \omega^2) = 0 + 0.3(0) = 0$.
$\beta_1^0 \leftarrow \beta_1^0 + p(\omega^2)\left((1-\lambda)\pi^1(\omega^2)^\top r(\omega^2) + \lambda\pi^1(\omega^2)^\top r(\omega^2)\right)$
$= 0 + 0.3((1-0.5)(0, 0, 105.556, 0, 0, 0, 0)^\top(0, 0, 0, 0, -20, -15, -15)$
$+ 0.5(0, 0, 105.556, 0, 0, 0, 0)^\top(0, 0, 0, 0, -20, -15, -15)) = 0$.
Compute $\beta_1 \leftarrow \beta_1 + p(\omega^2)\left((1-\lambda)T(\omega^2)^\top\pi^1(\omega^2) + \lambda(T(\omega^2)^\top\pi^1(\omega^2) - c^\top)\right)$

$$(-7.5, -4.5, 13.5833, -1.5)^\top + 0.3((1-0.5)(\begin{bmatrix} 1\,0\,0\,0 \\ 0\,1\,0\,0 \\ 0\,0\,1\,0 \\ 0\,0\,0\,1 \\ 0\,0\,0\,0 \\ 0\,0\,0\,0 \\ 0\,0\,0\,0 \end{bmatrix}^\top \quad 0, 0, 105.556,$$

$0, 0, 0, 0)^\top$

$$+0.5\begin{bmatrix}1&0&0&0\\0&1&0&0\\0&0&1&0\\0&0&0&1\\0&0&0&0\\0&0&0&0\\0&0&0&0\end{bmatrix}^{\top} 0, 0, 105.556, 0, 0, 0, 0)^{\top} - (50, 30, 15, 10))$$

$= (-22.5, -13.5, 40.75, -4.5)^{\top}$.

For $s = 3$, $\varphi^1(x^1, \omega^3) = \bar{Q}^1$:

Compute $\bar{\nu}^1 = \bar{\nu}^1 + p(\omega^3)\varphi^1(x^1, \omega^3) = 0 + 0.3(0) = 0$.

$\beta_1^0 \leftarrow \beta_1^0 + p(\omega^3)\left((1-\lambda)\pi^1(\omega^3)^{\top} r(\omega^3) + \lambda\pi^1(\omega^3)^{\top} r(\omega^3)\right)$

$= 0 + 0.3((1-0.5)(0, 0, 105.556, 0, 0, 0, 0)^{\top}(0, 0, 0, 0, -25, -20, -25)$

$+ 0.5(0, 0, 105.556, 0, 0, 0, 0)^{\top}(0, 0, 0, 0, -25, -20, -25)) = 0$.

Compute $\beta_1 \leftarrow \beta_1 + p(\omega^3)\left((1-\lambda)T(\omega^3)^{\top}\pi^1(\omega^3) + \lambda(T(\omega^3)^{\top}\pi^1(\omega^3) - c^{\top})\right)$

$$= (-22.5, -13.5, 40.75, -4.5)^{\top} + 0.3((1-0.5)\begin{bmatrix}1&0&0&0\\0&1&0&0\\0&0&1&0\\0&0&0&1\\0&0&0&0\\0&0&0&0\\0&0&0&0\end{bmatrix}^{\top} 0, 0, 105.556,$$

$0, 0, 0, 0)^{\top}$

$$+0.5\begin{bmatrix}1&0&0&0\\0&1&0&0\\0&0&1&0\\0&0&0&1\\0&0&0&0\\0&0&0&0\\0&0&0&0\end{bmatrix}^{\top} 0, 0, 105.556, 0, 0, 0, 0)^{\top} - (50, 30, 15, 10))$$

$= (-37.5, -22.5, 67.9167, -7.5)^{\top}$.

For $s = 4$, $\varphi^1(x^1, \omega^4) = \bar{Q}^1$:

Compute $\bar{\nu}^1 = \bar{\nu}^1 + p(\omega^4)\varphi^1(x^1, \omega^4) = 0 + 0.2(0) = 0$.

$\beta_1^0 \leftarrow \beta_1^0 + p(\omega^4)\left((1-\lambda)\pi^1(\omega^4)^{\top} r(\omega^4) + \lambda\pi^1(\omega^4)^{\top} r(\omega^4)\right)$

$= 0 + 0.2((1-0.5)(0, 0, 105.556, 0, 0, 0, 0)^{\top}(0, 0, 0, 0, -30, -25, -30)$

$+ 0.5(0, 0, 105.556, 0, 0, 0, 0)^{\top}(0, 0, 0, 0, -30, -25, -30)) = 0$.

Compute $\beta_1 \leftarrow \beta_1 + p(\omega^4)\left((1-\lambda)T(\omega^4)^\top \pi^1(\omega^4) + \lambda(T(\omega^4)^\top \pi^1(\omega^4) - c^\top)\right)$

$$= (-37.5, -22.5, 67.9167, -7.5)^\top + 0.3((1-0.5)\begin{bmatrix} 1\,0\,0\,0 \\ 0\,1\,0\,0 \\ 0\,0\,1\,0 \\ 0\,0\,0\,1 \\ 0\,0\,0\,0 \\ 0\,0\,0\,0 \\ 0\,0\,0\,0 \end{bmatrix}^\top 0, 0,$$

$$105.556, 0, 0, 0, 0)^\top$$

$$+ 0.5 \begin{bmatrix} 1\,0\,0\,0 \\ 0\,1\,0\,0 \\ 0\,0\,1\,0 \\ 0\,0\,0\,1 \\ 0\,0\,0\,0 \\ 0\,0\,0\,0 \\ 0\,0\,0\,0 \end{bmatrix}^\top 0, 0, 105.556, 0, 0, 0, 0)^\top - (50, 30, 15, 10))$$

$= (-47.5, -28.5, 86.0278, -9.5)^\top$.

For $s = 5$, $\varphi^1(x^1, \omega^5) = \bar{Q}^1$:

Compute $\bar{\nu}^1 = \bar{\nu}^1 + p(\omega^5)\varphi^1(x^1, \omega^5) = 0 + 0.2(0) = 0$.

$\beta_1^0 \leftarrow \beta_1^0 + p(\omega^5)\left((1-\lambda)\pi^1(\omega^5)^\top r(\omega^5) + \lambda \pi^1(\omega^5)^\top r(\omega^5)\right)$

$= 0 + 0.2((1-0.5)(0, 0, 105.556, 0, 0, 0, 0)^\top(0, 0, 0, 0, -10, -10, -10)$

$+ 0.5(0, 0, 105.556, 0, 0, 0, 0)^\top(0, 0, 0, 0, -10, -10, -10)) = 0$.

Compute $\beta_1 \leftarrow \beta_1 + p(\omega^5)\left((1-\lambda)T(\omega^5)^\top \pi^1(\omega^5) + \lambda(T(\omega^5)^\top \pi^1(\omega^5) - c^\top)\right)$

$$= (-47.5, -28.5, 86.0278, -9.5)^\top + 0.3((1-0.5)\begin{bmatrix} 1\,0\,0\,0 \\ 0\,1\,0\,0 \\ 0\,0\,1\,0 \\ 0\,0\,0\,1 \\ 0\,0\,0\,0 \\ 0\,0\,0\,0 \\ 0\,0\,0\,0 \end{bmatrix}^\top 0, 0,$$

$$105.556, 0, 0, 0, 0)^\top$$

$$+ 0.5 \begin{bmatrix} 1\,0\,0\,0 \\ 0\,1\,0\,0 \\ 0\,0\,1\,0 \\ 0\,0\,0\,1 \\ 0\,0\,0\,0 \\ 0\,0\,0\,0 \\ 0\,0\,0\,0 \end{bmatrix}^\top 0, 0, 105.556, 0, 0, 0, 0)^\top - (50, 30, 15, 10))$$

$= (-50, -30, 90.5556, -10)^\top$.

Generate an *optimality* cut: $\beta_0^\top x + \gamma \geq \beta_1^0 = (-50, -30, 90.5556, -10)^\top x + \gamma \geq 0$.

Set $u^1 \leftarrow c^\top x^1 + (1-\lambda)\bar{Q}^1 + \lambda\bar{\nu}^k = (50, 30, 15, 10)(-300, 0, 0, 0)^\top + (1-0.5)0 + 0.50 = -15{,}000$.
Set $u \leftarrow \min\{u^1, u\} = 0$.

Step 2. **Add Cut to Master Program and Solve.**

Add $\beta_0^\top x + \gamma \geq \beta_0^0$ to master problem:

$$
\begin{array}{llrrrrrl}
\nu^2 := \text{Min} & & & & & & \gamma & \\
\text{s.t.} & -x_1 & & & & & & \geq -300 \\
& & -x_2 & & & & & \geq -700 \\
& & & -x_3 & & & & \geq -600 \\
& & & & -x_4 & & & \geq -500 \\
& & -x_2 & -x_3 & -x_4 & & & \geq -1600 \\
& 141.667x_1 & -30x_2 & -15x_3 & -10x_4 & +\gamma & & \geq 0 \\
& -50x_1 & -30x_2 & +90.5556x_3 & -10x_4 & +\gamma & & \geq 0 \\
& x_1, & x_2, & x_3, & x_4 & & & \geq 0.
\end{array}
$$

Solve master to get $x^1 = (300, 0, 544.737, 0)^\top$, $\gamma = -45{,}914.5$ and $\nu^2 = -34{,}328.9$. Set $\ell \leftarrow \max\{\nu^2, \ell\} = \max\{-34{,}328.9, -\infty\} = -34{,}328.9$.

Step 3. **Termination.** Compute $u - \ell = 0 - (-34{,}328.9) = 34{,}328.9$ and $\epsilon|u| = 10^{-6}| - 34{,}328.9|) = 0.0343$. Since $u - \ell \geq \epsilon|u|$:

Set $k \leftarrow 1 + 1 = 2$. Return to step 1.

Return to Step 1.

After 18 iterations, the ASD-AGG algorithm terminates with optimal solution $x^* = (236.4, 690, 432, 318)^\top$ and objective value -2059.9. This solution is the same as the risk-neutral optimal solution. So in this case (true for all $\lambda \in [0, 1]$) the ASD risk measure performs the same as the risk-neutral case. However, we should point out that this is not always the case.

7.5 Separate Cut Decomposition for ASD

Let us now consider generating separate components of the subgradients on the expectation and deviation terms of the objective function, respectively. As we did in Sect. 7.3, let the optimality cut variable for the expectation term be γ and that for the deviation be ζ. Then we need to generate the left hand side coefficients β_k and the RHS β_k^0 for the expectation term. Similarly, we need to also generate the left hand side coefficients σ^k and the RHS σ_k^0 for the deviation term. Therefore, the ASD master program at iteration k is given as follows:

$$\begin{aligned} \ell_k := \text{Min } & (1-\lambda)c^\top x + (1-\lambda)\gamma + \lambda\zeta \\ \text{s.t. } & Ax \geq b \\ & \beta_\tau^\top x + \gamma \geq \beta_\tau^0,\ \tau = 1, \cdots, t \\ & \sigma_\tau^\top x + \zeta \geq \sigma_\tau^0,\ \tau = 1, \cdots, t \\ & x \geq 0. \end{aligned} \tag{7.17}$$

Let (x^k, γ^k, ζ^k) be the solution to the master problem. Then the subproblem for each scenario $\omega \in \Omega$ is as given in formulation (7.15) in the previous subsection.

7.5.1 *Subgradient Optimization Approach*

Let for all $\omega \in \Omega$, $y^k(\omega)$ solve Problem (7.15) and let $\pi^k(\omega)$ be the corresponding optimal dual multipliers associated with the subproblem constraints. The optimality cut for the expectation term can be derived as follows: Recall that $c - T(\omega)^\top \pi^k(\omega)$ is a subgradient of $f(., \omega)$. Therefore,

$$\beta_k(\omega) = T(\omega)^\top \pi^k(\omega)$$

and

$$\beta_k^0(\omega) = \pi^k(\omega)^\top r(\omega).$$

Thus the optimality cut coefficients β_t and RHS β_k^0 for the master problem (7.17) can be calculated as follows:

$$\beta_t = \sum_{\omega \in \Omega} p(\omega)\beta_k(\omega)$$

and

$$\beta_k^0 = \sum_{\omega \in \Omega} p(\omega)\beta_k^0(\omega).$$

The optimality cut to add to the master program is then $\beta_k^\top x + \gamma \geq \beta_k^0$. If a subproblem is *infeasible* for some scenario ω, then a dual extreme ray $\mu^k(\omega)$ is obtained and a feasibility cut generated as follows:

$$\beta_k = T(\omega)^\top \mu^k(\omega)$$

and

$$\beta_k^0 = \mu^k(\omega)^\top r(\omega).$$

Let us now turn to the optimality cut for the absolute semideviation term. We shall again have two cases to consider due to Equation (7.16):
Case 1: $\nu^k(\omega) = c^\top x^k + \varphi(x^k, \omega)$.
In this case we use the subgradient $c - T(\omega)^\top \pi^k(\omega)$ for $\omega \in \Omega$. Therefore, we can define $\sigma_k(\omega)$ and $\sigma_k^0(\omega)$ as follows:

$$\sigma_k(\omega) = T(\omega)^\top \pi^k(\omega) - c$$

and

$$\sigma_k^0(\omega) = \pi^k(\omega)^\top r(\omega).$$

Case 2: $\nu^k(\omega) = c^\top x^k + \sum_{\varpi \in \Omega} p(\varpi)\varphi(x^k, \varpi)$.
In this case, we use the expected subgradient $c - \sum_{\varpi \in \Omega} p(\varpi) T(\varpi)^\top \pi^k(\varpi)$ for $\omega \in \Omega$. Therefore define $\sigma_k(\omega)$ and $\sigma_k^0(\omega)$ as follows:

$$\sigma_k(\omega) = \sum_{\varpi \in \Omega} p(\varpi) T(\varpi)^\top \pi^k(\varpi) - c$$

and

$$\sigma_k^0(\omega) = \sum_{\varpi \in \Omega} p(\varpi) \pi^k(\varpi)^\top r(\varpi).$$

Consequently, the optimality cut coefficients σ_k and RHS σ_k^0 in (7.17) at iteration k can be calculated as follows, based on selecting the appropriate case for each scenario $\omega \in \Omega$:

$$\sigma_k = \sum_{\omega \in \Omega} p(\omega)\sigma_k(\omega)$$

and

$$\sigma_k^0 = \sum_{\omega \in \Omega} p(\omega)\sigma_k^0(\omega).$$

The optimality cut to add to the master program is then $\sigma_k^\top x + \zeta \geq \sigma_k^0$. Next, we give a formal statement of the separate cut algorithm.

7.5.2 ASD-SEP Algorithm

We can now state the SEP algorithm for ASD as follows:

Algorithm
begin

Step 0. **Initialization.** $k \leftarrow 0$, x^0 given or $x^0 \leftarrow \operatorname{argmin}\{c^\top x \mid Ax \geq b, x \geq 0\}$,
$\ell \leftarrow -\infty$, $u \leftarrow \infty$, $\epsilon > 0$, and $\lambda \in [0, 1]$.

Step 1. **Solve Subproblems.**

Set $\beta_k \leftarrow \mathbf{0}$, $\beta_k^0 \leftarrow 0$, $\sigma_k \leftarrow \mathbf{0}$, $\sigma_k^0 \leftarrow 0$, $\bar{Q}^k \leftarrow 0$, and $\bar{\nu}^k \leftarrow 0$.
For each $\omega \in \Omega$ solve subproblem (7.15):

Get and store optimal value $\varphi(x^k, \omega)$.
Get dual solution $\pi^k(\omega)$.
Compute $\bar{Q}^k \leftarrow \bar{Q}^k + p(\omega)\varphi(x^k, \omega)$.

If subproblem is *infeasible* for some ω:

Get dual extreme ray $\mu^k(\omega)$.
Compute $\beta_k \leftarrow T(\omega)^\top \mu^k(\omega)$.
Compute $\beta_k^0 \leftarrow \mu^k(\omega)^\top r(\omega)$.
Generate *feasibility* cut: $\beta_k^\top x \geq \beta_k^0$.
Go to step 2.

Else if *feasible* for all ω, for each $\omega \in \Omega$:

Compute $\beta_k \leftarrow \beta_k + p(\omega)T(\omega)^\top \pi^k(\omega)$.
Compute $\beta_k^0 \leftarrow \beta_k^0 + p(\omega)\pi^k(\omega)^\top r(\omega)$.
If $\varphi(x^k, \omega) \geq \bar{Q}^k$:

Compute $\bar{\nu}^k \leftarrow \bar{\nu}^k + p(\omega)\varphi(x^k, \omega)$.
Compute $\sigma_k \leftarrow \sigma_k + p(\omega)\left(T(\omega)^\top \pi^k(\omega) - c\right)$.
Compute $\sigma_k^0 \leftarrow \sigma_k^0 + p(\omega)\left(\pi^k(\omega)^\top r(\omega)\right)$.

Else:

Compute $\bar{\nu}^k \leftarrow \bar{\nu}^k + p(\omega)\bar{Q}^k$.
Compute $\sigma_k \leftarrow \sigma_k + p(\omega)\left(\sum_{\varpi \in \Omega} p(\varpi)T(\varpi)^\top \pi^k(\varpi) - c\right)$.
Compute $\sigma_k^0 \leftarrow \sigma_k^0 + p(\omega)\left(\sum_{\varpi \in \Omega} p(\varpi)\pi^k(\varpi)^\top r(\varpi)\right)$.

Generate *optimality* cuts: $\beta_k^\top x + \gamma \geq \beta_k^0$ and $\sigma_k^\top x + \zeta \geq \sigma_k^0$.
Compute *upper bound*:

Set $u^k \leftarrow c^\top x^k + (1 - \lambda)\bar{Q}^k + \lambda\bar{\nu}^k$.
Set $u \leftarrow \min\{u^k, u\}$.

If u is updated, set *incumbent* solution $x^* \leftarrow x^k$.

Step 2. **Add Cut to Master Program and Solve.** If some subproblem was infeasible:

Add feasibility cut to master problem (7.17).

Else:

Add optimality cut to master problem (7.17).

Solve master problem (7.17), and get optimal value ℓ_{k+1} and solution x^{k+1}.

Set $\ell \leftarrow \max\{\ell^{k+1}, \ell\}$.

Step 3. **Termination.** If $u - \ell \leq \epsilon|u|$:

Stop, solution x^* is ϵ-optimal.

Else:

Set $k \leftarrow k + 1$.

Return to Step 1.

End

As in the previous algorithms, to perform parametric optimization over λ values, ignore the initialization of λ in step 0 of the ASD-SEP algorithm and instead perform a Parametric ASD-SEP algorithm following the same steps as the Parametric ASD-AGG algorithm to determine and the value for λ for each algorithm run.

7.5.3 Numerical Example

Example 7.4 Apply two iterations of the ASD-SEP algorithm to the problem instance in Example 7.1.

Algorithm ASD-SEP
begin

Step 0. **Initialization.** Let $k \leftarrow 0$, choose $x^0 \in X$, and set $\ell \leftarrow -\infty$ and $u \leftarrow \infty$. Choose $\epsilon \leftarrow 10^{-6}$ and $x^0 \leftarrow \text{argmin}_{x\in X}\{50x_1+30x_2+15x_3+10x_4\} = (0, 0, 0, 0)^\top$ as the initial point , and $\lambda \leftarrow 0.5$.

For $k = 0$:

Step 1. **Solve Subproblems.** For $s = 1, \cdots, 5$, solve

$$
\begin{array}{rrrrl}
\varphi^0(x^0,\omega^s) := \text{Min} & -1150y_1 & -1525y_2 & -1900y_3 & \\
\text{s.t.} & -6y_1 & -8y_2 & -10y_3 & \geq 0 \\
& -20y_1 & -25y_2 & -28y_3 & \geq 0 \\
& -12y_1 & -15y_2 & -18y_3 & \geq 0 \\
& -8y_1 & -10y_2 & -14y_3 & \geq 0 \\
& -y_1 & & & \geq d_a^s \\
& & -y_2 & & \geq d_b^s \\
& & & -y_3 & \geq d_c^s \\
& y_1, & y_2, & y_3 & \geq 0.
\end{array}
$$

$s = 1$: feasible, $\varphi^0(x^0, \omega^1)$, $y_1^0{=}(0, 0, 0)^\top$, $\pi^0(\omega^1){=}(191.667, 0, 0, 0, 0, 0, 0)^\top$. Compute $\bar{Q}^0 \leftarrow \bar{Q}^0 + p(\omega^1)\varphi(x^0, \omega^1) = 0 + 0.15(0) = 0$.

$s = 2$: feasible, $\varphi^0(x^0, \omega^2) = 0$, $y_2^0 = (0, 0, 0)^\top$, $\pi^0(\omega^2) = (191.667, 0, 0, 0, 0, 0, 0)^\top$. Compute $\bar{Q}^0 \leftarrow \bar{Q}^0 + p(\omega^2)\varphi(x^0, \omega^2) = 0 + 0.3(0) = 0$.

$s = 3$: feasible, $\varphi^0(x^0, \omega^3) = 0$, $y_3^0 = (0, 0, 0)^\top$, $\pi^0(\omega^3) = (191.667, 0, 0, 0, 0, 0, 0)^\top$. Compute $\bar{Q}^0 \leftarrow \bar{Q}^0 + p(\omega^3)\varphi(x^0, \omega^3) = 0 + 0.3(0) = 0$.

$s = 4$: feasible, $\varphi^0(x^0, \omega^4) = 0$, $y_4^0 = (0, 0, 0)^\top$, $\pi^0(\omega^4) = (191.667, 0, 0, 0, 0, 0, 0)^\top$. Compute $\bar{Q}^0 \leftarrow \bar{Q}^0 + p(\omega^4)\varphi(x^0, \omega^4) = 0 + 0.2(0) = 0$.

$s = 5$: feasible, $\varphi^0(x^0, \omega^5) = 0$, $y_5^0 = (0, 0, 0)^\top$, $\pi^0(\omega^5) = (191.667, 0, 0, 0, 0, 0, 0)^\top$. Compute $\bar{Q}^0 \leftarrow \bar{Q}^0 + p(\omega^5)\varphi(x^0, \omega^5) = 0 + 0.05(0) = 0$.

For $s = 1$:

Compute $\beta_0 \leftarrow \beta_0 + p(\omega^1)(T(\omega^1)^\top \pi^0(\omega^1) - c)$

$$
= (0,0,0,0)^\top + 0.15(\begin{bmatrix} 1&0&0&0\\ 0&1&0&0\\ 0&0&1&0\\ 0&0&0&1\\ 0&0&0&0\\ 0&0&0&0\\ 0&0&0&0 \end{bmatrix}^\top (191.667, 0, 0, 0, 0, 0, 0)^\top - (50, 30, 15, 10))
$$

$= (21.25, -4.5, -2.25, -1.5)^\top$.

Compute $\beta_0^0 \leftarrow \beta_0^0 + p(\omega^1)\pi^0(\omega^1)^\top r(\omega^1)$

$= 0 + 0.15(191.667, 0, 0, 0, 0, 0, 0)(0, 0, 0, 0, -15, -10, -5)^\top = 0$.

$\varphi^0(x^0, \omega^1) = \bar{Q}^0$:

Compute $\bar{\nu}^0 = \bar{\nu}^0 + p(\omega^1)\varphi^0(x^0, \omega^1) = 0 + 0.15(0) = 0$.

$$\sigma_0 \leftarrow \sigma_0 + p(\omega^1)\left(T(\omega^1)^\top \pi^0(\omega^1) - c\right) = 0 + 0.15(\begin{bmatrix} 1\,0\,0\,0 \\ 0\,1\,0\,0 \\ 0\,0\,1\,0 \\ 0\,0\,0\,1 \\ 0\,0\,0\,0 \\ 0\,0\,0\,0 \\ 0\,0\,0\,0 \end{bmatrix}^\top$$

$(191.667, 0, 0, 0, 0, 0, 0)^\top$
$- (50, 30, 15, 10)) = (21.25, -4.5, -2.25, -1.5)^\top.$
Compute $\sigma_0^0 \leftarrow \sigma_0^0 + p(\omega^1)\left(\pi^0(\omega^1)^\top r(\omega^1)\right)$
$= 0 + 0.15(191.667, 0, 0, 0, 0, 0, 0)(0, 0, 0, 0, -15, -10, -5)^\top = 0.$
For $s = 2$:
Compute $\beta_0 \leftarrow \beta_0 + p(\omega^2)(T(\omega^2)^\top \pi^0(\omega^2) - c)$
$= (21.25, -4.5, -2.25, -1.5)^\top$

$$+0.3 \begin{bmatrix} 1\,0\,0\,0 \\ 0\,1\,0\,0 \\ 0\,0\,1\,0 \\ 0\,0\,0\,1 \\ 0\,0\,0\,0 \\ 0\,0\,0\,0 \\ 0\,0\,0\,0 \end{bmatrix}^\top (191.667, 0, 0, 0, 0, 0, 0)^\top - (50, 30, 15, 10))$$

$= (63.75, -13.5, -6.75, -4.5)^\top.$
Compute $\beta_0^0 \leftarrow \beta_0^0 + p(\omega^2)\pi^0(\omega^2)^\top r(\omega^2)$
$= 0 + 0.3(191.667, 0, 0, 0, 0, 0, 0)^\top(0, 0, 0, 0, -20, -15, -15) = 0.$
$\varphi^0(x^0, \omega^2) = \bar{Q}^0.$
Compute $\bar{\nu}^0 = \bar{\nu}^0 + p(\omega^2)\varphi^0(x^0, \omega^2) = 0 + 0.3(0) = 0.$
$\sigma_0^0 \leftarrow \sigma_0^0 + p(\omega^2)\pi^0(\omega^2)^\top r(\omega^2)$
$= 0 + 0.3((1 - 0.5)(191.667, 0, 0, 0, 0, 0, 0)^\top(0, 0, 0, 0, -20, -15, -15)$
$+ 0.5(191.667, 0, 0, 0, 0, 0, 0)^\top(0, 0, 0, 0, -20, -15, -15)) = 0.$
Compute $\sigma_0 \leftarrow \sigma_0 + p(\omega^2)\left(T(\omega^2)^\top \pi^0(\omega^2) - c\right) = (21.25, -4.5, -2.25, -1.5)^\top$

$$+0.3 \begin{bmatrix} 1\,0\,0\,0 \\ 0\,1\,0\,0 \\ 0\,0\,1\,0 \\ 0\,0\,0\,1 \\ 0\,0\,0\,0 \\ 0\,0\,0\,0 \\ 0\,0\,0\,0 \end{bmatrix}^\top (191.667, 0, 0, 0, 0, 0, 0)^\top - (50, 30, 15, 10))$$

$= (63.75, -13.5, -6.75, -4.5)^\top.$
For $s = 3$:
$\beta_0^0 \leftarrow \beta_0^0 + p(\omega^3)\pi^0(\omega^3)^\top r(\omega^3)$
$= 0 + 0.3(191.667, 0, 0, 0, 0, 0, 0)(0, 0, 0, 0, -25, -20, -25)^\top = 0.$

Compute $\beta_0 \leftarrow \beta_0 + p(\omega^3)(T(\omega^3)^\top \pi^0(\omega^3)) = (72.95, 0, 0, 0)^\top$

$$+ 0.3(\begin{bmatrix} 1\,0\,0\,0 \\ 0\,1\,0\,0 \\ 0\,0\,1\,0 \\ 0\,0\,0\,1 \\ 0\,0\,0\,0 \\ 0\,0\,0\,0 \\ 0\,0\,0\,0 \end{bmatrix}^\top (191.667, 0, 0, 0, 0, 0, 0)^\top$$

$- (50, 30, 15, 10)) = (106.25, -22.5, -11.25, -7.5)^\top$.

$\varphi^0(x^0, \omega^3) = \bar{Q}^0$:

Compute $\bar{\nu}^0 = \bar{\nu}^0 + p(\omega^3)\varphi^0(x^0, \omega^3) = 0 + 0.3(0) = 0$.

$\sigma_0^0 \leftarrow \sigma_0^0 + p(\omega^3)\pi^0(\omega^3)^\top r(\omega^3)$

$= 0 + 0.3(191.667, 0, 0, 0, 0, 0, 0)(0, 0, 0, 0, -25, -20, -25)^\top = 0$.

Compute $\sigma_0 \leftarrow \sigma_0 + p(\omega^3)(T(\omega^3)^\top \pi^0(\omega^3) - c) = (63.75, -13.5, -6.75, -4.5)^\top$

$$+ 0.3(\begin{bmatrix} 1\,0\,0\,0 \\ 0\,1\,0\,0 \\ 0\,0\,1\,0 \\ 0\,0\,0\,1 \\ 0\,0\,0\,0 \\ 0\,0\,0\,0 \\ 0\,0\,0\,0 \end{bmatrix}^\top (191.667, 0, 0, 0, 0, 0, 0)^\top$$

$- (50, 30, 15, 10)) = (106.25, -22.5, -11.25, -7.5)^\top$.

For $s = 4$:

$\beta_0^0 \leftarrow \beta_0^0 + p(\omega^4)\pi^0(\omega^4)^\top r(\omega^4)$

$= 0 + 0.2(191.667, 0, 0, 0, 0, 0, 0)(0, 0, 0, 0, -30, -25, -30)^\top = 0$.

Compute $\beta_0 \leftarrow \beta_0 + p(\omega^4)(T(\omega^4)^\top \pi^0(\omega^4) - c) = (106.25, -22.5,$
$-11.25, -7.5)^\top$

$$+ 0.2(\begin{bmatrix} 1\,0\,0\,0 \\ 0\,1\,0\,0 \\ 0\,0\,1\,0 \\ 0\,0\,0\,1 \\ 0\,0\,0\,0 \\ 0\,0\,0\,0 \\ 0\,0\,0\,0 \end{bmatrix}^\top (191.667, 0, 0, 0, 0, 0, 0)^\top - (50, 30, 15, 10))$$

$= (134.583, -28.5, -14.25, -9.5)^\top$.

$\varphi^0(x^0, \omega^4) = \bar{Q}^0$:

Compute $\bar{\nu}^0 = \bar{\nu}^0 + p(\omega^4)\varphi^0(x^0, \omega^4) = 0 + 0.2(0) = 0$.

$\sigma_0^0 \leftarrow \sigma_0^0 + p(\omega^4)\pi^0(\omega^4)^\top r(\omega^4)$

$= 0 + 0.2(191.667, 0, 0, 0, 0, 0, 0)(0, 0, 0, 0, -30, -25, -30)^\top = 0$.

Compute $\sigma_0 \leftarrow \sigma_0 + p(\omega^4)(T(\omega^4)^\top \pi^0(\omega^4) - c) = (106.25, -22.5, -11.25,$
$-7.5)$

$$+0.2(\begin{bmatrix}1\,0\,0\,0\\0\,1\,0\,0\\0\,0\,1\,0\\0\,0\,0\,1\\0\,0\,0\,0\\0\,0\,0\,0\\0\,0\,0\,0\end{bmatrix}^{\top}(191.667, 0, 0, 0, 0, 0, 0)^{\top} - (50, 30, 15, 10))$$

$= (134.583, -28.5, -14.25, -9.5)^{\top}$.

For $s = 5$:

$\beta_0^0 \leftarrow \beta_0^0 + p(\omega^5)\pi^0(\omega^5)^{\top} r(\omega^5)$

$= 0 + 0.05(191.667, 0, 0, 0, 0, 0, 0)(0, 0, 0, 0, -10, -10, -10)^{\top} = 0.$

Compute $\beta_0 \leftarrow \beta_0 + p(\omega^5)(T(\omega^5)^{\top}\pi^0(\omega^5) - c) = (134.583, -28.5, -14.25,$

$$-9.5)^{\top} + 0.3(\begin{bmatrix}1\,0\,0\,0\\0\,1\,0\,0\\0\,0\,1\,0\\0\,0\,0\,1\\0\,0\,0\,0\\0\,0\,0\,0\\0\,0\,0\,0\end{bmatrix}^{\top}(191.667, 0, 0, 0, 0, 0, 0)^{\top} - (50, 30, 15, 10))$$

$= (141.667, -30, -15, -10)^{\top}$.

$\varphi^0(x^0, \omega^5) = \bar{Q}^0$.

Compute $\bar{v}^0 = \bar{v}^0 + p(\omega^5)\varphi^0(x^0, \omega^5) = 0 + 0.2(0) = 0.$

$\sigma_0^0 \leftarrow \sigma_0^0 + p(\omega^5)\pi^0(\omega^5)^{\top} r(\omega^5)$

$= 0 + 0.2((191.667, 0, 0, 0, 0, 0, 0)(0, 0, 0, 0, -10, -10, -10)^{\top} = 0.$

Compute $\sigma_0 \leftarrow \sigma_0 + p(\omega^5)(T(\omega^5)^{\top}\pi^0(\omega^5) - c) = (134.583, -28.5, -14.25,$

$$-9.5)^{\top} + 0.3(\begin{bmatrix}1\,0\,0\,0\\0\,1\,0\,0\\0\,0\,1\,0\\0\,0\,0\,1\\0\,0\,0\,0\\0\,0\,0\,0\\0\,0\,0\,0\end{bmatrix}^{\top}(191.667, 0, 0, 0, 0, 0, 0)^{\top} - (50, 30, 15, 10))$$

$= (141.667, -30, -15, -10)^{\top}$.

Generate *optimality* cuts: $\beta_0^{\top} x + \gamma \geq \beta_0^0 = (141.667, -30, -15, -10)^{\top} x + \gamma \geq 0$, $\sigma_0^{\top} x + \zeta \geq \sigma_0^0 = (141.667, -30, -15, -10)^{\top} x + \zeta \geq 0$.

Set $u^0 \leftarrow c^{\top} x^0 + (1-\lambda)\bar{Q}^0 + \lambda\bar{v}^k = (50, 30, 15, 10)(0, 0, 0, 0)^{\top} + (1-0.5)0 + 0.50 = 0$.

Set $u \leftarrow \min\{u^0, u\} = 0$.

Step 2. **Add Cut to Master Program and Solve.**

Add $\beta_0^{\top} x + \gamma \geq \beta_0^0$ and $\sigma_0^{\top} x + \zeta \geq \sigma_0^0$ to master problem:

$$
\begin{array}{rlrrrrl}
\nu^1 := \text{Min} & (1-0.5)\gamma \; +0.5\zeta & & & & & \\
\text{s.t.} & -x_1 & & & & \geq & -300 \\
 & & -x_2 & & & \geq & -700 \\
 & & & -x_3 & & \geq & -600 \\
 & & & & -x_4 & \geq & -500 \\
 & & -x_2 & -x_3 & -x_4 & \geq & -1600 \\
 & 141.667x_1 & -30x_2 & -15x_3 & -10x_4 \; +\gamma & \geq & 0 \\
 & 141.667x_1 & -30x_2 & -15x_3 & -10x_4 \; +\zeta & \geq & 0 \\
 & x_1, & x_2, & x_3, & x_4 & \geq & 0.
\end{array}
$$

Solve master to get $x^1 = (300, 0, 0, 0)^\top$, $\gamma = -57{,}500$, $\zeta = -42{,}500$ and $\nu^1 = -42{,}500$. Set $\ell \leftarrow \max\{\nu^0, \ell\} = \max\{-42{,}500, -\infty\} = -42{,}500$.

Step 3. **Termination.** Compute $u - \ell = 0 - (-42{,}500) = 42{,}500$ and $\epsilon|u| = 10^{-6}| - 42{,}500|) = 0.0425$. Since $u - \ell \geq \epsilon|u|$:

Set $k \leftarrow 0 + 1 = 1$. Return to step 1.

Return to Step 1.

For $k = 1$:

Step 1. **Solve Subproblems.** For $s = 1, \cdots, 5$ solve

$$
\begin{array}{rrrrcl}
\varphi^1(x^1, \omega^s) := \text{Min} & -1150y_1 & -1525y_2 & -1900y_3 & & \\
\text{s.t.} & -6y_1 & -8y_2 & -10y_3 & \geq & 0 \\
 & -20y_1 & -25y_2 & -28y_3 & \geq & -300 \\
 & -12y_1 & -15y_2 & -18y_3 & \geq & 0 \\
 & -8y_1 & -10y_2 & -14y_3 & \geq & 0 \\
 & -y_1 & & & \geq & d_a^s \\
 & & -y_2 & & \geq & d_b^s \\
 & & & -y_3 & \geq & d_c^s \\
 & y_1, & y_2, & y_3 & \geq & 0.
\end{array}
$$

$s = 1$: feasible, $\varphi^1(x^1, \omega^1) = 0$, $y_1^{=}(0, 0, 0)^\top$, $\pi^1(\omega^1) = (0, 0, 105.556, 0, 0, 0, 0)^\top$. Compute $\bar{Q}^1 \leftarrow \bar{Q}^1 + p(\omega^1)\varphi(x^1, \omega^1) = 0 + 0.15(0) = 0$

$s = 2$: feasible, $\varphi^1(x^1, \omega^2) = 0$, $y_2^1 = (0, 0, 0)^\top$, $\pi^1(\omega^2) = (0, 0, 105.556, 0, 0, 0, 0)^\top$. Compute $\bar{Q}^1 \leftarrow \bar{Q}^1 + p(\omega^2)\varphi(x^1, \omega^2) = 0 + 0.15(0) = 0$

$s = 3$: feasible, $\varphi^1(x^1, \omega^3) = 0$, $y_3^1 = (0, 0, 0)^\top$, $\pi^1(\omega^3) = (0, 0, 105.556, 0, 0, 0, 0)^\top$. Compute $\bar{Q}^1 \leftarrow \bar{Q}^1 + p(\omega^3)\varphi(x^1, \omega^3) = 0 + 0.15(0) = 0$

$s = 4$: feasible, $\varphi^1(x^1, \omega^4) = 0$, $y_4^1 = (0, 0, 0)^\top$, $\pi^1(\omega^4) = (0, 0, 105.556, 0, 0, 0, 0)^\top$. Compute $\bar{Q}^1 \leftarrow \bar{Q}^1 + p(\omega^4)\varphi(x^1, \omega^4) = 0 + 0.15(0) = 0$

$s = 5$: feasible, $\varphi^1(x^1, \omega^5) = 0$, $y_5^1 = (0, 0, 0)^\top$, $\pi^1(\omega^5) = (0, 0, 105.556, 0, 0, 0, 0)^\top$. Compute $\bar{Q}^1 \leftarrow \bar{Q}^1 + p(\omega^5)\varphi(x^1, \omega^5) = 0 + 0.15(0) = 0$

For $s = 1$:

$\beta_1^0 \leftarrow \beta_1^0 + p(\omega^1)\pi^1(\omega^1)^\top r(\omega^1)$
$= 0 + 0.15(0, 0, 105.556, 0, 0, 0, 0)(0, 0, 0, 0, -15, -10, -5)^\top = 0.$

Compute $\beta_1 \leftarrow \beta_1 + p(\omega^1)(T(\omega^1)^\top \pi^1(\omega^1) - c) = (0, 0, 0, 0)^\top$

$$+0.15(\begin{bmatrix} 1\,0\,0\,0 \\ 0\,1\,0\,0 \\ 0\,0\,1\,0 \\ 0\,0\,0\,1 \\ 0\,0\,0\,0 \\ 0\,0\,0\,0 \\ 0\,0\,0\,0 \end{bmatrix}^\top (0, 0, 105.556, 0, 0, 0, 0)^\top - (50, 30, 15, 10))$$

$= (-7.5, -4.5, 13.5833, -1.5)^\top.$

$\varphi^1(x^1, \omega^1) = \bar{Q}^1$:

Compute $\bar{\nu}^1 = \bar{\nu}^1 + p(\omega^1)\varphi^1(x^1, \omega^1) = 0 + 0.15(0) = 0.$

$\sigma_1^0 \leftarrow \sigma_1^0 + p(\omega^1)\pi^1(\omega^1)^\top r(\omega^1)$
$= 0 + 0.15(0, 0, 105.556, 0, 0, 0, 0)(0, 0, 0, 0, -15, -10, -5)^\top = 0.$

Compute $\sigma_1 \leftarrow \sigma_1 + p(\omega^1)(T(\omega^1)^\top \pi^1(\omega^1) - c) = (0, 0, 0, 0)^\top$

$$+0.15(\begin{bmatrix} 1\,0\,0\,0 \\ 0\,1\,0\,0 \\ 0\,0\,1\,0 \\ 0\,0\,0\,1 \\ 0\,0\,0\,0 \\ 0\,0\,0\,0 \\ 0\,0\,0\,0 \end{bmatrix}^\top (0, 0, 105.556, 0, 0, 0, 0)^\top - (50, 30, 15, 10))$$

$= (-7.5, -4.5, 13.5833,$
$-1.5)^\top.$

For $s = 2$:

$\beta_1^0 \leftarrow \beta_1^0 + p(\omega^2)\pi^1(\omega^2)^\top r(\omega^2)$
$= 0 + 0.3(0, 0, 105.556, 0, 0, 0, 0)(0, 0, 0, 0, -20, -15, -15)^\top = 0.$

Compute $\beta_1 \leftarrow \beta_1 + p(\omega^2)(T(\omega^2)^\top \pi^1(\omega^2) - c) = (-7.5, -4.5, 13.5833,$

$$-1.5)^\top + 0.3(\begin{bmatrix} 1\,0\,0\,0 \\ 0\,1\,0\,0 \\ 0\,0\,1\,0 \\ 0\,0\,0\,1 \\ 0\,0\,0\,0 \\ 0\,0\,0\,0 \\ 0\,0\,0\,0 \end{bmatrix}^\top (0, 0, 105.556, 0, 0, 0, 0)^\top - (50, 30, 15, 10))$$

$= (-22.5, -13.5, 40.75, -4.5)^\top.$

$\varphi^1(x^1, \omega^2) = \bar{Q}^1$:

Compute $\bar{\nu}^1 = \bar{\nu}^1 + p(\omega^2)\varphi^1(x^1, \omega^2) = 0 + 0.3(0) = 0$.
$\sigma_1^0 \leftarrow \sigma_1^0 + p(\omega^2)\pi^1(\omega^2)^\top r(\omega^2)$
$= 0 + 0.3(0, 0, 105.556, 0, 0, 0, 0)(0, 0, 0, 0, -20, -15, -15)^\top = 0$.
Compute $\sigma_1 \leftarrow \sigma_1 + p(\omega^2)(T(\omega^2)^\top \pi^1(\omega^2) - c) = (-7.5, -4.5, 13.5833,$

$$- 1.5)^\top + 0.3((1 - 0.5)\begin{bmatrix} 1\,0\,0\,0 \\ 0\,1\,0\,0 \\ 0\,0\,1\,0 \\ 0\,0\,0\,1 \\ 0\,0\,0\,0 \\ 0\,0\,0\,0 \\ 0\,0\,0\,0 \end{bmatrix}^\top (0, 0, 105.556, 0, 0, 0, 0)^\top -$$

$(50, 30, 15, 10))$
$= (-22.5, -13.5, 40.75, -4.5)^\top$.
For $s = 3$:
$\beta_1^0 \leftarrow \beta_1^0 + p(\omega^3)\pi^1(\omega^3)^\top r(\omega^3)$
$= 0 + 0.3(0, 0, 105.556, 0, 0, 0, 0)(0, 0, 0, 0, -25, -20, -25)^\top = 0$.
Compute $\beta_1 \leftarrow \beta_1 + p(\omega^3)(T(\omega^3)^\top \pi^1(\omega^3) - c) = (-22.5, -13.5, 40.75,$

$$- 4.5)^\top + 0.3(\begin{bmatrix} 1\,0\,0\,0 \\ 0\,1\,0\,0 \\ 0\,0\,1\,0 \\ 0\,0\,0\,1 \\ 0\,0\,0\,0 \\ 0\,0\,0\,0 \\ 0\,0\,0\,0 \end{bmatrix}^\top (0, 0, 105.556, 0, 0, 0, 0)^\top - (50, 30, 15, 10))$$

$= (-37.5, -22.5, 67.9167, -7.5)^\top$.
$\varphi^1(x^1, \omega^3) = \bar{Q}^1$:
Compute $\bar{\nu}^1 = \bar{\nu}^1 + p(\omega^3)\varphi^1(x^1, \omega^3) = 0 + 0.3(0) = 0$.
$\sigma_1^0 \leftarrow \sigma_1^0 + p(\omega^3)\pi^1(\omega^3)^\top r(\omega^3)$
$= 0 + 0.3((0, 0, 105.556, 0, 0, 0, 0)(0, 0, 0, 0, -25, -20, -25)^\top = 0$.
Compute $\sigma_1 \leftarrow \sigma_1 + p(\omega^3)(T(\omega^3)^\top \pi^1(\omega^3) - c) = (-22.5, -13.5, 40.75,$

$$- 4.5)^\top + 0.3(\begin{bmatrix} 1\,0\,0\,0 \\ 0\,1\,0\,0 \\ 0\,0\,1\,0 \\ 0\,0\,0\,1 \\ 0\,0\,0\,0 \\ 0\,0\,0\,0 \\ 0\,0\,0\,0 \end{bmatrix}^\top (0, 0, 105.556, 0, 0, 0, 0)^\top - (50, 30, 15, 10))$$

$= (-37.5, -22.5, 67.9167, -7.5)^\top$.
For $s = 4$:
$\beta_1^0 \leftarrow \beta_1^0 + p(\omega^4)\pi^1(\omega^4)^\top r(\omega^4)$
$= 0 + 0.2(0, 0, 105.556, 0, 0, 0, 0)(0, 0, 0, 0, -30, -25, -30)^\top = 0$.

Compute $\beta_1 \leftarrow \beta_1 + p(\omega^4)(T(\omega^4)^\top \pi^1(\omega^4) - c) = (-37.5, -22.5, 67.9167,$

$$-7.5)^\top + 0.2(\begin{bmatrix} 1 & 0 & 0 & 0 \\ 0 & 1 & 0 & 0 \\ 0 & 0 & 1 & 0 \\ 0 & 0 & 0 & 1 \\ 0 & 0 & 0 & 0 \\ 0 & 0 & 0 & 0 \\ 0 & 0 & 0 & 0 \end{bmatrix}^\top (0, 0, 105.556, 0, 0, 0, 0)^\top - (50, 30, 15, 10))$$

$= (-47.5, -28.5, 86.0278, -9.5)^\top$.

$\varphi^1(x^1, \omega^4) = \bar{Q}^1$.

Compute $\bar{\nu}^1 = \bar{\nu}^1 + p(\omega^4)\varphi^1(x^1, \omega^4) = 0 + 0.2(0) = 0$.

$\sigma_1^0 \leftarrow \sigma_1^0 + p(\omega^4)\pi^1(\omega^4)^\top r(\omega^4)$

$= 0 + 0.2((0, 0, 105.556, 0, 0, 0, 0)(0, 0, 0, 0, -30, -25, -30)^\top = 0$.

Compute $\sigma_1 \leftarrow \sigma_1 + p(\omega^4)(T(\omega^4)^\top \pi^1(\omega^4) - c) = (-37.5, -22.5, 67.9167,$

$$-7.5)^\top + 0.2(\begin{bmatrix} 1 & 0 & 0 & 0 \\ 0 & 1 & 0 & 0 \\ 0 & 0 & 1 & 0 \\ 0 & 0 & 0 & 1 \\ 0 & 0 & 0 & 0 \\ 0 & 0 & 0 & 0 \\ 0 & 0 & 0 & 0 \end{bmatrix}^\top (0, 0, 105.556, 0, 0, 0, 0)^\top - (50, 30, 15, 10))$$

$= (-47.5, -28.5, 86.0278, -9.5)^\top$.

For $s = 5$:

$\beta_1^0 \leftarrow \beta_1^0 + p(\omega^5)\pi^1(\omega^5)^\top r(\omega^5)$

$+ 0.05(0, 0, 105.556, 0, 0, 0, 0)(0, 0, 0, 0, -10, -10, -10)^\top = 0$.

Compute $\beta_1 \leftarrow \beta_1 + p(\omega^5)(T(\omega^5)^\top \pi^1(\omega^5) - c) = (-47.5, -28.5, 86.0278,$

$$-9.5)^\top + 0.05(\begin{bmatrix} 1 & 0 & 0 & 0 \\ 0 & 1 & 0 & 0 \\ 0 & 0 & 1 & 0 \\ 0 & 0 & 0 & 1 \\ 0 & 0 & 0 & 0 \\ 0 & 0 & 0 & 0 \\ 0 & 0 & 0 & 0 \end{bmatrix}^\top (0, 0, 105.556, 0, 0, 0, 0)^\top - (50, 30, 15, 10))$$

$= (-50, -30, 90.5556, -10)^\top$.

$\varphi^1(x^1, \omega^5) = \bar{Q}^1$.

Compute $\bar{\nu}^1 = \bar{\nu}^1 + p(\omega^5)\varphi^1(x^1, \omega^5) = 0 + 0.2(0) = 0$.

$\sigma_1^0 \leftarrow \sigma_1^0 + p(\omega^5)\pi^1(\omega^5)^\top r(\omega^5)$

$= 0 + 0.05(0, 0, 105.556, 0, 0, 0, 0)(0, 0, 0, 0, -10, -10, -10)^\top = 0$.

Compute $\sigma_1 \leftarrow \sigma_1 + p(\omega^5)(T(\omega^5)^\top \pi^1(\omega^5) - c) = (-47.5, -28.5, 86.0278,$

$$-9.5)^\top + 0.05(\begin{bmatrix} 1\,0\,0\,0 \\ 0\,1\,0\,0 \\ 0\,0\,1\,0 \\ 0\,0\,0\,1 \\ 0\,0\,0\,0 \\ 0\,0\,0\,0 \\ 0\,0\,0\,0 \end{bmatrix}^\top (0, 0, 105.556, 0, 0, 0, 0)^\top - (50, 30, 15, 10))$$

$= (-50, -30, 90.5556, -10)^\top$.

Generate *optimality* cuts: $\beta_0^\top x + \gamma \geq \beta_1^0 = (-50, -30, 90.5556, -10)^\top x + \gamma \geq 0$ and $\sigma_0^\top x + \zeta \geq \sigma_1^0 = (-50, -30, 90.5556, -10)^\top x + \zeta \geq 0$.

Set $u^1 \leftarrow c^\top x^1 + (1-\lambda)\bar{Q}^1 + \lambda \bar{\nu}^k = (50, 30, 15, 10)(-300, 0, 0, 0)^\top + (1 - 0.5)0 + 0.50 = -15{,}000$.

Set $u \leftarrow \min\{u^1, u\} = 0$.

Step 2. Add Cut to Master Program and Solve.

Add $\beta_0^\top x + \gamma \geq \beta_0^0$ to master problem:

$$\begin{array}{llllllll}
\nu^1 := \text{Min } (1-0.5)\gamma & +0.5\zeta & & & & & & \\
\text{s.t.} & -x_1 & & & & & \geq & -300 \\
 & & -x_2 & & & & \geq & -700 \\
 & & & -x_3 & & & \geq & -600 \\
 & & & & -x_4 & & \geq & -500 \\
 & & -x_2 & -x_3 & -x_4 & & \geq & -1600 \\
 & 141.667x_1 & -30x_2 & -15x_3 & -10x_4 & +\gamma & \geq & 0 \\
 & 141.667x_1 & -30x_2 & -15x_3 & -10x_4 & +\zeta & \geq & 0 \\
 & -50x_1 & -30x_2 & 90.5556x_3 & -10x_4 & +\gamma & \geq & 0 \\
 & -50x_1 & -30x_2 & 90.5556x_3 & -10x_4 & +\zeta & \geq & 0 \\
 & x_1, & x_2, & x_3, & x_4 & & \geq & 0.
\end{array}$$

Solve master to get $x^2 = (300, 0, 544.737, 0)^\top$, $\gamma = -57{,}500$, $\zeta = -34{,}328.9$ and $\nu^2 = -34{,}328.9$. Set $\ell \leftarrow \max\{\nu^2, \ell\} = \max\{-34{,}328.9, -\infty\} = -34{,}328.9$.

Step 3. Termination. Compute $u - \ell = 0 - (-34{,}328.9) = 34{,}328.9$ and $\epsilon|u| = 10^{-6}| - 34{,}328.9|) = 0.0343$. Since $u - \ell \geq \epsilon|u|$:

Set $k \leftarrow 1 + 1 = 2$. Return to step 1.

Return to Step 1.

After 15 iterations, the *ASD*-SEP algorithm terminates with optimal solution $x^* = (236.4, 690, 432, 318)^\top$ and objective value -2059.9. This is the solution we obtained using the *ASD*-AGG algorithm, which took 18 iterations. As expected, the separate cut approach gives fewer iterations than the aggregated approach.

Bibliographic Notes

Recent work on MR-SLP algorithms includes convexity and decomposition of mean-risk stochastic programs [1], dissertation work on mean-risk portfolio optimization problems [14], and a computational study [3]. A thorough coverage of modern theory of MR-SLP approaches focusing on analysis of models, optimality theory, and duality can be found in Chapter 6 of the lectures on stochastic programming book by [27]. The stochastic programming book by [6] also has chapters on different aspects of risk-averse models and methods. A survey on modeling and optimization of risk [11] is also available.

Several works lay the foundation for MR-SLP: coherent risk measures [2], mean-absolute deviation [7], optimization of conditional value-at-risk and application to portfolio optimization [10, 20, 21], deviation measures [9], and optimization of convex risk function [22]. Algorithms for mean-risk stochastic integer programs [24] have been proposed including excess probabilities [26], deviation measures [13], and conditional value-at-risk [25]. Applications include optimization in energy [23] and for risk-averse power optimization in electricity networks [12]. Finally, the work in [15] establishes the relationship between stochastic dominance and the semideviation mean-risk model. The relation of stochastic dominance to related mean-risk models is studied in [16].

The methods presented in this chapter focus on two-stage MR-SLP. Methods have also been developed for risk-averse multistage stochastic programming (MSSP) for sequential decision-making over time. In risk-averse MSSP, the expected value is replaced by a risk measure [5]. The works [8, 18, 28] consider a nested formulation of the problem to include risk aversion based on a convex combination of CVaR and the expected value as the risk measure. In addition to CVaR, [28] also considers a mean-upper semideviation risk measure. A general coherent risk measure is used in [19]. Some form of the stochastic dual dynamic programming (SDDP) algorithm [4, 17] is used in these works to solve the resulting problem.

Problems

7.1 $\mathbb{D}$-AGG Algorithm

Perform two iterations of the $\mathbb{D}$-AGG algorithm towards solving the *abc*-Production Planning problem example instance for the following cases:

(a) CVaR with $\alpha := 0.95$ and $\lambda := 0.5$

(b) EE with target $\alpha := 0.95$, $\lambda := 5$ and target $\eta := -2000$

7.2 ASD-AGG Algorithm

Perform two iterations of the ASD-AGG algorithm towards solving the *abc*-Production Planning problem example instance with scenario probabilities

$p(\omega^s), s = 1, \cdots, 5$ given as follows: $p(\omega^1) := 0.01$, $p(\omega^2) := 0.30$, $p(\omega^3) := 0.30$, $p(\omega^4) := 0.20$, and $p(\omega^5) := 0.19$:

(a) Use $\lambda := 0.5$.
(b) Use $\lambda := 0.8$.
(c) Create the DEP and solve the instances in parts (a) and (b) to optimality. Compare and contrast the solutions obtained in parts (a) and (b) to the risk-neutral solution.

7.3 $\mathbb{D}$-AGG Algorithm Implementation

(a) Implement the $\mathbb{D}$-AGG algorithm using software and LP solver of your choice.
(b) Perform computational experiments based on selected standard test instances, and compare and contrast the solutions obtained under each of the risk measures QDEV, CVaR, and EE.
(c) For each risk measure, compare and contrast the solutions obtained under the different values of the risk factor λ.

7.4 $\mathbb{D}$-SEP Algorithm Implementation

(a) Implement the $\mathbb{D}$-SEP algorithm using software and LP solver of your choice.
(b) Perform computational experiments based on selected standard test instances, and compare and contrast the solutions obtained under each of the risk measures QDEV, CVaR, and EE.

7.5 $\mathbb{D}$-AGG Versus $\mathbb{D}$-SEP Algorithm

(a) Compare and contrast the computational performance of the $\mathbb{D}$-AGG and $\mathbb{D}$-SEP algorithms for each of the risk measures QDEV, CVaR, and EE.
(b) Which of the two algorithms do you prefer and why?

7.6 ASD-AGG Algorithm Implementation

(a) Implement the $\mathbb{D}$-AGG algorithm using software and LP solver of your choice and perform computational experiments based on standard test instances.
(b) Compare and contrast the solutions obtained under the different values of the risk factor λ.

7.7 ASD-SEP Algorithm Implementation

(a) Implement the $\mathbb{D}$-AGG algorithm using software and LP solver of your choice and perform computational experiments based on standard test instances.
(b) Compare and contrast the solutions obtained under the different values of the risk factor λ.

7.8 ASD-AGG Versus ASD-SEP Algorithm

(a) Compare and contrast the computational performance of the $\mathbb{D}$-AGG and $\mathbb{D}$-SEP algorithms for each of the risk measures QDEV, CVaR, and EE.
(b) Which of the two algorithms do you prefer and why?

7.9 Comparison of the Risk Measures

(a) Compare and contrast the risk measures QDEV, CVaR, EE, and ASD.
(b) How do these risk measures compare computationally?

7.10 Dantzig–Wolfe decomposition for MR-SLP with ASD

Instead of using subgradient optimization as done in this chapter, devise a Dantzig–Wolfe decomposition (column generation) method for tackling the MR-SLP problem with ASD.

7.11 Dual Decomposition Method for MR-SLP with ASD

Instead of *stage-wise* decomposition of the MR-SLP with ASD as done in this chapter, perform a *scenario* decomposition and devise a *dual* decomposition method for solving the problem.

7.12 Comparison of Decomposition Methods for MR-SLP with ASD

(a) Compare and contrast the different decomposition approaches for MR-SLP with ASD.
(b) Which approach do you prefer and why?

References

1. S. Ahmed. Convexity and decomposition of mean-risk stochastic programs. *Mathematical Programming*, 106(3):433–446, 2006.
2. P. Artzner, F. Delbean, J.M Eber, and D. Heath. Coherent measures of risk. *Mathematical Finance*, 9:203–228, 1999.
3. T.G. Cotton and L. Ntaimo. Computational study of decomposition algorithms for mean-risk stochastic linear programs. *Mathematical Programming Computation*, 7.4:471–499, 2015.
4. O. Dowson and L. Kapelevich. SDDP.jl: a Julia package for stochastic dual dynamic programming. *INFORMS Journal on Computing*, 2020. in press.
5. Tito Homem-de Mello and Bernardo K Pagnoncelli. Risk aversion in multistage stochastic programming: A modeling and algorithmic perspective. *European Journal of Operational Research*, 249(1):188–199, 2016.
6. G. Infanger, editor. *Stochastic Programming: The State of the Art in Honor of George B. Dantzig*. Springer, NY, NY, 2011.
7. H. Konno and H. Yamazaki. Mean-absolute deviation portfolio optimization model and its applications to Tokyo stock market. *Management Science*, 37(5):519–531, 1991.
8. Václav Kozmík and David P Morton. Evaluating policies in risk-averse multi-stage stochastic programming. *Mathematical Programming*, 152(1-2):275–300, 2015.
9. T. Kristoffersen. Deviation measures in linear two-stage stochastic programming. *Mathematical Methods of Operations Research*, 62(2):255–274, 2005.
10. P. Krokhmal, J. Palmquist, and S. Uryasev. Portfolio optimization with conditional value-at-risk objective and constraints. *Journal of Banking and Finance*, 4:43–68, 2002.
11. P. Krokhmal, M. Zabarankin, and S. Uryasev. Modeling and optimization of risk. *Surveys in Operations Research and Management Science*, 16:49–66, 2011.
12. S. Kuhn and R. Schultz. Risk neutral and risk averse power optimization in electricity networks with dispersed generation. *Mathematical Methods of Operations Research*, 69(2):353–367, 2009.

13. Andreas Märkert and Rüdiger Schultz. On deviation measures in stochastic integer programming. *Operations Research Letters*, 33(5):441–449, 2005.
14. N. Miller. *Mean-Risk Portfolio Optimization Problems with Risk-Adjusted Measures*. Dissertation, The State University of New Jersey, October 2008.
15. W. Ogryczak and A. Ruszcynski. From stochastic dominance to mean-risk model: Semideviations as risk measures. *European Journal of Operational Research*, 116:33–50, 1999.
16. W. Ogryczak and A. Ruszcynski. Dual stochastic dominance and related mean-risk models. *SIAM Journal on Optimization*, 13:60–78, 2002.
17. M.V. Pereira and L.M. Pinto. Multi-stage stochastic optimization applied to energy planning. *Mathematical Programming*, 52.1-3:359–375, 1991.
18. A. Philpott and V. De Matos. Dynamic sampling algorithms for multi-stage stochastic programs with risk aversion. *European Journal of Operational Research*, 218(2):470–483, 2012.
19. Andy Philpott, Vitor de Matos, and Erlon Finardi. On solving multistage stochastic programs with coherent risk measures. *Operations Research*, 61(4):957–970, 2013.
20. R. Rockafellar and S. Urysev. Conditional value-at-risk for general loss distributions. *Journal of Banking & Finance*, 26:1443–1471, 2002.
21. R.T. Rockafellar and S. Uryasev. Optimization of conditional value-at-risk. *The Journal at Risk*, 2:21–41, 2000.
22. A. Ruszcynski and A. Shapiro. Optimization of convex risk functions. *Mathematics of Operations Research*, 31(3):433–452, 2006.
23. R. Schultz and F. Neise. Algorithms for mean-risk stochastic integer programs in energy. In *Power Engineering Society General Meeting, 2006. IEEE*, page 8 pp., 2006.
24. R. Schultz and F. Neise. Algorithms for mean-risk stochastic integer programs in energy. *Revista Investigacion Operacional*, 28(1):4–16, 2007.
25. R. Schultz and S. Tiedemann. Conditional value-at-risk in stochastic programs with mixed-integer recourse. *Mathematical Programming*, 105:365–386, 2006.
26. Rüdiger Schultz and Stephan Tiedemann. Risk aversion via excess probabilities in stochastic programs with mixed-integer recourse. *SIAM J. on Optimization*, 14(1):115–138, 2003.
27. A. Shapiro, D. Dentcheva, and A. Ruszczyński. *Lectures on Stochastic Programming: Modeling and Theory*. SIAM, Philadelphia, PA., 2009.
28. Alexander Shapiro, Wajdi Tekaya, Joari Paulo da Costa, and Murilo Pereira Soares. Risk neutral and risk averse stochastic dual dynamic programming method. *European Journal of Operational Research*, 224(2):375–391, 2013.

Chapter 8
Sampling-Based Stochastic Linear Programming Methods

8.1 Introduction

In this chapter we explore statistical decomposition methods for two-stage mean-risk stochastic linear programs (MR-SLP). Dealing with instances of MR-SLP with very large number of outcomes can be very challenging. This is because the instance can be too large to even read into the computer, later on to solve using a direct solver or a decomposition method (e.g., those studied in Chaps. 6–7). Therefore, we need to find an alternative way to tackle such large-scale MR-SLP. This chapter explores two sampling-based techniques to give the student a flavor of statistically inspired methods for MR-SLP, *exterior* sampling and *interior* sampling. Exterior sampling is a Monte Carlo method [3] that involves taking a sample and solving an approximation problem. In this chapter, we study the basic sample average approximation (SAA) method [14] for MR-SLP. Interior sampling involves sequential sampling during the course of the algorithm to solve the approximation problem. This requires a streamlined design of the algorithm within which sequential sampling is done. We illustrate interior sampling with the basic stochastic decomposition (SD) method [7, 8] for MR-SLP.

To start, let us first restate the two-stage MR-SLP from Chap. 2 as follows:

$$\operatorname*{Min}_{x \in X} \; F(x) := \mathbb{E}[f(x, \tilde{\omega})] + \lambda \mathbb{D}[f(x, \tilde{\omega})], \tag{8.1}$$

where $\mathbb{E} : \mathscr{F} \mapsto \mathbb{R}$ denotes the expected value, $\mathbb{D} : \mathscr{F} \mapsto \mathbb{R}$ is the *risk measure*, and $\lambda \geq 0$ is a suitable weight factor that quantifies the trade-off between expected cost and risk. The problem is risk-neutral if $\lambda := 0$. Risk measure $\mathbb{D}$ is chosen so that the problem remains a *convex optimization problem*, allowing it to be solved using convex optimization methods. The set $X = \{Ax \geq b,\ x \geq 0\}$ is a nonempty polyhedron that defines the set of first-stage feasible solutions. The matrix $A \in \mathbb{R}^{m_1 \times n_1}$ and vector $b \in \mathbb{R}^{m_1}$ are the first-stage matrix and RHS vector, respectively. The family of real random cost variables $\{f(x, \tilde{\omega})\}_{x \in X} \subseteq \mathscr{F}$

L. Ntaimo, *Computational Stochastic Programming*, Springer Optimization and Its Applications 774, https://doi.org/10.1007/978-3-031-52464-6_8

are defined on $(\Omega, \mathscr{A}, \mathbb{P})$, where $\mathscr{F}$ is the space of all real random cost variables $f : \Omega \mapsto \mathbb{R}$ satisfying $\mathbb{E}[|f(\tilde{\omega})|] < \infty$. For a given $x \in X$ the real random cost variable $f(x, \tilde{\omega})$ is given by

$$f(x, \tilde{\omega}) := c^\top x + \varphi(x, \tilde{\omega}). \tag{8.2}$$

For a given realization ω of $\tilde{\omega}$, the *recourse function* $\varphi(x, \omega)$ is given by

$$\begin{aligned} \varphi(x, \omega) :=& \text{Min } q(\omega)^\top y(\omega) \\ & \text{s.t. } Wy(\omega) \geq r(\omega) - T(\omega)x \\ & \quad y(\omega) \geq 0, \end{aligned} \tag{8.3}$$

where $q(\omega) \in \mathbb{R}^{n_2}$ is the second-stage cost vector and $y(\omega) \in \mathbb{R}^{n_2}_+$ is the recourse decision. The matrix $W \in \mathbb{R}^{m_2 \times n_2}$ is the *recourse matrix*, $T(\omega) \in \mathbb{R}^{m_2 \times n_1}$ is the *technology matrix*, and $r(\omega) \in \mathbb{R}^{m_2}$ is the RHS vector. By *scenario* ω we mean the realization of the stochastic problem data, i.e., $\omega := (q(\omega), T(\omega), r(\omega))$.

To ensure that Problem (8.1) is well-defined for sampling and computational purposes, we shall make the following assumptions:

(A1) The multivariaterandom variable $\tilde{\omega}$ is discretely distributed with finitely many scenarios $\omega \in \Omega$, each with probability of occurrence $p(\omega)$.
(A2) The first-stage feasible set X is compact.
(A3) Problem (8.1) has *complete* or *relatively complete recourse*, i.e., for all x or $x \in X$, $\{Wy(\omega) \geq r(\omega) - T(\omega)x,\ y(\omega) \geq 0\} \neq \emptyset$.
(A4) The risk measure $\mathbb{D}$ is convexity preserving so that when $\lambda > 0$, function $f(x, \tilde{\omega})$ is convex and continuous.

Even though statistical methods can handle continuous distributions, we make assumption (A1) to stay in line with the previous chapters as many problems of interest in practice can be modeled using discrete probability distributions. We can formulate Problem (8.1) to attain assumption (A2) to ensure that we have a limit point in X. Assumption (A3) guarantees the feasibility of the second-stage problem for every x or $x \in X$. This assumption implies that $\varphi(x, \omega) < \infty$ with probability one for all $x \in X$. The last assumptions ensure that we are dealing with a "nice" recourse function that can be approximated using sampling techniques.

To solve Problem (8.1), we need to evaluate $\mathbb{E}[f(x, \tilde{\omega})] + \lambda\mathbb{D}[f(x, \tilde{\omega})]$, which involves multidimensional summations (or integration for $\tilde{\omega}$ with continuous distribution). This requires evaluating, at least implicitly, $f(x, \omega)$ for all scenarios $\omega \in \Omega$. These functions evaluations can be prohibitively expensive if f is hard to evaluate and/or there is a huge number of scenarios. Because $\tilde{\omega}$ is a random variable, $f(x, \tilde{\omega})$ is also a random variable, for all x. Therefore, the objective function involves the expected value (mean) of a random variable. This implies that the natural thing one can do is to estimate both $\mathbb{E}[f(x, \tilde{\omega})]$ and $\mathbb{D}[f(x, \tilde{\omega})]$ *statistically*. Next, we present an example MR-SLP instance that we later use to illustrate sampling, the SAA scheme, and the SD method.

8.2 Example Numerical Instance

To help illustrate the sampling based techniques we study in this chapter, we shall use an instance of the *abc*-Production Planning MR-SLP problem from Chap. 3 which we describe below.

Example 8.1 (The *abc*-Production Planning Problem)
The *abc*-Production Planning problem involves maximizing the profit of producing three different products (*Product-a, Product-b, Product-c*) given a limited supply of raw materials (*Material-abc*) and the available number of processing hours involving three different processes (*Process-1, Process-2,* and *Process-3*) required for making the products. The first-stage decision variables are as follows:

x_1: number of units of *Material-abc* purchased
x_2: number of hours of *Process-1* purchased
x_3: number of hours of *Process-2* purchased
x_4: number of hours of *Process-3* purchased

The second-stage decision variables are as follows:

$y_1(\omega)$: number of units of *Product-a* produced under scenario ω
$y_2(\omega)$: number of units of *Product-b* produced under scenario ω
$y_3(\omega)$: number of units of *Product-c* produced under scenario ω

Each of the three products has five *independent* demand outcomes: *Low, Moderate, High, Extreme, and Rare*. The distribution of the product demand is a collection of demand scenarios and corresponding probabilities. This forms a multivariate probability distribution given in Table 8.1 with the corresponding scenario tree shown in Fig. 8.1.

Table 8.1 The *abc*-Production Planning Problem product marginal demand distribution

Random variable	Demand	Probability	Demand type
1. Product-a (d_a^s)	15	0.15	Low
	20	0.30	Moderate
	25	0.30	High
	30	0.20	Extreme
	10	0.05	Rare
2. Product-b (d_b^s)	10	0.15	Low
	15	0.30	Moderate
	20	0.30	High
	25	0.20	Extreme
	10	0.05	Rare
3. Product-c (d_c^s)	5	0.15	Low
	15	0.30	Moderate
	25	0.30	High
	30	0.20	Extreme
	10	0.05	Rare

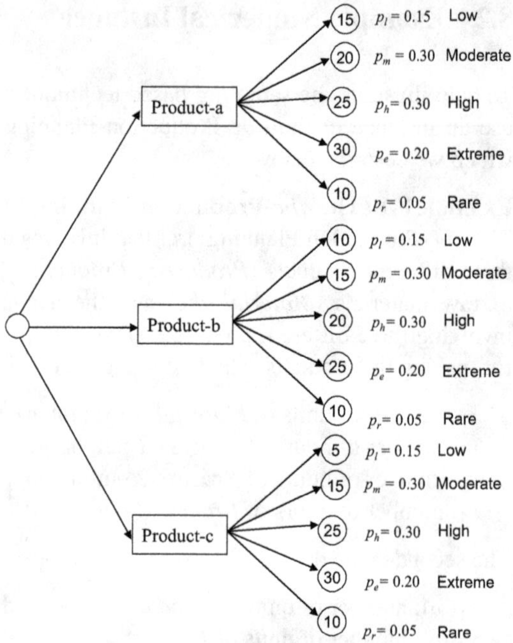

Fig. 8.1 Demand scenario tree for the *abc*-Production Planning Problem instance

Given the marginal distributions for each product from Table 8.1, the total number of scenarios is $5^3 = 125$. Let the sample space Ω be given as $\Omega = \{\omega^1, \omega^2, \cdots, \omega^{125}\}$. Then the product demand distribution can be given as shown in Table 8.2 in which we show a partial list of the scenarios for illustration purposes.

Let us now write the *abc*-Production Planning formulation in the form of Problem (8.1). Let $x = (x_1, x_2, x_3, x_4)^\top$ and $y = (y_1, y_2, y_3)^\top$ and specify the instance data as follows:

First-stage:

Cost vector: $c = (50, 30, 15, 10)^\top$.

$$\text{Matrix: } A = \begin{bmatrix} -1 & 0 & 0 & 0 \\ 0 & -1 & 0 & 0 \\ 0 & 0 & -1 & 0 \\ 0 & 0 & 0 & -1 \\ 0 & -1 & -1 & -1 \end{bmatrix}.$$

RHS vector: $b = (-300, -800, -600, -500, -1600)^\top$.

Second-stage:

Sample space: $\Omega = \{\omega^1, \omega^2, \cdots, \omega^{125}\}$.

Probabilities of occurrence: $\{p(\omega^1), p(\omega^2), \cdots, p(\omega^{125})\}$ are given in Table 8.2.

Cost vector: $q = (-1150, -1525, -1900)^\top$.

Table 8.2 Product demand distribution for the *abc*-Production Planning problem instance

	Demand			
Scenario	*Product-a*	*Product-b*	*Product-c*	Probability
ω^1	15	10	5	0.003375
ω^2	15	10	15	0.00675
ω^3	15	10	25	0.00675
ω^4	15	10	30	0.00450
ω^5	15	10	10	0.001125
ω^6	15	15	5	0.006750
ω^7	15	15	15	0.0135
ω^8	15	15	25	0.0135
ω^9	15	15	30	0.00900
ω^{10}	15	15	10	0.00225
ω^{11}	15	20	5	0.00675
ω^{12}	15	20	15	0.0135
ω^{13}	15	20	25	0.0135
ω^{14}	15	20	30	0.009
ω^{15}	15	20	10	0.00225
ω^{16}	15	25	5	0.0045
ω^{17}	15	25	15	0.009
ω^{18}	15	25	25	0.009
ω^{19}	15	25	30	0.006
ω^{20}	15	25	10	0.0015
ω^{21}	15	10	5	0.001125
ω^{22}	15	10	15	0.00225
ω^{23}	15	10	25	0.00225
ω^{24}	15	10	30	0.0015
ω^{25}	15	10	10	0.000375
ω^{26}	20	10	5	0.00675
...	...	...	...	...
...	...	...	...	...
...	...	...	...	...
ω^{125}	10	10	10	0.000125
				$\sum_{s=1}^{125} p_{\omega^s} = 1$

$$\text{Recourse matrix: } W = \begin{bmatrix} -6 & -8 & -10 \\ -20 & -25 & -28 \\ -12 & -15 & -18 \\ -8 & -10 & -14 \\ -1 & 0 & 0 \\ 0 & -1 & 0 \\ 0 & 0 & -1 \end{bmatrix}.$$

$$\text{Technology matrix: } T = \begin{bmatrix} 1\,0\,0\,0 \\ 0\,1\,0\,0 \\ 0\,0\,1\,0 \\ 0\,0\,0\,1 \\ 0\,0\,0\,0 \\ 0\,0\,0\,0 \\ 0\,0\,0\,0 \end{bmatrix}.$$

RHS vector:

$r(\omega^1) = (0, 0, 0, 0, -15, -10, -5)^\top.$
$r(\omega^2) = (0, 0, 0, 0, -15, -10, -15)^\top.$
$r(\omega^3) = (0, 0, 0, 0, -15, -10, -25)^\top.$
$r(\omega^4) = (0, 0, 0, 0, -15, -10, -30)^\top.$
$r(\omega^5) = (0, 0, 0, 0, -15, -10, -10)^\top.$
$r(\omega^6) = (0, 0, 0, 0, -15, -15, -5)^\top.$
$r(\omega^7) = (0, 0, 0, 0, -15, -15, -15)^\top.$
$r(\omega^8) = (0, 0, 0, 0, -15, -15, -25)^\top.$
$r(\omega^9) = (0, 0, 0, 0, -15, -15, -30)^\top.$
$r(\omega^{10}) = (0, 0, 0, 0, -15, -15, -10)^\top.$
$\cdots$
$r(\omega^{125}) = (0, 0, 0, 0, -10, -10, -10)^\top.$

Since we are considering the risk-neutral setting, $\lambda := 0$, otherwise $\mathbb{D}$ has to be specified. Using the given instance data, the first-stage feasible set X is given as follows:

$$\begin{aligned} X = \{ -x_1 & & & & \geq -300, \\ & -x_2 & & & \geq -800, \\ & & -x_3 & & \geq -600, \\ & & & -x_4 & \geq -500, \\ & -x_2 & -x_3 & -x_4 & \geq -1600, \\ x_1, & x_2, & x_3, & x_4 & \geq 0\}. \end{aligned}$$

Now for an outcome $\omega \in \Omega$,

$$\begin{aligned} f(x, \omega) &:= c^\top x + \varphi(x, \omega) \\ &= 50x_1 + 30x_2 + 15x_3 + 10x_4 + \varphi(x, \omega) \end{aligned} \tag{8.4}$$

and the explicit formulation can be given as follows:

$$\underset{x \in X}{\text{Min}}\ 50x_1 + 30x_2 + 15x_3 + 10x_4 + \sum_{\omega \in \Omega} p(\omega)\varphi(x, \omega),$$

where for each $x \in X$ and $\omega \in \Omega$, $\varphi(x,\omega)$ is given as

$$\begin{array}{rrrrl}
\varphi(x,\omega) := \text{Min} & -1150y_1(\omega) & -1525y_2(\omega) & -1900y_3(\omega) & \\
\text{s.t.} & -6y_1(\omega) & -8y_2(\omega) & -10y_3(\omega) & \geq -x_1 \\
& -20y_1(\omega) & -25y_2(\omega) & -28y_3(\omega) & \geq -x_2 \\
& -12y_1(\omega) & -15y_2(\omega) & -18y_3(\omega) & \geq -x_3 \\
& -8y_1(\omega) & -10y_2(\omega) & -14y_3(\omega) & \geq -x_4 \\
& -y_1(\omega) & & & \geq d_a(\omega) \\
& & -y_2(\omega) & & \geq d_b(\omega) \\
& & & -y_3(\omega) & \geq d_c(\omega) \\
& y_1(\omega), & y_2(\omega), & y_3(\omega) & \geq 0.
\end{array}$$

The values $d_a(\omega)$, $d_b(\omega)$, and $d_c(\omega)$ correspond to the negative of the demand for Product-a, Product-b, and Product-c, respectively. Let the dual multipliers associated with the constraints in the subproblem be denoted $\pi^\omega = (\pi_1^\omega, \pi_2^\omega, \cdots, \pi_7^\omega)^\top$. Then the dual to the second-stage subproblem can be written as follows:

$$\begin{array}{llllllll}
\text{Max} & -x_1\pi_1^\omega & -x_2\pi_2^\omega & -x_3\pi_3^\omega & -x_4\pi_4^\omega & +d_a(\omega)\pi_5^\omega & +d_b(\omega)\pi_6^\omega & +d_c(\omega)\pi_7^\omega \\
\text{s.t.} & -6\pi_1^\omega & -20\pi_2^\omega & -12\pi_3^\omega & -8\pi_4^\omega & -\pi_5^\omega & & \leq -1150 \\
& -8\pi_1^\omega & -25\pi_2^\omega & -15\pi_3^\omega & -10\pi_4^\omega & & -\pi_6^\omega & \leq -1525 \\
& -6\pi_1^\omega & -20\pi_2^\omega & -12\pi_3^\omega & -8\pi_4^\omega & & & -\pi_7^\omega \leq -1900 \\
& \pi_1^\omega, & \pi_2^\omega, & \pi_3^\omega, & \pi_4^\omega, & \pi_5^\omega, & \pi_6^\omega, & \pi_7^\omega \geq \quad 0.
\end{array}$$

From the dual problem formulation, we can see that the dual feasible space Π is given as

$$\begin{array}{lllllllll}
\Pi = \{ & -6\pi_1^\omega & -20\pi_2^\omega & -12\pi_3^\omega & -8\pi_4^\omega & -\pi_5^\omega & & & \leq -1150, \\
& -8\pi_1^\omega & -25\pi_2^\omega & -15\pi_3^\omega & -10\pi_4^\omega & & -\pi_6^\omega & & \leq -1525, \\
& -6\pi_1^\omega & -20\pi_2^\omega & -12\pi_3^\omega & -8\pi_4^\omega & & & -\pi_7^\omega & \leq -1900, \\
& \pi_1^\omega, & \pi_2^\omega, & \pi_3^\omega, & \pi_4^\omega, & \pi_5^\omega, & \pi_6^\omega, & \pi_7^\omega & \geq \quad 0\}.
\end{array}$$

Thus the dual problem takes the form

$$\underset{\pi\in\Pi}{\text{Max}} -x_1\pi_1^\omega - x_2\pi_2^\omega - x_3\pi_3^\omega - x_4\pi_4^\omega + d_a(\omega)\pi_5^\omega + d_b(\omega)\pi_6^\omega + d_c(\omega)\pi_7^\omega.$$

Observe that the dual feasible space Π is fixed for all $(x,\omega) \in X \times \Omega$ and the only change is in the objective function. We shall apply the SAA scheme and the SD algorithm to this instance later and will see that the SD method exploits this property of MR-SLP with fixed recourse matrix W and fixed object function coefficient vector q. Next, we review the basic statistics of how to generate a random sample on a computer.

8.3 Generating Random Samples

In this section we review how to generate random samples for MR-SLP. This is an important step in sampling algorithms for MR-SLP. Recall from statistics that we can generate a random sample using the uniform distribution, $\mathscr{U}(0, 1)$, by transforming the given distribution. If $\tilde{\omega}$ is a random variable with cumulative distribution function (CDF) $\mathbb{P}$, i.e., $\tilde{\omega} \sim \mathbb{P}$, then $\mathbb{P}(\tilde{\omega})$ can be transformed into $\mathscr{U}(0, 1)$, i.e., $\mathbb{P}(\tilde{\omega}) \sim \mathscr{U}(0, 1)$ as described next. Let $u \in [0, 1]$, then

$$\begin{aligned}
\mathbb{P}(\mathbb{P}(\tilde{\omega}) \leq u) &= \mathbb{P}(\mathbb{P}^{-1}(\mathbb{P}(\tilde{\omega})) \leq \mathbb{P}^{-1}(u)) \\
&= \mathbb{P}(\tilde{\omega} \leq \mathbb{P}^{-1}(u)) \\
&= \mathbb{P}(\mathbb{P}^{-1}(u)) \\
&= u.
\end{aligned}$$

This implies that we can use $\mathscr{U}(0, 1)$ to generate a random variable that is distributed differently.

The generation of a $\mathscr{U}(0, 1)$ by computer is a well-understood process and programming languages usually have standard libraries that include *pseudorandom number generators*. For example, C/C++ language standard libraries have functions *rand*() and *srand*(), with the latter function allow for specifying different "seeds" to change the sequence in which the numbers are generated. Therefore, we simply need to have a pseudorandom number generator for [0, 1] in order to generator independent and identically distributed (IID) samples. Recall that "identically distributed" means that all outcomes in the sample are taken from the same distribution, while "independent" means that the outcomes are all independent events. To accomplish this with, we use different *seeds* for the pseudorandom number generator.

Let $\tilde{\omega}$ be discretely distribution with sample space $\Omega = \{\omega^s\}_{s=1}^N$, where $N = |\Omega|$, and cumulative mass function $\{p_{\omega^s}\}_{\omega^s \in \Omega}$. Then given $u \in [0, 1]$, we need to compute $\mathbb{P}^{-1}(u) = \omega^{s(u)}$, where

$$s(u) = \min\{s \mid \sum_{i=1}^{s} p_{\omega^i} \geq u\}.$$

Next, we give some numerical examples to illustrate the foregoing discussion.

We illustrate generating samples with three different examples. The first one illustrates generating samples from a continuous distribution, while the second example uses a discrete distribution. The third example shows how to generate random samples based on the STOCH file for the MR-SLP instance.

Table 8.3 Random samples of $\tilde{\omega}$ using $\mathscr{U}(0, 1)$

s	u	$\omega^s := G^{-1}(u)$
1	0.19562	
2	0.20499	
3	0.38151	
4	0.46682	
5	0.53370	
6	0.65197	
7	0.77441	
8	0.82813	
9	0.88431	
10	0.95502	

8.3.1 Numerical Example 1: Continuous Distribution

Example 8.2 (Using the Uniform Distribution to Generate Random Samples from a Continuous Distribution) Consider the random variable $\tilde{\omega}$ with probability density function (PDF), $g(\omega) = 3\omega^2,\ 0 < \omega < 1$.

(a) Derive the CDF $G(\omega)$ and its inverse, $G^{-1}(u)$.
(b) Use the given $\mathscr{U}(0, 1)$ outcomes (column u) in Table 8.3 to generate 10 IID samples of $\tilde{\omega}$, ω^s, $s = 1, \cdots, 10$ for each of the u values listed in the Table.

Solution:

(a) Given PDF: $g(\omega) = 3\omega^2,\ 0 < \omega < 1$.

$$CDF : G(\omega) = \int_{v=0}^{\omega} 3v^2 dv = v^3|_{v=0}^{\omega}$$
$$= \omega^3.$$

$$\therefore G(\omega) = \omega^3.$$

Given $u = \mathscr{U}(0, 1)$,

$$G(\omega) = \omega^3 = u \Rightarrow \omega := G^{-1}(u) = u^{\frac{1}{3}}.$$

(b) Plugging the u values in $G^{-1}(u)$ in Table 8.3 we get the outcomes shown in Table 8.4.

Table 8.4 Random samples of $\tilde{\omega}$ using $\mathcal{U}(0, 1)$

s	u	$\omega^s := G^{-1}(u)$
1	0.19562	0.58050
2	0.20499	0.58963
3	0.38151	0.72528
4	0.46682	0.77574
5	0.53370	0.81114
6	0.65197	0.86712
7	0.77441	0.91831
8	0.82813	0.93908
9	0.88431	0.95985
10	0.95502	0.98477

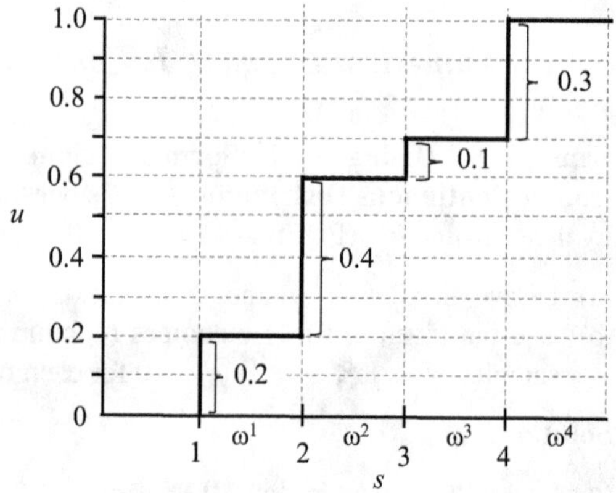

Fig. 8.2 The CDF for Example 8.3 problem

8.3.2 *Numerical Example 2: Discrete Distribution*

Example 8.3 (Using the Uniform Distribution to Generate Random Samples from Discrete Distribution) Consider discretely distributed random variable $\tilde{\omega}$ with sample space $\Omega = \{\omega^s\}_{s=1}^4$ and probabilities of occurrence are $\{p_{\omega^s}\}_{s=1}^4 = \{0.2, 0.4, 0.1, 0.3\}$. Determine and plot the CDF to use for randomly generating ω^s's using $\mathcal{U}(0, 1)$.

Solution:
For each of the outcomes $\{\omega^1, \omega^2, \omega^3, \omega^4\}$, the cumulative probabilities are $\{0.2, 0.2+0.4, 0.2+0.4+0.1, 0.2+0.4+0.1+0.3\} = \{0.2, 0.6, 0.7, 1.0\}$ with the following probability ranges for each: $[0, 0.2)$, $[0.2, 0.6)$, $[0.6, 0.7)$, and $[0.7, 1.0]$. A plot of the CDF is given in Fig. 8.2. Observe that generating a $u \in [0, 1]$ (vertical axis) using a pseudorandom number generator would determine an outcome ω^s (horizontal axis) whose occurrence follows the original distribution.

Table 8.5 STOCH file data for an instance of the *abc*-Production Planning problem in INDEPENDENT format

Column	Row	Value	Prob
RHS	ROW25	−15	0.15
RHS	ROW25	−20	0.3
RHS	ROW25	−25	0.3
RHS	ROW25	−30	0.2
RHS	ROW25	−10	0.05
RHS	ROW26	−10	0.15
RHS	ROW26	−15	0.3
RHS	ROW26	−20	0.3
RHS	ROW26	−25	0.2
RHS	ROW26	−10	0.05
RHS	ROW27	−5	0.15
RHS	ROW27	−15	0.3
RHS	ROW27	−25	0.3
RHS	ROW27	−30	0.2
RHS	ROW27	−10	0.05

8.3.3 Numerical Example 3: STOCH File

Example 8.4 (Using the Uniform Distribution to Generate Random Samples Based on the STOCH File) This example illustrates how to sample scenarios based on marginal distributions from a STOCH file, which is a standard input file for stochastic programs (see REF). Consider the STOCH file data given in Table 8.5 in *independent* format for the *abc*-Production Planning problem. The columns of the table correspond to the columns (decision variables), rows (constraints), value (outcome), and probability of occurrence of each outcome. *Independent* means that the random variables represented in the file are independent, and in this case, the RHS of the SLP is random.

(a) What are the random variables in this instance?
(b) How many outcomes does each random variable have? State each random variable and its outcomes and their probabilities of occurrence.
(c) What is the total number of scenarios in this instance? List the scenarios and their probabilities of occurrence.
(d) Determine the (marginal) cumulative distribution for each random variable to use for generating random samples (scenarios) using $\mathscr{U}(0, 1)$ and plot it.
(e) What will be the random sample when a pseudorandom number generator independently generates the following values, denoted u_1, u_2, u_3, respectively, for the first, second, and third random variables:

(i) $u_1 = 0.5666$, $u_2 = 0.7541$, and $u_3 = 0.0666$.
(ii) $u_1 = 0.2005$, $u_2 = 0.4415$, and $u_3 = 0.8047$.
(iii) $u_1 = 0.9909$, $u_2 = 0.7541$, and $u_3 = 0.6666$.

Solution:

(a) The random variables in this instance are the RHS constraints ROW25, ROW26, and ROW27. Thus, we have three random variables whose outcomes will be denoted $\omega_{i,j}$, where $i = 1, 2, 3$ is the index for i-th random variable and j is its j-th component.

(b) ROW25 has five outcomes

$$\{\omega_{1,1}, \omega_{1,2}, \omega_{1,3}, \omega_{1,4}, \omega_{1,5}\} = \{-15, -20, -25, -30, -10\},$$

with the following corresponding probabilities of occurrence:

$$\{0.15, 0.3, 0.3, 0.2, 0.05\}.$$

ROW26 has five outcomes

$$\{\omega_{2,1}, \omega_{2,2}, \omega_{2,3}, \omega_{2,4}, \omega_{2,5}\} = \{-10, -15, -20, -25, -10\},$$

with the following corresponding probabilities of occurrence:

$$\{0.15, 0.3, 0.3, 0.2, 0.05\}.$$

ROW27 has five outcomes

$$\{\omega_{3,1}, \omega_{3,2}, \omega_{3,3}, \omega_{3,4}, \omega_{3,5}\} = \{-5, -15, -25, -30, -10\},$$

with the following corresponding probabilities of occurrence:

$$\{0.15, 0.3, 0.3, 0.2, 0.05\}.$$

(c) The total number of scenarios is $5 \times 5 \times 5 = 125$. The scenarios for this SLP instance are partially listed in Table 8.6. Since the random variables are independent, the probabilities are calculated as a product, e.g., $p_{\omega^1} = 0.15 \times 0.15 \times 0.15 = 0.003375$. Also, verify that $\sum_{s=1}^{125} p_{\omega^s} = 1.0$ as expected.

(d) The marginal distributions of the three random variables are given in Table 8.7 and are plotted in Fig. 8.3.

(e) Using Table 8.7 or Fig. 8.3 to map the given u on the vertical axis to the outcome generates the following random samples:

(i) $u_1 = 0.1466 \mapsto \omega_{1,1} := -15$, $u_2 = 0.7541 \mapsto \omega_{2,4} := -25$, and $u_3 = 0.0667 \mapsto \omega_{3,1} := -5$. This corresponds to scenario ω^{16} in Table 8.6.

(ii) $u_1 = 0.2015 \mapsto \omega_{1,2} := -20$, $u_2 = 0.1415 \mapsto \omega_{2,1} := -10$, and $u_3 = 0.0045 \mapsto \omega_{3,1} = -5$. This corresponds to scenario ω^{26} in Table 8.6.

(iii) $u_1 = 0.0919 \mapsto \omega_{1,1} := -15$, $u_2 = 0.9641 \mapsto \omega_{2,5} := -10$, and $u_3 = 0.6655 \mapsto \omega_{3,3} := -25$. This corresponds to scenario ω^{23} in Table 8.6.

Table 8.6 A partial list of scenarios for the *abc*-Production Planning problem instance

Scenario ω^s	ROW25	ROW26	ROW27	Probability
ω^1	−15	−10	−5	0.003375
ω^2	−15	−10	−15	0.00675
ω^3	−15	−10	−25	0.00675
ω^4	−15	−10	−30	0.0045
ω^5	−15	−10	−10	0.001125
ω^6	−15	−15	−5	0.00675
ω^7	−15	−15	−15	0.0135
ω^8	−15	−15	−25	0.0135
ω^9	−15	−15	−30	0.009
ω^{10}	−15	−15	−10	0.00225
ω^{11}	−15	−20	−5	0.00675
ω^{12}	−15	−20	−15	0.0135
ω^{13}	−15	−20	−25	0.0135
ω^{14}	−15	−20	−30	0.009
ω^{15}	−15	−20	−10	0.00225
ω^{16}	−15	−25	−5	0.0045
ω^{17}	−15	−25	−15	0.009
ω^{18}	−15	−25	−25	0.009
ω^{19}	−15	−25	−30	0.006
ω^{20}	−15	−25	−10	0.0015
ω^{21}	−15	−10	−5	0.001125
ω^{22}	−15	−10	−15	0.00225
ω^{23}	−15	−10	−25	0.00225
ω^{24}	−15	−10	−30	0.0015
ω^{25}	−15	−10	−10	0.000375
ω^{26}	−20	−10	−5	0.00675
…	…	…	…	…
…	…	…	…	…
…	…	…	…	…
ω^{125}	−10	−10	−10	0.000125
				$\sum_{s=1}^{125} p_{\omega^s} = 1$

8.4 Exterior Sampling

Exterior sampling based methods involve taking a random sample and then solving approximation problem using a particular algorithm, e.g., the L-shaped method. This process is repeated several times (replications) and then statistical bounds on key solution quantities are calculated. There are several papers and books on exterior sampling for SLP. For example, Chapter 10 of [3] provides a study of Monte Carlo methods for stochastic programming while [9] gives a comprehensive survey of Monte Carlo sampling-based methods for stochastic optimization problems. In this

Table 8.7 Marginal distributions of the three random variables for the *abc*-Production Planning problem instance

	ROW25
Outcomes	$\{\omega_{1,1}, \omega_{1,2}, \omega_{1,3}, \omega_{1,4}, \omega_{1,5}\} = \{-15, -20, -25, -30, -10\}$
Cumulative probabilities	{0.15, 0.45,0.75,0.95 1.0}
Probability ranges	[0, 0.15), [0.15, 0.45), [0.45, 0.75], [0.75, 0.95], [0.95, 1.0]
	ROW26
Outcomes	$\{\omega_{2,1}, \omega_{2,2}, \omega_{2,3}, \omega_{2,4}, \omega_{2,5}\} = \{-10, -15, -20, -25, -10\}$
Cumulative probabilities	{0.15, 0.45,0.75,0.95 1.0}
Probability ranges	[0, 0.15), [0.15, 0.45), [0.45, 0.75], [0.75, 0.95], [0.95, 1.0]
	ROW27
Outcomes	$\{\omega_{3,1}, \omega_{3,2}, \omega_{3,3}, \omega_{3,4}, \omega_{3,5}\} = \{-5, -15, -25, -30, -10\}$
Cumulative probabilities	{0.15, 0.45,0.75,0.95 1.0}
Probability ranges	[0, 0.15), [0.15, 0.45), [0.45, 0.75], [0.75, 0.95], [0.95, 1.0]

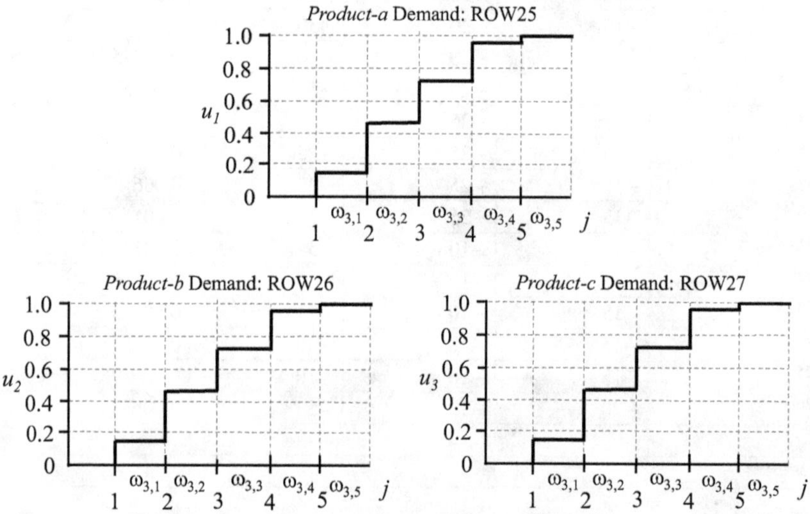

Fig. 8.3 Plot of the product demand marginal CDFs for the *abc*-Production Planning problem instance

section, we explore the SAA scheme based on the theory of the SAA in Chapter 6 of [14].

8.4.1 Sample Average Approximation

The SAA scheme was originally derived for Problem (8.1) with $\lambda = 0$, which is can be stated as

$$z := \underset{x \in X}{\text{Min}} \; F(x) := \mathbb{E}[f(x, \tilde{\omega})]. \tag{8.5}$$

Instead of solving Problem (8.5), the idea of SAA is to solve an approximating problem. Let $\{\omega^s\}_{s=1}^N$ be a random sample from $\tilde{\omega}$ using the procedure described in the previous subsection, where ω^s's are IID observations of $\tilde{\omega}$. Then $F(x)$ can be estimated using the following SAA problem:

$$\hat{z}_N := \underset{x \in X}{\text{Min}} \; \hat{F}_N(x) := \frac{1}{N}\sum_{s=1}^{N} f(x, \omega^s). \tag{8.6}$$

Observe that Problem (8.6) is an SLP with $\tilde{\omega}$ replaced by $\tilde{\omega}_N$ whose distribution is the empirical distribution of $\{\omega^s\}_{s=1}^N$. Let x_N^* be the optimal solution to Problem (8.6). Then x_N^* is not necessarily optimal to Problem (8.1) but is feasible for all $\{\omega^s\}_{s=1}^N$ due to assumption (A3), i.e., there exists $y(\omega) \in \mathbb{R}_+^{n_2}$ such that

$$Wy(\omega) \geq r(\omega) - T(\omega)x_N^*, \; y(\omega) \geq 0, \;\; \forall \omega \in \{\omega^s\}_{s=1}^N.$$

We also have that x_N^* is a random variable since $x_N^* \in \text{argmin}_{x \in X} \frac{1}{N}\sum_{s=1}^N f(x, \omega^s)$. Because the ω^s's are drawn from $\tilde{\omega}$ using $\mathbb{P}$, $\mathbb{E}[\hat{z}_N]$ is an *unbiased* estimator of the mean $\mathbb{E}[f(x, \tilde{\omega})]$. Next, we need to establish the lower bound and upper bound on the optimal value z.

Let $x^* \in X$ be the optimal solution to Problem (8.5) (original problem) and let x_N^* be the optimal solution to Problem (8.6) (SAA problem). Then for $x_N^* \in X$ we have

$$\begin{aligned} &\underset{x \in X}{\text{Min}} \; \mathbb{E}[f(x, \tilde{\omega})] \leq \mathbb{E}[f(x_N^*, \tilde{\omega})] \\ &\Rightarrow \mathbb{E}[f(x^*, \tilde{\omega})] \leq \mathbb{E}[f(x_N^*, \tilde{\omega})] \\ &\therefore \; z \leq \mathbb{E}[f(x_N^*, \tilde{\omega})] \end{aligned}$$

The second relation is true since x_N^* is simply a feasible solution to the original problem, and therefore $\mathbb{E}[f(x_N^*, \tilde{\omega})]$ is an upper bound on the optimal value z. But since x^* is a feasible solution to the SAA problem, the following is also true:

$$\frac{1}{N}\sum_{s=1}^{N} f(x_N^*, \omega^s) \leq \frac{1}{N}\sum_{s=1}^{N} f(x^*, \omega^s).$$

Taking expectations on both sides we have the following:

$$\mathbb{E}[\frac{1}{N}\sum_{s=1}^{N} f(x_N^*, \omega^s)] \leq \mathbb{E}[\frac{1}{N}\sum_{s=1}^{N} f(x^*, \omega^s]).$$

The left hand side of the above inequality becomes

$$\mathbb{E}[\frac{1}{N}\sum_{s=1}^{N} f(x_N^*, \omega^s)] = \mathbb{E}[\hat{z}_N].$$

Similarly, the RHS can be expressed as follows:

$$\mathbb{E}[\frac{1}{N}\sum_{s=1}^{N} f(x^*, \omega^s)] = \mathbb{E}[f(x^*, \tilde{\omega}])] = z.$$

Putting the two together it follows that

$$\mathbb{E}[\hat{z}_N] \leq z.$$

From the above results, we now have the following "sandwich" inequality:

$$\mathbb{E}[\frac{1}{N}\sum_{s=1}^{N} f(x_N^*, \omega^s)] \;\leq\; \mathbb{E}[f(x^*, \tilde{\omega})] \;\leq\; \mathbb{E}[f(x_N^*, \tilde{\omega})$$

$$\Rightarrow \mathbb{E}[\hat{z_N}] \leq z \leq \mathbb{E}[f(x_N^*, \tilde{\omega})].$$

This inequality implies that $\mathbb{E}[\hat{z_N}]$ is a lower bound and $\mathbb{E}[f(x_N^*, \tilde{\omega})]$ is an upper bound on the optimal value z. Assuming that the sample mean is an *unbiased* estimator of the population mean, then $\mathbb{E}[\hat{z}_N]$ is an *unbiased* estimator for z. However, $\hat{z}_N$ is a *biased* estimator for z.

Let us now extend the above results to the risk-averse case assuming that assumption (A4) holds, i.e., the risk measure $\mathbb{D}$ is convexity preserving for all $\lambda > 0$ and that $f(x, \tilde{\omega})$ is convex and continuous. We defined different risk measures in Chap. 2 and used some of them in the decomposition algorithms in Chap. 7. Here we consider three risk measures, quantile deviation (QDEV), conditional value-at-risk (CVaR), and expected excess (EE). For continuity of the ongoing discourse, let us restate the definitions of these risk measures. QDEV is a two-sided risk measure and reflects the expectation of the deviation above and below the α-quantile of the cumulative distribution of $f(x, \tilde{\omega})$. Given $x \in X$ and $\alpha \in (0, 1)$, QDEV is mathematically given as follows:

$$\phi_{QDEV_\alpha}(x) := \underset{\eta\in\mathbb{R}}{\text{Min}}\; \mathbb{E}[\varepsilon_1(\eta - f(x, \tilde{\omega}))_+ + \varepsilon_2(f(x, \tilde{\omega}) - \eta)_+],$$

where $\varepsilon_1 > 0$ and $\varepsilon_2 > 0$ are such that $\alpha = \varepsilon_2/(\varepsilon_1 + \varepsilon_2)$. Problem (8.1) with $\mathbb{D} := \phi_{QDEV_\alpha}$ and $\lambda \geq 0$ can be written as follows:

$$z := \underset{x\in X}{\text{Min}}\; F(x) := \mathbb{E}[f(x, \tilde{\omega})] + \lambda\phi_{QDEV_\alpha}(x). \tag{8.7}$$

Given $\lambda \in [0, 1/\varepsilon_1]$ under assumption (A1), the DEP for Problem (8.7) is given as follows:

$$\begin{aligned}
\text{Min} \quad & (1-\lambda\varepsilon_1)c^\top x + \lambda\varepsilon_1\eta + (1-\lambda\varepsilon_1)\sum_{\omega\in\Omega} p(\omega)q(\omega)^\top y(\omega) \\
& + \lambda(\varepsilon_1+\varepsilon_2)\sum_{\omega\in\Omega} p(\omega)v(\omega) \\
\text{s.t.} \quad & T(\omega)x + Wy(\omega) \geq r(\omega), \ \forall\omega\in\Omega \\
& -c^\top x - q(\omega)^\top y(\omega) + \eta + v(\omega) \geq 0, \ \forall\omega\in\Omega \\
& x\in X, \ \eta\in\mathbb{R}, \ y(\omega)\in\mathbb{R}_+^{n_2}, \ v(\omega)\in\mathbb{R}_+, \forall\omega\in\Omega.
\end{aligned}$$

Given a sample $\{\omega^s\}_{s=1}^N$, we need to solve the following SAA problem for QDEV instead of solving Problem (8.7):

$$\hat{z}_N := \underset{x\in X}{\text{Min}} \ \hat{F}_N(x) := \frac{1}{N}\sum_{s=1}^N f(x,\omega^s) + \lambda\hat{\phi}_{QDEV_\alpha}(x), \tag{8.8}$$

where $\hat{\phi}_{QDEV_\alpha}(x)$ is given as

$$\hat{\phi}_{QDEV_\alpha}(x) := \underset{\eta\in\mathbb{R}}{\text{Min}} \ \frac{1}{N}\sum_{s=1}^N \big(\varepsilon_1(\eta - f(x,\omega^s))_+ + \varepsilon_2(f(x,\omega^s)-\eta)_+\big).$$

Given an optimal solution $x^* \in X$ to Problem (8.7) and an optimal solution $x_N^* \in X$ to SAA Problem (8.8), we have the following "sandwich" inequality for QDEV:

$$\begin{aligned}
\mathbb{E}[\frac{1}{N}\sum_{s=1}^N f(x_N^*,\omega^s) + \lambda\hat{\phi}_{QDEV_\alpha}(x_N^*)] &\leq \mathbb{E}[f(x^*,\tilde{\omega})] + \lambda\phi_{QDEV_\alpha}(x^*) \\
&\leq \mathbb{E}[f(x_N^*,\tilde{\omega})] + \lambda\phi_{QDEV_\alpha}(x_N^*) \\
\Rightarrow \ \mathbb{E}[\hat{z_N}] \leq z &\leq \mathbb{E}[f(x_N^*,\tilde{\omega})] + \lambda\phi_{QDEV_\alpha}(x_N^*).
\end{aligned}$$

This inequality implies that $\mathbb{E}[\hat{z_N}]$ is a lower bound on z, while $\mathbb{E}[f(x_N^*,\tilde{\omega})] + \lambda\phi_{QDEV_\alpha}(x_N^*)$ is an upper bound.

CVaR reflects the expectation of the $(1-\alpha).100\%$ worst outcomes for a given probability level $\alpha \in (0, 1)$. Given $x \in X$ and $\alpha \in (0, 1)$, CVaR can be expressed as follows:

$$\phi_{CVaR_\alpha}(x) := \underset{\eta\in\mathbb{R}}{\text{Min}} \ \{\eta + \frac{1}{1-\alpha}\mathbb{E}[(f(x,\tilde{\omega})-\eta)_+]\}.$$

CVaR can also be expressed in terms of $\phi_{QDEV_\alpha}(x)$ as follows:

$$\phi_{CVaR_\alpha}(x) := \mathbb{E}[f(x, \tilde{\omega})] + \frac{1}{\varepsilon_1}\phi_{QDEV_\alpha}(x).$$

For any $\lambda \geq 0$, an MR-SLP with $\mathbb{D} := \phi_{CVaR_\alpha}$ can be written as follows:

$$z := \underset{x \in X}{\text{Min}}\ F(x) := \mathbb{E}[f(x, \tilde{\omega})] + \lambda \phi_{CVaR_\alpha}(x). \tag{8.9}$$

An alternative formulation, for $0 \leq \lambda \leq 1$, is coherent and is given as

$$z := \underset{x \in X}{\text{Min}}\ F(x) := (1-\lambda)\mathbb{E}[f(x, \tilde{\omega})] + \lambda \phi_{CVaR_\alpha}(x). \tag{8.10}$$

The DEP for this model can be given as follows:

$$\begin{aligned}
\text{Min} \quad & (1-\lambda)c^\top x + (1-\lambda)\sum_{\omega \in \Omega} p(\omega) q(\omega)^\top y(\omega) \ + \lambda \eta + \frac{\lambda}{1-\alpha}\sum_{\omega \in \Omega} p(\omega) v(\omega) \\
\text{s.t.} \quad & T(\omega)x + W y(\omega) \geq r(\omega),\ \forall \omega \in \Omega \\
& -c^\top x - q(\omega)^\top y(\omega) + \eta + v(\omega) \geq 0,\ \forall \omega \in \Omega \\
& x \in X,\ \eta \in \mathbb{R},\ y(\omega) \in \mathbb{R}_+^{n_2},\ v(\omega) \in \mathbb{R}_+, \forall \omega \in \Omega.
\end{aligned}$$

Given a sample $\{\omega^s\}_{s=1}^N$, we need to solve the following SAA problem for CVaR instead of Problem (8.10):

$$\hat{z}_N := \underset{x \in X}{\text{Min}}\ \hat{F}_N(x) := (1-\lambda)\frac{1}{N}\sum_{s=1}^N f(x, \omega^s) + \lambda \hat{\phi}_{CVaR_\alpha}(x), \tag{8.11}$$

where $\hat{\phi}_{CVaR_\alpha}(x)$ is given as

$$\hat{\phi}_{CVaR_\alpha}(x) := \underset{\eta \in \mathbb{R}}{\text{Min}}\ \{\eta + \frac{1}{(1-\alpha)N}\sum_{s=1}^N (f(x, \omega^s) - \eta)_+\}.$$

An SAA problem for formulation (8.9) can also be written in a similar manner. Given an optimal solution $x^* \in X$ to Problem (8.10) and an optimal solution $x_N^* \in X$ to SAA Problem (8.11), we have the following "sandwich" inequality for CVaR:

$$\begin{aligned}
\mathbb{E}[(1-\lambda)\frac{1}{N}\sum_{s=1}^N f(x_N^*, \omega^s) + \lambda \hat{\phi}_{CVaR_\alpha}(x_N^*)] &\leq (1-\lambda)\mathbb{E}[f(x^*, \tilde{\omega})] + \lambda \phi_{CVaR_\alpha}(x^*) \\
&\leq (1-\lambda)\mathbb{E}[f(x_N^*, \tilde{\omega})] + \lambda \phi_{CVaR_\alpha}(x_N^*) \\
\Rightarrow \mathbb{E}[\hat{z_N}] \leq z &\leq (1-\lambda)\mathbb{E}[f(x_N^*, \tilde{\omega})] + \lambda \phi_{CVaR_\alpha}(x_N^*).
\end{aligned}$$

This inequality implies that $\mathbb{E}[\hat{z_N}]$ is a lower bound on z, while $(1-\lambda)\mathbb{E}[f(x_N^*, \tilde{\omega})]+\lambda\phi_{CVaR_\alpha}(x_N^*)$ is an upper bound.

EE is the expected value of the excess over a given target η. Given $x \in X$, $\eta \in \mathbb{R}$ and $\lambda \geq 0$, EE is defined as

$$\phi_{EE_\eta}(x) := \mathbb{E}[f(x, \tilde{\omega}) - \eta]_+.$$

An MR-SLP with $\mathbb{D} := \phi_{EE_\eta}$ is given as

$$z := \underset{x \in X}{\text{Min}} \ F(x) := \mathbb{E}[f(x, \tilde{\omega})] + \lambda\phi_{EE_\eta}(x), \tag{8.12}$$

while the DEP to this problem can be given as follows:

$$\begin{aligned}
\text{Min} \quad & c^\top x + \sum_{\omega \in \Omega} p(\omega) q(\omega)^\top y(\omega) + \lambda \sum_{\omega \in \Omega} p(\omega) v(\omega) \\
\text{s.t.} \quad & T(\omega)x + Wy(\omega) \geq r(\omega), \ \forall \omega \in \Omega \\
& -c^\top x - q(\omega)^\top y(\omega) + v(\omega) \geq -\eta, \ \forall \omega \in \Omega \\
& x \in X, \ \ y(\omega) \in \mathbb{R}_+^{n_2}, \ v(\omega) \in \mathbb{R}_+, \ \forall \omega \in \Omega.
\end{aligned}$$

For a sample $\{\omega^s\}_{s=1}^N$, the SAA problem for EE formulation (8.12) we need to solve is

$$\hat{z}_N := \underset{x \in X}{\text{Min}} \ \hat{F}_N(x) := \frac{1}{N} \sum_{s=1}^{N} f(x, \omega^s) + \lambda \hat{\phi}_{EE_\eta}(x), \tag{8.13}$$

where $\hat{\phi}_{EE_\eta}(x)$ is given as

$$\hat{\phi}_{EE_\eta}(x) := \frac{1}{N} \sum_{s=1}^{N} \left(f(x, \omega^s) - \eta\right)_+.$$

Given an optimal solution $x^* \in X$ to Problem (8.12) and an optimal solution $x_N^* \in X$ to SAA Problem (8.13), we have the following "sandwich" inequality for EE:

$$\begin{aligned}
\mathbb{E}[\frac{1}{N} \sum_{s=1}^{N} f(x_N^*, \omega^s) + \lambda \hat{\phi}_{EE_\eta}(x_N^*)] \ & \leq \ \mathbb{E}[f(x^*, \tilde{\omega})] + \lambda \hat{\phi}_{EE_\eta}(x^*) \\
& \leq \ \mathbb{E}[f(x_N^*, \tilde{\omega})] + \lambda \phi_{EE_\eta}(x_N^*). \\
& \Rightarrow \mathbb{E}[\hat{z_N}] \leq z \leq \mathbb{E}[f(x_N^*, \tilde{\omega})] + \lambda \phi_{EE_\eta}(x_N^*).
\end{aligned}$$

This inequality implies that $\mathbb{E}[\hat{z_N}]$ is a lower bound on z, while $\mathbb{E}[f(x_N^*, \tilde{\omega})] + \lambda\phi_{EE_\eta}(x_N^*)$ is an upper bound.

8.4.2 The Sample Average Approximation Scheme

We are now in a position to state the steps of the SAA scheme to compute the lower and upper bound, respectively, on the optimal value to the original problem. We start with the SAA scheme for calculating an estimate for the lower bound $\mathbb{E}[\hat{z}_N]$.

Scheme *SAA–LowerBound*
begin

Step 0. **Initialization.** Choose sample (batch) size N and number of replications M. Set $i \leftarrow 1$ and lower bound $L_{MN} \leftarrow 0$.
Step 1. **Sample Generation.** Generate IID sample (batch) of size N: $\{\omega^{s,i}\}_{s=1}^N$.
Step 2. **Solve SAA Problem.** Solve the SAA problem

$$\hat{z}_N^i := \operatorname*{Min}_{x \in X} \frac{1}{N} \sum_{s=1}^{N} f(x, \omega^{s,i})$$

to get optimal solution $\hat{x}_N^{i*}$.
Step 3. **Compute Lower Bound.** Calculate

$$L_{MN} \leftarrow L_{MN} + \frac{1}{M} \hat{z}_N^i.$$

Step 4. **Termination.** If $i == M$, stop and report $\hat{z}_N^i, \hat{x}_N^{i*}$, for $i = 1, \cdots, M$, and lower bound L_{MN}. Otherwise, set $i \leftarrow i + 1$ and return to step 1.

---

end

In step 2 of SAA–LowerBound, depending on the size of N and instance difficulty, one can use a direct solver or a decomposition algorithm of choice (e.g., L-shaped algorithm). Step 3 of the scheme computes the lower bound $L_{MN} := \frac{1}{M} \sum_{i=1}^{M} \hat{z}_N^i$.

Next, we state the SAA scheme for estimating the upper bound on the optimal value. Given a feasible solution $\hat{x}$, $\mathbb{E}[f(x^*, \tilde{\omega})] \le \mathbb{E}[f(\hat{x}, \tilde{\omega})]$. The RHS is clearly an upper bound and we can estimate as follows:

Scheme *SAA–UpperBound*
begin

Step 0. **Initialization.** Set feasible solution $\hat{x} \leftarrow \hat{x}_N^{i*}$, for some $i \in \{1, \cdots, M\}$. Choose sample (batch) size $\bar{N} > N$ and number of replications $\bar{M}$. Set $j \leftarrow 1$ and upper bound $U_{\bar{N}\bar{M}} \leftarrow 0$.

Step 1. **Sample Generation.** Generate IID sample (batch) of size $\bar{N}$: $\{\omega^{s,j}\}_{s=1}^{\bar{N}}$.
Step 2. **Solve SAA Problem.** Solve the SAA problem

$$\hat{z}^j_{\bar{N}} = \underset{x \in X}{\text{Min}} \ \frac{1}{\bar{N}} \sum_{s=1}^{N} f(x, \omega^{s,j}).$$

Step 3. **Compute Lower Bound.** Calculate

$$U_{\bar{M}\bar{N}} \leftarrow L_{\bar{M}\bar{N}} + \frac{1}{\bar{M}} \hat{z}^i_{\bar{N}}.$$

Step 4. **Termination.** If $j == \bar{M}$, stop and report upper bound $U_{\bar{M}\bar{N}}$. Otherwise, set $j \leftarrow j + 1$ and return to step 1.

---

end

Confidence Interval on Lower Bound
Given $L_{MN} = \frac{1}{M} \sum_{t=1}^{M} \hat{z}^i_N$:

- ◇ The estimate L_{MN} is an unbiased estimate of $\mathbb{E}[\hat{z}_N]$.
- ◇ Provide a statistical *lowerbound* for z (*true* optimal value).
- ◇ When the M batches (of sample size N) are IID, by the central limit theorem (CLT) it follows that: $\sqrt{M}\left(L_{MN} - \mathbb{E}[\hat{z}_N]\right) \rightarrow \mathcal{N}(0, \sigma_L^2)$ *as* $M \rightarrow \infty$, where $\mathcal{N}$ denotes the normal distribution.
- ◇ The *sample variance* estimator of σ_L^2 is

$$s_L^2(M) = \frac{1}{M-1} \sum_{i=1}^{M} \left(\hat{z}^i_N - L_{MN}\right)^2.$$

- ◇ Define z_α to satisfy $\mathbb{P}(\mathcal{N}(0, 1) \leq z_\alpha) = 1 - \alpha$.
- ◇ Replace σ_L by $s_L(M)$.
- ◇ We can then obtain an approximate $(1 - \alpha)$-confidence interval (CI) for $\mathbb{E}[\hat{z}_N]$ as follows:

$$\left[L_{MN} - z_{\frac{\alpha}{2}} \frac{s_L(M)}{\sqrt{M}},\ L_{MN} + z_{\frac{\alpha}{2}} \frac{s_L(M)}{\sqrt{M}}\right].$$

Confidence Interval on UB
Given $U_{\bar{M}\bar{N}}(\hat{x}) = \frac{1}{\bar{M}} \sum_{j=1}^{\bar{M}} \hat{z}^j_{\bar{N}}(\hat{x})$:

- ◇ The estimate $U_{\bar{M}\bar{N}}(\hat{x})$ is an unbiased estimate of $\mathbb{E}[f(\hat{x}, \tilde{\omega})]$.
- ◇ Provide a statistical UB for z (*true* optimal value).
- ◇ When the $\bar{M}$ batches (of sample size $\bar{N}$) are IID we have by the CLT that

$$\sqrt{\bar{M}}\left(U_{\bar{M}\bar{N}} - \mathbb{E}[\hat{z}_{\bar{N}}(\hat{x})]\right) \rightarrow \mathcal{N}(0, \sigma_U^2(\hat{x}))\ as\ \bar{M} \rightarrow \infty.$$

◇ The *sample variance* estimator of $\sigma^2_{\bar{U}}(\hat{x})$ is

$$s^2_{\bar{U}}(\hat{x}, \bar{M}) = \frac{1}{\bar{M}-1}\sum_{j=1}^{\bar{M}} \left(\hat{z}^j_{\bar{N}} - U_{\bar{M}\bar{N}}(\hat{x})\right)^2.$$

◇ Define z_α to satisfy $\mathbb{P}(\mathcal{N}(0,1) \leq z_\alpha) = 1 - \alpha$.
◇ Replace σ_U by $s_U(\hat{x}, \bar{M})$.
◇ Obtain an approximate $(1-\alpha)$-CI for $\mathbb{E}[\hat{z}_{\bar{N}}(\hat{x})]$ as follows:

$$\left[U_{\bar{M}\bar{N}}(\hat{x}) - z_{\frac{\alpha}{2}}\frac{s_U(\hat{x}, \bar{M})}{\sqrt{\bar{M}}},\ U_{\bar{M}\bar{N}}(\hat{x}) + z_{\frac{\alpha}{2}}\frac{s_U(\hat{x}, \bar{M})}{\sqrt{\bar{M}}}\right].$$

Next, we give example of detailed numerical results for the SAA scheme.

8.4.3 Numerical Examples

We are now ready to illustrate applying the SAA scheme to using two numerical examples. The first example is based on Example 8.5 from the previous section, while the second example involves the *abc-Production Planning* problem from Sect. 8.2.

Example 8.5 (SAA Based on Random Samples from a Continuous Distribution) Consider the following two-stage MR-SLP:

$$z^* = \underset{x \geq 4}{\text{Min}}\ F(x) := \mathbb{E}[f(x, \tilde{\omega})],$$

where for outcome ω of $\tilde{\omega}$, $f(x, \omega) := x + \varphi(x, \omega)$ and

$$\varphi(x, \omega) = \underset{y \geq 0}{\text{Min}}\ \{y \mid y \geq x,\ y \geq 5\omega\}.$$

Supposed that the random variable $\tilde{\omega}$ has PDF $g(\omega) = 3\omega^2,\ 0 < \omega < 1$ as given in Example 8.2. Use the outcomes, $\omega^s, s = 1, \cdots, 10$, you listed in Table 8.3 to answer the following questions.

(a) Create an SAA problem for the 10 samples ($N = 10$) in Table 8.3. State your SAA problem formulation explicitly.
(b) Solve your SAA problem in part (a) and state its optimal value $\hat{z}^*_N$ and solution x^*_N.

Solution:

(a) We can state the SAA problem as follows:

$$\hat{z}^*_N = \underset{x \geq 4}{\text{Min}}\{\hat{F}_N(x)\},$$

where

$$\hat{F}_N(x) = x + \frac{1}{10}\sum_{s=1}^{10}\varphi(x,\omega^s)$$

$$\varphi(x,\omega^s) = \text{Min}\,\{y^s \mid y^s \ge x,\, y^s \ge 5\omega^s,\, y^s \ge 0\}.$$

Alternatively, we can write the DEP formulation as follows:

$$\begin{aligned}
\hat{z}_N^* = \text{Min}\; & x + \frac{1}{10}\sum_{s=1}^{10} y^s \\
\text{s.t.}\quad & x \ge 0 \\
& -x + y^s \ge 0,\; s = 1, \cdots, 10 \\
& y^s \ge 5\omega^s,\; s = 1, \cdots, 10 \\
& x,\; y^s \ge 0,\; s = 1, \cdots, 10.
\end{aligned}$$

(b) Substituting the ω^s values in Table 8.4 and solving the instance we get the sample optimal value $z_N^* = 8.3401$ and sample solution $x_N^* = 4$. The interested reader can perform several replications M to compute a CI on the LB and perform replications $\bar{M}$ to compute a CI on the UB. We leave this as an exercise at the end of this chapter.

Example 8.6 (Applying SAA to the *abc*-Production Planning Problem Instance) Apply the SAA scheme based on the L-shaped algorithm to the *abc*-Production Planning problem given in Example 8.1 and compute the $1-\alpha$-confidence interval (CI) on L_{MN} and $L_{\bar{M}\bar{N}}$, respectively. Use $\alpha = 0.05$, i.e., 95% CI. Set the number of samples N for each run of the L-shaped algorithm and the number of replications M as well as $\bar{N}$ and $\bar{M}$ as follows:

(a) $N = 25$ and $M = 10$ to compute the lower bound L_{MN}, and $\bar{N} = 50$ and $\bar{M} = 20$ to compute the upper bound $L_{\bar{M}\bar{N}}$.
(b) $N = 50$ and $M = 10$ to compute the lower bound L_{MN}, and $\bar{N} = 75$ and $\bar{M} = 20$ to compute the upper bound $L_{\bar{M}\bar{N}}$.

Solution:
The results for case (a) are shown in Table 8.8. The columns of the table are as follows: Replication number (Rep i), first-stage solution ($\hat{x}_N^{i*}$), objective value ($\hat{z}_N^i$), sample variance of the upper bound ($s_U^2(\hat{x}_N^{i*}, \bar{M})$), upper bound value ($U_{\bar{M}\bar{N}}(\hat{x}_N^{i*})$), and the 95% CI on the upper bound (UB CI). At the bottom of the table we provide the lower bound L_{MN}, sample variance of the lower bound ($s_L^2(M)$), and the 95% CI on the lower bound (LB CI).

Table 8.8 SAA results for the *abc*-Production instance: $N = 25$, $M = 10$, $\bar{N} = 50$, $\bar{M} = 20$, and $\alpha = 0.05$

Rep i	$(\hat{x}_N^{i*})^\top$	$\hat{z}_N^i$	$s_U^2(\hat{x}_N^{i*}, \bar{M})$	$U_{\bar{M}\bar{N}}(\hat{x}_N^{i*})$	UB CI
1	(204.40, 590.00, 372.00, 278.00)	−2134.19	12,460.80	−2034.70	[−2183.12, −2085.27]
2	(246.43, 690.00, 443.57, 345.00)	−2520.64	46,468.40	−2358.80	[−2453.28, −2264.32]
3	(236.40, 690.00, 432.00, 318.00)	−2250.10	12,134.00	−2328.73	[−2377.01, −2280.46]
4	(246.68, 690.00, 432.00, 319.33)	−2334.67	15,288.50	−2282.81	[−2337.00, −2228.62]
5	(246.43, 690.00, 443.57, 345.00)	−2674.96	52,991.30	−2291.83	[−2392.72, −2190.93]
6	(237.14, 690.00, 432.86, 320.00)	−2095.63	26,613.60	−2270.50	[−2342.00, −2199.00]
7	(236.40, 690.00, 432.00, 318.80)	−1871.70	16,056.30	−2289.38	[−2344.92, −2233.84]
8	(236.67, 690.00, 432.00, 319.33)	−2498.30	18,567.90	−2325.43	[−2385.16, −2265.71]
9	(236.40, 690.00, 432.00, 318.00)	−2261.20	16,425.20	−2240.71	[−2296.88, −2184.53]
10	(246.43, 690.00, 443.57, 345.00)	−2363.44	67,998.40	−2307.54	[−2421.83, −2193.25]
L_{MN}	−2290.53				
$s_L^2(M)$	59,168.60				
LB CI	[−2441.31, −2139.76]				

Using the L-shaped algorithm, we obtain the optimal solution to the instance to be $x^* = (240.714, 700.000, 439.286, 325.000)^\top$ with an optimal value of $-2,374.04$. Note that the SAA scheme is able to compute the LB CI $[-2,441.31, -2,139.76]$ which includes the optimal value as expected. Similarly, the UB CIs for each replication provides an interval with valid upper bounds on the optimal value.

The results for case (b) are shown in Table 8.9. Note that in this case $N = 50$ and $\bar{N} = 75$ are increased and we again see that the SAA scheme generates the LB CI $[-2,550.33, -2,279.83]$ which includes the optimal value $-2,374.04$. In this case, the CI is tighter than that generated in case (a) and also, the sample variance on the LB is smaller. This is expected since the number of samples N in increased from 25 to 50. Furthermore, increasing the number of replications M could also reduce the sample variance. The UB CIs for each replication provides valid upper bounds on the optimal value.

Two important topics in Monte Carlo sampling based methods for stochastic optimization that have not been covered in this chapter are *solution quality* and

Table 8.9 SAA results for the *abc*-Production instance: $N = 50$, $M = 10$, $\bar{N} = 75$, $\bar{M} = 20$, and $\alpha = 0.05$

Rep i	$(\hat{x}_N^{i*})^\top$	$\hat{z}_N^i$	$s_U^2(\hat{x}_N^{i*}, \bar{M})$	$U_{\bar{M}\bar{N}}(\hat{x}_N^{i*})$	UB CI
1	(237.14, 690.00, 432.86, 320.00)	−2188.02	17,369.80	−2327.32	[−2385.08, −2269.56]
2	(236.40, 690.00, 432.00, 318.00)	−2209.15	8297.27	−2356.89	[−2396.82, −2316.97]
3	(246.43, 690.00, 443.57, 345.00)	−2321.84	28,433.90	−2259.61	[−2333.52, −2185.71]
4	(250.00, 700.00, 450.00, 350.00)	−2553.20	51,170.00	−2236.87	[−2336.01, −2137.72]
5	(250.00, 700.00, 450.00, 350.00)	−2522.40	42,336.00	−2328.26	[−2418.44, −2238.08]
6	(246.43, 690.00, 443.57, 345.00)	−2481.14	28,304.60	−2196.73	[−2270.47, −2122.99]
7	(240.71, 700.00, 439.29, 325.00)	−2315.01	18,612.70	−2345.65	[−2405.45, −2285.86]
8	(250.00, 700.00, 450.00, 350.00)	−2926.40	71,768.50	−2213.72	[−2331.14, −2096.30]
9	(240.71, 700.00, 439.29, 325.00)	−2325.29	9031.69	−2329.63	[−2371.28, −2287.97]
10	(237.14, 690.00, 432.86, 320.00)	−2308.33	13,196.10	−2342.29	[−2392.63, −2291.94]
L_{MN}	−2415.08				
$s_L^2(M)$	47,615.00				
LB CI	[−2550.33, −2279.83]				

variance reduction. We refer the interested reader to [9, 12] for variance reduction techniques for sequential sampling and to [1, 2] for sequential sampling and solution quality assessment in stochastic programs.

8.5 Interior Sampling

In contrast to *exterior* sampling methods like the SAA that take a set of samples and then solve an approximation problem, *interior* sampling involves sampling during the course of the algorithm. This requires a streamlined design of the algorithm to embed sequential sampling into it. The number of samples needed to solve the approximation problem is not known a priori, and typically, algorithm termination is based on some statistical stopping criterion that has to be met. Examples of interior sampling methods from the literature include the L-shaped with sampling method [10], stochastic quasi-gradient methods [11], successive sample mean optimization

(SSMO) [16], and stochastic decomposition (SD) [7, 8]. In this chapter, we review the latter method and provide illustrative results based on Example 8.1 in Sect. 8.2.

8.5.1 Stochastic Decomposition

The SD method was originally derived for Problem (8.1) with $\lambda := 0$, which we state as follows:

$$z := \operatorname*{Min}_{x \in X} \ F(x) := \mathbb{E}[f(x, \tilde{\omega})], \tag{8.14}$$

where for a given $x \in X$ the real random cost variable $f(x, \tilde{\omega}) := c^\top x + \varphi(x, \tilde{\omega})$ and given scenario $\omega \in \Omega$ of $\tilde{\omega}$, the *recourse function* $\varphi(x, \omega)$ is given by

$$\begin{aligned} \varphi(x, \omega) :=& \text{Min } q^\top y(\omega) \\ &\text{s.t. } Wy(\omega) \geq r(\omega) - T(\omega)x \\ &\quad y(\omega) \geq 0, \end{aligned} \tag{8.15}$$

where $q \in \mathbb{R}^{n_2}$ is the second-stage cost vector and is fixed for all scenarios $\omega \in \Omega$. A *scenario* ω gives the realization of the stochastic problem data, i.e., $\omega := (T(\omega), r(\omega))$. In addition to assumptions (A1)–(A4), to ensure that problem (8.14–8.15) is well-defined for the SD method, the following assumption is made:

(A5) A lower bound $\mathscr{L} \leq \varphi(x, \omega) \ \ \forall (x, \omega) \in X \times \Omega$ is known.

The lower bound $\mathscr{L}$ is needed to ensure that the probability that an optimal point is cut off by an optimality cut is asymptotically zero. In general, Problem (8.14) can be formulated such that $\mathscr{L} = 0$ provides a lower bound on φ. So we shall assume that $\mathscr{L} = 0$ and therefore $\varphi(x, \omega) \geq 0$.

Instead of solving Problem (8.14–8.15), the basic idea of SD is to solve an approximating problem in which F is recursively approximated using a piecewise linear function F_k. This is accomplished by sequentially generating one new sample ω^k at each iteration k within an optimization algorithm. The basic SD algorithm uses Benders decomposition to determine a piecewise linear approximation F_k at each iteration. Therefore, each iteration k of the algorithm a master program of the following form is solved to get the point x^{k+1}:

$$\operatorname*{Min}_{x \in X} \ F_k(x) \tag{8.16}$$

Given $x^k \in X$, the subproblem to solve at iteration k for sample ω^k takes the following form:

$$\varphi(x^k, \omega^k) := \text{Min } q^\top y$$

$$\text{s.t. } Wy \geq r(\omega^k) - T(\omega^k)x^k \tag{8.17}$$
$$y \geq 0.$$

Notice that even though the recourse decision $y(\omega)$ depends on ω, we conveniently write it as y as shown since the problem data $(r(\omega^k), T(\omega^k))$ is what changes at each iteration k for each sample ω^k.

Now let the dual multipliers associated with constraints (8.17) be denoted by π^k and define set $\Pi = \{W^\top \pi \leq q,\ \pi \geq 0\}$. Then the dual to the subproblem takes the following form:

$$\varphi(x^k, \omega^k) := \operatorname*{Max}_{\pi \in \Pi}\ \pi^\top \left(r(\omega^k) - T(\omega^k)x^k \right). \tag{8.18}$$

Observe that the dual solution set Π remains fixed for all ω^k due to both the recourse matrix W and the second-stage cost vector q being independent of ω. Let ω^k denote the optimal dual solution to subproblem (8.18) at iteration k. Then π^k together with previously generated dual solutions can be used to create an approximation F_k of the function F. Since π^k depends on (x^k, ω^k), in stating the SD algorithm we shall write π^k as $\pi^k(x^k, \omega^k)$ when it is necessary to show what (x^k, ω^k) led to that π^k. We shall use the set V_k to store the dual solutions seen up to iteration k and the set $\mathcal{T}_k$ to store the iteration indices. We should point out that the current iteration k of the SD algorithm is also the number of observations of $\tilde{\omega}$ that are currently available.

The SD algorithm exploits the fixed dual problem feasible set Π by solving only one subproblem per iteration and reusing previously generated dual solutions to create the approximation F_k of the function F. Computationally, doing this can save a significant amount of solution time, especially for large-scale instances. Next, we give a formal statement of the SD algorithm.

8.5.2 *Stochastic Decomposition Algorithm*

In stating the SD algorithm, we shall denote by t the algorithm iteration at which a cutting-plane was generated, while the current iteration will be denoted by k. We use $k + 1$ to denote the next iteration. A basic SD algorithm for Problem (8.14) can be stated as follows:

Algorithm *Basic SD*
begin

Step 0. **Initialization.** Set $k \leftarrow 0$, $V_0 \leftarrow \emptyset$, $\mathcal{T}_0 \leftarrow \emptyset$, $F_0(x) = -\infty$, and choose $x^1 \in X$. Compute lower bound $\mathcal{L}$.

Step 1. **Generate New Sample.** Set $k \leftarrow k + 1$. Randomly generate a sample of $\tilde{\omega}$, ω^k, independent of any previously generated samples.

Step 2 **Update Approximation F_k and V_k, $\mathcal{T}_k$.** Determine $F_k(x)$, a piecewise linear approximation of $F(x)$:

(a) Solve subproblem 8.17 for sample ω^k and update V_k. Get

$$\pi^k(x^k, \omega^k) \in \underset{\pi \in \Pi}{\operatorname{argmax}}\{\pi^\top r(\omega^k) - T(\omega^k)x^k)\}.$$

$$\text{Set } V_k \leftarrow V_{k-1} \cup \{\pi^k(x^k, \omega^k), \pi^k(\bar{x}^{k-1}, \omega^k)\}.$$

(b) Determine the coefficients of the k-th cut to add to the master program.

$$\alpha_k^k + \beta_k^k x \equiv \frac{1}{k}\sum_{t=1}^{k}(\pi_t^k)^\top(r(\omega^t) - T(\omega^t)x),$$

where

$$\pi_t^k \in \underset{\pi \in V_k}{\operatorname{argmax}}\{\pi^\top(r(\omega^t) - T(\omega^t)x^k)\}.$$

(c) Update the coefficients of all previously generated cuts. For $t = 1, \cdots, k-1$, set

$$\alpha_t^k \leftarrow \frac{k-1}{k}\alpha_t^{k-1} + \frac{1}{k}\mathcal{L}, \quad \beta_t^k \leftarrow \frac{k-1}{k}\beta_t^{k-1}.$$

(d) Update the approximation $F_k(x)$. Set $\mathcal{T}_k \leftarrow \mathcal{T}_{k-1} \cup \{k\}$ and

$$F_k(x) = c^\top x + \text{Max}_{t \in \mathcal{T}_k} \{\alpha_t^k + \beta_t^k x\}.$$

Step 3. **Solve Master Program.** Solve $\text{Min}_{x \in X}\ F_k(x)$ to get x^{k+1}.

Step 4. **Termination.** If *termination criterion* is satisfied, stop. Otherwise, repeat from step 1.

Terminating the SD algorithm in step 4 requires identifying a convergent subsequence, and therefore *termination criterion* must be based on the statistical nature of the algorithm. For the given basic SD algorithm, one can consider specifying a set number of samples as a termination criterion. So next we give some illustrative numerical example results. We discuss termination criteria for the SD algorithm in Sect. 8.5.4 and provide an extension of the algorithm for stabilizing.

8.5.3 Numerical Example

We are now ready to illustrate the results of applying the SD algorithm to the *abc-Production Planning* problem described in Sect. 8.2.

Example 8.7 (Applying SD to the *abc*-Production Planning Instance) Apply the SD algorithm to the *abc*-Production Planning problem given in Example 8.1. Reformulate the problem to satisfy assumption (A5) and then apply the SD algorithm using the following number of samples k for each run: 25, 50, 75, 100, 125, 150, 175, 200. Conduct two experiment runs using different initial seeds for the random number generator for each experiment. Report the solution at termination, lower bound (LB), upper bound (UB), and the percentage gap.

Solution:
We can see from the SAA results that the optimal objective function value of the example instance is negative. Therefore, to satisfy assumption (A5), i.e., the requirement that $\mathcal{L} \geq 0$, we can reformulate the problem as follows: Introduce a nonnegative decision variable, denoted v, and add it to the second-stage objective function $-1150y_1(\omega) - 1525y_2(\omega) - 1900y_3(\omega)$ to translate it so that it remains nonnegative for all $x \in X$. Given the possible values that $y_1(\omega)$, $y_2(\omega)$ and $y_3(\omega)$ can take, we can determine a value for $v \geq 0$. From the second-stage constraints, we see that the second-stage decision variables are bounded as follows:

$$0 \leq y_1(\omega) \leq d_a(\omega),\ 0 \leq y_2(\omega) \leq d_b(\omega),\ 0 \leq y_3(\omega) \leq d_c(\omega).$$

Therefore, we can set the second-stage objective function to

$$-1150y_1(\omega) - 1525y_2(\omega) - 1900y_3(\omega) + v$$

for $v \geq 0$ so that $\mathcal{L} = 0$. For example, $v = 140{,}000$ would satisfy this requirement and the problem formulation can be rewritten as follows:

$$\operatorname*{Min}_{x\in X}\ 50x_1 + 30x_2 + 15x_3 + 10x_4 + \sum_{\omega\in\Omega} p(\omega)\varphi(x,\omega),$$

where for each $x \in X$ and $\omega \in \Omega$, $\varphi(x, \omega)$ is given as

Table 8.10 SD results for transformed *abc*-Production Planning instance for case (a)

k	$(x^k)^\top$	LB	UB	% Gap
25	(202.24, 700.00, 410.17, 345.13)	130,069	142,031	8.42%
50	(209.69, 700.00, 403.79, 294.89)	134,707	140,431	4.08%
75	(236.21, 700.00, 407.59, 297.38)	135,406	139,691	3.07%
100	(227.57, 700.00, 425.52, 324.06)	135,932	138,656	1.97%
125	(249.58, 700.00, 469.62, 329.24)	135,806	138,282	1.79%
150	(249.79, 700.00, 438.82, 357.67)	136,136	137,998	1.34%
175	(253.88, 700,00, 440.28, 343.02)	136,378	138,018	1.19%
200	(245.48, 700.00, 456.66, 328.89)	136,295	137,958	1.21%

$$
\begin{aligned}
\varphi(x,\omega) := \text{Min } & -1150y_1(\omega) & -1525y_2(\omega) & -1900y_3(\omega) + v & & \\
\text{s.t.} & -6y_1(\omega) & -8y_2(\omega) & -10y_3(\omega) & \geq & -x_1 \\
& -20y_1(\omega) & -25y_2(\omega) & -28y_3(\omega) & \geq & -x_2 \\
& -12y_1(\omega) & -15y_2(\omega) & -18y_3(\omega) & \geq & -x_3 \\
& -8y_1(\omega) & -10y_2(\omega) & -14y_3(\omega) & \geq & -x_4 \\
& -y_1(\omega) & & & \geq & -d_a(\omega) \\
& & -y_2(\omega) & & \geq & -d_b(\omega) \\
& & & -y_3(\omega) & \geq & -d_c(\omega) \\
& & & v & = & 140{,}000. \\
& y_1(\omega), & y_2(\omega), & y_3(\omega),\ v & \geq & 0.
\end{aligned}
$$

Applying the basic SD algorithm to the transformed instance, we obtain the results for case (a) shown in Table 8.10. The columns of the table are as follows: number of samples (k), first-stage solution (x^k), lower bound (LB), upper bound (UB), and the algorithm percentage gag (% Gap) at termination. Recall that the optimal solution to the original instance is $x^* = (240.714, 700.000, 439.286, 325.000)^\top$ with an optimal value of -2374.04. Solving the transformed instance results in the same optimal solution, but with an optimal value of 13,7626. Translating this value back, i.e., 137,626.96 – 140,000 $= -2374.04$, we get the optimal value to the original instance. Looking at Table 8.8 we see that the SD algorithm is able to get the lower and upper bounds that include the optimal value, with the gap improving as the number of samples k increases as expected. Table 8.11 shown similar results for case (b). With a statistically based termination condition, the SD algorithm can identify a convergent subsequence of the iterates x^k whose limit point is the optimal. We provide details of the termination condition in the next subsection.

8.5.4 Stabilizing the Stochastic Decomposition Algorithm

The optimal number of samples required to solve an instance of Problem (8.14) is typically not known. Therefore, to stop the SD algorithm we can use the basic

Table 8.11 SD results for transformed *abc*-Production Planning instance for case (b)

k	$(x^k)^\top$	LB	UB	% Gap
25	(202.29, 700.00, 415.94, 345.02)	130,156	142,115	8.42%
50	(253.04, 700.00, 401.59, 301.97)	134,176	140,888	4.76%
75	(226.85, 700.00, 401.77, 312.96)	135,143	139,607	3.20%
100	(234.88, 700.00, 467.43, 315.98)	135,667	138,439	2.00%
125	(248.09, 700.00, 434.45, 352.54)	135,751	138,287	1.84%
150	(238.52, 700.00, 448.04, 311.72)	135,857	138,274	1.75%
175	(238.39, 700,00, 493.27, 336.46)	136,374	138,739	1.70%
200	(248.33, 700.00, 455.80, 336.24)	136,659	137,995	0.97%

asymptotic properties of the algorithm to identify a convergent subsequence of the iterates x^k. We summarize the asymptotic properties of the SD algorithm in the following Theorem:

Theorem 8.1 *Suppose that the given assumptions (A1–A5) hold. Let* $F_k(x) = c^\top x + \text{Max}_{t \in \mathcal{T}_k} \{\alpha_t^k + \beta_t^k x\}$ *denote the sequence of approximations and* $\{x^k\}$ *denote the sequence of iterates generated by the SD algorithm. Then*

(a) $\lim_k \to \infty\ F_k(x) - F_{k-1}(x) = 0$ *with probability one.*
(b) There exists a subsequence of iterations indexed by $\mathcal{K}$ *such that every limit point of* $\{x^k\}_{\mathcal{K}}$ *is an optimal solution with probability one.*

Theorem 8.1 provides a way to terminate the SD algorithm by identifying $\{x^k\}_{\mathcal{K}}$. But before we can arrive at this termination criterion, we first need to address one shortcoming of the basic SD algorithm: the iterates x^k can be "unstable" since the algorithm adds one sample at a time. Therefore, to *stabilize* the algorithm a penalty or *regularization* term $\frac{\sigma}{2}||x - \bar{x}^k||^2$ is added to the objective. Thus, problem approximation (8.16) becomes

$$\underset{x \in X}{\text{Min}}\ F_k(x) + \frac{\sigma}{2}||x - \bar{x}^k||^2, \tag{8.19}$$

where σ is a penalty and $\bar{x}^k$ is the incumbent solution (best solution seen so far) at iteration k. Since the objective function values generated by the SD algorithm are based on statistical estimates, as iterations proceed the objective function value estimates based on the incumbent solution become more accurate and the asymptotic optimality of the incumbent can be attained.

The SD algorithm will now require updating the incumbent solution by performing an *incumbent test* before Step 3. This will be done as follows: Let i_k denote the iteration at which the incumbent $\bar{x}^k$ was identified. Set $i_0 \leftarrow 0$ at initialization. Also, let $\gamma \in (0, 1)$ and $\sigma > 0$ be given. The incumbent test to perform is as follows:

If

$$|F_k(x^k) - F_k(\bar{x}^{k-1})| < \gamma |F_{k-1}(x^k) - F_{k-1}(\bar{x}^{k-1})| \tag{8.20}$$

then

$$\bar{x}^k \leftarrow x^k, \; i_k \leftarrow k$$

else

$$\bar{x}^k \leftarrow \bar{x}^{k-1}, \; i_k \leftarrow i_{k-1}.$$

By construction, $F_k(x^k)$ and $F_k(\bar{x}^{k-1})$ are estimates of $F(x^k)$ and $F(\bar{x}^{k-1})$, respectively. The RHS of inequality (8.20) is at most zero since $x^k \in \operatorname{argmin} F_{k-1}(x^k)$. The incumbent test implies that if the objective function value at x^k is sufficiently lower than that at $\bar{x}^{k-1}$, then x^k is adopted as the new incumbent.

Let $\{\bar{x}^k\}_{k=1}^{\infty}$ be the subsequence of incumbents, which accumulate at optimal solutions. Let $\{k_n\}_{n=1}^{\infty}$ be the subsequence of iterations at which the incumbent changes. Let $\delta^k = F_{k-1}(x^k) - F_{k-1}(\bar{x}^{k-1})$. Then, if N^* is the index set such that

$$n \in N^* \Leftrightarrow \delta^{k_n} \geq \frac{1}{n} \sum_{m=1}^{n} \delta^{k_m},$$

then N^* is an infinite set and every accumulation point of $\{\bar{x}^{k_n}\}_{n \in N^*}$ is optimal with probability one. This means that $\lim_{n \in N^*} \delta^{k_n} = 0$ with probability one. Let us now characterize a subsequence of incumbent solutions for which all accumulation points are optimal with probability one. Given that i_k is the iteration at which the current incumbent was identified, then $k - i_k$ be the number of iterations that have passed without changing the incumbent solution. If the incumbent solution changes only finitely often, then $k - i_k$ will eventually increase without bound. Otherwise, the incumbent solution changes infinitely often. Let $\epsilon > 0$ be small enough to serve as a surrogate for zero. Then, the following stopping rule can be used for the SD algorithm:

Stopping Rule: If k is "large enough," stop the SD algorithm if $\delta^k > -\epsilon$ and either $k - i_k$ is large or $i_k \in N^*$.

The index k is required to be "large enough" and ϵ to be "small enough" (both are instance dependent) to prevent premature termination. Next, we give a formal statement of the SD algorithm with regularization.

Algorithm *SD with Regularization*
begin

Step 0. **Initialization.** Set $k \leftarrow 0$, $V_0 \leftarrow \emptyset$, $\mathcal{T}_0 \leftarrow \emptyset$, $F_0(x) = -\infty$, $x^1 \in X$, incumbent solution $\bar{x}^0 \leftarrow x^1$, and $i_0 \leftarrow 0$. Compute lower bound $\mathcal{L}$ and choose $\sigma > 0$ and $\gamma \in (0, 1)$.

Step 1. **Generate New Sample.** Set $k \leftarrow k + 1$. Randomly generate a sample of $\tilde{\omega}$, ω^k, independent of any previously generated samples.

Step 2 **Update Approximation F_k and V_k, $\mathscr{T}_k$.** Determine $F_k(x)$, a piecewise linear approximation of $F(x)$:

(a) Solve subproblem 8.17 for sample ω^k and update V_k. Get

$$\pi^k(x^k, \omega^k) \in \underset{\pi \in \Pi}{\operatorname{argmax}}\{\pi^\top r(\omega^k) - T(\omega^k)x^k)\}.$$

$$\text{Set } V_k \leftarrow V_{k-1} \cup \{\pi^k(x^k, \omega^k), \pi^k(\bar{x}^{k-1}, \omega^k)\}.$$

(b) Determine the coefficients of the k-th cut to add to the master program.

$$\alpha_k^k + \beta_k^k x \equiv \frac{1}{k}\sum_{t=1}^{k}(\pi_t^k)^\top(r(\omega^t) - T(\omega^t)x),$$

where

$$\pi_t^k \in \underset{\pi \in V_k}{\operatorname{argmax}}\{\pi^\top(r(\omega^t) - T(\omega^t)x^k)\}.$$

(c) Update the coefficients of the cut indexed by i_{k-1}.

$$\alpha_{i_{k-1}}^k + \beta_{i_{k-1}}^k x \equiv \frac{1}{k}\sum_{t=1}^{k}(\bar{\pi}_t^k)^\top(r(\omega^t) - T(\omega^t)x),$$

where

$$\bar{\pi}_t^k \in \underset{\pi \in V_k}{\operatorname{argmax}}\{\pi^\top(r(\omega^t) - T(\omega^t)\bar{x}^{k-1})\}.$$

(d) Update $\mathscr{T}_k$ and remaining cuts. Set

$$\mathscr{T}_k \leftarrow J_{k-1} \cup \{k\}.$$

For $t \in \mathscr{T}_k \setminus \{i_{k-1}, k\}$, set

$$\alpha_t^k \leftarrow \frac{k-1}{k}\alpha_t^{k-1} + \frac{1}{k}\mathscr{L}, \ \beta_t^k \leftarrow \frac{k-1}{k}\beta_t^{k-1}.$$

(d) Set

$$F_k(x) = c^\top x + \underset{t \in \mathscr{T}_k}{\operatorname{Max}}\{\alpha_t^k + \beta_t^k x\}.$$

Step 3. **Perform Incumbent Test.** If

$$F_k(x^k) - F_k(\bar{x}^{k-1}) < \gamma \left(F_{k-1}(x^k) - F_{k-1}(\bar{x}^{k-1}) \right),$$

then

$$\bar{x}^k \leftarrow x^k, \; i_k \leftarrow k.$$

Otherwise,

$$\bar{x}^k \leftarrow \bar{x}^{k-1}, \; i_k \leftarrow i_{k-1}.$$

Step 4. **Solve Master Program.** Solve

$$\underset{x \in X}{\text{Min}} \; F_k(x) + \frac{\sigma}{2} ||x - \bar{x}^k||^2,$$

to get x^{k+1}.

Step 5. **Termination.** If *Stopping Rule* is satisfied, stop. Otherwise, repeat from step 1.

Let n_k denote the number of incumbents that have been generated up to current iteration k and k_n be the iteration at which the n-th incumbent is generated. In order to determine which iteration indices belong to N^*, the quantity $\frac{1}{n_k} \sum_{n=1}^{n_k} \delta^{k_n}$ must be updated at all iterations at which the incumbent changes. Other stopping rules have been proposed, including those based on objective value estimates associated with a sequence of incumbents, statistical error bound estimates, and optimality conditions. We refer the reader to [8] for details on stopping rules as well as the convergence results and their proofs.

Bibliographic Notes

The sampling-based methods presented in this chapter focus on two-stage SP. Sample-based methods have also been developed for multistage SP (MSP) for sequential decision-making over time. These include the stochastic dual dynamic programming (SDDP) algorithm [13], originally developed for solving hydrothermal scheduling problems. A Julia package for the implementation of the SDDP algorithm was developed by [6]. The algorithm has been widely used to deal with problems with a large number of stages [4, 5]. A mathematical analysis of the convergence properties of the SDDP method is available in [15].

Table 8.12 Random samples of $\tilde{\omega}$ using $\mathscr{U}(0, 1)$

s	u	$\omega^s := G^{-1}(u)$
1	0.19562	
2	0.20499	
3	0.38151	
4	0.46682	
5	0.53370	
6	0.65197	
7	0.77441	
8	0.82813	
9	0.88431	
10	0.95502	

Problems

8.1 (Generating Random Samples Using the Uniform Distribution) Consider the following two-stage MR-SLP:

$$z^* = \underset{x \geq 8}{\text{Min}} \; F(x) := \mathbb{E}[f(x, \tilde{\omega})],$$

where for outcome ω of $\tilde{\omega}$, $f(x, \omega) := x + \varphi(x, \omega)$ and

$$\varphi(x, \omega) = \underset{y \geq 0}{\text{Min}} \; \{y \mid y \geq x, \; y \geq 10\omega\}.$$

Supposed that the random variable $\tilde{\omega}$ has a PDF, $g(\omega) = \frac{1}{4}\omega^3, \; 0 < \omega < 2$.

(a) Determine the cumulative distribution function (CDF), denoted $G(\omega)$. What is its inverse, $G^{-1}(u)$.
(b) Use the given $\mathscr{U}(0, 1)$ outcomes (column u) in Table 8.3 to generate 10 IID samples of $\tilde{\omega}$, ω^s, $s = 1, \cdots, 10$, for each of the u values listed in the Table.
(c) Create an SAA problem for your samples ($N = 10$) in Table 8.12. State your SAA problem formulation stagewise and write the DEP, explicitly.
(d) Solve your SAA problem in part (c) and state its optimal value $\hat{z}_N^*$ and solution x_N^*.

8.2 (Generating Random Samples from a STOCH File) Consider the STOCH file data given in Table 8.13 in *independent* format. The columns of the table Column, Row, Value and Prob correspond to the decision variable, constraint, outcome, and probability of occurrence of the outcome, respectively. Answer the following questions:

(a) What are the random variables in this instance?
(b) How many outcomes does each random variable have? State each random variable, its outcomes and their probabilities of occurrence.

Table 8.13 STOCH file data for an SLP instance in INDEPENDENT format

Column	Row	Value	Prob
RHS	ROW21	5	0.3
RHS	ROW21	10	0.4
RHS	ROW21	15	0.3
RHS	ROW22	20	0.5
RHS	ROW22	40	0.5
RHS	ROW23	10	0.2
RHS	ROW23	20	0.2
RHS	ROW23	30	0.4
RHS	ROW23	40	0.2

(c) What is the total number of scenarios in this instance? List the scenarios and their probabilities of occurrence.

(d) Determine the (marginal) cumulative distribution for each random variable to use for generating random samples (scenarios) using $\mathcal{U}(0, 1)$ and plot it.

(e) What will be the random sample (scenario) when a pseudorandom number generator independently generates the following values, denoted u_1, u_2, u_3, respectively, for the first, second and third random variables:

(i) $u_1 = 0.5666$, $u_2 = 0.7541$, and $u_3 = 0.0666$.
(ii) $u_1 = 0.2005$, $u_2 = 0.4415$, and $u_3 = 0.8047$.
(iii) $u_1 = 0.9909$, $u_2 = 0.7541$, and $u_3 = 0.6666$.

8.3 Exterior Sampling—Sample Average Approximation
Consider Problem 8.1 and apply SAA based on $N = 10$ as follows: Repeat Problem 8.1(b) four more times to have five ($M = 5$) independent replications. For each replication i, record your optimal sample solution $\hat{x}_N^{i*}$ and value $\hat{z}_N^{i*}$ as in Table 8.14. Calculate the sample lower bound L_{MN} and a 95% CI on the lower bound (LB). Set $\hat{x} := \hat{x}_N^{i*}$, where $\hat{x}_N^{i*}$ corresponds to your sample solution with the lowest $\hat{z}_N^{i*}$ value (you may use any of your sample solutions if you prefer). Next, perform $\bar{M} = 5$ replications to compute a CI on the upper bound (UB) using $\bar{N} = 20$ independent samples. Report your sample upper bound $U_{\bar{M}\bar{N}}(\hat{x})$ and sample variance $s_U^2(\hat{x}, \bar{M})$ as in Table 8.14.

8.4 Exterior Sampling—Sample Average Approximation
Implement (code) the *SAA Scheme* based on the L-shaped algorithm using your favorite programming language and LP/MIP solver. Test your code using the *abc*-Production Planning problem instance in Example 8.1 as well as your own MR-SLP instances to make sure that the code is working correctly.

8.5 Exterior Sampling—Sample Average Approximation
Apply your implementation of the SAA scheme to the *abc*-Production Planning problem instance given in Example 8.1 and compute a 95%-confidence interval (CI) based on the following experiments:

Table 8.14 SAA results for $N = 10$, $M = 5$, $\bar{N} = 20$, $\bar{M} = 5$, and $\alpha = 0.05$

Replication i	$\hat{x}_N^{i*}$	$\hat{z}_N^i$
1		
2		
3		
4		
5		
L_{MN}		
$s_L^2(M)$		
LB CI		
$U_{\bar{M}\bar{N}}(\hat{x})$		
$s_U^2(\hat{x}, \bar{M})$		
UB CI		

(a) $N = 65$ and $M = 20$ to compute the lower bound L_{MN}, and $\bar{N} = 100$ and $\bar{M} = 20$ to compute the upper bound $L_{\bar{M}\bar{N}}$.
(b) $N = 75$ and $M = 20$ to compute the lower bound L_{MN}, and $\bar{N} = 100$ and $\bar{M} = 20$ to compute the upper bound $L_{\bar{M}\bar{N}}$.

Report your results in a table as in Example 8.6. Compare and contrast your results in part (a) and (b).

8.6 Exterior Sampling—Sample Average Approximation
Apply your implementation of the *SAA Scheme* to standard MR-SLP/SLP test instances from the literature (those available online) and compare the solutions, objective function values and CIs generated by your algorithm to those reported in the literature.

8.7 Interior Sampling—Stochastic Decomposition
Compare and contrast the SAA and SD methods. Which of the two methods do you prefer and why? Under what circumstances would you prefer one of the methods over the other?

8.8 Interior Sampling—Stochastic Decomposition
Implement (code) the *SD with Regularization* algorithm outlined in Sect. 8.5.2 using your favorite programming language and LP/MIP solver. Test your code using the *abc*-Production Planning problem instance in Example 8.1 as well as your own MR-SLP instances to make sure that the code is working correctly.

8.9 Interior Sampling—Stochastic Decomposition
Apply your implementation of the *SD with Regularization* algorithm to the *abc*-Production Planning problem instance given in Example 8.1. Reformulate the problem to satisfy assumption (A5) using $\upsilon = 100{,}000$ and then apply the algorithm using your choice of termination criteria. Repeat your experiment four times using different initial seeds for the pseudorandom number generator and report

the solution and objective value. Compare the solution and objective value generated by your algorithm to the true optimal solution and value.

8.10 Interior Sampling—Stochastic Decomposition
Apply your implementation of the *SD with Regularization* algorithm to standard MR-SLP/SLP test instances from the literature (those available online) and compare the solutions and objective function values generated by your algorithm to those reported in the literature.

References

1. G. Bayraksan and D.P. Morton. Assessing solution quality in stochastic programs via sampling. In *Decision Technologies and Applications*, pages 102–122. INFORMS, 2009.
2. G. Bayraksan and D.P. Morton. A sequential sampling procedure for stochastic programming. *Operations Research*, 59(4):898–913, 2011.
3. J.R. Birge and F. V. Louveaux. *Introduction to Stochastic Programming*. Springer, New York, 1997.
4. Santiago Cerisola, Jesus M Latorre, and Andres Ramos. Stochastic dual dynamic programming applied to nonconvex hydrothermal models. *European Journal of Operational Research*, 218(3):687–697, 2012.
5. Vitor L De Matos, Andy B Philpott, and Erlon C Finardi. Improving the performance of stochastic dual dynamic programming. *Journal of Computational and Applied Mathematics*, 290:196–208, 2015.
6. O. Dowson and L. Kapelevich. SDDP.jl: a Julia package for stochastic dual dynamic programming. *INFORMS Journal on Computing*, 2020. in press.
7. J.L. Higle and S. Sen. Stochastic decomposition: An algorithm for two-stage stochastic linear programs with recourse. *Mathematics of Operational Research*, 16:650–669, 1991.
8. J.L. Higle and S. Sen. *Stochastic Decomposition*. Kluwer Academic Publishers, 101 Phillip Drive, Norwell, MA 02061, 1996.
9. T. Homem-de Mello and G. Bayraksan. Monte Carlo sampling-based methods for stochastic optimization. *Surveys in Operations Research and Management Science*, 19(1):56–85, 2014.
10. G. Infanger. Monte Carlo (importance) sampling within a benders decomposition algorithm for stochastic linear programs. *Annals of Operations Research*, 39(1):69–95, 1992.
11. V.I. Norkin, Y.M. Ermoliev, and A. Ruszczyński. On optimal allocation of indivisibles under uncertainty. *Operations Research*, 46(3):381–395, 1998.
12. J. Park, R. Stockbridge, and G. Bayraksan. Variance reduction for sequential sampling in stochastic programming. *Annals of Operations Research*, 300(1):171–204, 2021.
13. M.V. Pereira and L.M. Pinto. Multi-stage stochastic optimization applied to energy planning. *Mathematical Programming*, 52.1-3:359–375, 1991.
14. A. Ruszczyn'ski and A. Shapiro, editors. *Handbooks in Operations Research and Management Science. Volume 10: Stochastic Programming*. Elsevier, New York, 2003.
15. Alexander Shapiro. Analysis of stochastic dual dynamic programming method. *European Journal of Operational Research*, 209(1):63–72, 2011.
16. R.J-B. Wets. Epi-consistency of convex stochastic programs. *Stochastic and Statistics Reports*, 34:83–92, 1989.

Chapter 9
Stochastic Mixed-Integer Programming Methods

9.1 Introduction

Modeling and derivation of algorithms for two-stage stochastic mixed-integer programming (SMIP) require a good understanding of the structural properties and the nature of the mixed-integer recourse function. The structural properties of SMIP follow those of deterministic MIPs and thus, decomposition methods for this class of problems are based on MIP and stochastic linear programming (SLP) methods. MIP methods include cutting-planes, branch-and-bound (BAB), and branch-and-cut (BAC). Pure cutting-plane methods and BAB methods may not necessarily solve the problem to optimality in practice. Therefore, in such cases BAC methods that combine cutting-planes in a BAB setting are often employed. Several SMIP methods are based on the Benders or Lagrangian decomposition (Chap. 4) and thus adopt the L-shaped method or dual decomposition (see Chap. 5) framework. In this chapter, we begin by reviewing the basic structural properties of SMIP. We illustrate the key properties with both graphical and numerical examples. Next, we discuss designing decomposition algorithms for SMIP and explore three algorithms with detailed numerical examples: *binary first-stage* (BFS), *disjunctive decomposition* (D^2), and *Fenchel decomposition* (FD) algorithms.

As with SLP, given an instance with a relatively large number of scenarios, the deterministic equivalent problem (DEP) may be impossible to solve using direct MIP solvers. This motivates researchers in this field to exploit the *problem structure* and decompose the problem into manageable subproblems that can be solved in some coordination scheme. A typical solution method follows the basic idea of Benders decomposition or Lagrangian decomposition (see Chap. 4) that incorporates a cutting-plane generation, BAB or BAC. The idea is to: a) decompose the problem into a master program and subproblems, b) relax the master and/or subproblems, c) employ a coordination/iterative scheme to solve the master and subproblem relaxations, d) generate cutting-planes as needed at every iteration, and/or perform BAB until some termination criterion is satisfied.

L. Ntaimo, *Computational Stochastic Programming*, Springer Optimization and Its Applications 774, https://doi.org/10.1007/978-3-031-52464-6_9

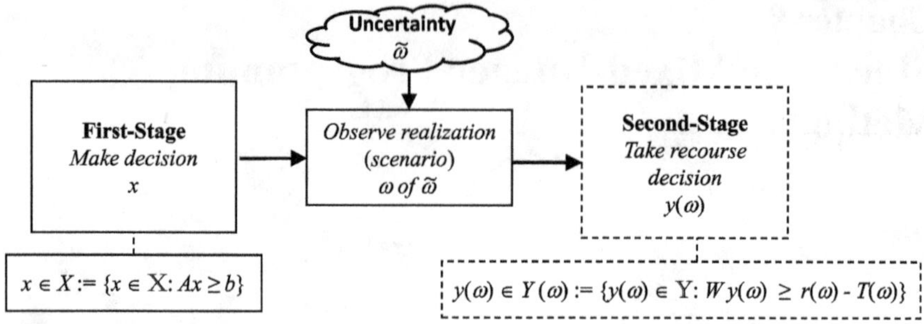

Fig. 9.1 The two-stage recourse decision-making process

In *stage-wise* decomposition of SMIP, the first stage involves a here-and-now decision variable vector $x \in \mathbb{X} \subset \mathbb{R}_+^{n_1}$, while the second-stage has the recourse decision variable vector $y(\tilde{\omega}) \in \mathbb{Y} \subset \mathbb{R}_+^{n_2}$. The sets $\mathbb{X}$ and $\mathbb{Y}$ impose integrality (binary or integer) requirements on some or all components of x and y, respectively. We shall denote the binary (0, 1) restrictions by $\mathbb{B}$, i.e., $\mathbb{B} = \{0, 1\}$. Similarly, we shall denote the general integer restrictions by $\mathbb{Z}_+$, i.e., $\mathbb{Z}_+ = \{0, 1, 2, \cdots\}$. The multivariate random variable $\tilde{\omega}$ is defined on a probability space $(\Omega, \mathscr{A}, \mathbb{P})$. The realization (scenario) of $\tilde{\omega}$ is denoted by ω, $\omega \in \Omega$, and the random cost based on $\tilde{\omega}$ is represented by the random cost function $f(x, \tilde{\omega})$. The decision-making process in this setting is illustrated in Fig. 9.1. In the first stage, decision x has to be made here-and-now without full information on the future realization ω of $\tilde{\omega}$. Then in the second stage, a recourse or corrective action $y(\omega)$ is taken based on both the decision x that was made in the first stage and scenario ω, which only becomes known in the second stage. Thus, the decision $y(\omega)$ adapts to a given scenario ω.

Formally, a two-stage mean-risk SMIP (MR-SMIP) with recourse can be stated as follows:

$$\underset{x \in X \cap \mathbb{X}}{\text{Min}} \quad \mathbb{E}[f(x, \tilde{\omega})] + \lambda \mathbb{D}[f(x, \tilde{\omega})], \tag{9.1}$$

where $\mathbb{E} : \mathscr{F} \mapsto \mathbb{R}$ denotes the expected value, $\mathbb{D} : \mathscr{F} \mapsto \mathbb{R}$ is the *risk measure*, and $\lambda \geq 0$ is a suitable weight factor that quantifies the trade-off between expected cost and risk. The problem is risk-neutral if $\lambda := 0$. Risk measure $\mathbb{D}$ is chosen so that it satisfies nice properties such as convexity. In Chap. 6, we study two main classes of mean-risk measures: *quantile* and *deviation* risk measures. We consider three quantile risk measure: excess probability (EP), quantile deviation (QDEV), and conditional value-at-risk (CVaR); and two deviation risk measures: expected excess (EE) and absolute semideviation (ASD). The set $X = \{x \in \mathbb{R}_+^{n_1} : Ax \geq b\}$ is nonempty and $X \cap \mathbb{X}$ defines the first-stage feasible solutions, where $\mathbb{X}$ imposes integrality restrictions on x. The matrix $A \in \mathbb{R}^{m_1 \times n_1}$ and vector $b \in \mathbb{R}^{m_1}$ are the first-stage matrix and right hand side vector, respectively. The family of real random cost variables $\{f(x, \tilde{\omega})\}_{x \in X \cap \mathbb{X}} \subseteq \mathscr{F}$ are defined on $(\Omega, \mathscr{A}, \mathbb{P})$, where $\mathscr{F}$ is the space of all real random cost variables $f : \Omega \mapsto \mathbb{R}$ satisfying $\mathbb{E}[|f(\tilde{\omega})|] < \infty$. For a given $x \in X \cap \mathbb{X}$ the real random cost variable $f(x, \tilde{\omega})$ is given by

$$f(x,\tilde{\omega}) := c^\top x + \varphi(x,\tilde{\omega}). \tag{9.2}$$

If $x \notin X \cap \mathbb{X}$, $f(x,\tilde{\omega}) = \infty$. For a given realization ω of $\tilde{\omega}$ the *recourse function* $\varphi(x,\omega)$ is given by

$$\begin{aligned} \varphi(x,\omega) :=& \text{Min } q(\omega)^\top y(\omega) \\ & \text{s.t. } Wy(\omega) \geq r(\omega) - T(\omega)x \\ & y(\omega) \in \mathbb{Y}, \end{aligned} \tag{9.3}$$

where $q(\omega) \in \mathbb{R}^{n_2}$ is the second-stage cost vector and $y(\omega)$ is the recourse decision. The matrix $W \in \mathbb{R}^{m_2 \times n_2}$ is the *recourse matrix*, $T(\omega) \in \mathbb{R}^{m_2 \times n_1}$ is the *technology matrix*, and $r(\omega) \in \mathbb{R}^{m_2}$ is the second-stage right hand side vector. By *scenario* ω we mean the realization of the stochastic problem data, i.e., $\omega := (q(\omega), T(\omega), r(\omega))$.

To ensure that Problem (9.1–9.3) is well-defined for computational purposes, we make the following assumptions:

(A1) The multivariate random variable $\tilde{\omega}$ is discretely distributed with finitely many scenarios $\omega \in \Omega$, each with probability of occurrence $p(\omega)$.
(A2) For every $x \in X \cap \mathbb{X}$ and $\omega \in \Omega$, we have $-\infty < f(x,\tilde{\omega}) < \infty$.
(A3) The first-stage feasible set X is nonempty and bounded.
(A4) The components of A, b, and W are rational and for every $\omega' \in \Omega$, the probability $\mathbb{P}\{\omega = \omega'\}$ is rational.

Assumption (A2) is the *relatively complete recourse* assumption and it guarantees the feasibility of the second-stage problem for every $x \in X \cap \mathbb{X}$ and $\omega \in \Omega$. In practice, one can formulate the problem so that it satisfies assumption (A2). The second-stage feasible set for a given $(x,\omega) \in X \cap \mathbb{X} \times \Omega$ can be defined as follows:

$$Y(x,\omega) = \{y(\omega) \in \mathbb{Y} : Wy(\omega) \geq r(\omega) - T(\omega)x\}.$$

As in Chap. 6, we shall continue to make the assumption that the risk measure $\mathbb{D}$ is convex and the definitions of the risk measures we consider are given in Chap. 2. Next, we define notation and review some basic properties needed for the decomposition algorithms.

9.2 Basic Structural Properties

Let us begin with defining some notation we shall use in describing the basic structural properties of SMIP. Let $\lceil a \rceil$ denote the rounding up function, i.e., smallest integer greater than or equal to a. Similarly, let $\lfloor a \rfloor$ denote the rounding down function, i.e., largest integer less than or equal to a, and define $\lfloor a \rfloor = -\lceil -a \rceil$. Also, for $a \in \mathbb{R}$ define $(a)_+ = \max\{a, 0\}$, where max is the *maximum* operator. For an

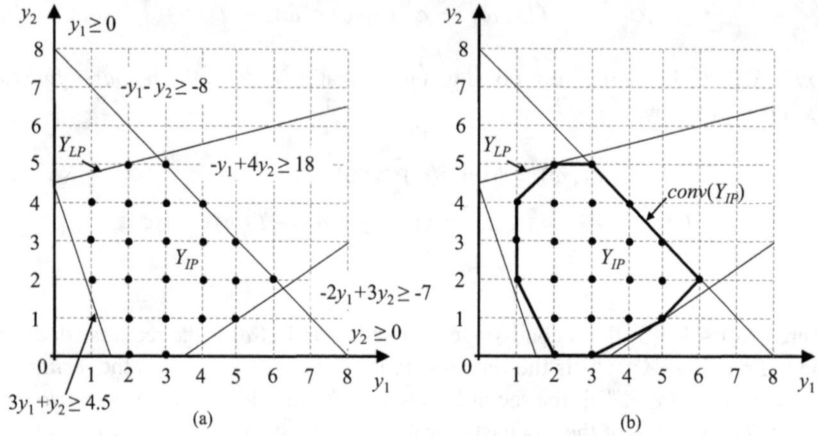

Fig. 9.2 An illustration of the convex hull of integer points: (**a**) feasible sets; (**b**) $conv(Y_{IP})$

integer set Y_{IP}, let $conv(Y_{IP})$ denote the convex hull of set Y_{IP}. We illustrate this graphically in Fig. 9.2.

Definition 9.1 For a function $f : Y_{IP} \mapsto \mathbb{R} \cup \{\infty\}$, we define the *convex envelope* $co(f) : conv(Y_{IP}) \mapsto \mathbb{R}$ and its closed convex envelope $\bar{co}(f) : conv(Y_{IP}) \mapsto \mathbb{R}$ as the pointwise supremum of all convex, respectively affine, functions majorized by f.

A key result from MIP literature [6, 7] is that the recourse function $f(x, \omega)$ is the value function of an MIP and is in general nonconvex, discontinuous,and lower semicontinuous. The structure of the expectation model, i.e., MR-SMIP with $\lambda := 0$, has been studied in the literature [20, 21, 29] and can be summarized by the following result:

Theorem 9.1 *Assuming (A1)–(A4), the expected recourse function* $\mathbb{E}[f(x, \tilde{\omega})]$ *is real-valued and lower semicontinuous on* $\mathbb{R}^{n_1}$.

If assumption (A1) does not hold and $\tilde{\omega}$ has an absolutely continuous density, then $\mathbb{E}[f(x, \omega)]$ is continuous on $\mathbb{R}^{n_1}$. MR-SMIP for different specifications of $\mathbb{D}$ (EP, QDEV, CVaR, EE,ASD) are nonconvex nonlinear optimization problems in general. We summarize a fundamental structural result from the literature [23] as follows:

Theorem 9.2 *Assuming (A1)–(A4) and that* $\mathbb{D} \in$ *{EP, QDEV, CVaR, EE, ASD}, the function* $\mathbb{E}[f(x, \tilde{\omega})] + \lambda\mathbb{D}[f(x, \tilde{\omega})]$ *is real-valued and lower semicontinuous on* $\mathbb{R}^{n_1}$.

Both Theorems 9.1 and 9.2 imply that to solve Problem (9.1) requires devising algorithms that approximate the nonconvex objective function, which is challenging in general. In fact, algorithms for SMIP typically imbed MIP techniques into lower-bounding procedures using a cutting-plane, BAB or BAC scheme. Even with alleviating numerical integration challenges by using discrete probability distributions, the problem remains to find a global minimizer in a *nonconvex* optimization problem. Analytical properties of $\mathbb{E}[f(x, \tilde{\omega})] + \lambda\mathbb{D}[f(x, \tilde{\omega})]$ are particularly poor. The lack of smoothness and continuity makes local subgradient-based descent approaches to solve SMIP less promising. In Chap. 6, we derived deterministic equivalent problem (DEP) formulations for mean-risk SLP. Applying general-purpose MIP solvers to solve the DEP is often prohibitive, thus the need for devising decomposition methods.

Let us now use a numerical example to illustrate the basic structural properties of SMIP summarized in Theorems 9.1 and 9.2.

Example 9.1 Consider the following MR-SMIP:

$$\underset{x \in X}{\text{Min}} \ \mathbb{E}[f(x, \tilde{\omega})] + \lambda\mathbb{D}[f(x, \tilde{\omega})],$$

where $f(x, \tilde{\omega}) := c^\top x + \varphi(x, \tilde{\omega})$ and for an outcome $\omega \in \Omega$ of $\tilde{\omega}$

$$\begin{aligned} \varphi(x, \omega) := &\text{Min } q^\top y(\omega) \\ &\text{s.t. } Wy(\omega) \geq r(\omega) - Tx \\ &\quad y(\omega) \in \mathbb{Y}, \end{aligned}$$

with the problem data given below. Also, define the LP relaxation

$$f_{LP}(x, \tilde{\omega}) := c^\top x + \varphi_{LP}(x, \tilde{\omega}),$$

where

$$\begin{aligned} \varphi_{LP}(x, \omega) := &\text{Min } q^\top y(\omega) \\ &\text{s.t. } Wy(\omega) \geq r(\omega) - Tx \\ &\quad y(\omega) \geq 0. \end{aligned}$$

First-stage:
$c = (1)^\top$, $A = [-1]$, $b = (-3)^\top$, and $X = \{x \in \mathbb{R}_+ : -x \geq -3\}$.
Second-stage:
$\Omega = \{\omega^1, \omega^2\}$ and $p(\omega^1) = p(\omega^2) = 0.5$.
$q = (2, 1)^\top$.
$T = \begin{bmatrix} -1 \\ 1 \end{bmatrix}$, $W = \begin{bmatrix} 1 & 0 \\ 0 & 1 \end{bmatrix}$, $r(\omega^1) = (-1, 1)^\top$, and $r(\omega^2) = (-2, 2)^\top$.
$\mathbb{D} := EE$ (Expected Excess) with target $\eta := 1.5$.
$\mathbb{Y} := \mathbb{Z}^2_+$.
Answer the following questions:

1. Derive closed-form expressions for the functions $\varphi(x, \omega)$ and $f(x, \omega)$ for ω^1 and ω^2, respectively.

2. Graph the following functions on the same axes for each case and then give a brief description of the functions:

   (*i*) $\varphi(x, \omega^1)$ and $\varphi_{LP}(x, \omega^1)$; and $f(x, \omega^1)$ and $f_{LP}(x, \omega^1)$.
   (*ii*) $\varphi(x, \omega^2)$ and $\varphi_{LP}(x, \omega^2)$; and $f(x, \omega^2)$ and $f_{LP}(x, \omega^2)$.

3. Write closed-form expressions for $\mathbb{E}[\varphi(x, \tilde{\omega})]$ and $\mathbb{E}[f(x, \tilde{\omega})]$.
4. Assuming $\mathbb{D} := EE$ (Expected Excess), write a closed-form expression for $\mathbb{D}[f(x, \tilde{\omega})]$ for $\lambda = 1$.
5. Graph the following functions on the same axes for each case and then give a brief description of the functions:

   (*i*) $\varphi(x, \omega^1)$, $\varphi(x, \omega^2)$ and $\mathbb{E}[\varphi(x, \tilde{\omega})]$; and $f(x, \omega^1)$, $f(x, \omega^2)$ and $\mathbb{E}[f(x, \tilde{\omega})]$.
   (*ii*) $\mathbb{E}[\varphi(x, \tilde{\omega})]$, $\mathbb{E}[\varphi_{LP}(x, \tilde{\omega})]$, $\mathbb{E}[\varphi(x, \tilde{\omega})]+ \mathbb{D}[\varphi(x, \tilde{\omega})]$, and $\mathbb{E}[\varphi_{LP}(x, \tilde{\omega})]+ \mathbb{D}[\varphi_{LP}(x, \tilde{\omega})]$;
   and
   $\mathbb{E}[f(x, \tilde{\omega})]$, $\mathbb{E}[f_{LP}(x, \tilde{\omega})]$, $\mathbb{E}[f(x, \tilde{\omega})]+ \mathbb{D}[f(x, \tilde{\omega})]$ and $\mathbb{E}[f_{LP}(x, \tilde{\omega})]+ \mathbb{D}[f_{LP}(x, \tilde{\omega})]$.

Solution

1. **Derivation of closed-form expressions for $\varphi(x, \omega)$ and $f(x, \omega)$ for ω^1 and ω^2:**
   Consider ω_1:

$$
\begin{aligned}
\varphi(x, \omega^1) := \text{Min } & 2y_1 + y_2 \\
\text{s.t. } & y_1 \geq -1 + x \\
& y_2 \geq 1 - x \\
& y_1, \; y_2 \in \mathbb{Z}_+.
\end{aligned}
$$

For $0 \leq x \leq 1$, the optimal second-stage solution is $y_1 = 0$, $y_2 = \lceil 1 - x \rceil$. For $1 < x \leq 3$, the optimal second-stage solution is $y_1 = \lceil x - 1 \rceil$, $y_2 = 0$. Hence,

$$\varphi(x, \omega^1) := \max\{2(\lceil x - 1 \rceil), \lceil 1 - x \rceil\}.$$

$$f(x, \omega^1) = c^\top x + \varphi(x, \omega^1) := x + \max\{2(\lceil x - 1 \rceil), \lceil 1 - x \rceil\}.$$

Similarly, for ω^2:

$$
\begin{aligned}
\varphi(x, \omega^2) := \text{Min } & 2y_1 + y_2 \\
\text{s.t. } & y_1 \geq -2 + x \\
& y_2 \geq 2 - x \\
& y_1, \; y_2 \in \mathbb{Z}_+.
\end{aligned}
$$

For $0 \leq x \leq 2$, the optimal second-stage solution is $y_1 = 0$, $y_2 = \lceil 2 - x \rceil$. For $2 < x \leq 3$, the optimal second-stage solution is $y_1 = \lceil x - 2 \rceil$, $y_2 = \lceil 2 - x \rceil$. Hence,

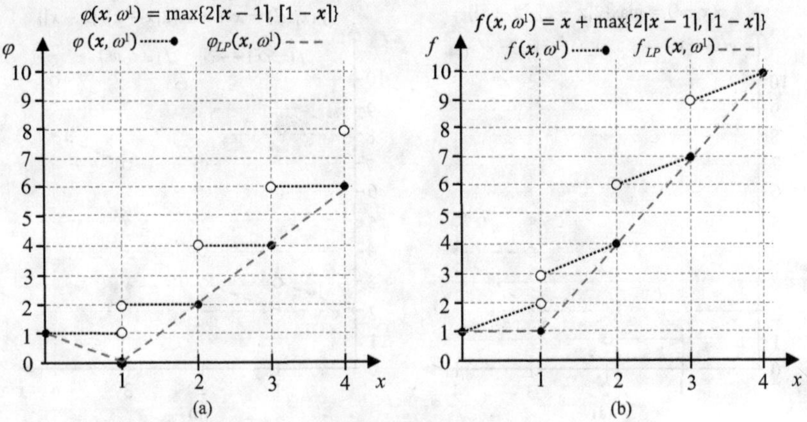

Fig. 9.3 Graph showing functions: (**a**) $\varphi(x, \omega^1)$ and $\varphi_{LP}(x, \omega^1)$; (**b**) $f(x, \omega^1)$ and $f_{LP}(x, \omega^1)$

$$\varphi(x, \omega^2) := \max\{2(\lceil x - 2 \rceil), \lceil 2 - x \rceil\}.$$

$$f(x, \omega^2) = c^\top x + \varphi(x, \omega^2) := x + \max\{2(\lceil x - 2 \rceil), \lceil 2 - x \rceil\}.$$

2. **Graphing and Describing the Functions:**

   (*i*) The functions $\varphi(x, \omega^1)$ and $f(x, \omega^1)$ are plotted in Fig. 9.3a, b, respectively. As can be seen in the figures, both functions are typical Gomory functions; they are *nonconvex* and *discontinuous* at $x = 1, 2, 3, 4$. Furthermore, these functions are *lower semicontinuous*; they are continuous along the x-axis but have discontinuities or "jumps" along the vertical axis. The functions $\varphi_{LP}(x, \omega^1)$ and $f_{LP}(x, \omega^1)$, on the contrary, are *continuous* and *piecewise linear*. They provide a *lower convex envelope* of the original function.

   (*ii*) The functions $\varphi(x, \omega^2)$ and $f(x, \omega^2)$ are plotted in Fig. 9.4a, b, respectively. Both functions are Gomory functions; they are nonconvex and lower semicontinuous. The functions $\varphi_{LP}(x, \omega^2)$ and $f_{LP}(x, \omega^2)$ are continuous, piecewise linear, and provide a lower convex envelope of the original function.

3. **Closed-form expressions for $\mathbb{E}[\varphi(x, \tilde{\omega})]$ and $\mathbb{E}[f(x, \tilde{\omega})]$:**

$$\begin{aligned}
\mathbb{E}[\varphi(x, \tilde{\omega})] &= 0.5\varphi(x, \omega^1) + 0.5\varphi(x, \omega^2) \\
&= 0.5 \max\{2(\lceil x - 1 \rceil), \lceil 1 - x \rceil\} + 0.5 \max\{2(\lceil x - 2 \rceil), \lceil 2 - x \rceil\}. \\
\mathbb{E}[f(x, \tilde{\omega})] &= c^\top x + \mathbb{E}[\varphi(x, \tilde{\omega})] \\
&= x + 0.5 \max\{2(\lceil x - 1 \rceil), \lceil 1 - x \rceil\} + 0.5 \max\{2(\lceil x - 2 \rceil), \lceil 2 - x \rceil\}.
\end{aligned}$$

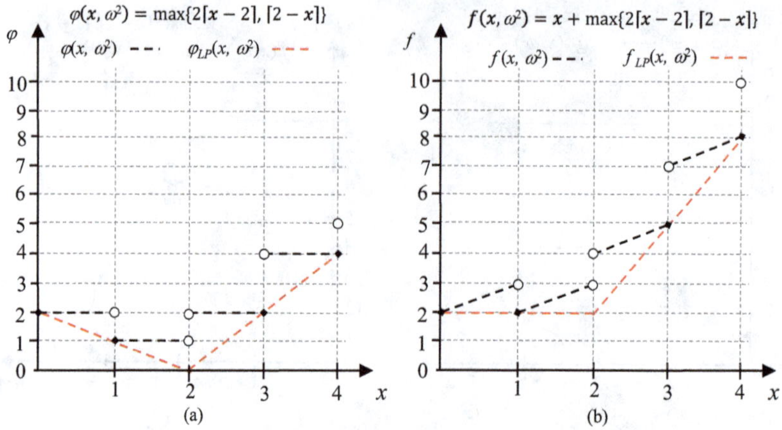

Fig. 9.4 Graph showing functions: (**a**) $\varphi(x, \omega^2)$ and $\varphi_{LP}(x, \omega^2)$; (**b**) $f(x, \omega^2)$ and $f_{LP}(x, \omega^2)$

4. **Closed-form expression for $\mathbb{D}[f(x, \tilde{\omega})]$ for $\lambda = 1$ (Expected Excess):** Recall from Chap. 2 that for the expected excess risk measure, $\mathbb{D}[f(x, \tilde{\omega})] = \mathbb{E}[(f(x, \tilde{\omega}) - \eta)_+]$.

$$\begin{aligned}
\mathbb{D}[f(x, \tilde{\omega})] &= \mathbb{E}[(f(x, \tilde{\omega}) - \eta)_+] \\
&= 0.5(f(x, \omega^1) - \eta)_+ + 0.5(f(x, \omega^2) - \eta)_+ \\
&= 0.5(x + \max\{2(\lceil x - 1\rceil), \lceil 1 - x\rceil\} - \eta)_+ + \\
&\quad 0.5(x + \max\{2(\lceil x - 2\rceil), \lceil 2 - x\rceil\} - \eta)_+.
\end{aligned}$$

5. **Graphing and Describing the Functions:**
   (*i*) The functions $\varphi(x, \omega^1)$, $\varphi(x, \omega^2)$ and $\mathbb{E}[\varphi(x, \tilde{\omega})]$ are plotted in Fig. 9.5a while the functions $f(x, \omega^1)$, $f(x, \omega^2)$ and $\mathbb{E}[f(x, \tilde{\omega})]$ are shown in Fig. 9.5b. Clearly, these functions are nonconvex, discontinuous, and lower semicontinuous.
   (*ii*) The functions $\mathbb{E}[\varphi(x, \tilde{\omega})]$, $\mathbb{E}[\varphi_{LP}(x, \tilde{\omega})]$, $\mathbb{E}[\varphi(x, \tilde{\omega})] + \mathbb{D}[\varphi(x, \tilde{\omega})]$, and $\mathbb{E}[\varphi_{LP}(x, \tilde{\omega})] + \mathbb{D}[\varphi_{LP}(x, \tilde{\omega})]$ are given in Fig. 9.6a while the functions $\mathbb{E}[f(x, \tilde{\omega})]$, $\mathbb{E}[f_{LP}(x, \tilde{\omega})]$, $\mathbb{E}[f(x, \tilde{\omega})] + \mathbb{D}[f(x, \tilde{\omega})]$ are shown in Fig. 9.6b. $\mathbb{E}[f_{LP}(x, \tilde{\omega})] + \mathbb{D}[f_{LP}(x, \tilde{\omega})]$. The functions $\mathbb{E}[\varphi(x, \tilde{\omega})] + \mathbb{D}[\varphi(x, \tilde{\omega})]$ and $\mathbb{E}[f(x, \tilde{\omega})] + \mathbb{D}[f(x, \tilde{\omega})]$ are nonconvex, discontinuous, and lower semicontinuous. However, the functions $\mathbb{E}[\varphi_{LP}(x, \tilde{\omega})] + \mathbb{D}[\varphi_{LP}(x, \tilde{\omega})]$ and $\mathbb{E}[f_{LP}(x, \tilde{\omega})] + \mathbb{D}[f_{LP}(x, \tilde{\omega})]$ are continuous and piecewise linear and provide a lower convex envelope of the original function.

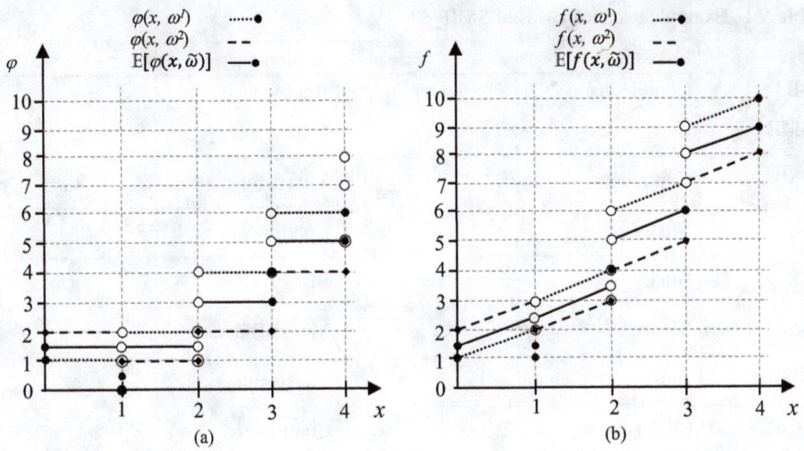

Fig. 9.5 Graph showing the following functions: **(a)** $\varphi(x, \omega^1)$, $\varphi(x, \omega^2)$ and $\mathbb{E}[\varphi(x, \tilde{\omega})]$; and **(b)** $f(x, \omega^1)$, $f(x, \omega^2)$ and $\mathbb{E}[f(x, \tilde{\omega})]$

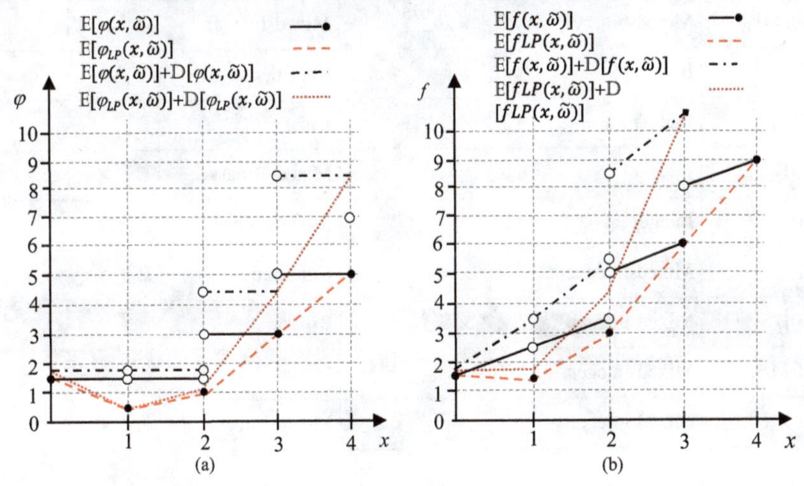

Fig. 9.6 Graph of showing the following functions: **(a)** $\mathbb{E}[\varphi(x, \tilde{\omega})]$, $\mathbb{E}[\varphi_{LP}(x, \tilde{\omega})]$, $\mathbb{E}[\varphi(x, \tilde{\omega})] + \mathbb{D}[\varphi(x, \tilde{\omega})]$, and $\mathbb{E}[\varphi_{LP}(x, \tilde{\omega})] + \mathbb{D}[\varphi_{LP}(x, \tilde{\omega})]$; **(b)** $\mathbb{E}[f(x, \tilde{\omega})]$, $\mathbb{E}[f_{LP}(x, \tilde{\omega})]$, $\mathbb{E}[f(x, \tilde{\omega})] + \mathbb{D}[f(x, \tilde{\omega})]$ and $\mathbb{E}[f_{LP}(x, \tilde{\omega})] + \mathbb{D}[f_{LP}(x, \tilde{\omega})]$

9.3 Designing Algorithms for SMIP

Algorithms for SMIP are designed to solve specific cases of SMIP based on the nature of the decision variables in both stages. The sets $\mathbb{X}$ and $\mathbb{Y}$ define whether the decision variables in each stage are continuous (C), binary (B), general integer (I), mixed-binary (MB), or mixed-integer (MI). Clearly, there are several combinations of these types of decision variables one can encounter for different applications. We

Table 9.1 Example instance cases of SMIP

Case	First-stage	$\mathbb{X}$	Second-stage	$\mathbb{Y}$
C-B	Continuous	$\mathbb{R}_+^{n_1}$	Binary	$\mathbb{B}_+^{n_2}$
C-MB	Continuous	$\mathbb{R}_+^{n_1}$	Mixed-Binary	$\mathbb{B}_+^{n_2'} \times \mathbb{R}_+^{n_2-n_2'}$
C-MI	Continuous	$\mathbb{R}_+^{n_1}$	Mixed-Integer	$\mathbb{Z}_+^{n_2'} \times \mathbb{R}_+^{n_2-n_2'}$
B-C	Binary	$\mathbb{B}_+^{n_1}$	Continuous	$\mathbb{R}_+^{n_2}$
B-B	Binary	$\mathbb{B}_+^{n_1}$	Binary	$\mathbb{B}_+^{n_2}$
B-MB	Binary	$\mathbb{B}_+^{n_1}$	Mixed-Binary	$\mathbb{B}_+^{n_2'} \times \mathbb{R}_+^{n_2-n_2'}$
B-MI	Binary	$\mathbb{B}_+^{n_1}$	Mixed-Integer	$\mathbb{Z}_+^{n_2'} \times \mathbb{R}_+^{n_2-n_2'}$
MB-C	Mixed-Binary	$\mathbb{B}_+^{n_1'} \times \mathbb{R}_+^{n_1-n_1'}$	Continuous	$\mathbb{R}_+^{n_2}$
MB-B	Mixed-binary	$\mathbb{B}_+^{n_1'} \times \mathbb{R}_+^{n_1-n_1'}$	Binary	$\mathbb{B}_+^{n_2'}$
MB-MB	Mixed-binary	$\mathbb{B}_+^{n_2'} \times \mathbb{R}_+^{n_2-n_2'}$	Mixed-Binary	$\mathbb{B}_+^{n_2'} \times \mathbb{R}_+^{n_2-n_2'}$
MB-MI	Mixed-binary	$\mathbb{B}_+^{n_1'} \times \mathbb{R}_+^{n_1-n_1'}$	Mixed-Integer	$\mathbb{Z}_+^{n_2'} \times \mathbb{R}_+^{n_2-n_2'}$
I-C	Integer	$\mathbb{Z}_+^{n_1}$	Continuous	$\mathbb{R}_+^{n_2}$
I-B	Integer	$\mathbb{Z}_+^{n_1}$	Binary	$\mathbb{B}_+^{n_2}$
I-MB	Integer	$\mathbb{Z}_+^{n_1}$	Mixed-Binary	$\mathbb{B}_+^{n_2'} \times \mathbb{R}_+^{n_2-n_2'}$
I-MI	Integer	$\mathbb{Z}_+^{n_1}$	Mixed-Integer	$\mathbb{Z}_+^{n_2'} \times \mathbb{R}_+^{n_2-n_2'}$
MI-C	Mixed-Integer	$\mathbb{Z}_+^{n_1'} \times \mathbb{R}_+^{n_1-n_1'}$	Continuous	$\mathbb{R}_+^{n_2}$
MI-B	Mixed-integer	$\mathbb{Z}_+^{n_1'} \times \mathbb{R}_+^{n_1-n_1'}$	Binary	$\mathbb{B}_+^{n_2}$
MI-MB	Mixed-integer	$\mathbb{Z}_+^{n_1'} \times \mathbb{R}_+^{n_1-n_1'}$	Mixed-Binary	$\mathbb{B}_+^{n_2'} \times \mathbb{R}_+^{n_2-n_2'}$
MI-MI	Mixed-integer	$\mathbb{Z}_+^{n_1'} \times \mathbb{R}_+^{n_1-n_1'}$	Mixed-Integer	$\mathbb{Z}_+^{n_2'} \times \mathbb{R}_+^{n_2-n_2'}$

give some example cases in Table 9.1. We should point out that SMIP cases with continuous second-stage are still amenable to Benders decomposition and the L-shaped algorithm can be applied since the recourse function is convex. Nevertheless, even for such cases the SMIP instances can be very challenging to solve.

9.4 Example Instance

To numerically illustrate the algorithms in this chapter, we use a simple SMIP instance described below. This instance has one first-stage decision variable x and two second-stage decision variables y_1 and y_2. This instance was designed to make it relatively easy to illustrate the three algorithms studied in this chapter.

Example 9.2 Consider the following MR-SMIP:

$$\underset{x \in X \cap \mathbb{B}^{n_1}}{\text{Min}} \quad \mathbb{E}[f(x, \tilde{\omega})] + \lambda \mathbb{D}[f(x, \tilde{\omega})], \tag{9.4}$$

where $f(x, \tilde{\omega}) := c^\top x + \varphi(x, \tilde{\omega})$ and for an outcome $\omega \in \Omega$ of $\tilde{\omega}$

$$\begin{aligned} \varphi(x, \omega) :=& \text{Min } q^\top y \\ & \text{s.t. } Wy \geq r(\omega) - T(\omega)x \\ & y \in \mathbb{Y}, \end{aligned} \tag{9.5}$$

Let the problem data for an instance of Problem (9.4–9.5) be given as follows:

First-stage:
$c = (-2)^\top$, $A = [-1]$, $b = (-1)^\top$.
Second-stage:
$\Omega = \{\omega^1, \omega^2\}$ with probabilities of occurrence $p(\omega^1) = p(\omega^2) = 0.5$.
$q(\omega^1) = q(\omega^2) = (-2, -2)^\top$.

Recourse matrix: $W = \begin{bmatrix} 1 & -1 \\ -1 & -1 \\ -1 & 0 \\ 0 & -1 \end{bmatrix}$.

Technology matrix: $T(\omega^1) = T(\omega^2) = \begin{bmatrix} 0 \\ -0.5 \\ 0 \\ 0 \end{bmatrix}$.

Right hand side:

$$r(\omega^1) = (-0.6, -1.2, -1, -1)^\top,$$

$$r(\omega^2) = (-0.6, -1.4, -1, -1)^\top.$$

$\lambda := 0$, we consider the risk-neutral setting. Otherwise, $\mathbb{D}$ has to be specified.
$\mathbb{Y} = \{0, 1\}^2$.
Let us rewrite the example instance in a more explicit form as follows:

$$\underset{x \in X}{\text{Min}} \; -2x + \sum_{s=1}^{2} p(\omega^s)\varphi(x, \omega^s),$$

where $X = \{x \in \{0, 1\} \mid -x \geq -1\}$. For a given $x \in X$ and scenario ω^1, the second-stage subproblem is given as follows:

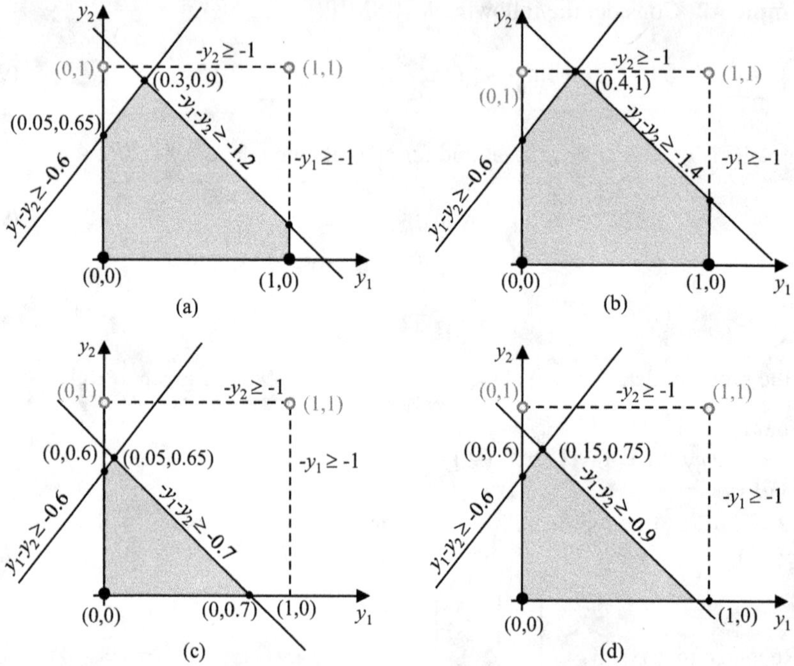

Fig. 9.7 Subproblem feasible region of Example 9.2 instance. (**a**) Scenario $\omega^1, x = 0$. (**b**) Scenario $\omega^2, x = 0$. (**c**) Scenario $\omega^1, x = 1$. (**d**) Scenario $\omega^2, x = 1$

$$
\begin{array}{rrrrl}
\varphi(x,\omega^1) := \text{Min} & -2y_1 & -2y_2 & & \\
\text{s.t.} & y_1 & -y_2 & \geq & -0.6 \\
 & -y_1 & -y_2 & \geq & -1.2 - (-0.5)x \\
 & -y_1 & & \geq & -1 \\
 & & -y_2 & \geq & -1 \\
 & y_1, & y_2 & \in & \{0, 1\}.
\end{array}
$$

Similarly, for a given $x \in X$ and scenario ω^2, the second-stage subproblem is given as follows:

$$
\begin{array}{rrrrl}
\varphi(x,\omega^2) := \text{Min} & -2y_1 & -2y_2 & & \\
\text{s.t.} & y_1 & -y_2 & \geq & -0.6 \\
 & -y_1 & -y_2 & \geq & -1.4 - (-0.5)x \\
 & -y_1 & & \geq & -1 \\
 & & -y_2 & \geq & -1 \\
 & y_1, & y_2 & \in & \{0, 1\}.
\end{array}
$$

We plot the subproblem feasible regions for $x = 0$ and $x = 1$ for each scenario in Fig. 9.7.

We are now ready to present and illustrate the SMIP algorithms. So in each of the next three subsections, we first derive the algorithm and then give a detailed numerical illustration of the algorithm using Example 9.2.

9.5 Binary First-Stage

One of the early algorithms for two-stage SMIP with recourse was derived by Laporte and Louveaux [15] for pure binary first-stage decision variables. This particular algorithm was designed for Problem (9.1) with $\lambda = 0$ (risk-neutral), pure binary first-stage, and arbitrary (continuous, binary, mixed-binary, integer, mixed-integer, etc.) second-stage. According to Table 9.1, the algorithm can be applied to cases B-C, B-B, B-MB, and B-MI. We shall refer to this algorithm as the *Binary First-Stage* (BFS) algorithm and we shall apply it to SMIP problems of the following form:

$$\operatorname*{Min}_{x \in X \cap \mathbb{B}^{n_1}} \; c^\top x + \mathbb{E}[\varphi(x, \tilde{\omega})], \tag{9.6}$$

where $X = \{Ax \geq b, x \geq 0\}$ and for an outcome $\omega \in \Omega$ of $\tilde{\omega}$,

$$\begin{aligned} \varphi(x, \omega) = &\operatorname{Min} q(\omega)^\top y \\ &\text{s.t. } Wy \geq r(\omega) - T(\omega)x \\ &\quad y \in \mathbb{Y}. \end{aligned} \tag{9.7}$$

The set $\mathbb{Y}$ imposes possible integral restrictions on some or all components of y.

In addition to assumptions (A1)–(A4) in Sect. 9.1, the BFS algorithm requires the following assumption on Problem (9.6–9.7):

(A5) A lower bound $\mathscr{L}$ on $\mathbb{E}[\varphi(x, \tilde{\omega})]$ is known, i.e.,

$$\mathscr{L} \leq \operatorname*{Min}_{x \in X \cap \mathbb{B}^{n_1}} \mathbb{E}[\varphi(x, \tilde{\omega})].$$

It is desirable to have $\mathscr{L}$ as large as possible because it plays a significant role in the convergence of the BFS algorithm to an optimal solution. The value of $\mathscr{L}$ affects the number of iterations of the algorithm. Therefore, $\mathscr{L}$ must be carefully selected based on the instance under consideration. Computational experience with the BFS algorithm shows that it is very sensitive to the value of $\mathscr{L}$. The reason for this is because $\mathscr{L}$ appears in the BFS algorithm optimality cut, which we shall refer to as the "L^2 cut." L^2 stands for Laporte and Louveaux [15], who derived the cut.

To derive the L^2 cut, let k denote the algorithm iteration index and $F(x^k) = \mathbb{E}[\varphi(x^k, \tilde{\omega})]$. In addition, for an $x^k \in X \cap \mathbb{B}^{n_1}$, define the set

$$\mathbb{S}^k = \{j \mid x_j^k = 1\}.$$

We should make it clear that while x is a decision variable, x^k is not, it is a specification of x at iteration k and x_j^k is the j-th component of x^k. Thus, $\mathbb{S}^k$ is a well-defined index set and can be empty. We define an L^2 cut in the following theorem:

Theorem 9.3 *The L^2 cut, defined as*

$$\eta \geq (F(x^k) - \mathscr{L})(\sum_{j \in \mathbb{S}^k} x_j - \sum_{j \notin \mathbb{S}^k} x_j - |\mathbb{S}^k| + 1) + \mathscr{L},$$

is a valid optimality cut for $F(x)$, $x \in X \cap \mathbb{B}^{n_1}$. This cut is valid in the sense that it is tight (active) at x^k and the inequality holds for all $x^k \in X \cap \mathbb{B}^{n_1}$.

Proof Consider the quantity

$$d = \sum_{j \in \mathbb{S}^k} x_j - \sum_{j \notin \mathbb{S}^k} x_j.$$

Then $d \leq |\mathbb{S}^k|$. Observe that $d = |\mathbb{S}^k|$ if and only if $\mathbb{S}^k$ is the set based on x^k. If $d = |\mathbb{S}^k|$, then the cut is

$$\eta \geq F(x^k).$$

If $d < |\mathbb{S}^k|$, then x^k is not the solution on which $\mathbb{S}^k$ is based. In this case, $d < |\mathbb{S}^k|$ implying that $d \leq |\mathbb{S}^k| - 1$. Thus

$$(\sum_{j \in \mathbb{S}^k} x_j - \sum_{j \notin \mathbb{S}^k} x_j - |\mathbb{S}^k| + 1) \leq 0.$$

Setting

$$M(x, x^k) = (F(x^k) - \mathscr{L})(\sum_{j \in \mathbb{S}^k} x_j - \sum_{j \notin \mathbb{S}^k} x_j - |\mathbb{S}^k| + 1) \leq 0$$

we get the cut

$$\eta \geq M(x, x^k) + \mathscr{L}.$$

Clearly, with $M(x, x^k) \leq 0$, the cut is valid for x^k. ■

The L^2 cut in Theorem 9.3 can be rewritten as follows:

$$\eta \geq (F(x^k) - \mathscr{L})(\sum_{j \in \mathbb{S}^k} x_j - \sum_{j \notin \mathbb{S}^k} x_j) + (F(x^k) - \mathscr{L})(1 - |\mathbb{S}^k|) + \mathscr{L}$$

$$\Rightarrow (F(x^k)-\mathscr{L}) \sum_{j \notin \mathbb{S}^k} x_j - (F(x^k)-\mathscr{L}) \sum_{j \in \mathbb{S}^k} x_j + \eta \geq (F(x^k)-\mathscr{L})(1 - |\mathbb{S}^k|) + \mathscr{L}.$$

Setting the cut coefficient $\beta'_k = (F(x^k)-\mathscr{L})$ and the right hand side $\alpha'_k = (F(x^k)-\mathscr{L})(1 - |\mathbb{S}^k|) + \mathscr{L}$, the cut becomes

$$\beta'_k \sum_{j \notin \mathbb{S}^k} x_j - \beta'_k \sum_{j \in \mathbb{S}^k} x_j + \eta \geq \alpha'_k. \tag{9.8}$$

We shall use the form of the L^2 cut in (9.8) in the BFS algorithm. It should be pointed out that the L^2 cut is a "weak" cut in general and recommend using it together with other lower-bounding cuts (e.g., Benders optimality cut) to be effective. We consider a basic BFS algorithm that, in addition to the L^2 cut, generates and adds a Benders optimality cut at each iteration of the algorithm.

Using the L-shaped algorithm framework, let the master program at iteration k of the BFS algorithm be given as follows:

$$z^{k+1} := \text{Min } c^\top x + \eta$$

$$\text{s.t. } Ax \geq bn$$

$$(\beta_t)^\top x + \eta \geq \alpha_t,\ t = 1, \cdots, k \tag{9.9a}$$

$$\beta'_t \sum_{j \notin \mathbb{S}^t} x_j - \beta'_t \sum_{j \in \mathbb{S}^t} x_j + \eta \geq \alpha'_k,\ t = 1, \cdots, k \tag{9.9b}$$

$$x \in \mathbb{B}^{n_1}, \eta \text{ free},$$

where constraints (9.9a) are the Benders optimality cuts and constraints (9.9b) are the L^2 cuts. Recall from Chap. 6 that to generate a Benders optimality cut given x^k, we need to solve the following LP relaxation of subproblem (9.7) for all $\omega \in \Omega$:

$$\varphi_{LP}(x^k, \omega) = \text{Min } q(\omega)^\top y$$

$$\text{s.t. } Wy \geq r(\omega) - T(\omega)x^k \tag{9.10}$$

$$y \geq 0.$$

Let the optimal dual multipliers associated with constraints (9.10) be denoted by $\pi_k(\omega)$. Then a Benders optimality cut is calculated as follows:

$$\eta \geq \sum_{\omega \in \Omega} p(\omega)\{\pi_k(\omega)^\top (r(\omega) - T(\omega)x)\},\ \forall x \in X.$$

This simplifies to

$$(\beta_k)^\top x + \eta \geq \alpha_k,$$

where

$$(\beta_k)^\top = \sum_{\omega \in \Omega} p(\omega)\pi_k(\omega)^\top T(\omega) \text{ and } \alpha_k = \sum_{\omega \in \Omega} p(\omega)\pi_k(\omega)^\top r(\omega).$$

We maintain the *relatively complete recourse* and, thus, do not include feasibility cuts in the master program.

9.5.1 *BFS Algorithm*

A basic BFS algorithm for Problem (9.6) can be stated as follows:

Algorithm BFS
begin

Step 0. **Initialization.** Set $k \leftarrow 1$, $L^1 \leftarrow -\infty$, and $U^1 \leftarrow \infty$. Choose $\epsilon > 0$, $x^1 \in X \cap \{0,1\}^{n_1}$, and compute $\mathscr{L}$. Set incumbent solution $x_\epsilon^* \leftarrow x^1$ and objective value $z_\epsilon^* \leftarrow \infty$.

Step 1. **Solve Subproblems.**

(a) **Solve Subproblem MIPs.** Solve subproblems MIPs for all $\omega \in \Omega$ and compute $F(x^k) \leftarrow \mathbb{E}[f(x, \tilde{\omega})]$. Update upper bound $U^{k+1} \leftarrow \min\{c^\top x + F(x^k), U^k\}$. If U^{k+1} is updated, set $x_\epsilon^* \leftarrow x^k$ with objective value $z_\epsilon^* \leftarrow U^{k+1}$.

(b) **Solve Subproblem LPs.** Solve subproblems LPs for all $\omega \in \Omega$ to get dual solutions $\pi_k(\omega)$. Compute $(\beta_k')^\top$ and α_k' for a Benders optimality cut.

Step 2 **Update and Solve the Master Program.** Using $\mathscr{L}$, x^k, and $F(x^k)$, compute $\beta_k = (F(x^k) - \mathscr{L})$ and $\alpha_k = (F(x^k) - \mathscr{L})(1 - |\mathbb{S}^k|) + \mathscr{L}$ for L^2 optimality cut and append cut to master program. Add a Benders optimality cut to master program and solve to get optimal solution x^{k+1} and optimal value z^{k+1}. Set $L^{k+1} \leftarrow \max\{z^{k+1}, L^k\}$.

Step 3. **Termination.** If $U^{k+1} - L^{k+1} \leq \epsilon|U^{k+1}|$, stop and declare x_ϵ^* to be ϵ-optimal with objective value z_ϵ^*. Otherwise, set $k \leftarrow k+1$ and go to step 1.

We should note that computing $\mathscr{L}$ may not be trivial unless you know something about the instance. Observe that $\mathscr{L}$ can be set to be equal to the expected recourse function value for the LP relaxation, but this may not necessarily provide a good lower bound.

9.5.2 Numerical Illustration of the BFS Algorithm

Example 9.3 We are now in a position to use the SMIP instance given in Example 9.2 to illustrate the BFS algorithm. To apply the algorithm, we need to first compute the lower bound on the expected recourse function value, $\mathscr{L} \leq \underset{x \in X \cap \mathbb{B}^{n_1}}{\text{Min}} \ \mathbb{E}[\varphi(x, \tilde{\omega})]$. To do this, let us consider the LP relaxation of the second-stage expected recourse function, denoted $\mathbb{E}[\varphi_{LP}(x, \tilde{\omega})]$, and compute

$$\mathscr{L} = \underset{x \in X \cap \mathbb{B}^{n_1}}{\text{Min}} \ \mathbb{E}[\varphi_{LP}(x, \tilde{\omega})],$$

which is a valid lower bound on $\mathbb{E}[\varphi(x, \tilde{\omega})]$. Since $x \in X \cap \mathbb{B}^{n_1} = \{0, 1\}$, for $x = 0$,

$$\varphi_{LP}(0, \omega^1) = -2.4, \ \varphi_{LP}(0, \omega^2) = -2.8 \text{ and } \mathbb{E}[\varphi_{LP}(0, \tilde{\omega})] = -2.6.$$

When $x = 1$ we have

$$\varphi_{LP}(1, \omega^1) = -1.4, \ \varphi_{LP}(1, \omega^2) = -1.8 \text{ and } \mathbb{E}[\varphi_{LP}(1, \tilde{\omega})] = -1.6.$$

Therefore, we set $\mathscr{L} \leftarrow \min\{-1.6, -2.6\} = -2.6$.

Algorithm BFS
begin

Step 0. Initialization. Set $k \leftarrow 1$, $U^1 \leftarrow \infty$, and $L^1 \leftarrow -\infty$. Choose $\epsilon \leftarrow 10^{-6}$, $x^1 \leftarrow 1$ and set $\mathscr{L} \leftarrow -2.6$. Initialize incumbent solution $x^*_\epsilon \leftarrow x^1$ and objective value $z^*_\epsilon \leftarrow \infty$.

Step 1.

(a) Solve Subproblem MIPs. For scenario ω^1,

$$\begin{aligned}
\varphi(x^1, \omega^1) := \text{Min} \ & -2y_1 - 2y_2 \\
\text{s.t.} \ & y_1 - y_2 \geq -0.6 \\
& -y_1 - y_2 \geq -1.2 - (-0.5)x^1 \\
& -y_1 \geq -1 \\
& -y_2 \geq -1 \\
& y_1, \ y_2 \in \{0, 1\}.
\end{aligned}$$

The optimal solution is $y_1(\omega^1) = 0$, $y_2(\omega^1) = 0$ (see Fig. 9.8a) with $\varphi(x^1, \omega^1) = 0$. For scenario ω^2,

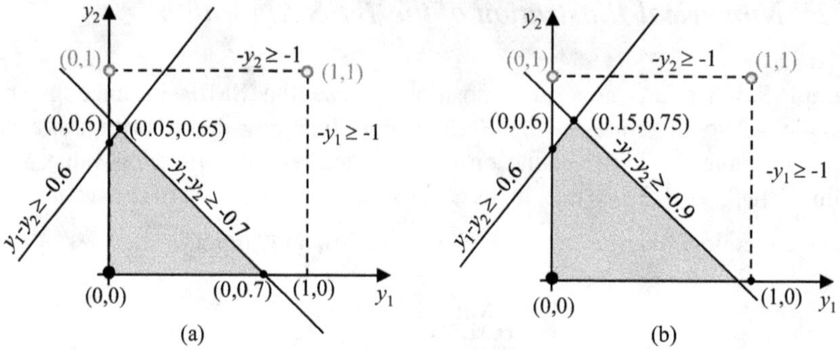

Fig. 9.8 Subproblem feasible region of Example 9.2 instance for $x = 1$. (**a**) Scenario $\omega^1, x = 1$. (**b**) Scenario ω^2, $x = 1$

$$
\begin{array}{rllll}
\varphi(x^1, \omega^2) := \text{Min} & -2y_1 & -2y_2 & & \\
\text{s.t.} & y_1 & -y_2 & \geq & -0.6 \\
& -y_1 & -y_2 & \geq & -1.4 - (-0.5)x^1 \\
& -y_1 & & \geq & -1 \\
& & -y_2 & \geq & -1 \\
& y_1, & y_2 & \in & \{0, 1\}.
\end{array}
$$

The optimal solution is $y_1(\omega^2) = 0$, $y_2(\omega^2) = 0$ with $\varphi(x^1, \omega^2) = 0$ (see Fig. 9.8b). Therefore, $F(x^1) \leftarrow \mathbb{E}[f(x^1, \tilde{\omega})] = 0.5(0) + 0.5(0) = 0$. Update upper bound $U^2 \leftarrow \min\{c^\top x^1 + F(x^1), U^1\} = \min\{0, \infty\} = 0$. Since U^2 is updated, set $x^*_\epsilon \leftarrow x^1 = 1$ and $z^*_\epsilon \leftarrow U^2 = 0$.

(b) Solve Subproblems (LP). For scenario ω^1,

$$
\begin{array}{rllll}
\varphi_{LP}(x^1, \omega^1) := \text{Min} & -2y_1 & -2y_2 & & \\
\text{s.t.} & y_1 & -y_2 & \geq & -0.6 \\
& -y_1 & -y_2 & \geq & -1.2 - (-0.5)x^1 \\
& -y_1 & & \geq & -1 \\
& & -y_2 & \geq & -1 \\
& y_1, & y_2 & \geq & 0.
\end{array}
$$

The optimal solution is $y^1(\omega^1) = 0.05$, $y^2(\omega^1) = 0.65$ (see Fig. 9.8a) with $\varphi_{LP}(x^1, \omega^1) = -1.4$. Get dual solution $\pi_1(\omega^1) = (0, 2, 0, 0)^\top$. The resulting cut is

$$
\eta_1 \geq (0, 2, 0, 0)(\begin{pmatrix} -0.6 \\ -1.2 \\ -1 \\ -1 \end{pmatrix} - \begin{bmatrix} 0 \\ -0.5 \\ 0 \\ 0 \end{bmatrix} x)
$$

$$
\Rightarrow \eta_1 \geq -2.4 + x.
$$

For scenario ω^2,

$$\begin{aligned}\varphi_{LP}(x^1,\omega^2) := \text{Min } & -2y_1 \; -2y_2 \\ \text{s.t.} \quad & \;\; y_1 \; -y_2 \geq -0.6 \\ & -y_1 \; -y_2 \geq -1.4 - (-0.5)x^1 \\ & -y_1 \qquad \;\; \geq \;\; -1 \\ & \qquad \;\; -y_2 \geq \;\; -1 \\ & \;\; y_1, \quad \; y_2 \geq \;\; 0.\end{aligned}$$

The optimal solution is $y_1(\omega^2) = 0.15$, $y_2(\omega^2) = 0.75$ (see Fig. 9.8b) with $\varphi_{LP}(x^1,\omega^2) = -1.8$. Get dual solution $\pi_1(\omega^2) = (0,2,0,0)^\top$. The resulting cuts are

$$\eta_2 \geq (0,2,0,0,)(\begin{pmatrix}-0.6\\-1.4\\-1\\-1\end{pmatrix} - \begin{bmatrix}0\\-0.5\\0\\0\end{bmatrix} x)$$

$$\Rightarrow \eta_2 \geq -2.8 + x.$$

Step 2: Update and Solve the Master Program. Since the two scenarios are equally likely, the expected values associated with the cut coefficients yield $-x + \eta \geq -2.6$, and we add the optimality cut to the master program. Using $\mathscr{L}$, x^1, and $F(x^1)$, compute

$$\beta_1 = (F(x^1) - \mathscr{L}) \leftarrow (0 - (-2.6)) = 2.6$$

and

$$\alpha_1 = (F(x^1) - \mathscr{L})(1 - |\mathbb{S}^1|) + \mathscr{L} \leftarrow 0.6(1-1) + (-2.6) = -2.6.$$

Then the L^2 cut is $-2.6x + \eta \geq -2.6$. Appending the cut to master program (see Fig. 9.9), we get

$$\begin{aligned} z^2 := \text{Min } & -2x + \eta \\ \text{s.t. } & -x \geq -1 \\ & -x + \eta \geq -2.6 \\ & -2.6x + \eta \geq -2.6 \\ & x \in \{0,1\},\; \eta \text{ free.}\end{aligned}$$

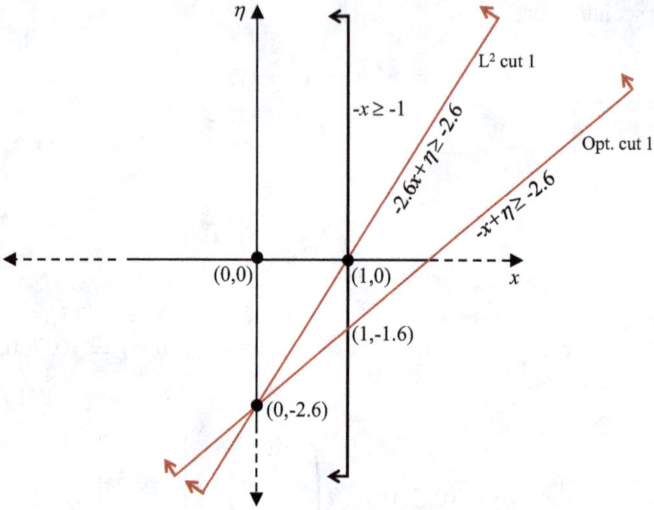

Fig. 9.9 *BFS* algorithm master program feasible region after iteration 1

Solving the master program, we get $x^2 = 0$, $\eta = -2.6$ and objective value $z^2 = -2.6$. Therefore, the lower bound $L^2 \leftarrow \max\{z^2, L^1\} = \{-2.6, -\infty\} = -2.6$. This completes the first iteration of the algorithm.

Step 3. Termination. We have $U^2 - L^2 = 0 - (-2.6) = 2.6$. Since $U^2 - L^2 > \epsilon U^2 = 2.6 \times 10^{-6}$, set $k \leftarrow 2$ and go to step 1.

Iteration $k = 2$:

Step 1.

(a) Solve Subproblem MIPs. For scenario ω^1

$$
\begin{array}{rrrrl}
\varphi(x^2, \omega^1) := \text{Min} & -2y_1 & -2y_2 & & \\
\text{s.t.} & y_1 & -y_2 & \geq & -0.6 \\
 & -y_1 & -y_2 & \geq & -1.2 - (-0.5)x^2 \\
 & -y_1 & & \geq & -1 \\
 & & -y_2 & \geq & -1 \\
 & y_1, & y_2 & \in & \{0, 1\}.
\end{array}
$$

The optimal solution is $y^1(\omega^1) = 1$ and $y_2(\omega^1) = 0$ (see Fig. 9.10a) with $\varphi(x^2, \omega^1) = -2$. For scenario ω^2,

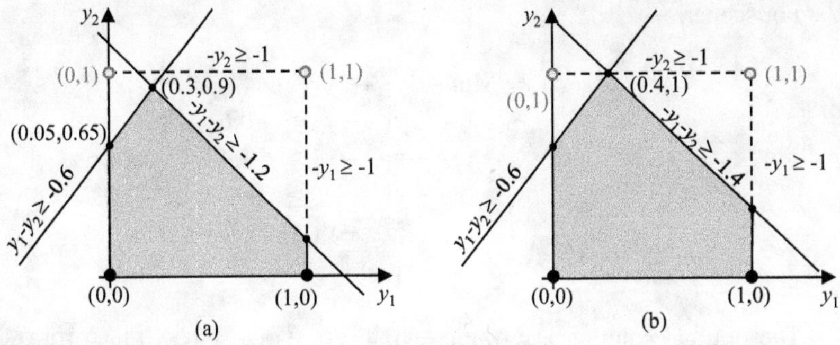

Fig. 9.10 Subproblem feasible region of Example 9.2 instance for $x = 0$. (**a**) Scenario ω^1, $x = 0$. (**b**) Scenario ω^2, $x = 0$

$$\begin{array}{rrrll}
\varphi(x^2, \omega^1) := \text{Min} & -2y_1 & -2y_2 & & \\
\text{s.t.} & y_1 & -y_2 \geq & -0.6 & \\
& -y_1 & -y_2 \geq & -1.4 & -(-0.5)x^2 \\
& -y_1 & \geq & -1 & \\
& & -y_2 \geq & -1 & \\
& y_1, & y_2 \in & \{0, 1\}. &
\end{array}$$

The optimal solution is $y_1(\omega^2) = 1$ and $y_2(\omega^2) = 0$ (see Fig. 9.10b) with $\varphi(x^2, \omega^1) = -2$. Therefore, $F(x^2) \leftarrow \mathbb{E}[f(x^2, \tilde{\omega})] = 0.5(-2)+0.5(-2) = -2$. Update upper bound $U^3 \leftarrow \min\{c^\top x^2 + F(x^2), U^2\} = \min\{-2, 0\} = -2$. Since U^2 is updated, set $x_\epsilon^* \leftarrow x^2 = 1$ and $z_\epsilon^* \leftarrow U^3 = -2$.

(b) Solve Subproblem LPs. For scenario ω^1

$$\begin{array}{rrrll}
\varphi_{LP}(x^2, \omega^1) := \text{Min} & -2y_1 & -2y_2 & & \\
\text{s.t.} & y_1 & -y_2 \geq & -0.6 & -0 \\
& -y_1 & -y_2 \geq & -1.2 & -(-0.5)x^2 \\
& -y_1 & \geq & -1 & -0 \\
& & -y_2 \geq & -1 & -0 \\
& y_1, & y_2 \geq & 0. &
\end{array}$$

The optimal solution is $y_1(\omega^1) = 0.3$, $y_2(\omega^1) = 0.9$ (see Fig. 9.10a) with $\varphi_{LP}(x^2, \omega^1) = -2.4$. Get dual solution $\pi_2(\omega^1) = (0, 2, 0, 0)^\top$. The resulting cut is

$$\eta_1 \geq (0, 2, 0, 0)(\begin{pmatrix} -0.6 \\ -1.2 \\ -1 \\ -1 \end{pmatrix} - \begin{bmatrix} 0 \\ -0.5 \\ 0 \\ 0 \end{bmatrix} x)$$

$$\Rightarrow \eta_1 \geq -2.4 + x.$$

For scenario ω^2,

$$\begin{array}{rrrrl} \varphi_{LP}(x^2,\omega^2) := \text{Min} & -2y_1 & -2y_2 & & \\ \text{s.t.} & y_1 & -y_2 & \geq & -0.6 \\ & -y_1 & -y_2 & \geq & -1.4 - (-0.5)x^2 \\ & -y_1 & & \geq & -1 \\ & & -y_2 & \geq & -1 \\ & y_1, & y_2 & \geq & 0. \end{array}$$

The optimal solution is $y_1(\omega^2) = 0.4$, $y_2(\omega^2) = 1$ (see Fig. 9.10b) with $\varphi_{LP}(x^2, \omega^1) = -2.8$. Therefore, $F(x^2) \leftarrow \mathbb{E}[f(x^2, \tilde{\omega})] = 0.5(-2) + 0.5(-2) = -2$. Get dual solution $\pi_2(\omega^2) = (0, 2, 0, 0)^\top$. The resulting cut is

$$\eta_2 \geq (0, 2, 0, 0,)(\begin{pmatrix} -0.6 \\ -1.4 \\ -1 \\ -1 \end{pmatrix} - \begin{bmatrix} 0 \\ -0.5 \\ 0 \\ 0 \end{bmatrix} x)$$

$$\Rightarrow \eta_2 \geq -2.8 + x.$$

Step 2. Update and Solve the Master Program. Since the two scenarios are equally likely, the expected values associated with the cut coefficients yield $-x + \eta \geq -2.6$, and we add the optimality cut to the master program. Using $\mathscr{L}$, x^2, and $F(x^2)$, compute

$$\beta_1 = (F(x^2) - \mathscr{L}) = (-2 - (-2.6)) = 0.6$$

and

$$\alpha_1 = (F(x^2) - \mathscr{L})(1 - |\mathbb{S}^1|) + \mathscr{L} = 0.6(1) + (-2.6) = -2.$$

Then the L^2 cut is $0.6x + \eta \geq -2$. Appending the cut to the master program (see Fig. 9.11), we get

$$\begin{aligned} z^3 := \text{Min}\ & -2x + \eta \\ \text{s.t.}\ & -x \geq -1 \\ & -x + \eta \geq -2.6 \\ & -2.6x + \eta \geq -2.6 \\ & 0.6x + \eta \geq -2 \\ & x \in \{0, 1\},\ \eta \text{ free}. \end{aligned}$$

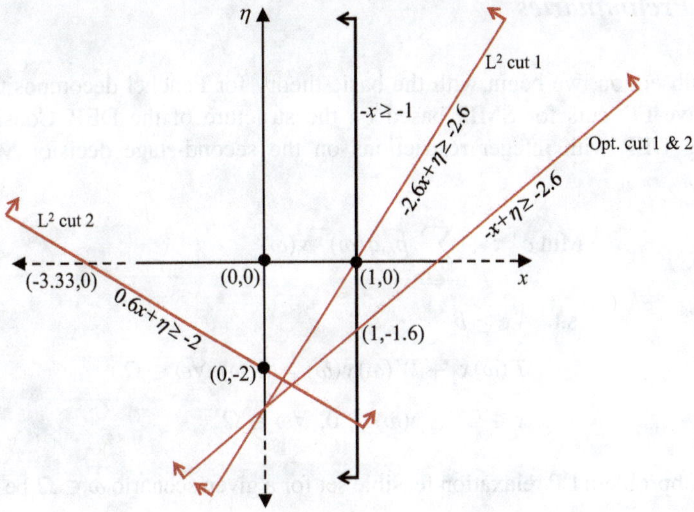

Fig. 9.11 *BFS* algorithm master program feasible region after iteration 2

Solving the master program, we get $x^3 = 0$, $\eta = -2$ and objective value $z^3 = -2$. Therefore, the lower bound $L^3 \leftarrow \max\{z^3, L^2\} = \{-2, -2.6\} = -2$. This completes the second iteration of the algorithm.

Step 3. **Termination.** Since $U^3 - L^3 = 0$, we terminate the algorithm and declare the optimal solution to be $x^*_\epsilon = 1$ with optimal value $z^*_\epsilon = -2$.

---

end

9.6 Fenchel Decomposition

We shall now study the Fenchel decomposition (FD) method, which is a cutting-plane method for SMIP with arbitrary first- and second-stage. As with the BFS method, FD adopts a Benders decomposition framework allowing for the LP relaxation of the SMIP to be solved using the L-shaped algorithm. The FD method we present keeps the master problem as an integer program while the second stage is relaxed and cutting-planes (called *FD cuts*) are generated and added at each iteration of the algorithm. FD cuts are designed to separate non-integer points from the convex hull of integer points of the feasible space. Specifically, the FD method generates and uses FD cuts in the second stage of the relaxed problem to approximate the second-stage MIP feasible set. In turn, these cuts allow for approximating the recourse function via the dual solutions as in Benders decomposition. Next, we provide the foundations of FD and then move on to the derivation of the FD method.

9.6.1 Preliminaries

In this subsection, we begin with the basic theory for Fenchel decomposition and then derive FD cuts for SMIP based on the structure of the DEP. Consider the following DEP with integer restrictions on the second-stage decision variables relaxed:

$$\begin{aligned} \text{Min } & c^\top x + \sum_{\omega \in \Omega} p_\omega q(\omega)^\top y(\omega) \\ \text{s.t. } & Ax \geq b \\ & T(\omega)x + W(\omega)y(\omega) \geq h(\omega) \;\; \forall \omega \in \Omega \\ & x \in \mathbb{B}^{n_1}, \;\; y(\omega) \geq 0, \;\; \forall \omega \in \Omega. \end{aligned} \tag{9.11}$$

Let the subproblem LP relaxation feasible set for a given scenario $\omega \in \Omega$ be defined as

$$Y_{LP}(\omega) = \left\{(x, y(\omega)) \in \mathbb{B}^{n_1} \times \mathbb{R}_+^{n_2} \mid Ax \geq b, \; T(\omega)x + W(\omega)y(\omega) \geq h(\omega)\right\}.$$

Then the set of subproblem feasible solutions for ω can be given by

$$Y_{IP}(\omega) = \left\{(x, y(\omega)) \in Y_{LP}(\omega) \mid y_j(\omega) \in \mathbb{B}, \; \forall j \in J\right\}.$$

Also, let $Y_{IP}^c(\omega)$ denote the convex hull of $Y_{IP}(\omega)$, denoted $conv(Y_{IP}(\omega))$. A key result from FD is summarized in the following theorem:

Theorem 9.4 *Let $(\hat{x}, \hat{y}(\omega)) \in Y_{LP}(\omega)$ be given. Define $g(\omega, \alpha(\omega), \beta(\omega)) = Max \{\alpha(\omega)^\top x + \beta(\omega)^\top y(\omega) \mid (x, y(\omega)) \in Y_{IP}^c(\omega)\}$ and let $\delta(\omega, \alpha(\omega), \beta(\omega)) = \alpha(\omega)^\top \hat{x} + \beta(\omega)^\top \hat{y}(\omega) - g(\omega, \alpha(\omega), \beta(\omega))$. Then there exists vectors $\alpha(\omega)$ and $\beta(\omega)$ for which $\delta(\omega, \alpha(\omega), \beta(\omega)) > 0$ if and only if $(\hat{x}, \hat{y}(\omega)) \notin Y_{IP}^c(\omega)$.*

Theorem 9.4 follows from a basic result for generating a *Fenchel cut* in IP [8]. The theorem enables generating valid inequalities for $Y_{IP}^c(\omega)$ for a given scenario $\omega \in \Omega$. The inequality $\alpha(\omega)^\top x + \beta(\omega)^\top y(\omega) \leq g(\omega, \alpha(\omega), \beta(\omega))$ is valid for $Y_{IP}^c(\omega)$ and separates $Y_{IP}^c(\omega, x)$ from $(\hat{x}, \hat{y}(\omega))$ if and only if $\delta(\omega, \alpha(\omega), \beta(\omega)) > 0$ and is then referred to as an *FD cut*.

The function $\delta(\omega, \alpha(\omega), \beta(\omega))$ is piecewise linear and concave since the value $\alpha(\omega)^\top \hat{x} + \beta(\omega)^\top \hat{y}(\omega)$ is a linear function of $\alpha(\omega)$ and $\beta(\omega)$ and the function $g(\omega, \alpha(\omega), \beta(\omega))$ is piecewise linear and convex. Thus, we can use subgradient optimization to maximize $\delta(\omega, \alpha(\omega), \beta(\omega))$ if $\alpha(\omega)$ and $\beta(\omega)$ are chosen from some convex set Π. We know that for a concave function $f : \mathbb{R}^m \mapsto \mathbb{R}$, a subgradient $s \in \mathbb{R}^m$ of f at $\bar{x} \in \mathbb{R}^m$ must satisfy $f(x) \leq f(\bar{x}) + s^\top(x - \bar{x})$. Using the following result, we can compute the subgradient of $\delta(\omega, \alpha(\omega), \beta(\omega))$.

Proposition 9.1 *Let $(\hat{x}, \hat{y}(\omega)) \in Y_{LP}(\omega)$ be given and let $(\bar{x}, \bar{y}(\omega)) \in Y^c_{IP}(\omega)$ satisfy $g(\omega, \bar{\alpha}(\omega), \bar{\beta}(\omega)) = \bar{\alpha}(\omega)^\top \bar{x} + \bar{\beta}(\omega)^\top \bar{y}(\omega)$. Then $[(\hat{x} - \bar{x})^\top; (\hat{y}(\omega) - \bar{y}(\omega))^\top]^\top$ is a subgradient of $\delta(\omega, \alpha(\omega), \beta(\omega))$ at $(\bar{\alpha}(\omega), \bar{\beta}(\omega))$.*

The fundamental idea of FD is to sequentially generate (at least partially) $Y^c_{IP}(\omega)$ for each $\omega \in \Omega$. Because $Y^c_{IP}(\omega)$ appears in the definition of $g(\omega, \alpha(\omega), \beta(\omega))$ in Theorem 9.4, we deal with $conv\{y(\omega) \mid y(\omega) \in Y_{IP}(\omega)\}$ instead. For a given non-integer point $(\hat{x}, \hat{y}(\omega)) \in Y_{LP}(\omega)$, we can solve the following problem to generate a valid inequality for $Y_{IP}(\omega)$:

$$\delta(\omega) = \underset{(\alpha(\omega),\beta(\omega))\in\Pi}{\text{Max}} \left\{\alpha(\omega)^\top \hat{x} + \beta(\omega)^\top \hat{y}(\omega) - g(\omega, \alpha(\omega), \beta(\omega))\right\}, \tag{9.12}$$

where

$$g(\omega, \alpha(\omega), \beta(\omega)) = \text{Max}\left\{\alpha(\omega)^\top x + \beta(\omega)^\top y \mid (x, y(\omega)) \in Y^c_{IP}(\omega)\right\}.$$

According to Theorem 9.4, the non-integer point $(\hat{x}, \hat{y}(\omega)) \in Y_{LP}(\omega)$ is cut off by the FD cut $\alpha(\omega)^\top x + \beta(\omega)^\top y(\omega) \leq g(\omega, \alpha(\omega), \beta(\omega))$ if $\delta(\omega) > 0$. And in this case, we append the FD cut to $Y_{LP}(\omega)$. Note that the FD cut is generated so that it cuts off the largest distance (based on Π) between the point $(\hat{x}, \hat{y}(\omega))$ and the cut itself. However, to generate FD cuts Problem (9.12) has to be solved several times in general. Therefore, it is desirable to have a linearly constrained domain for Π to speed up computation time. The L^1 unit sphere, $\Pi = \{(\alpha(\omega), \beta(\omega)) \in \mathbb{R}^{n_1+n_2}_+ \mid 0 \leq \alpha(\omega) \leq 1, 0 \leq \beta(\omega) \leq 1\}$, provides such a choice.

9.6.2 FD Algorithm

We shall now devise a basic *FD* algorithm based on the Benders decomposition [4] framework so that the LP relaxation (9.11) can be solved using the L-shaped method [28]. In our LP relaxation, we keep the master problem as an MIP and only relax the second-stage problem. Now let k denote the current iteration of the *FD* algorithm and t denote previous iterations. Furthermore, let $\mathcal{K}_k(\omega)$ denote the set of iteration indices up to k where an FD cut is generated and added to the subproblem for scenario ω. Then the master program at iteration k can be given as follows:

$$\begin{aligned} &\text{Min}\ c^\top x + \eta \\ &\text{s.t.}\ Ax \geq b \\ &\qquad \sigma_t^\top x + \eta \leq \nu_t,\ t = 1, \cdots, k \\ &\qquad x \in \mathbb{B}^{n_1},\ \eta \text{ free}. \end{aligned} \tag{9.13}$$

The second set of constraints in Problem (9.13) is the Benders *optimality* cuts. We do not include feasibility cuts because of the *relatively complete recourse* assumption (A2). For a given solution x^k from the master program and outcome $\omega \in \Omega$, the relaxed subproblem is

$$\begin{aligned} \varphi_c^k(x^k, \omega) = \text{Min } & q(\omega)^\top y \\ \text{s.t. } & W(\omega)y \geq h(\omega) - T(\omega)x^k \\ & \beta(\omega)^\top y \leq g(\omega, \alpha(\omega), \beta(\omega)) - \alpha(\omega)^\top x^k, \ t \in \mathcal{K}_k(\omega) \\ & y \geq 0, \end{aligned} \tag{9.14}$$

where the second set of constraints in Problem (9.14) is the FD cuts generated and added to the scenario problem by iteration k.

A basic FD algorithm can stated as follows:

Algorithm FD
begin

Step 0. Initialization. Set $k \leftarrow 1$, $L^1 \leftarrow -\infty$, $U^1 \leftarrow \infty$, $\mathcal{K}_1(\omega) \leftarrow \emptyset, \forall \omega \in \Omega$, and choose $\epsilon > 0$ and $x^1 \in X \cap \mathbb{B}^{n_1}$. Initialize incumbent solution $x_\epsilon^* \leftarrow x^1$ and optimal value $z_\epsilon^* \leftarrow \infty$.

Step 1. Solve LP Relaxation. Solve LP relaxation (9.11) to get solution $(x^k, \{y^k(\omega)\}_{\omega \in \Omega})$ and objective value z^k. Set $L^k \leftarrow z_k$. If $(x^k, \{y^k(\omega)\}_{\omega \in \Omega})$ satisfy the integer restrictions, set $x_\epsilon^* \leftarrow x^k$ and $z_\epsilon^* \leftarrow z^k$, $U^k \leftarrow z^k$ and stop, x_ϵ^* is optimal. Otherwise, set $k \leftarrow k+1$ and go to step 3.

Step 2. Solve Subproblem LPs. Solve subproblem (9.14) for all $\omega \in \Omega$. If $(x^k, \{y^k(\omega)\}_{\omega \in \Omega})$ satisfy integer restrictions, set $U^{k+1} \leftarrow \min\{c^\top x^k + \mathbb{E}[\varphi_c^k(x^k, \tilde{\omega})], U^k\}$. If U^{k+1} is updated, set incumbent solution $x_\epsilon^* \leftarrow x^k$ and $z_\epsilon^* \leftarrow U^{k+1}$, and go to step 5.

Step 3. Solve FD Cut Generation Subproblems and Add Cuts. For all $\omega \in \Omega$ such that $(x^k, y^k(\omega))$ is non-integer, form and solve (9.12) to obtain $\alpha^k(\omega)$, $\beta^k(\omega)$ and $g(\omega, \alpha^k(\omega), \beta^k(\omega))$. Form an FD cut

$$\alpha^k(\omega)^\top x + \beta^k(\omega)^\top y \leq g(\omega, \alpha^k(\omega), \beta^k(\omega))$$

and append to subproblem (9.14) for ω. Update the set $\mathcal{K}_{k+1}(\omega) \leftarrow \mathcal{K}_k(\omega) \cup \{k\}$.

Step 4. Re-Solve Subproblem LPs.
Re-solve subproblem (9.14) for all $\omega \in \Omega$ for which an FD cut was added. If $(x^k, \{y^k(\omega)\}_{\omega \in \Omega})$ satisfy integer restrictions, set $U^{k+1} \leftarrow \min\{c^\top x^k + \mathbb{E}[\varphi_c^k(x^k, \omega)], U^k\}$. If U^{k+1} is updated, set incumbent solution $x_\epsilon^* \leftarrow x^k$ and $z_\epsilon^* \leftarrow U^{k+1}$. Go to step 5.

Step 5. Update and Solve the Master Problem.
Compute an *optimality cut* using the dual multipliers from the most recently solved subproblem and add to the master Problem (9.13). Solve the master

problem to get optimal solution x^{k+1} and optimal value z^{k+1}. Set $L^{k+1} \leftarrow \max\{z^{k+1},\ L^k\}$. If $U^{k+1} - L^{k+1} \leq \epsilon$, stop and declare x^*_ϵ ϵ-optimal with value and z^*_ϵ. Otherwise, $k \leftarrow k + 1$ and repeat from step 2.

---

end

In step 1 of the FD algorithm, one has a choice of algorithms to use for solving the SLP such as a decomposition algorithm (e.g., L-shaped) or a direct solver applied to the DEP to the SLP.

9.6.3 FD Cut Generation

Let us now move on to how to generate an FD cut in Step 3 of the FD algorithm. We need to solve Problem (9.12). One way is to use a cutting-plane decomposition approach based on Proposition 9.1. To that end, for a given non-integer solution $(x^k, y^k(\omega)) \in Y_{LP}(\omega)$, $\delta^k(\omega)$ can be maximized by solving

$$\begin{aligned} \delta^k(\omega) = \quad & \underset{(\alpha(\omega),\beta(\omega))\in\Pi}{\text{Max}} \quad \theta \\ & \text{s.t.} \quad -\theta + (x^k - x)^\top \alpha(\omega) + (y^k(\omega) - y(\omega))^\top \beta(\omega) \geq 0 \\ & \quad (x, y(\omega)) \in Y^c_E(\omega), \end{aligned} \tag{9.15}$$

where $Y^c_E(\omega)$ is the set of extreme points of $Y^c_{IP}(\omega)$. The free variable θ in Problem (9.15) gives the maximum distance (based on Π) between the non-integer point $(x^k, y^k(\omega))$ and the FD cut. We shall solve Problem (9.15) using a cutting-plane method as follows: Let τ and τ' be iteration indices for the FD cut generation subroutine. Then given $(x^k, y^k(\omega)) \in Y_{LP}(\omega)$ at iteration k of the FD algorithm, the subroutine master problem at iteration τ is given as follows:

$$\begin{aligned} \delta^k_\tau(\omega) = \quad & \underset{(\alpha(\omega),\beta(\omega))\in\Pi}{\text{Max}} \quad \theta \\ \text{s.t.} \quad & -\theta + (x^k - x^\tau(\omega))^\top \alpha(\omega) + (y^k(\omega) - y^\tau(\omega))^\top \beta(\omega) \geq 0, \\ & \qquad \tau = 1, \cdots, \tau'. \end{aligned} \tag{9.16}$$

Given a solution $(\theta^\tau, \alpha^\tau(\omega), \beta^\tau(\omega))$ to (9.16) at iteration τ, $(x^\tau, y^\tau(\omega))$ is the optimal solution to the following subproblem:

$$\begin{aligned} g(\omega, \alpha^\tau(\omega), \beta^\tau(\omega)) = \text{Max} \;\; & \alpha^\tau(\omega)^\top x + \beta^\tau(\omega)^\top y(\omega) \\ \text{s.t.} \;\; & (x, y(\omega)) \in Y_{IP}(\omega). \end{aligned} \tag{9.17}$$

We can now state a basic FD cut generation subroutine for step 3 of the FD algorithm as follows:

Algorithm *FD Cut Generation Subroutine*
begin

Step 3.0. **Initialization.** Set $\tau \leftarrow 0$, $\ell^0 \leftarrow -\infty$, $u^0 \leftarrow \infty$, choose $\epsilon' > 0$ and $(\alpha^0(\omega), \beta^0(\omega)) \in \Pi$.

Step 3.1. **Solve Subproblem and Compute Lower Bound.** Use $(\alpha^\tau(\omega), \beta^\tau(\omega))$ to form and solve subproblem (9.17) to get solution $(x^\tau, y^\tau(\omega))$ and objective value $g(\omega, \alpha^\tau(\omega), \beta^\tau(\omega))$. Let

$$d_\tau = (x^k)^\top \alpha^\tau(\omega) + y^k(\omega)^\top \beta^\tau(\omega) - g(\omega, \alpha^\tau(\omega), \beta^\tau(\omega)).$$

Set $\ell^{\tau+1} \leftarrow \max\{d_\tau,\ \ell^\tau\}$. If $\ell^{\tau+1}$ is updated, set incumbent solution $(\alpha^*(\omega), \beta^*(\omega), g^*(\omega, \alpha^*(\omega), \beta^*(\omega))) \leftarrow (\alpha^\tau(\omega), \beta^\tau(\omega), g(\omega, \alpha^\tau(\omega), \beta^\tau(\omega)))$.

Step 3.2. **Solve Master Problem.** Use current non-integer solution $(x^k, y^k(\omega))$ and subproblem (9.17) solution $(x^\tau, y^\tau(\omega))$ to form and add constraint (9.16) to master problem. Solve master problem to get an optimal solution $(\theta^{\tau+1}, \alpha^{\tau+1}(\omega), \beta^{\tau+1}(\omega))$. Set $u^{\tau+1} \leftarrow \min\{\theta^{\tau+1},\ u^\tau\}$. If $u^{\tau+1} - \ell^{\tau+1} \le \epsilon'$, stop and declare incumbent solution ϵ'-optimal. Otherwise, set $\tau \leftarrow \tau + 1$ and go to step 1.

---

end

Generating an FD cut can be expensive because subproblem (9.17) is an IP and is difficult to solve in general. Therefore, the FD method is better suited for problems with special structure or where the structure of the subproblem can be exploited so that generating an FD cut is not computationally expensive. Knapsack problems with nonnegative left hand side coefficients are a good example of problems from the literature with special structure where the FD has worked very well.

A few remarks regarding the finite termination of the FD algorithm are now in order. Finite termination of the FD algorithm to an optimal solution can be guaranteed, in theory, if Π is chosen to be an n_1+n_2-dimensional set containing the origin in its strict interior so that $\delta(\omega)$ attains a positive value in $\mathbb{R}^{n_1+n_2}$ if and only if it achieves a positive value on Π [9]. However, the rate of convergence can vary based on the type of norm used to define Π. Furthermore, cutting-plane methods tend to be more effective if used in a BAC setting, i.e., in combination with BAB.

9.6.4 Numerical Illustration of the FD Algorithm

Example 9.4 We are now ready to illustrate the basic FD algorithm using the instance given in Example 9.2. The FD algorithm can be applied to the example instance as follows:

Algorithm *FD*
begin

Step 0. **Initialization.** Set $k \leftarrow 1$, $L^1 \leftarrow -\infty$, $U^1 \leftarrow \infty$, $\mathscr{K}_1(\omega^1) = \mathscr{K}_1(\omega^2) \leftarrow \emptyset$, and choose $\epsilon \leftarrow 10^{-6}$ and $x^1 \leftarrow 1$. Initialize incumbent solution $x^*_\epsilon \leftarrow x^1$ and optimal value $z^*_\epsilon \leftarrow \infty$.

Step 1. **Solve LP Relaxation.** We can use the L-shaped algorithm to solve the LP relaxation, but since we only have two scenarios, we can instead simply formulate the DEP relaxation with decision variables y_1^s and y_2^s for $s = 1, 2$ as follows:

$$
\begin{array}{llll}
\text{Min} - 2x & -y_1^1 - y_2^1 & -y_1^2 - y_2^2 & \\
\text{s.t.} & y_1^1 - y_2^1 & & \geq -0.6 \\
- 0.5x & -y_1^1 - y_2^1 & & \geq -1.2 \\
& -y_1^1 & & \geq -1 \\
& - y_2^1 & & \geq -1 \\
& & y_1^2 - y_2^2 & \geq -0.6 \\
- 0.5x & & -y_1^2 - y_2^2 & \geq -1.4 \\
& & -y_1^2 & \geq -1 \\
& & - y_2^2 & \geq -1 \\
x \in \{0, 1\}, & y_1^1, y_2^1, & y_1^2, y_2^2 & \geq 0.
\end{array}
$$

Solving the relaxation and we get $x^1 = 1$, $y_1^1 = 0.05$, $y_2^1 = 0.65$, $y_1^2 = 0.15$, $y_2^2 = 0.75$ and objective value $z^1 = -3.6$. This means that $x^k = x^1 = 1$, $y^k(\omega_1) = y^1(\omega_1) = (0.05, 0.65)^\top$, and $y^k(\omega_2) = y^1(\omega_2) = (0.15, 0.75)^\top$. Set $L^1 \leftarrow \max\{z^1, L^1\} = -3.6$. Since the solution does not satisfy integer restrictions, set $k \leftarrow 2$ and go to step 3.

Step 3. **Solve FD Cut Generation Subproblems and Add Cuts.**
Form and solve (9.12) using the cut generation subroutine.

Scenario ω^1:

Algorithm *FD Cut Generation Subroutine*
begin

Step 3.0. **Initialization.** Set $\tau \leftarrow 0$, $\ell^0 \leftarrow -\infty$, $u^0 \leftarrow \infty$ and choose $\epsilon' \leftarrow 10^{-6}$, $\alpha^0(\omega^1) \leftarrow 1$, $\beta^0(\omega^1) \leftarrow (1, 1)^\top$.

Step 3.1. Solve Subproblem and Compute Lower Bound.

$$\begin{aligned} g(\omega^1, \alpha^0(\omega^1), \beta^0(\omega^1)) = \text{Max} \quad & \alpha^0(\omega^1)x + \beta_1^0(\omega^1)y_1 + \beta_2^0(\omega^1)y_2 \\ \text{s.t.} \quad & y_1 - y_2 \geq -0.6 \\ & -0.5x - y_1 - y_2 \geq -1.2 \\ & -y_1 \geq -1 \\ & -y_2 \geq -1 \\ & x,\ y_1, y_2 \in \{0, 1\}. \end{aligned}$$

Solve to get solution $(x^0, y_1^0(\omega^1), y_2^0(\omega^1)) = (0, 1, 0)$ with objective value 1. Let $d_0 = x^1\alpha^0(\omega^1) + y^1(\omega^1)^\top \beta^0(\omega^1) - 1 = 1 + 0.05 + 0.65 - 1 = 0.7$, $\ell^1 \leftarrow \max\{0.7, \ell^0\} = 0.7$, and set incumbent solution to $\alpha^0(\omega^1) \leftarrow 1, \beta^0(\omega^1) \leftarrow (1, 1)$ and $g(\omega^1, \alpha^0(\omega^1), \beta^0(\omega^1)) \leftarrow 1$.

Step 3.2. Solve Master Problem.

$$\begin{aligned} \delta_1^1(\omega^1) = & \text{Max } \theta \\ & \text{s.t. } -\theta + (1-0)\alpha(\omega^1) + (0.05-1)\beta_1(\omega^1) + (0.65-0)\beta_2(\omega^1) \geq 0. \\ & 0 \leq \alpha(\omega^1) \leq 1,\ 0 \leq \beta_1(\omega^1), \beta_2(\omega^1) \leq 1. \end{aligned}$$

Solve to get $\theta^1 = 1.65, \alpha^1(\omega^1) = 1, \beta^1(\omega^1) = (0, 1)^\top$. Upper bound $u^1 \leftarrow \min\{1.65, u^0\} = 1.65$. Since $u^1 - \ell^1 = 1.65 - 0.7 = 0.95 > \epsilon'$, set $\tau \leftarrow \tau + 1 = 1$ and return to step 1.

Step 3.1. Solve Subproblem and Compute Lower Bound.

$$\begin{aligned} g(\omega^1, \alpha^1(\omega^1), \beta^1(\omega^1)) = \text{Max } & \alpha^1(\omega^1)x + \beta_1^1(\omega^1)y_1 + \beta_2^1(\omega^1)y_2 \\ \text{s.t.} \quad & y_1 - y_2 \geq -0.6 \\ & -0.5x - y_1 - y_2 \geq -1.2 \\ & -y_1 \geq -1 \\ & -y_2 \geq -1 \\ & x,\ y_1, y_2 \in \{0, 1\}. \end{aligned}$$

Solve to get solution $(x^1, y_1^1(\omega^1), y_2^1(\omega^1)) = (1, 0, 0)$ with objective value 1. Let $d_1 = x^1\alpha^1(\omega^1) + y^1(\omega^1)^\top \beta^1(\omega^1) - 1 = 1 + 0 + 0.65 - 1 = 0.65$.
$\ell^2 \leftarrow \max\{0.7, 0.65\} = 0.7$, and the incumbent solution is $\alpha^1(\omega^1) \leftarrow 1$, $\beta^1(\omega^1) \leftarrow (0, 1)$ and $g(\omega^1, \alpha^1(\omega^1), \beta^1(\omega^1)) \leftarrow 1$.

Step 3.2. **Solve Master Problem.**

$$\delta_2^1(\omega^1) = \text{Max } \theta$$

$$\text{s.t. } -\theta + (1-0)\alpha(\omega^1) + (0.05-1)\beta_1(\omega^1) + (0.65-0)\beta_2(\omega^1) \geq 0$$

$$-\theta + (1-1)\alpha(\omega^1) + (0.05-0)\beta_1(\omega^1) + (0.65-0)\beta_2(\omega^1) \geq 0$$

$$0 \leq \alpha(\omega^1) \leq 1,\ 0 \leq \beta_1(\omega^1), \beta_2(\omega^1) \leq 1.$$

Solving to get $\theta^2 = 0.7, \alpha^2 = 1, \beta_1^2(\omega^1) = \beta_2^2(\omega^1) = 1$. Upper bound $u^2 = \min\{0.7, 1.65\} = 0.7$. Since $u^2 - \ell^2 = 0.7 - 0.7 = 0$, stop the cut generation subroutine. Add FD cut $x + y_1 + y_2 \leq 1$ to the subproblem for scenario ω^1 and update $\mathcal{K}_1(\omega^1) \leftarrow \{1\}$.

Scenario ω^2:

Step 3.0. **Initialization.** Set $\tau \leftarrow 1, \ell^0 \leftarrow -\infty, u^0 \leftarrow \infty$ and choose $\epsilon' \leftarrow 10^{-6}, \alpha^0(\omega^2) \leftarrow 1, \beta^0(\omega^2) \leftarrow (1,1)^\top$.

Step 3.1. **Solve Subproblem and Compute Lower Bound.**

$$g(\omega^2, \alpha^0(\omega^2), \beta^0(\omega^2)) = \text{Max } \alpha^0(\omega^2)x + \beta_1^0(\omega^2)y_1 + \beta_2^0(\omega^2)y_2$$

$$\begin{aligned} \text{s.t.} \quad & y_1 - y_2 \geq -0.6 \\ & -0.5x - y_1 - y_2 \geq -1.4 \\ & -y_1 \geq -1 \\ & -y_2 \geq -1 \\ & x, y_1, y_2 \in \{0,1\}. \end{aligned}$$

Solving the subproblem we get $(x^0, y_1^0(\omega^2), y_2^0(\omega^2)) = (0, 1, 0)$ with objective value $= 1$. Let $d_0 = x^1\alpha^0(\omega^2) + y^1(\omega^2)^\top \beta^0(\omega^2) - 1 = 1 + 0.15 + 0.75 - 1 = 0.9$, $\ell^1 \leftarrow \max\{0.9, \ell^0\} = 0.9$, and the incumbent solution is $\alpha^0(\omega^2) \leftarrow 1$, $\beta^0(\omega^2) \leftarrow (1,1)^\top$ and $g(\omega^2, \alpha^0(\omega^2), \beta^0(\omega^2)) \leftarrow 1$.

Step 3.2. **Solve Master Problem.**

$$\delta_1^1(\omega^2) = \text{Max } \theta$$

$$\text{s.t. } -\theta + (1-0)\alpha(\omega^2) - 0.85\beta_1(\omega^2) + 0.75\beta_2(\omega^2) \geq 0.$$

$$0 \leq \alpha(\omega^2) \leq 1,\ 0 \leq \beta_1(\omega^2), \beta_2(\omega^2) \leq 1.$$

Solving the master program we obtain $\theta^1 = 1.75, \alpha^1(\omega^2) = 1, \beta_1^1(\omega^2) = 0$, and $\beta_2^1(\omega^2) = 1$. Upper bound $u^1 \leftarrow \min\{1.75, u^0\} = 1.75$. Since $u^1 - \ell^1 = 1.75 - 0.9 = 0.85 > \epsilon'$, set $\tau \leftarrow \tau + 1 = 1$.

Step 3.1. Solve Subproblem and Compute Lower Bound.

$$g(\omega^2, \alpha^1(\omega^2), \beta^1(\omega^2)) = \text{Max } \alpha^1(\omega^2)x + \beta_1^1(\omega^2)y_1 + \beta_2^1(\omega^2)y_2$$
$$\text{s.t.} \quad y_1 - y_2 \geq -0.6$$
$$-0.5x - y_1 - y_2 \geq -1.4$$
$$-y_1 \geq -1$$
$$-y_2 \geq -1$$
$$x,\ y_1, y_2 \in \{0, 1\}.$$

Solve and we get solution $(x^1, y_1^1(\omega^2), y_2^1(\omega^2)) = (1, 0, 0)$ with objective value 1. Let $d_1 = x^1\alpha^1(\omega^2) + y^1(\omega^2)^\top \beta^1(\omega^2) - 1 = 1 + 0 + 0.75 - 1 = 0.75$. $\ell^2 \leftarrow \max\{0.75, 0.9\} = 0.9$, and the incumbent solution is $\alpha^1(\omega^2) = 1$, $\beta^1(\omega^2) = (0, 1)^\top$ and $g(\omega^2, \alpha^1(\omega^2), \beta^1(\omega^2)) = 1$.

Step 3.2. Solve Master Problem.

$$\delta_2^1(\omega^2) = \text{Max } \theta$$
$$\text{s.t} \ -\theta + (1-0)\alpha(\omega^2) + (0.15-1)\beta_1(\omega^2) + (0.75-0)\beta_2(\omega^2) \geq 0$$
$$-\theta + (1-1)\alpha(\omega^2) + (0.15-0)\beta_1(\omega^2) + (0.75-0)\beta_2(\omega^2) \geq 0.$$
$$0 \leq \alpha(\omega^2) \leq 1,\ 0 \leq \beta_1(\omega^2), \beta_2(\omega^2) \leq 1.$$

Solving the master program we get $\theta^2 = 0.9$, $\alpha^2(\omega^2) = 1$, $\beta_1^2(\omega^2) = \beta_2^2(\omega^2) = 1$. Upper bound $u^2 \leftarrow \min\{0.9, u^1\} = 0.9$. Since $u^2 - \ell^2 = 0.9 - 0.9 = 0$, stop the cut generation subroutine. Add $x + y_1 + y_2 \leq 1$ to the subproblem for scenario ω^2 and update $\mathcal{K}_1(\omega^2) \leftarrow \{1\}$. See Fig. 9.12 for a plot of the feasible region for each scenario after adding the FD cuts.

end

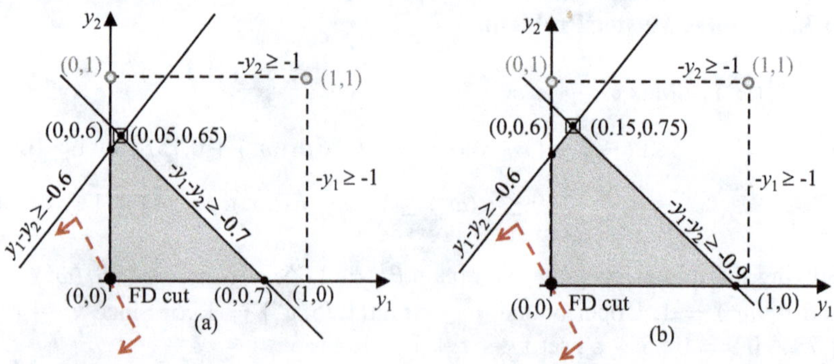

Fig. 9.12 Subproblem feasible region for each scenario after adding FD cuts at iteration 1. (**a**) Scenario ω^1. (**b**) Scenario ω^2

Step 4. Resolve Subproblem LPs

Scenario ω^1:

$$
\begin{aligned}
\varphi_{LP}(x^1, \omega^1) = \text{Min} \quad & -2y_1 - 2y_2 \\
\text{s.t.} \quad & y_1 - y_2 \geq -0.6 \\
& -y_1 - y_2 \geq -1.2 - (-0.5)x^1 \\
& -y_1 \geq -1 \\
& -y_2 \geq -1 \\
& -y_1 - y_2 \geq -1 + x^1 \\
& y_1, y_2 \geq 0.
\end{aligned}
$$

The optimal solution is $x = 1$ and $y^1(\omega^1) = (0, 0)^\top$ with corresponding dual solution $d(\omega^1) = (0, 0, 0, 0, 2)^\top$.

Scenario ω^2:

$$
\begin{aligned}
\varphi_{LP}(x^1, \omega^2) = \text{Min} \quad & -2y_1 - 2y_2 \\
\text{s.t.} \quad & y_1 - y_2 \geq -0.6 \\
& -y_1 - y_2 \geq -1.4 - (-0.5)x^1 \\
& -y_1 \geq -1 \\
& -y_2 \geq -1 \\
& -y_1 - y_2 \geq -1 + x^1 \\
& y_1, y_2 \geq 0.
\end{aligned}
$$

The optimal solution is $x = 1$ and $y^1(\omega^2) = (0, 0)^\top$ with corresponding dual solution $d(\omega^2) = (0, 0, 0, 0, 2)^\top$. Since the integer restrictions are satisfied, we update $U^2 \leftarrow \min\{-2, U^1\} = -2$, $x_\epsilon^* \leftarrow 1$ and objective value $z_\epsilon^* \leftarrow U^2 = -2$, and go to step 5.

Step 5. Update and Solve the Master Problem

Using the dual solution for each subproblem from Step 4, formulate the optimality cut: $\eta \geq \sum_{s=1}^{2} p(\omega^s) d(\omega^s)^\top (r(\omega^s) - (T)^\top(\omega^s)x)$.

Scenario ω^1:

$$
\eta_1 \geq (0, 0, 0, 0, 2)\left(\begin{pmatrix} -0.6 \\ -1.2 \\ -1 \\ -1 \\ -1 \end{pmatrix} - \begin{bmatrix} 0 \\ -0.5 \\ 0 \\ 0 \\ -1 \end{bmatrix} x\right)
$$

$$
\Rightarrow \eta_1 \geq -2 + 2x
$$

Scenario ω^2:

$$\eta_2 \geq (0, 0, 0, 0, 2)(\begin{pmatrix} -0.6 \\ -1.4 \\ -1 \\ -1 \\ -1 \end{pmatrix} - \begin{bmatrix} 0 \\ -0.5 \\ 0 \\ 0 \\ -1 \end{bmatrix} x)$$

$$\Rightarrow \eta_2 \geq -2 + 2x.$$

Since the two scenarios are equally likely, the expected values associated with the cut coefficients yield $-2x + \eta \geq -2$. Adding the optimality cut to the master program (see Fig. 9.13), we get:

$$\begin{aligned} z^2 := \text{Min} \quad & -2x + \eta \\ \text{s.t.} \quad & -x \geq -1 \\ & -2x + \eta \geq -2 \\ & x \in \{0, 1\}, \eta \text{ free.} \end{aligned}$$

Solving the master program, we obtain $x^2 = 1$ and $\eta = 0$ with objective value $z^2 = -2$. Therefore, the lower bound becomes $L^2 \leftarrow \max\{-2, L^1\} = -2$. This completes the first iteration of the algorithm. Since $U^2 - L^2 = 0 < \epsilon$, we terminate the algorithm and declare the optimal solution to be $x^*_\epsilon = 1$ with optimal value $z^*_\epsilon = -2$. Notice that $x^*_\epsilon = 0$ is an alternative optimal solution.

---

end

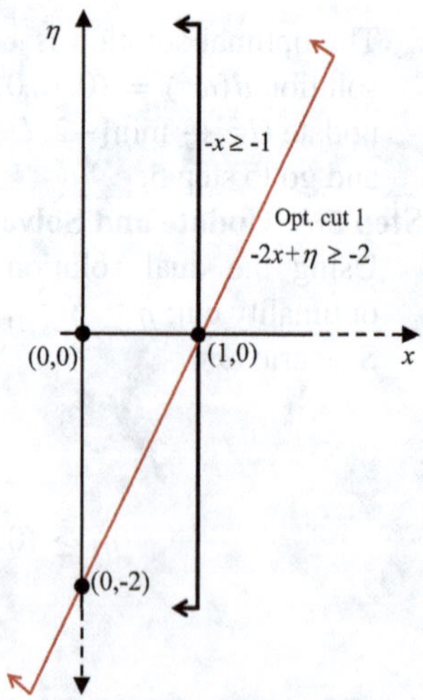

Fig. 9.13 *FD* algorithm master program feasible region after iteration 1

9.7 Disjunctive Decomposition

In this section, we study the disjunctive decomposition (D^2) method, a cutting-plane method for SMIP with pure binary first stage and mixed-binary second stage. As with FD, the D^2 method adopts a Benders decomposition framework allowing for the LP relaxation of the SMIP to be solved using the L-shaped method. D^2requires the first-stage solution to be an extreme point of the first-stage feasible set and therefore, the master problem is solved as a binary program. The second stage is relaxed and cutting-planes are generated and added at each iteration of the algorithm. The D^2 method is based on disjunctive programming, a branch of mathematical programming that deals with the characterization of disjunctions or unions of sets. Specifically, the D^2 method is a cutting-plane method that uses disjunctive cuts (called D^2 cuts) in the second stage of the relaxed problem to sequentially approximate the second-stage MIP feasible set. In turn, these cuts allow for approximating the recourse function via the dual solutions as in Benders decomposition.

The form of Problem (9.1) suitable for the D^2 method can be stated as follows:

$$\underset{x \in X \cap \mathbb{B}^{n_1}}{\text{Min}} \quad \mathbb{E}[f(x, \tilde{\omega})] + \lambda \mathbb{D}[f(x, \tilde{\omega})], \tag{9.18}$$

where the set $X = \{x \in \mathbb{R}_+^{n_1} \mid Ax \geq b\}$ is nonempty and includes the inequality $-x \geq -1$. For a given $x \in X \cap \mathbb{B}^{n_1}$, the real random cost variable $f(x, \tilde{\omega})$ is given by $f(x, \tilde{\omega}) := c^\top x + \varphi(x, \tilde{\omega})$ and for a given realization ω of $\tilde{\omega}$, the recourse function $\varphi(x, \omega)$ is given by

$$\begin{aligned} \varphi(x, \omega) :=& \text{Min } q(\omega)^\top y \\ &\text{s.t. } Wy \geq r(\omega) - T(\omega)x \\ &\qquad y \in \mathbb{B}^{n_z} \times \mathbb{R}_+^{n_2 - n_z}. \end{aligned} \tag{9.19}$$

In Problem (9.19), the inequality $Wy \geq r(\omega) - T(\omega)x$ includes $-y_j \geq -1, \ j = 1, \cdots, n_z$. For a given $(x, \omega) \in X \times \Omega$, the second-stage integer feasible set is given as

$$Y(x, \omega) = \{y \geq 0 \mid Wy(\omega) \geq r(\omega) - T(\omega)x, \ y_j \in \{0, 1\}, \ j = 1, \cdots, n_z\}.$$

Thus, we have

$$\varphi(x, \omega) := \underset{Y(x,\omega)}{\text{Min}} \quad q(\omega)^\top y.$$

Letting $y = (y_z, y_u)$ and $q(\omega) = (q_z(\omega), q_u(\omega))$, we can alternatively write Problem (9.19) as follows:

$$\varphi(x,\omega) := \text{Min } q_z(\omega)^\top y_z + q_u(\omega)^\top y_z \tag{9.20}$$
$$\text{s.t. } W_z y_z + W_u y_u \geq r(\omega) - T(\omega)x$$
$$y_z \in \mathbb{B}^{n_z}, \; y_u \in \mathbb{R}_+^{n_u},$$

where $n_u = n_2 - n_z$. For the remainder of this section, we shall continue to make the assumption that the risk measure $\mathbb{D}$ is convex and focus our algorithm derivation on the risk-neutral case, i.e., for $\lambda := 0$.

9.7.1 Preliminaries

Now that we have formally stated the appropriate problem for the D^2 method , let us start with stating the fundamental concepts needed for the derivation of the D^2 algorithm. The reader familiar with disjunctive programming may skip this subsection.

Definition 9.2 (Disjunctive Set) Consider the sets

$$S_h = \{y \in \mathbb{R}_+^{n_2} \mid G_h y \geq r_h\}, h \in H,$$

where set h has two atoms, i.e., $H = \{0, 1\}$. Then the feasible set S, expressed as

$$S = \cup_{h \in H} S_h,$$

is a disjunctive set.

A *disjunctive set* is a union of sets. In our case, we shall focus on the union of polyhedral sets of the form S_h without integer requirements.

Definition 9.3 (Valid Inequality) An inequality $\pi^\top y \geq \pi_0$ is *valid* for S if it is satisfied by all $y \in S$, i.e., $S \subseteq \{y \in \mathbb{R}_+^{n_2} \mid \pi^\top y \geq \pi_0\}$.

By definition, the inequality $\pi^\top y \geq \pi_0$ is valid for $S = S_0 \cup S_1$ if it does not cut off any feasible points in S_0 and S_1 as illustrated in Fig. 9.14. As a cutting-plane method, the D^2approach focuses on the generation of such valid inequalities. Note that the union of the convex sets S_h, $h \in H$, is not convex, i.e., S is not convex. Consequently, to maintain a convex approach, we will need to derive a convexification of S using *reverse* convex programming. The fundamental theoretical result that will enable us to generate valid inequalities for S is the *disjunctive cut principle*, which we state in the following theorem.

Theorem 9.5 (Disjunctive Cut Principle)

(a) ***Forward Part*** *[3]. Assume that S and $S_h, h \in H$, are defined as before. If we can find $\lambda_h \geq 0, \forall h \in H$, then*

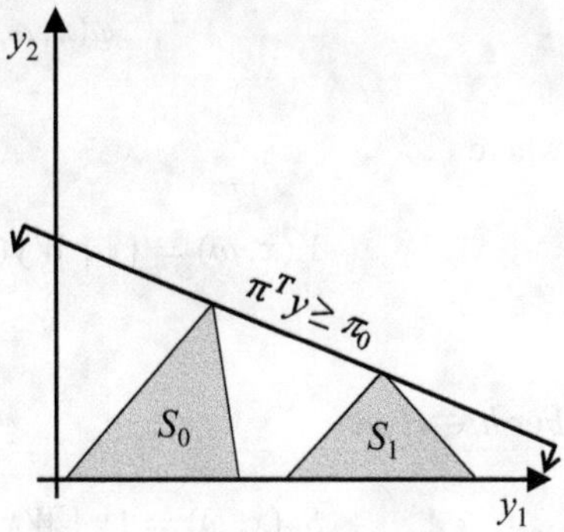

Fig. 9.14 A valid inequality for $S = S_0 \cup S_1$

$$\sum_j \{\max_{h \in H} \lambda_h^\top G_{h_j}\} y_j \geq \min_{h \in H} \{\lambda_h^\top r_h\} \tag{9.21}$$

is valid for $S = \cup_{h \in H} S_h$.

(b) ***Reverse Part*** *[5]. Suppose* $\pi^\top y \geq \pi_0$ *is valid for* S *and* $H^* = \{h \in H \mid S_h \neq \emptyset\}$, *then there exists* $\{\lambda_h \geq 0\}_{h \in H}$ *such that*

$$\pi_j \geq \max_{h \in H^*} \{\lambda_h^\top G_{h_j}\}, \forall j \text{ and } \pi_0 \leq \min_{h \in H^*} \{\lambda_h^\top r_h\}. \tag{9.22}$$

Note that π_0 in (9.22) is a piecewise linear concave function since it is a minimum of affine functions. An important set we need is the set that contains all π and π_0 that form all valid inequalities for disjunctive set S. This set is referred to as the *reverse polar*.

Definition 9.4 (Reverse Polar) The reverse polar of the set S, denoted $S^{\#}$, is defined as

$$S^{\#} = \{(\pi, \pi_0) \mid \exists \{\lambda_h \geq 0\}_{h \in H} \text{ such that relation (9.22) is satisfied}\}.$$

When π_0 is fixed, we can denote the reverse polar by $S^{\#}(\pi_0)$. We shall assume that S is *full-dimensional* and $S_h \neq \emptyset,\ h \in H$. Recall that a polytope is full-dimensional if it is an n-dimensional object in $\mathbb{R}^n$. Then $\pi^\top y \geq \pi_0$, with $\pi_0 \neq 0$ is a *facet* of $cl(conv(S))$ if and only if (π, π_0) is an extreme point of $S^{\#}(\pi_0)$. We shall use this concept later to convexify $S = \cup_{h \in H} S_h$.

9.7.2 D^2 Cut Generation

We shall now apply the preliminary results we have outlined so far to derive valid inequalities for our second-stage problem:

$$\varphi(x, \omega) := \underset{Y(x,\omega)}{\text{Min}} \ q(\omega)^\top y,$$

where

$$Y(x, \omega) = \{y \mid Wy(\omega) \geq r(\omega) - T(\omega)x,$$
$$y \geq 0, \ y_j \in \{0, 1\}, \ j = 1, \cdots, n_z\}.$$

For $h \in H$, let

$$S_h(x, \omega) = \{y \mid Wy \geq r(\omega) - T(\omega)x,$$
$$C_h^\top y \geq d_h,$$
$$y \geq 0, -y_j \geq -1, \ j = 1, \cdots, n_z\},$$

where $C_h^\top y \geq d_h$ denotes a set of sequentially generated valid inequalities. Letting $W_h = [W^\top; C_h^\top]^\top$ and $r_h(\omega) = (r(\omega); d_h)$, we can rewrite $S_h(x, \omega)$ as follows:

$$S_h(x, \omega) = \{y \mid W_h y(\omega) \geq r_h(\omega) - T_h(\omega)x,$$
$$y \geq 0, \ y_j \in \{0, 1\}, \ j = 1, \cdots, n_z\}.$$

We shall now consider the disjunction $S(x, \omega) = \cup_{h \in H} S_h(x, \omega)$.

Theorem 9.6 (The Common-Cut-Coefficients (C^3) Theorem [25]) *Let $(\bar{x}, \bar{\omega})$ be given (after solving the first stage and choosing some scenario $\bar{\omega} \in \Omega$) and suppose that $S_h(\bar{x}, \bar{\omega}) \neq \emptyset, \forall h \in H$, and $\pi^\top y \geq \pi_0(\bar{x}, \bar{\omega})$ is valid for $S(\bar{x}, \bar{\omega})$. Then there exists a function $\pi_0 : X \times \Omega \rightarrow \mathbb{R}$ such that for any $(x, \omega) \in X \times \Omega$, $\pi^\top y \geq \pi_0(x, \omega)$ is a valid inequality for $S(x, \omega)$.*

The C^3 Theorem 9.6 allows for a valid inequality for $S(\bar{x}, \bar{\omega})$ of the form $\pi^\top y \geq \pi_0(\bar{x}, \bar{\omega})$ to be translated to $\pi^\top y \geq \pi_0(x, \omega)$, which is valid for $S(x, \omega)$ for all $(x, \omega) \in X \times \omega$.

The theorem is called C^3 due to the fact that one (common) π is generated at each iteration of the algorithm and used for all $\omega \in \Omega$. This saves on computation time because computing a π for each scenario is very expensive as it requires solving an LP that is more than twice the size (number of variables and constraints) of the subproblem LP. Thus, the key idea of the D^2 method is to generate a π for all $\omega \in \Omega$ and then compute $\pi_0(x, \omega)$ for each $\omega \in \Omega$ as needed. This means that at each iteration we append $\pi^\top y \geq \pi_0(x, \omega)$ to the subproblem LP relaxation as follows:

$$\varphi_0(x, \omega) := \text{Min } q(\omega)^\top y$$
$$\text{s.t. } Wy \geq r(\omega) - T(\omega)x$$

$$\pi^\top y \geq \pi_0(x, \omega)$$
$$y \geq 0.$$

We are interested in using $\varphi_0(x, \omega)$ to generate lower bounds on the recourse function $\varphi(x, \omega)$ at each iteration of the algorithm. This can be accomplished via Benders decomposition by using the dual multipliers to the above LP. However, $\pi_0(x, \omega)$, which is part of the right hand side, is a piecewise linear *concave* function. Therefore, we need to convexify $\pi_0(x, \omega)$ to get $\pi_c(x, \omega)$ to have a convex approximation of $\varphi(x, \omega)$ so that

$$\begin{aligned}\varphi(x, \omega) \geq \varphi_{LP}(x, \omega) := &\text{Min } q(\omega)^\top y \\ &\text{s.t. } Wy \geq r(\omega) - T(\omega)x \\ &\quad \pi^\top y \geq \pi_c(x, \omega) \\ &\quad y \geq 0.\end{aligned}$$

We shall now address how to generate the left hand side coefficient π. Let k denote the D^2 algorithm iteration counter. Then, at iteration k we have

$$\begin{aligned}\varphi_{LP}(x^k, \omega) := &\text{Min } q(\omega)^\top y \\ &\text{s.t. } W^k y \geq r^k(\omega) - T^k(\omega)x^k \\ &\quad y \geq 0,\end{aligned} \tag{9.23}$$

where for $k = 1$, $W^1 := W$, $T^1(\omega) := T(\omega)$, and $r^1(\omega) := r(\omega)$, and as iterations progress, i.e., for $k \geq 1$, $(\pi^k)^\top y \geq \pi_c(x^k, \omega)$ is generated and added to the subproblem LP relaxation. In other words, π^k is appended to W^{k-1} and $\pi_c(x^k, \omega)$ is appended to $r^{k-1}(\omega) - T^{k-1}(\omega)x^k$. The C^3 theorem ensures that with a translation, a valid inequality $(\pi^k)^\top y \geq \pi_c(\bar{x}^k, \bar{\omega})$ derived for one pair $(\bar{x}, \bar{\omega}) \in X \times \Omega$ is used to derive a valid inequality for other pair $(x, \omega) \in X \times \Omega$.

The goal of generating cutting-planes is to cut off (eliminate) fractional (non-integer) solutions for each scenario $\omega \in \Omega$ subproblem LP. For binary (0–1) programs, the closure of the convex hull of 0–1 integer points can be recovered by sequentially generating convex hulls using one disjunction at a time. We illustrate this concept using Fig. 9.15. In the figure, we have the LP relaxation feasible set shown in (a). Sequentially, in (b) we generate the disjunctive cut $(\pi^1)^\top y \geq \pi_c^1$ based on y_1 and add it to the LP relaxation feasible region to cut off the non-integer point (0.5, 1). In part (c), we generate the disjunctive cut $(\pi^2)^\top y \geq \pi_c^2$ based on y_2 and add it to the LP relaxation feasible region to cut off the non-integer point $(1, 0.5)$. Observe that through this sequential process, in (c) we fully recover the convex hull of the 0–1 integer points!

Let at iteration k the solution to the subproblem LP relaxation (9.23) be given as $y^k(\omega)$. Denote by $j(k)$ the binary component (index) of $y^k(\omega)$ with a fractional

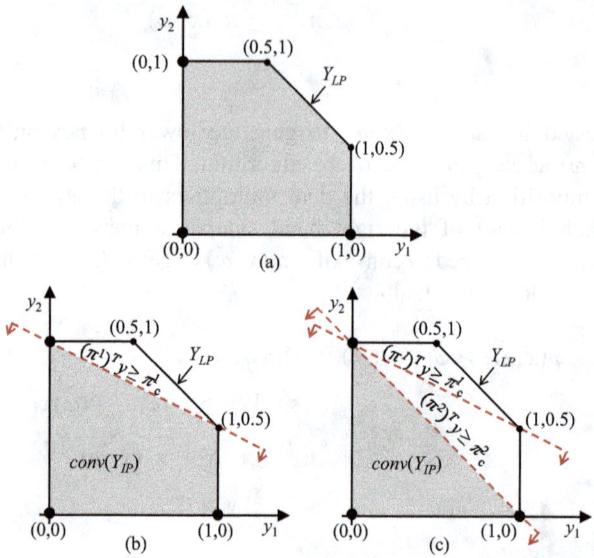

Fig. 9.15 Illustration of sequentially recovering the convex hull of binary points: (**a**) feasible region; (**b**) disjunction on y_1; and (**c**) disjunction on y_2

value, i.e., $0 < y_j^k(\omega) < 1$. At each iteration k we shall choose a fractional component, indexed $j(k)$, to use as a basis for creating a disjunction to generate a D^2 cut to cut off the fractional solution. We refer to $j(k)$ as the *disjunction index* and to $y_j^k(\omega)$ as the *disjunction variable*. Various rules can be used to select the disjunction variable. Here are some examples of such rules:

(R1) Choose $j(k)$ such that $y_j^k(\omega)$ is the *first* fractional solution component among all $\omega \in \Omega$.

(R2) Choose $j(k)$ such that $y_j^k(\omega)$ is the *most* fractional solution component among all $\omega \in \Omega$, i.e.,

$$j(k) := \underset{j=1,\cdots,n_z}{\text{argmin}} \{|y_j^k(\omega) - 0.5| : \forall \omega \in \Omega\}.$$

(R3) Choose $j(k)$ such that $y_j^k(\omega)$ is the *smallest* fractional solution component among all $\omega \in \Omega$, i.e.,

$$j(k) := \underset{j=1,\cdots,n_z}{\text{argmin}} \{y_j^k(\omega) : \forall \omega \in \Omega\}.$$

(R4) Choose $j(k)$ such that $y_j^k(\omega)$ is the *largest* fractional solution component among all $\omega \in \Omega$, i.e.,

Table 9.2 Example subproblem LP (9.23) solution at iteration k

Scenario	Solution	
	$j(k) = 1$	$j(k) = 2$
ω^1	$y_1^k(\omega_1) = 0.30$	$y_2^k(\omega_1) = 0.90$
ω^2	$y_1^k(\omega_2) = 0.40$	$y_2^k(\omega_2) = 1.00$

$$j(k) := \underset{j=1,\cdots,n_z}{\text{argmax}} \ \{y_j^k(\omega) : \ \forall \omega \in \Omega\}.$$

(R5) Choose $j(k)$ as the component j with the *largest number of scenarios* with $0 < y_j^k(\omega) < 1$.

Example 9.5 Consider the subproblem LP (9.23) solution given in Table 9.2 at iteration k of the D^2 algorithm for some instance with two scenarios, ω^1 and ω^2.

(a) What is $j(k)$ when disjunction variable rule (R1) is applied?
(b) What is $j(k)$ when disjunction variable rule (R2) is applied?
(c) What is $j(k)$ when disjunction variable rule (R3) is applied?
(d) What is $j(k)$ when disjunction variable rule (R4) is applied?
(e) What is $j(k)$ when disjunction variable rule (R5) is applied?

Solution

(a) Rule (R1): The *first* fractional value in Table 9.2 is $y_1^k(\omega_2) = 0.30$. Therefore, the disjunction variable index $j(k) = 1$.
(b) Rule (R2): The component with the most fractional value must correspond to $y_j^k(\omega)$ closest to 0.5. From Table 9.2, we can see that $y_1^k(\omega_2) = 0.40$ is the closest to 0.5. Therefore, the disjunction variable index $j(k) = 1$.
(c) Rule (R3): The solution with the smallest fractional value in Table 9.2 is $y_1^k(\omega_1) = 0.30$. Therefore, the disjunction variable index $j(k) = 1$.
(d) Rule (R4): We can see from Table 9.2 that the solution with the largest fractional component value is $y_2^k(\omega_1) = 0.90$. Therefore, the disjunction variable index $j(k) = 2$.
(e) Rule (R5): The solution component in Table 9.2 with a fractional value in both scenarios is $j = 1$ with $y_1^k(\omega_1) = 0.30$ and $y_2^k(\omega_1) = 0.40$. Therefore, the disjunction variable index $j(k) = 1$.

Generating the Common-Cut-Coefficients π

After choosing a disjunctive index $j(k)$ at iteration k of the algorithm, we need to create a disjunction to form the reverse polar to obtain $(\pi, \pi_0(\omega))$ based on the first-stage solution x^k. Recall that $H = \{0, 1\}$ and let

$$\begin{aligned} S(x^k, \omega) &= \cup_{h \in H} S_{h, j(k)}(x^k, \omega) \\ &= S_{0, j(k)}(x^k, \omega) \cup S_{1, j(k)}(x^k, \omega), \end{aligned} \tag{9.24}$$

where

$$S_{0,j(k)}(x^k,\omega) := \{y \geq 0 \mid W^k y \geq r^k(\omega) - T^k(\omega)x^k, \tag{9.25}$$

$$- y_j \geq -\lfloor y_j^k \rfloor \} \tag{9.26}$$

and

$$S_{1,j(k)}(x^k,\omega) := \{y \geq 0 \mid W^k y \geq r^k(\omega) - T^k(\omega)x^k, \tag{9.27}$$

$$y_j \geq \lceil y_j^k \rceil \}. \tag{9.28}$$

Notice that since $0 < y_j^k < 1$, we have $\lfloor y_j^k \rfloor = 0$ in constraint (9.26) and $\lceil y_j^k \rceil = 1$ in constraint (9.28). Let vector λ_{01} and scalar λ_{02} be nonnegative multipliers associated with constraints (9.25) and (9.26), respectively. Similarly, let λ_{11} and λ_{12} be nonnegative multipliers associated with constraints (9.27) and (9.28), respectively. We assume that both vectors λ_{01} and λ_{11} are appropriately dimensioned. Let us now define

$$I_j^k = \begin{cases} 0, & \text{if } j \neq j(k) \\ 1, & \text{otherwise} \end{cases}$$

and let $S^\#$ denote the (scaled) reverse polar. Applying the C^3 Theorem 9.6, $S^\#$ can be written as follows:

$$\begin{aligned}
S^\# := \{\pi \in \mathbb{R}^{n_2}, \pi_0(\omega) \in \mathbb{R},\ \forall \omega \in \Omega \mid \\
& \pi_j - \lambda_{01}^\top W^k + I_j^k \lambda_{02} \geq 0,\ \forall j = 1, \cdots, n_2 \\
& \pi_j - \lambda_{11}^\top W^k - I_j^k \lambda_{12} \geq 0,\ \forall j = 1, \cdots, n_2 \\
& - \pi_0(\omega) + \lambda_{01}^\top (r^k(\omega) - T^k(\omega)x^k) - \lambda_{02} \lfloor y_j^k(\omega) \rfloor \geq 0,\ \forall \omega \in \Omega \\
& - \pi_0(\omega) + \lambda_{11}^\top (r^k(\omega) - T^k(\omega)x^k) + \lambda_{12} \lceil y_j^k(\omega) \rceil \geq 0,\ \forall \omega \in \Omega \\
& - 1 \leq \pi_j \leq 1,\ \forall j = 1, \cdots, n_2 && (9.29a) \\
& - 1 \leq \pi_0(\omega) \leq 1,\ \forall \omega \in \Omega && (9.29b) \\
& \lambda_{01}, \lambda_{02}, \lambda_{11}, \lambda_{12} \geq 0\}.
\end{aligned}$$

The reverse polar $S^\#$ is scaled by normalizing (π, π_0) based on the L_∞-norm in constraints (9.29a) and (9.29b). This normalization is necessary because we want to generate (π, π_0) so that $\pi^\top y \geq \pi_0(\omega)$ cuts off the non-integer solution $y^k(\omega)$ for $\omega \in \Omega$. Therefore, to obtain the desired coefficients, we need to maximize the slackness of this inequality, i.e., maximize the amount by which the fractional solution $y^k(\omega)$ violates $\pi^\top y \geq \pi_0(\omega)$ for $\omega \in \Omega$. This can be accomplished by solving the following SLP:

$$\underset{(\pi,\pi_0)\in S^\#}{\text{Max}} \quad \mathbb{E}[\pi_0(\tilde{\omega})] - \mathbb{E}[y^k(\tilde{\omega})^\top]\pi. \tag{9.30}$$

We shall refer to Problem (9.30) as the C^3-SLP since the π generated from this SLP is independent of ω, i.e., it is common to all $\omega \in \Omega$. We should also point out that problem (9.30) has *simple recourse*.

Remark 9.1 The objective function in C^3-SLP is simply one of the possible alternatives for generating a D^2 cut. The expectation over all scenarios may not necessarily cut off the fractional solution $y^k(\omega)$. Alternatively, one can use conditional expectation based on the scenarios for which the disjunctive variable has a fractional solution.

Convexifying $\pi_0(x, \omega)$

Let us now return to the convexification of $\pi_0(x, \omega)$. Solving C^3-SLP (9.30), we obtain the following solution:

$$\pi^k, \pi_0^k(\omega),\ \forall\omega \in \Omega,\ \lambda_{01}^k, \lambda_{02}^k, \lambda_{11}^k, \text{ and } \lambda_{12}^k.$$

Using this solution, we compute the following parameters:

$$\begin{aligned}
\hat{\nu}_0^k(\omega) &= (\lambda_{01}^k)^\top r^k(\omega) - \lambda_{02}^k \lfloor y_j^k(\omega) \rfloor, \\
\hat{\nu}_1^k(\omega) &= (\lambda_{11}^k)^\top r^k(\omega) + \lambda_{12}^k \lceil y_j^k(\omega) \rceil, \\
\hat{\gamma}_0^k(\omega)^\top &= (\lambda_{01}^k)^\top T^k(\omega), \text{ and} \\
\hat{\gamma}_1^k(\omega)^\top &= (\lambda_{11}^k)^\top T^k(\omega).
\end{aligned}$$

Recall that from the *disjunctive cut principle*, we can write

$$\begin{aligned}
\pi_0(x, \omega) &\le \min_{h\in H} (\lambda_h^k)^\top (r^k(\omega) - T^k(\omega)x) \\
&= \min_{h\in H} \{\hat{\nu}_h^k(\omega) - \hat{\gamma}_h^k(\omega)^\top x\},
\end{aligned}$$

where $\hat{\nu}_h^k(\omega) = (\lambda_h^k)^\top r(\omega)$ and $\hat{\gamma}_h^k(\omega) = (\lambda_h^k)^\top T(\omega)$. Since $H = \{0, 1\}$, we have

$$\pi_0(x, \omega) \le \min \{\hat{\nu}_0^k(\omega) - \hat{\gamma}_0^k(\omega)^\top x,\ \hat{\nu}_1^k(\omega) - \hat{\gamma}_1^k(\omega)^\top x\}.$$

Clearly, $\pi_0(x, \omega)$ is the minimum of affine functions and is, therefore, piecewise linear concave (see Fig. 9.16). This means that we can convexify it.

Define $\Pi_X(\omega)$ to be the epigraph of $\pi_0^k(\cdot, \omega)$ for $\omega \in \Omega$ restricted to $x \in X$ as

$$\Pi_X(\omega) := \{(\theta, x) \mid \theta \ge \pi_0(x, \omega), x \in X\},$$

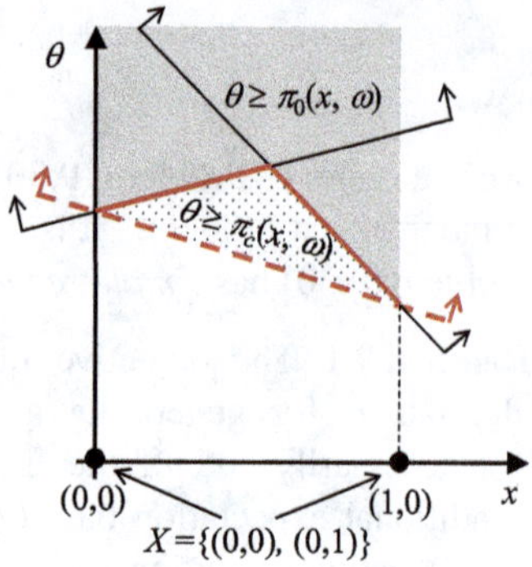

Fig. 9.16 Convexification of $\theta \geq \pi_0(x, \omega)$ to $\theta \geq \pi_c(x, \omega)$ over X

where $X = \{Ax \geq b, x \geq 0\}$. We can write $\Pi_X(\omega)$ as a disjunction as follows:

$$\begin{aligned}\Pi_X(\omega) &= \cup_{h \in H}\, E_h(\omega)\\ &= E_0(\omega) \cup E_1(\omega),\end{aligned}$$

where

$$E_0(\omega) := \{(\theta, x) \mid \theta \geq \hat{\nu}_0(\omega) - \hat{\gamma}_0(\omega)^\top x,\ Ax \geq b, x \geq 0\} \tag{9.31}$$

and

$$E_1(\omega) := \{(\theta, x) \mid \theta \geq \hat{\nu}_1(\omega) - \hat{\gamma}_1(\omega)^\top x,\ Ax \geq b, x \geq 0\}. \tag{9.32}$$

We can now apply the *disjunctive cut principle* to convexify $\Pi_0(\omega)$ over X using the reverse polar, denoted $\Pi_X(\omega)$. We assume that X is bounded and for all $x \in X$, $\theta \geq 0$ in both (9.31) and (9.32). Let $\tau_{0,h} \in \mathbb{R}_+$ and $\tau_h \in \mathbb{R}_+^{m_1}$ be the multipliers associated with the constraints in (9.31) for $h = 0$, and with the constraints in (9.32) for $h = 1$. The scaled reverse polar, $\Pi_X^\#(\omega)$, can be written as follows:

$$\begin{aligned}\Pi_X^\# := \{&\sigma_0(\omega) \in \mathbb{R}, \sigma(\omega) \in \mathbb{R}^{n_1}, \delta(\omega) \in \mathbb{R} \mid\\ &\sigma_o(\omega) - \tau_{00} \geq 0\\ &\sigma_o(\omega) - \tau_{01} \geq 0\\ &\sigma_j(\omega) - \hat{\gamma}_0(\omega)\tau_{00} - \tau_0^\top A_j \geq 0,\ \forall j = 1, \cdots, n_1\\ &\sigma_j(\omega) - \hat{\gamma}_1(\omega)\tau_{01} - \tau_1^\top A_j \geq 0,\ \forall j = 1, \cdots, n_1\\ &-\delta(\omega) + \hat{\nu}_0(\omega)\tau_{00} + b^\top \tau_0 \geq 0\\ &-\delta(\omega) + \hat{\nu}_1(\omega)\tau_{01} + b^\top \tau_1 \geq 0\\ &\tau_{00} + \tau_{01} = 1\\ &\tau_{00}, \tau_{01}, \tau_0, \tau_1 \geq 0\}.\end{aligned} \tag{9.33}$$

Notice that reverse polar $\Pi_X^{\#}(\omega)$ is scaled via Eq. (9.33). This also ensures that $\sigma_0(\omega) \neq 0$ so that θ has a nonzero coefficient. A feasible solution to $\Pi_X^{\#}(\omega)$, $\sigma_0(\omega), \sigma(\omega), \delta(\omega), \tau_{00}, \tau_0, \tau_{01}, \tau_1$, yields

$$\sigma_0(\omega)\theta + \sigma(\omega)^\top x \geq \delta(\omega).$$

This inequality is normalized by dividing by the entire cut $\sigma_0(\omega)$ to get

$$\theta + \frac{\sigma(\omega)^\top}{\sigma_0(\omega)} x \geq \frac{\delta(\omega)}{\sigma_0(\omega)},$$

where $\hat{\nu}(\omega) = \frac{\delta(\omega)}{\sigma_0(\omega)}$ and $\hat{\gamma}(\omega) = \frac{\sigma(\omega)}{\sigma_0(\omega)}$. Therefore, the convexification of $\pi_0(x, \omega)$ yields

$$\pi_c(x, \omega) = \hat{\nu}(\omega) - \hat{\gamma}(\omega)^\top x, \ x \in X. \tag{9.34}$$

Note that $\pi_0(x, \omega) = \pi_c(x, \omega)$ whenever $x \in X$ is an extreme point of X.

Turning to the algorithm context, given $(x^k, \omega) \in X \times \Omega$ at iteration k together with the parameters $\hat{\nu}_0^k(\omega)$, $\hat{\nu}_1^k(\omega)$, $\hat{\gamma}_0^k(\omega)^\top$, and $\hat{\gamma}_1^k(\omega)^\top$ from the C^3-SLP, we form and solve

$$\underset{\Pi_X^{\#}(\omega)}{\text{Max}} \ \delta(\omega) - \sigma_0(\omega) - (x^k)^\top \sigma(\omega) \tag{9.35}$$

to get an optimal solution $\delta^k(\omega)$, $\sigma_0^k(\omega)$, and $\sigma^k(\omega)$, as well as τ_{00}^k, τ_{01}^k, τ_0^k, and τ_1^k. LP (9.35) is referred to as the right hand side LP or RH-SLP for $\omega \in \Omega$. The objective maximizes the distance between $\pi_c^k(x, \omega)$ and $\pi_0^k(x, \omega)$. At iteration k, we generate π^k from C^3-SLP (9.30) and $(\hat{\nu}^k(\omega), \hat{\gamma}^k(\omega))$ from RHS-LP (9.35) for all $\omega \in \Omega$ and then form the *D^2 cut:*

$$\begin{aligned} &(\pi^k)^\top y \geq \pi_c^k(x, \omega) \\ \Rightarrow\ &(\pi^k)^\top y \geq \hat{\nu}^k(\omega) - \hat{\gamma}^k(\omega)^\top x. \end{aligned} \tag{9.36}$$

Inequality (9.36) is added to each subproblem at iteration k with a goal of cutting off the non-integer solution $y^k(\omega)$ for scenario $\omega \in \Omega$.

9.7.3 D^2 Algorithm

We continue to maintain that $X = \{Ax \geq b, x \geq 0\}$. Then at iteration k of the D^2 algorithm, we have the following LP relaxation of the (risk-neutral) problem:

$$\underset{x \in X \cap \mathbb{B}^{n_1}}{\text{Min}} \ c^\top x + \mathbb{E}[\varphi_{LP}(x, \tilde{\omega})]. \tag{9.37}$$

For a given solution $x^k \in X \cap \mathbb{B}^{n_1}$ and $\omega \in \Omega$,

$$\begin{aligned} \varphi_{LP}(x^k, \omega) := &\text{Min } q(\omega)^\top y(\omega), \\ &\text{s.t. } W^k y(\omega) \geq r^k(\omega) - T^k(\omega)x^k \\ &\quad y(\omega) \geq 0. \end{aligned} \tag{9.38}$$

Adopting the Benders decomposition [4] setting for Problem (9.37–9.38), we define the master program at iteration k as follows:

$$\begin{aligned} &\text{Min } c^\top x + \eta \\ &\text{s.t. } Ax \geq b \\ &\quad \sigma_t^\top x + \eta \leq \nu_t, \; t = 1, \cdots, k \\ &\quad x \in \mathbb{B}^{n_1}, \; \eta \geq 0. \end{aligned} \tag{9.39}$$

The second set of constraints in Problem (9.39) is the Benders *optimality* cuts. The optimality cut variable η is a piecewise linear approximation of the translated value of $\mathbb{E}[\varphi_{LP}(x, \tilde{\omega})]$ in the k-th iteration, which in turn is an approximation of $\mathbb{E}[\varphi(x, \tilde{\omega})]$. A basic D^2 algorithm can be stated as follows:

Algorithm *D2*
begin

Step 0. Initialization. Set $k \leftarrow 1$, $x^1 \in X \cap \mathbb{B}^{n_1}$, $U^1 \leftarrow \infty$, $L^1 \leftarrow -\infty$, $\epsilon > 0$, $W^1 \leftarrow W$, $T^1(\omega) \leftarrow T(\omega)$, and $r^1(\omega) = r(\omega)$, for all $\omega \in \Omega$. Initialize incumbent solution $x^*_\epsilon \leftarrow x^1$ and objective value $z^*_\epsilon \leftarrow \infty$.

Step 1. Solve Subproblem LPs. For each $\omega \in \Omega$, use the matrix W^k and right hand side vector $r^k(\omega) - T^k(\omega)x^k$ to solve (9.38). If $\{y^k(\omega)\}_{\omega \in \Omega}$ satisfy the integer restrictions, update $U^{k+1} \leftarrow \min\{c^\top x^k + \mathbb{E}[\varphi_{LP}(x^k, \tilde{\omega})], U^k\}$. Set incumbent solution $x^*_\epsilon \leftarrow x^k$ and $z^*_\epsilon \leftarrow U^{k+1}$ if U^{k+1} is updated and go to step 4.

Step 2. Solve Cut Generation LPs and Perform Updates.
Choose a disjunction index $j(k)$ based on $\{y^k(\omega)\}_{\omega \in \Omega}$ using a disjunction variable rule of choice.

(a) Formulate and solve C^3-SLP (9.30) to obtain $\pi^k(\omega)$ and multipliers ($\lambda^k_{0,1}$, λ^k_{02}, λ^k_{11}, λ^k_{12}). Define $W^{k+1} = [(W^k)^\top; \pi^k]^\top$.

(b) Use the multipliers ($\lambda^k_{0,1}$, λ^k_{02}, λ^k_{11}, λ^k_{12}) obtained in (a) to form and solve RHS-LP (9.35) for each outcome $\omega \in \Omega$. Use the solution to define $\nu^k(\omega)$ and $\gamma^k(\omega)$ and update $r^{k+1}(\omega) = [r^k(\omega), \nu^k(\omega)]$ and $T^{k+1}(\omega) = [(T^k(\omega))^\top; \gamma^k(\omega)]^\top$.

Step 3. **Update and Solve Subproblem LPs.** For each $\omega \in \Omega$, solve (9.38) using W^{k+1} and $r^{k+1}(\omega) - T^{k+1}(\omega)x^k$. If $y^k(\omega)$ satisfies the integer restrictions for all $\omega \in \Omega$, update $U^{k+1} \leftarrow \min\{c^\top x^k + \mathbb{E}[\varphi_{LP}(x^k, \tilde{\omega})], U^k\}$. Set incumbent solution $x^*_\epsilon \leftarrow x^k$ and $z^*_\epsilon \leftarrow U^{k+1}$ if U^k is updated. Otherwise, $U^{k+1} \leftarrow U^k$.

Step 4. **Update and Solve the Master Problem.** Compute an optimality cut using the dual multipliers from the most recently solved subproblem and add to the master Problem (9.39). Solve the master program to get x^{k+1} and let v_{k+1} denote the optimal value. Set $L^{k+1} \leftarrow \max\{v_{k+1}, L^k\}$. If $U^{k+1} - L^{k+1} \leq \epsilon$, stop and declare x^*_ϵ ϵ-optimal with objective value z^*_ϵ. Otherwise, $k \leftarrow k+1$ and repeat from Step 1.

---

end

Remark 9.2 Note that in the basic D^2 algorithm we avoid computing the upper bound U^k since it requires the solution of all scenario subproblem MIPs, which is computationally expensive in general. However, it may be necessary to compute U^k in Step 2 at some iteration k when the first-stage solution x^k stabilizes (stops changing). At this point, we recommend generating the L^2 optimality cut (as described in Sect. 9.5) to add to the master program to aid in closing the LP gap.

9.7.4 Numerical Illustration of the D^2 Algorithm

Example 9.6 We shall now apply the D^2 algorithm to solve the instance given in Example 9.2. Referring to the instance, we observe that for binary values of the second-stage variables, a valid lower bound on the objective function $-2y_1 - 2y_2$ is -4. In order to be consistent with the requirement that the lower bound on the second-stage objective value must be zero, we translate the second-stage objective function by adding 4, thereby ensuring nonnegativity values after translation.

Algorithm D^2
begin

Step 0. **Initialization.** Set $k \leftarrow 1$, $x^1 \leftarrow 0$, $U^1 \leftarrow \infty$, $L^1 \leftarrow -\infty$, $\epsilon \leftarrow 10^{-6}$, $W^1 \leftarrow W$, $T^1(\omega) \leftarrow T(\omega)$, and $r^1(\omega) = r(\omega)$, for all $\omega \in \Omega$:

$$W^1 = W = \begin{bmatrix} 1 & -1 \\ -1 & -1 \\ -1 & 0 \\ 0 & -1 \end{bmatrix}, \quad T^1(\omega^1) = T^1(\omega^2) = T = \begin{bmatrix} 0 \\ -0.5 \\ 0 \\ 0 \end{bmatrix},$$

$$r^1(\omega^1) = (-0.6, -1.2, -1, -1)^\top, \text{ and}$$

$$r^1(\omega^2) = (-0.6, -1.4, -1, -1)^\top.$$

Initialize incumbent solution $x^*_\epsilon \leftarrow x^1 := 0$ and objective value $z^*_\epsilon \leftarrow \infty$.

Step 1. Solve Subproblem LPs. Solve the LP relaxation of the subproblems.
Scenario ω^1:

$$\begin{aligned}
\varphi_{LP}(x^1, \omega^1) := \text{Min} \quad & -y_1 - y_2 \\
\text{s.t.} \quad & y_1 - y_2 \geq -0.6 \\
& -y_1 - y_2 \geq -1.2 - (-0.5)x^1 \\
& -y_1 \geq -1 \\
& -y_2 \geq -1 \\
& y_1, y_2 \geq 0.
\end{aligned}$$

Solving, we get optimal solution $y(\omega^1) = (0.3, 0.9)^\top$.
Scenario ω^2:

$$\begin{aligned}
\varphi_{LP}(x^1, \omega^2) := \text{Min} \quad & -y_1 - y_2 \\
\text{s.t.} \quad & y_1 - y_2 \geq -0.6 \\
& -y_1 - y_2 \geq -1.4 - (-0.5)x^1 \\
& -y_1 \geq -1 \\
& y_1 - y_2 \geq -1 \\
& y_1, y_2 \geq 0.
\end{aligned}$$

The optimal solution for this subproblem is $y(\omega^2) = (0.4, 1)^\top$.

Step 2. Solve Cut Generation LPs and Perform Updates. Since y_1 has the largest fractional value compared to y_2, set $j(k) = j(1) = 1$, i.e., choose y_1 to be the *disjunction variable*.

(a) Form and solve C^3-SLP (9.30):

$$\begin{aligned}
\text{Max } & 0.5\pi_0(\omega^1) + 0.5\pi_0(\omega^2) - 0.5(0.3, 0.9)(\pi_1, \pi_2)^\top - 0.5(0.4, 1)(\pi_1, \pi_2)^\top \\
\text{s.t. } & \pi_1 - \lambda_{01}^\top(1, -1, -1, 0)^\top + \lambda_{02} \geq 0 \\
& \pi_2 - \lambda_{01}^\top(-1, -1, 0, -1)^\top + 0\lambda_{02} \geq 0 \\
& \pi_1 - \lambda_{11}^\top(1, -1, -1, 0)^\top - \lambda_{12} \geq 0 \\
& \pi_2 - \lambda_{11}^\top(-1, -1, 0, -1)^\top - 0\lambda_{12} \geq 0 \\
& \pi_0(\omega^1) - \lambda_{01}^\top(-0.6, -1.2, -1, -1)^\top + 0\lambda_{02} \geq 0 \\
& \pi_0(\omega^2) - \lambda_{01}^\top(-0.6, -1.4, -1, -1)^\top + 0\lambda_{02} \geq 0
\end{aligned}$$

$$\pi_0(\omega^1) - \lambda_{11}^\top(-0.6, -1.2, -1, -1)^\top - \lambda_{12} \geq 0$$
$$\pi_0(\omega^2) - \lambda_{11}^\top(-0.6, -1.4, -1, -1)^\top - \lambda_{12} \geq 0$$
$$-1 \leq \pi_1,\ \pi_2,\ \pi_0(\omega^1),\ \pi_0(\omega^2) \leq 1.$$

Solve to get $\pi^1 = (-0.2, -1)^\top$, $\pi_0(\omega^1) = -0.6$, $\pi_0(\omega^2) = -0.6$, $\lambda_{01} = (1, 0, 0, 0)^\top$, $\lambda_{11} = (0, 1, 0, 0)^\top$, $\lambda_{02} = 1.2$, and $\lambda_{12} = 0.8$.

Perform updates:

$$W^2 \leftarrow [W^1; (\pi^1)^\top] = \begin{bmatrix} 1 & -1 \\ -1 & -1 \\ -1 & 0 \\ 0 & -1 \\ -0.2 & -1 \end{bmatrix}.$$

$\hat{\nu}_0^1(\omega^1) = (1, 0, 0, 0)(-0.6, -1.2, -1, -1)^\top = -0.6.$
$\hat{\nu}_1^1(\omega^1) = (0, 1, 0, 0)(-0.6, -1.2, -1, -1)^\top + 0.8 = -0.4.$
$\hat{\gamma}_0^1(\omega^1) = (1, 0, 0, 0)(0, -0.5, 0, 0)^\top = 0.$
$\hat{\gamma}_1^1(\omega^1) = (0, 1, 0, 0)(0, -0.5, 0, 0)^\top = -0.5.$

$\hat{\nu}_0^1(\omega^2) = (1, 0, 0, 0)(-0.6, -1.4, -1, -1)^\top = -0.6.$
$\hat{\nu}_1^1(\omega^2) = (0, 1, 0, 0)(-0.6, -1.4, -1, -1)^\top + 0.8 = -0.6.$
$\hat{\gamma}_0^1(\omega^2) = (1, 0, 0, 0)(0, -0.5, 0, 0)^\top = 0.$
$\hat{\gamma}_1^1(\omega^2) = (0, 1, 0, 0)(0, -0.5, 0, 0)^\top = -0.5.$

Therefore, we have

$$\begin{aligned}\pi_0(x, \omega^1) &= \min_{h \in H}\{\hat{\nu}_h(\omega^1) - \hat{\gamma}_h(\omega^1)^\top x\} \\ &= \min\{-0.6 - 0x, -0.4 - (-0.5)x\}.\end{aligned}$$

(b) Form and solve RHS-LP (9.35) for scenario ω^1:

$$\begin{aligned}\text{Max}\ & \delta(\omega^1) - \sigma_0(\omega^1) - x^1\sigma_1(\omega^1) \\ \text{s.t.}\ & \sigma_0(\omega^1) - \tau_{00} \geq 0 \\ & \sigma_0(\omega^1) - \tau_{01} \geq 0 \\ & \tau_{00} + \tau_{01} = 1 \\ & \sigma_1(\omega^1) - (-1)\tau_0 - 0\tau_{00} \geq 0\end{aligned}$$

$$
\begin{aligned}
&\sigma_1(\omega^1) - (-1)\tau_1 - (-0.5)\tau_{01} \geq 0 \\
&\delta(\omega^1) - (-1)\tau_0 - (-0.6)\tau_{00} \geq 0 \\
&\delta(\omega^1) - (-1)\tau_1 - (-0.4)\tau_{01} \geq 0 \\
&\sigma_0(\omega^1), \delta(\omega^1), \sigma(\omega^1) \text{ free} \\
&\tau_0, \tau_1, \tau_{00}, \tau_{01} \geq 0.
\end{aligned}
$$

The optimal solution is

$$\sigma_0(\omega^1) = 0.5, \sigma_1(\omega^1) = 0, \delta(\omega^1) = -0.3, \tau_{00} = 0.5, \tau_{01} = 0.5.$$

Therefore, $\nu^1(\omega^1) = -0.3/0.5 = -0.6$ and $\gamma^1(\omega^1) = 0/0.5 = 0$. We obtain $r^2(\omega^1)$ by appending $\nu^1(\omega^1)$ to $r^1(\omega^1)$ and obtain $T^2(\omega^1)$ by appending $\gamma^1(\omega^1)$ to $T^1(\omega^1)$:

$$r^2(\omega^1) = (-0.6, -1.2, -1, -1, -0.6)^\top$$

and

$$T^2(\omega^1) = (0, -0.5, 0, 0, 0)^\top.$$

Form and solve RHS-LP (9.35) for scenario ω^2:

$$
\begin{aligned}
\text{Max } & \delta(\omega^2) - \sigma_0(\omega^2) - x^1\sigma_1(\omega^2) \\
\text{s.t. } & \sigma_0(\omega^2) - \tau_{00} \geq 0 \\
& \sigma_0(\omega^2) - \tau_{01} \geq 0 \\
& \tau_{00} + \tau_{01} = 1 \\
& \sigma_1(\omega^2) - (-1)\tau_0 - 0\tau_{00} \geq 0 \\
& \sigma_1(\omega^2) - (-1)\tau_1 - (-0.5)\tau_{01} \geq 0 \\
& \delta(\omega^2) - (-1)\tau_0 - (-0.6)\tau_{00} \geq 0 \\
& \delta(\omega^2) - (-1)\tau_1 - (-0.6)\tau_{01} \geq 0 \\
& \sigma_0(\omega^2), \delta(\omega^2), \sigma(\omega^2) \text{ free} \\
& \tau_0, \tau_1, \tau_{00}, \tau_{01} \geq 0.
\end{aligned}
$$

The optimal solution is

$$\sigma_0(\omega^2) = 0.5, \sigma_1(\omega^2) = 0, \delta(\omega^2) = -0.3, \tau_0 = 0, \tau_1 = 0, \tau_{00} = 0.5, \tau_{01} = 0.5.$$

Therefore, $\nu^1(\omega^2) = -0.3/0.5 = -0.6$ and $\gamma^1(\omega^2) = 0/0.5 = 0$. We obtain $r^2(\omega^2)$ by appending ν^1 to $r^1(\omega^2)$ and obtain $T^2(\omega^2)$ by appending γ^1 to $T^1(\omega^2)$:

$$r^2(\omega^2) = (-0.6, -1.4, -1, -1, -0.6)^\top$$

and

$$T^2(\omega^2) = (0, -0.5, 0, 0, 0)^\top.$$

Step 3. Update and Solve Subproblem LPs. For each $\omega \in \Omega$, solve subproblem LP using W^{k+1} and $r^{k+1}(\omega) - T^{k+1}(\omega)x^k$:

$$\begin{aligned} \varphi_{LP}(x^1, \omega^1) := \text{Min} \quad & -y_1 - y_2 \\ \text{s.t.} \quad & y_1 - y_2 \geq -0.6 \\ & -y_1 - y_2 \geq -1.2 - (-0.5)x^1 \\ & -y_1 \geq -1 \\ & -y_2 \geq -1 \\ & -0.2y_1 - y_2 \geq -0.6 \\ & y_1, y_2 \geq 0. \end{aligned}$$

Solving the subproblem, we get $y(\omega^1) = (0.75, 0.45)^\top$ (see Fig. 9.17a), and dual multipliers $d(\omega^1) = (0, 2, 0, 0, 0)^\top$.

$$\begin{aligned} \varphi_{LP}(x^1, \omega^2) := \text{Min} \quad & -y_1 - y_2 \\ \text{s.t.} \quad & y_1 - y_2 \geq -0.6 \\ & -y_1 - y_2 \geq -1.4 - (-0.5)x^1 \\ & -y_1 \geq -1 \\ & -y_2 \geq -1 \\ & -0.2y_1 - y_2 \geq -0.6 \\ & y_1, y_2 \geq 0. \end{aligned} \tag{9.40}$$

Solving the subproblem, we get $y(\omega^2) = (1, 0.4)^\top$ (see Fig. 9.17b), and dual multipliers $d(\omega^2) = (0, 2, 0, 0, 0)^\top$. Since the integer restrictions are *not* satisfied, we set $U^2 \leftarrow U^1 = \infty$.

Step 4. Update and Solve Master Problem. Using the dual solution for each subproblem from Step 3, construct Benders optimality cuts of the form $\eta_s \geq d(\omega^s)^\top (r^2(\omega^s) - (T^2)^\top(\omega^s)x)$:

For scenario ω^1, the optimality cut is

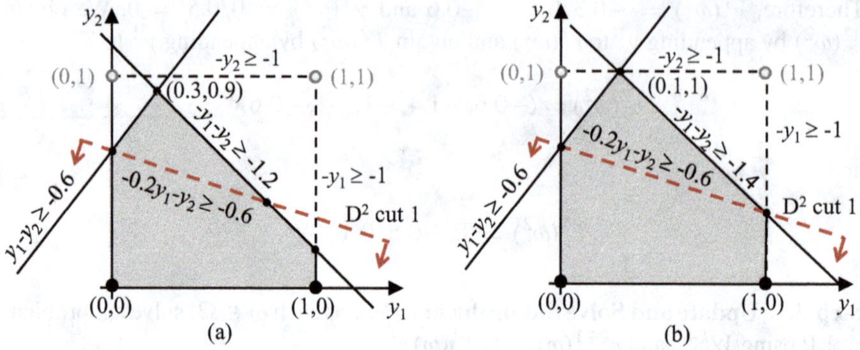

Fig. 9.17 Subproblem feasible region at iteration 1 after adding the D^2 cuts. **(a)** Scenario ω^1. **(b)** Scenario ω^2

$$\eta_1 \geq (0, 2, 0, 0, 0)(\begin{pmatrix} -0.6 \\ -1.2 \\ -1 \\ -1 \\ -0.6 \end{pmatrix} - \begin{bmatrix} 0 \\ -0.5 \\ 0 \\ 0 \\ 0 \end{bmatrix} x)$$

$$\Rightarrow \eta_1 \geq -2.4 + x.$$

For scenario ω^2, the optimality cut is

$$\eta_2 \geq (0, 2, 0, 0, 0)(\begin{pmatrix} -0.6 \\ -1.4 \\ -1 \\ -1 \\ -0.6 \end{pmatrix} - \begin{bmatrix} 0 \\ -0.5 \\ 0 \\ 0 \\ 0 \end{bmatrix} x)$$

$$\Rightarrow \eta_2 \geq -2.8 + x.$$

Since the two scenarios are equally likely, the expected values associated with the cut coefficients yield $\eta - x \geq -2.6$. Applying the translation $\theta = \eta + 4$, we get $\theta - x \geq 1.4$ as the optimality cut to add to the master program:

$$\begin{aligned} z^2 := \text{Min} \quad & -2x + \theta \\ \text{s.t.} \quad & -x \geq -1 \\ & -x + \theta \geq 1.4 \\ & x \in \{0, 1\}, \theta \geq 0. \end{aligned}$$

Solving the master program, we get $x^2 = 1, \theta = 2.4$ and an objective value $z^2 = 0.4$. Therefore, the lower bound becomes $L^2 \leftarrow \max\{z^2, L^1\} = \max\{0.4, -\infty\} = 0.4$. This completes the first iteration of the algorithm. Since $U^2 - L^2 = \infty - 0.4 > \epsilon$, set $k \leftarrow 2$ and go to the next iteration.

Iteration k $= 2$:

Step 1. Solve Subproblem LPs. We solve the LP relaxation of the subproblems based on x^2:

$$
\begin{array}{rrrrl}
\varphi_{LP}(x^2, \omega^1) := \text{Min} & -2y_1 & -2y_2 & & \\
\text{s.t.} & y_1 & -y_2 & \geq & -0.6 \\
 & -y_1 & -y_2 & \geq & -1.2 - (-0.5)x^2 \\
 & -y_1 & & \geq & -1 \\
 & & -y_2 & \geq & -1 \\
 & -0.2y_1 & -y_2 & \geq & -0.6 \\
 & y_1, & y_2 & \in & \{0, 1\}.
\end{array}
$$

The optimal solution for $\varphi_{LP}(x^2, \omega^1)$ is $y(\omega^1) = (0.125, 0.575)^\top$.

$$
\begin{array}{rrrrl}
\varphi_{LP}(x^2, \omega^2) := \text{Min} & -2y_1 & -2y_2 & & \\
\text{s.t.} & y_1 & -y_2 & \geq & -0.6 \\
 & -y_1 & -y_2 & \geq & -1.4 - (-0.5)x^2 \\
 & -y_1 & & \geq & -1 \\
 & & -y_2 & \geq & -1 \\
 & -0.2y_1 & -y_2 & \geq & -0.6 \\
 & y_1, & y_2 & \in & \{0, 1\}.
\end{array}
$$

The optimal solution for $\varphi_{LP}(x^2, \omega^2)$ is $y(\omega^2) = (0.375, 0.525)^\top$.

Step 2. Solve Cut Generation LPs and Perform Updates. Since y_1 has the 'most' fractional value compared to y_1, set $j(k) = j(2) = 1$, i.e., choose y_1 to be the *disjunction variable*.

(a) Form and solve C^3-SLP (9.30):

$$
\begin{aligned}
&\text{Max } 0.5\pi_0(\omega^1) + 0.5\pi_0(\omega^2) - 0.5(0.125, 0.575)(\pi_1, \pi_2)^\top \\
&\qquad - 0.5(0.375, 525)(\pi_1, \pi_2)^\top \\
&\text{s.t. } \pi_1 - \lambda_{01}^\top(1, -1, -1, 0, -0.2)^\top + \lambda_{02} \geq 0 \\
&\qquad \pi_2 - \lambda_{01}^\top(-1, -1, 0, -1, -1)^\top + 0\lambda_{02} \geq 0 \\
&\qquad \pi_1 - \lambda_{11}^\top(1, -1, -1, 0, -0.2)^\top - \lambda_{12} \geq 0 \\
&\qquad \pi_2 - \lambda_{11}^\top(-1, -1, 0, -1, -1)^\top - 0\lambda_{12} \geq 0
\end{aligned}
$$

$$\pi_0(\omega^1) - \lambda_{01}^\top(-0.6, -0.7, -1, -1, -0.6)^\top + 0\lambda_{02} \ge 0$$
$$\pi_0(\omega^2) - \lambda_{01}^\top(-0.6, -0.9, -1, -1, -0.6)^\top + 0\lambda_{02} \ge 0$$
$$\pi_0(\omega^1) - \lambda_{11}^\top(-0.6, -0.7, -1, -1, -0.6)^\top - \lambda_{12} \ge 0$$
$$\pi_0(\omega^2) - \lambda_{11}^\top(-0.6, -0.9, -1, -1, -0.6)^\top - \lambda_{12} \ge 0$$
$$-1 \le \pi_1,\ \pi_2,\ \pi_0(\omega^1),\ \pi_0(\omega^2) \le 1.$$

Solve to get $\pi^2 = (-1, 0)^\top$, $\pi_0(\omega^1) = 0$, $\pi_0(\omega^2) = 0$ and $\lambda_{01} = (0, 0, 0, 0, 0)^\top$, $\lambda_{11} = (0, 10, 0, 0, 0)^\top$, $\lambda_{02} = 0$, $\lambda_{12} = 9$. We obtain W^3 by appending π^2.

$$W^3 \leftarrow [W^2; (\pi^2)^\top] = \begin{bmatrix} 1 & -1 \\ -1 & -1 \\ -1 & 0 \\ 0 & -1 \\ -0.2 & -1 \\ -1 & 0 \end{bmatrix}.$$

$$\hat{\nu}_0^2(\omega^1) = (0, 0, 0, 0, 0)(-0.6, -1.2, -1, -1, -0.6)^\top = 0$$
$$\hat{\nu}_1^2(\omega^1) = (0, 10, 0, 0, 0)(-0.6, -1.2, -1, -1, -0.6)^\top + 9 = -3$$
$$\hat{\gamma}_0^2(\omega^1) = (0, 0, 0, 0, 0)(0, -0.5, 0, 0, 0)^\top = 0$$
$$\hat{\gamma}_1^2(\omega^1) = (0, 10, 0, 0, 0)(0, -0.5, 0, 0, 0)^\top = -5$$

$$\hat{\nu}_0^2(\omega^2) = (0, 0, 0, 0, 0)(-0.6, -1.4, -1, -1, -0.6)^\top = 0$$
$$\hat{\nu}_1^2(\omega^2) = (0, 10, 0, 0, 0)(-0.6, -1.4, -1, -1, -0.6)^\top + 9 = -5$$
$$\hat{\gamma}_0^2(\omega^2) = (0, 0, 0, 0, 0)(0, -0.5, 0, 0, 0)^\top = 0$$
$$\hat{\gamma}_1^2(\omega^2) = (0, 10, 0, 0, 0)(0, -0.5, 0, 0, 0)^\top = -5$$

(b) Form and solve RHS-LP (9.35) for scenario ω^1:

$$\begin{aligned} \text{Max } & \delta(\omega^1) - \sigma_0(\omega^1) - x^2\sigma_1(\omega^1) \\ \text{s.t. } & \sigma_0(\omega^1) - \tau_{00} \ge 0 \\ & \sigma_0(\omega^1) - \tau_{01} \ge 0 \\ & \tau_{00} + \tau_{01} = 1 \\ & \sigma_1(\omega^1) - (-1)\tau_0 - 0\tau_{00} \ge 0 \\ & \sigma_1(\omega^1) - (-1)\tau_1 - (-5)\tau_{01} \ge 0 \\ & \delta(\omega^1) - (-1)\tau_0 - 0\tau_{00} \ge 0 \\ & \delta(\omega^1) - (-1)\tau_1 - (-3)\tau_{01} \ge 0 \end{aligned}$$

$$\sigma_0(\omega^1), \delta(\omega^1), \sigma(\omega^1) \text{ free}$$

$$\tau_0, \tau_1, \tau_{00}, \tau_{01} \geq 0.$$

The optimal solution for scenario ω^1 is: $\sigma_0(\omega^1) = 0.5$, $\sigma_1(\omega^1) = -1.5$, $\delta(\omega^1) = -1.5$, $\tau_0 = 1.5$, $\tau_1 = 0$, $\tau_{00} = 0.5$, $\tau_{01} = 0.5$. Therefore, $\nu^2(\omega^1) = -1.5/0.5 = -3$ and $\gamma^2(\omega^1) = -1.5/0.5 = -3$. We obtain $r^3(\omega^1)$ by appending $\nu^2(\omega^1)$ to $r^2(\omega^1)$ and obtain $T^3(\omega^1)$ by appending $\gamma^2(\omega^1)$ to $T^2(\omega^1)$:

$$r^3(\omega^1) = (-0.6, -1.2, -1, -1, -0.6, -3)^\top$$

and

$$T^3(\omega^1) = (0, -0.5, 0, 0, 0, -3)^\top.$$

Form and solve RHS-LP (9.35) for scenario ω^2:

$$\begin{aligned}
\text{Max } & \delta(\omega^2) - \sigma_0(\omega^2) - x^2\sigma(\omega^2) \\
\text{s.t. } & \sigma_0(\omega^2) - \tau_{00} \geq 0 \\
& \sigma_0(\omega^2) - \tau_{01} \geq 0 \\
& \tau_{00} + \tau_{01} = 1 \\
& \sigma_1(\omega^2) - (-1)\tau_0 - 0\tau_{00} \geq 0 \\
& \sigma_1(\omega^2) - (-1)\tau_1 - (-5)\tau_{01} \geq 0 \\
& \delta(\omega^2) - (-1)\tau_0 - 0\tau_{00} \geq 0 \\
& \delta(\omega^2) - (-1)\tau_1 - (-5)\tau_{01} \geq 0 \\
& \sigma_0(\omega^2), \delta(\omega^2), \sigma(\omega^2) \text{ free} \\
& \tau_0, \tau_1, \tau_{00}, \tau_{01} \geq 0.
\end{aligned}$$

The optimal solution for scenario ω^2 is: $\sigma_0(\omega^2) = -2.5$, $\sigma_1(\omega^2) = -2.5$, $\delta(\omega^2) = 0.5$, $\tau_0 = 2.5$, $\tau_1 = 0$, $\tau_{00} = 0.5$, $\tau_{01} = 0.5$. Therefore, $\nu^2(\omega^2) = -2.5/0.5 = -5$ and $\gamma^2(\omega^2) = -2.5/0.5 = -5$. We obtain $r^3(\omega^2)$ by appending ν^2 to $r^2(\omega^2)$ and obtain $T^3(\omega^2)$ by appending γ^2 to $T^2(\omega^2)$:

$$r^3(\omega^2) = (-0.6, -1.4, -1, -1, -0.6, -5)^\top$$

and

$$T^3(\omega^2) = (0, -0.5, 0, 0, 0, -5)^\top.$$

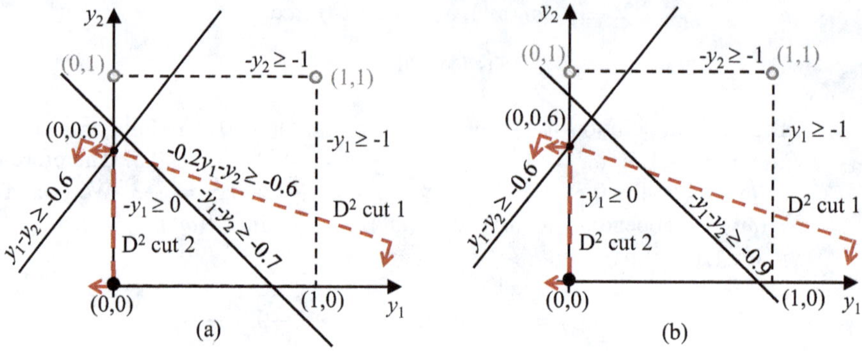

Fig. 9.18 Subproblem feasible region at iteration 2 after adding the D^2 cuts. (**a**) Scenario ω^1. (**b**) Scenario ω^2

Step 3. Update and Solve Subproblem LPs.

$$\begin{array}{llrrl}
\varphi_{LP}(x^2,\omega^1) := \text{Min} & -2y_1 & -2y_2 & & \\
\text{s.t.} & y_1 & -y_2 & \geq -0.6 & \\
 & -y_1 & -y_2 & \geq -1.2 & -(-0.5)x^2 \\
 & -y_1 & & \geq \ \ -1 & \\
 & & -y_2 & \geq \ \ -1 & \\
 & -0.2y_1 & -y_2 & \geq -0.6 & \\
 & -y_1 & & \geq \ \ -3 & -(-3)x^2 \\
 & y_1, & y_2 \geq 0. & &
\end{array}$$

Solving the subproblem, we get $y(\omega^1) = (0, 0.6)^\top$ (see Fig. 9.18a), and dual multipliers $d(\omega^1) = (2, 0, 0, 0, 0, 4)^\top$.

$$\begin{array}{llrrl}
\varphi_{LP}(x^2,\omega^2) := \text{Min} & -y_1 & -y_2 & & \\
\text{s.t.} & y_1 & -y_2 \geq & -0.6 & \\
 & -y_1 & -y_2 \geq & -1.4 & -(-0.5)x^2 \\
 & -y_1 & \geq & -1 & \\
 & & -y_2 \geq & -1 & \\
 & -0.2y_1 & -y_2 \geq & -0.6 & \\
 & -y_1 & \geq & -5 & -(-5)x^2 \\
 & y_1, & y_2 \in & \{0,1\}. &
\end{array}$$

Solving the subproblem, we get $y(\omega^2) = (0, 0.6)^\top$ (see Fig. 9.18b), and dual multipliers $d(\omega^2) = (2, 0, 0, 0, 0, 4)^\top$.

Since the integer restrictions are not satisfied, we set $U^3 \leftarrow U^2 = \infty$.

Step 4. Update and Solve Master Problem. Using the dual solution for each subproblem from Step 3, construct Benders optimality cuts of the form $\eta_s \geq d(\omega^s)^\top(r^2(\omega^s) - (T^2(\omega^s))^\top(\omega^s)x)$:

For scenario ω^1, the cut is

$$\eta_1 \geq (2, 0, 0, 0, 0, 4)(\begin{pmatrix} -0.6 \\ -1.2 \\ -1 \\ -1 \\ -0.6 \\ -3 \end{pmatrix} - \begin{bmatrix} 0 \\ -0.5 \\ 0 \\ 0 \\ 0 \\ -3 \end{bmatrix} x)$$

$$\Rightarrow \eta_1 \geq -13.2 + 12x.$$

For scenario ω^2, the cut is

$$\eta_2 \geq (2, 0, 0, 0, 0, 4)(\begin{pmatrix} -0.6 \\ -1.4 \\ -1 \\ -1 \\ -0.6 \\ -5 \end{pmatrix} - \begin{bmatrix} 0 \\ -0.5 \\ 0 \\ 0 \\ 0 \\ -5 \end{bmatrix} x)$$

$$\Rightarrow \eta_2 \geq -21.2 + 20x.$$

Since the two scenarios are equally likely, the expected values associated with the cut coefficients yield $\eta - 16x \geq -17.2$. Applying the translation $\theta = \eta + 4$, we get $\theta - 16x \geq -13.2$ as the optimality cut to add to the master program:

$$\begin{aligned} z^3 := \text{Min} \quad & -2x + \theta \\ \text{s.t.} \quad & -x \geq -1 \\ & -x + \theta \geq 1.4 \\ & -16x + \theta \geq -13.2 \\ & x \in \{0, 1\}, \theta \geq 0. \end{aligned}$$

Solving the master program, we get $x^3 = 1, \theta = 2.8$ and an objective value of $z^3 \leftarrow 0.8$. Therefore, the lower bound becomes $L^3 \leftarrow \max\{z^3, L^2\} = \max\{0.8, 0.4\} = 0.8$. This completes the second iteration of the algorithm. Since $U^3 - L^3 = \infty > \epsilon$, set $k \leftarrow 3$, and we begin the next iteration.

Iteration $k = 3$:

Step 1. Since y_2 does not satisfy the integer restrictions, we choose y_2 as the *disjunction variable* and create the disjunction on $y_2 \leq 0$ and $y_2 \geq 1$.

Step 2. Solve Cut Generation LPs and Perform Updates.

(a) Form and solve C^3-SLP (9.30):

$$\text{Max } 0.5\pi_0(\omega^1) + 0.5\pi_0(\omega^2) - 0.5(0, 0.6)(\pi_1, \pi_2)^\top - 0.5(0, 0.6)(\pi_1, \pi_2)^\top$$

$$\begin{aligned}
\text{s.t. } & \pi_1 - \lambda_{01}^\top(1, -1, -1, 0, -0.2, -1)^\top + 0\lambda_{02} \geq 0 \\
& \pi_2 - \lambda_{01}^\top(-1, -1, 0, -1, -1, 0,)^\top + \lambda_{02} \geq 0 \\
& \pi_1 - \lambda_{11}^\top(1, -1, -1, 0, -0.2, -1)^\top - 0\lambda_{12} \geq 0 \\
& \pi_2 - \lambda_{11}^\top(-1, -1, 0, -1, -1, 0)^\top - \lambda_{12} \geq 0 \\
& \pi_0(\omega^1) - \lambda_{01}^\top(-0.6, -0.7, -1, -1, -0.6, -3)^\top + 0\lambda_{02} \geq 0 \\
& \pi_0(\omega^2) - \lambda_{01}^\top(-0.6, -0.9, -1, -1, -0.6, -5)^\top + 0\lambda_{02} \geq 0 \\
& \pi_0(\omega^1) - \lambda_{11}^\top(-0.6, -0.7, -1, -1, -0.6, -3)^\top - \lambda_{12} \geq 0 \\
& \pi_0(\omega^2) - \lambda_{11}^\top(-0.6, -0.9, -1, -1, -0.6, -5)^\top - \lambda_{12} \geq 0 \\
& -1 \leq \pi_1, \pi_2, \pi_0(\omega^1), \pi_0(\omega^2) \leq 1.
\end{aligned}$$

Solve to get $\pi^3 = (1, -1)$, $\pi_0(\omega^1) = 0$, $\pi_0(\omega^2) = 0$ and $\lambda_{01} = (0, 0, 0, 0, 0, 0)$, $\lambda_{11} = (0, 10, 0, 0, 0, 0)$, $\lambda_{02} = 1$, $\lambda_{12} = 9$. We obtain W^4 by appending π^3.

$$W^4 = \begin{bmatrix} 1 & -1 \\ -1 & -1 \\ -1 & 0 \\ 0 & -1 \\ -0.2 & -1 \\ -1 & 0 \\ 1 & -1 \end{bmatrix}.$$

$$\begin{aligned}
\hat{\nu}_0^3(\omega^1) &= (0, 0, 0, 0, 0, 0)(-0.6, -1.2, -1, -1, -0.6, -3)^\top = 0 \\
\hat{\nu}_1^3(\omega^1) &= (0, 10, 0, 0, 0, 0)(-0.6, -1.2, -1, -1, -0.6, -3)^\top + 9 = -3 \\
\hat{\gamma}_0^3(\omega^1) &= (0, 0, 0, 0, 0, 0)(0, -0.5, 0, 0, 0, -3)^\top = 0 \\
\hat{\gamma}_1^3(\omega^1) &= (0, 10, 0, 0, 0, 0)(0, -0.5, 0, 0, 0, -3)^\top = -5
\end{aligned}$$

$$\begin{aligned}
\hat{\nu}_0^3(\omega^2) &= (0, 0, 0, 0, 0, 0)(-0.6, -1.4, -1, -1, -0.6, -5)^\top = 0 \\
\hat{\nu}_1^3(\omega^2) &= (0, 10, 0, 0, 0, 0)(-0.6, -1.4, -1, -1, -0.6, -5)^\top + 9 = -5 \\
\hat{\gamma}_0^3(\omega^2) &= (0, 0, 0, 0, 0, 0)(0, -0.5, 0, 0, 0, -5)^\top = 0 \\
\hat{\gamma}_1^3(\omega^2) &= (0, 10, 0, 0, 0, 0)(0, -0.5, 0, 0, 0, -5)^\top = -5
\end{aligned}$$

Form and solve RHS-LP (9.35) for scenario ω^1:

$$\begin{aligned}
\text{Max } & \delta(\omega^1) - \sigma_0(\omega^1) - (x^3)^\top \sigma_1(\omega^1) \\
\text{s.t. } & \sigma_0(\omega^1) - \tau_{00} \geq 0 \\
& \sigma_0(\omega^1) - \tau_{01} \geq 0 \\
& \tau_{00} + \tau_{01} = 1 \\
& \sigma_1(\omega^1) - (-1)\tau_0 - 0\tau_{00} \geq 0 \\
& \sigma_1(\omega^1) - (-1)\tau_1 - (-5)\tau_{01} \geq 0 \\
& \delta(\omega^1) - (-1)\tau_0 - 0\tau_{00} \geq 0 \\
& \delta(\omega^1) - (-1)\tau_1 - (-3)\tau_{01} \geq 0 \\
& \sigma_0(\omega^1), \delta(\omega^1).\sigma(\omega^1) \text{ free} \\
& \tau_0, \tau_1, \tau_{00}, \tau_{01} \geq 0.
\end{aligned}$$

The optimal solution for scenario ω^1 is: $\sigma_0(\omega^1) = 0.5$, $\sigma_1(\omega^1) = -1.5$, $\delta(\omega^1) = -1.5$, $\tau_0 = 1.5$, $\tau_1 = 0$, $\tau_{00} = 0.5$, $\tau_{01} = 0.5$. Therefore, $\nu^3(\omega^1) = -1.5/0.5 = -3$ and $\gamma^3(\omega^1) = -1.5/0.5 = -3$. We obtain $r^4(\omega^1)$ by appending $\nu^3(\omega^1)$ to $r^3(\omega^1)$ and obtain $T^4(\omega^1)$ by appending $\gamma^3(\omega^1)$ to $T^3(\omega^1)$:

$$r^4(\omega^1) = (-0.6, -1.2, -1, -1, -0.6, -3, -3)^\top$$

and

$$T^4(\omega^1) = (0, -0.5, 0, 0, 0, -3, -3)^\top.$$

Form and solve RHS-LP (9.35) for scenario ω^2:

$$\begin{aligned}
\text{Max } & \delta(\omega^2) - \sigma_0(\omega^2) - (x^3)^\top \sigma(\omega^2) \\
\text{s.t. } & \sigma_0(\omega^2) - \tau_{00} \geq 0 \\
& \sigma_0(\omega^2) - \tau_{01} \geq 0 \\
& \tau_{00} + \tau_{01} = 1 \\
& \sigma_1(\omega^2) - (-1)\tau_0 - 0\tau_{00} \geq 0 \\
& \sigma_1(\omega^2) - (-1)\tau_1 - (-5)\tau_{01} \geq 0 \\
& \delta(\omega^2) - (-1)\tau_0 - 0\tau_{00} \geq 0 \\
& \delta(\omega^2) - (-1)\tau_1 - (-5)\tau_{01} \geq 0 \\
& \sigma_0(\omega^2), \delta(\omega^2), \sigma(\omega^2) \text{ free} \\
& \tau_0, \tau_1, \tau_{00}, \tau_{01} \geq 0.
\end{aligned}$$

The optimal solution for scenario ω^2 is: $\sigma_0(\omega^2) = 0.5, \sigma_1(\omega^2) = -2.5, \delta(\omega^2) = -2.5, \tau_0 = 2.5, \tau_1 = 0, \tau_{00} = 0.5, \tau_{01} = 0.5$. Therefore, $\nu^3(\omega^2) = -2.5/0.5 = -5$ and $\gamma^3(\omega^2) = -2.5/0.5 = -5$. We obtain $r^4(\omega^2)$ by appending ν^3 to $r^3(\omega^2)$ and obtain $T^4(\omega^2)$ by appending γ^3 to $T^3(\omega^2)$:

$$r^4(\omega^2) = (-0.6, -1.4, -1, -1, -0.6, -5, -5)^\top$$

and

$$T^4(\omega^2) = (0, -0.5, 0, 0, 0, -5, -5)^\top.$$

Step 3. Update and Solve Subproblem LPs.

$$\begin{array}{rllll}
\varphi_{LP}(x^3, \omega^1) := \text{Min} & -2y_1 & -2y_2 & & \\
\text{s.t.} & y_1 & -y_2 & \geq -0.6 & \\
 & -y_1 & -y_2 & \geq -1.2 & -(-0.5)x^3 \\
 & -y_1 & & \geq -1 & \\
 & & -y_2 & \geq -1 & \\
 & -0.2y_1 & -y_2 & \geq -0.6 & \\
 & -y_1 & & \geq -3 & -(-3)x^3 \\
 & y_1 & -y_2 & \geq -3 & -(-3)x^3 \\
 & y_1, & y_2 \geq 0. & &
\end{array}$$

Solving the subproblem, we get $y(\omega^1) = (0, 0)^\top$ (see Fig. 9.19a), and dual multipliers $d(\omega^1) = (0, 0, 0, 0, 0, 4, 2)^\top$.

$$\begin{array}{rllll}
\varphi_{LP}(x^3, \omega^2) := \text{Min} & -2y_1 & -2y_2 & & \\
\text{s.t.} & y_1 & -y_2 & \geq -0.6 & \\
 & -y_1 & -y_2 & \geq -1.4 & -(-0.5)x^3 \\
 & -y_1 & & \geq -1 & \\
 & & -y_2 & \geq -1 & \\
 & -0.2y_1 & -y_2 & \geq -0.6 & \\
 & -y_1 & & \geq -5 & -(-5)x^3 \\
 & y_1 & -y_2 & \geq -5 & -(-5)x^3 \\
 & y_1, & y_2 \geq 0. & &
\end{array}$$

Solving the subproblem, we get $y(\omega^2) = (0, 0)^\top$ (see Fig. 9.19b), and dual multipliers $d(\omega^2) = (0, 0, 0, 0, 0, 4, 2)^\top$. Since the integer restrictions are satisfied, we update $U^4 \leftarrow \min\{c^\top x^3 + \mathbb{E}[\varphi_{LP}(x^3, \tilde{\omega})] + 4, U^3\} = \min\{-2 + 0 + 4, \infty\} = 2$. Notice that we add 4 to the objective value to translate it back. We can now update the incumbent solution $x^*_\epsilon \leftarrow x^3 = 1$ and objective value $z^*_\epsilon \leftarrow 2$.

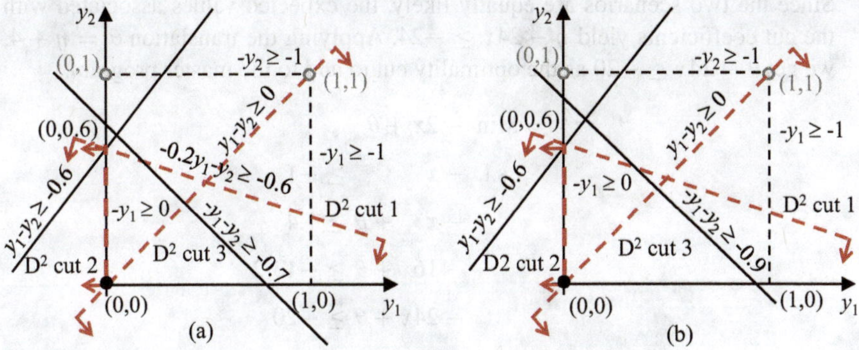

Fig. 9.19 Subproblem feasible region at iteration 3 after adding the D^2 cuts. (**a**) Scenario ω^1. (**b**) Scenario ω^2

Step 4. Update and Solve Master Problem. Using the dual solution for each subproblem from Step 3, construct Benders optimality cuts of the form $\eta_s \geq d(\omega^s)^\top(r^2(\omega^s) - (T^2(\omega^s))^\top(\omega^s)x)$:

The resulting cuts are

$$\eta_1 \geq (0, 0, 0, 0, 0, 4, 2)\left(\begin{pmatrix} -0.6 \\ -1.2 \\ -1 \\ -1 \\ -0.6 \\ -3 \\ -3 \end{pmatrix} - \begin{bmatrix} 0 \\ -0.5 \\ 0 \\ 0 \\ 0 \\ -3 \\ -3 \end{bmatrix} x\right)$$

$$\eta_1 \geq -18 + 18x.$$

The resulting cuts are

$$\eta_2 \geq (0, 0, 0, 0, 0, 4, 2)\left(\begin{pmatrix} -0.6 \\ -1.4 \\ -1 \\ -1 \\ -0.6 \\ -5 \\ -5 \end{pmatrix} - \begin{bmatrix} 0 \\ -0.5 \\ 0 \\ 0 \\ 0 \\ -5 \\ -5 \end{bmatrix} x\right)$$

$$\eta_2 \geq -30 + 30x.$$

Since the two scenarios are equally likely, the expected values associated with the cut coefficients yield $\eta - 24x \geq -24$. Applying the translation $\theta = \eta + 4$, we get $\theta - 24x \geq -20$ as the optimality cut to add to the master program:

$$\begin{aligned} z^4 := \text{Min} \quad & -2x + \theta \\ \text{s.t.} \quad & -x \geq -1 \\ & -x + \theta \geq 1.4 \\ & -16x + \theta \geq -13.2 \\ & -24x + \theta \geq -20 \\ & x \in \{0, 1\}, \theta \geq 0. \end{aligned}$$

Solving the master program, we get $x^4 = 0, \theta = 1.4$ and objective value $z^4 = 1.4$. Therefore, the lower bound becomes $L^4 \leftarrow \max\{z^4, L^3\} = \max\{1.4, 0.8\} = 1.4$. This completes the third iteration of the algorithm. Since $U^4 - L^4 = 0.6 > \epsilon$, set $k \leftarrow 4$, and we begin the next iteration.

Iteration $k = 4$:

Step 1. Solve Subproblem LPs. We solve the LP relaxation of the subproblems. For scenario ω^1 we have:
We start the second iteration by solving the following updated subproblems with $x^4 = 0$:

$$\begin{aligned} \varphi_{LP}(x^4, \omega^1) := \text{Min} \quad & -2y_1 - 2y_2 \\ \text{s.t.} \quad & y_1 - y_2 \geq -0.6 \\ & -y_1 - y_2 \geq -1.2 \\ & -y_1 \geq -1 \\ & -y_2 \geq -1 \\ & -0.2y_1 - y_2 \geq -0.6 \\ & -y_1 \geq -3 \\ & y_1 - y_2 \geq -3 \\ & y_1, \quad y_2 \geq 0. \end{aligned}$$

Solving the subproblem, we get $y(\omega^1) = (0.75, 0.45)^\top$, and dual multipliers $d(\omega^1) = (0, 2, 0, 0, 0, 0, 0)^\top$.

$$\begin{aligned} \varphi_{LP}(x^4, \omega^2) := \text{Min} \quad & -2y_1 - 2y_2 \\ \text{s.t.} \quad & y_1 - y_2 \geq -0.6 \\ & -y_1 - y_2 \geq -1.4 \\ & -y_1 \geq -1 \\ & -y_2 \geq -1 \\ & -0.2y_1 - y_2 \geq -0.6 \\ & -y_1 \geq -5 \\ & y_1 - y_2 \geq -5 \\ & y_1, \quad y_2 \geq 0. \end{aligned}$$

Solving the subproblem, we get $y(\omega^2) = (1, 0.4)^\top$, and dual multipliers $d(\omega^2) = (0, 2, 0, 0, 0, 0, 0)^\top$.

Step 2. Solve Cut Generation LPs and Perform Updates. Since $y_2(\omega^1)$ does not satisfy the integer restrictions, we choose $y_2(\omega^1)$ as the *disjunction variable* and create the disjunction on $y_2(\omega^1) \leq 0$ and $y_2(\omega^1) \geq 1$.

(a) Form and solve C^3-SLP (9.30):

$$
\begin{aligned}
\text{Max } & 0.5\pi_0(\omega^1) + 0.5\pi_0(\omega^2) - 0.5(0.75, 0.45)(\pi_1, \pi_2)^\top - 0.5(1, 0.4)(\pi_1, \pi_2)^\top \\
\text{s.t. } & \pi_1 - \lambda_{01}^\top(1, -1, -1, 0, -0.2, -1, 0)^\top + 0\lambda_{02} \geq 0 \\
& \pi_2 - \lambda_{01}^\top(-1, -1, 0, -1, -1, 0, -1)^\top + \lambda_{02} \geq 0 \\
& \pi_1 - \lambda_{11}^\top(1, -1, -1, 0, -0.2, -1, 0)^\top - 0\lambda_{12} \geq 0 \\
& \pi_2 - \lambda_{11}^\top(-1, -1, 0, -1, -1, 0, -1)^\top - \lambda_{12} \geq 0 \\
& \pi_0(\omega^1) - \lambda_{01}^\top(-0.6, -1.2, -1, -1, -0.6, -3, -3)^\top + 0\lambda_{02} \geq 0 \\
& \pi_0(\omega^2) - \lambda_{01}^\top(-0.6, -1.4, -1, -1, -0.6, -5, -5)^\top + 0\lambda_{02} \geq 0 \\
& \pi_0(\omega^1) - \lambda_{11}^\top(-0.6, -1.2, -1, -1, -0.6, -3, -3)^\top - \lambda_{12} \geq 0 \\
& \pi_0(\omega^2) - \lambda_{11}^\top(-0.6, -1.4, -1, -1, -0.6, -5, -5)^\top - \lambda_{12} \geq 0 \\
& -1 \leq \pi_1,\ \pi_2,\ \pi_0(\omega^1),\ \pi_0(\omega^2) \leq 1.
\end{aligned}
$$

Solve to get $\pi^4 = (0, -1)$, $\pi_0(\omega^1) = 0$, $\pi_0(\omega^2) = 0$ and $\lambda_{01} = (0, 0, 0, 0, 0, 0, 0)$, $\lambda_{11} = (0, 0, 0, 0, 2.5, 0, 0)$, $\lambda_{02} = 1$, $\lambda_{12} = 1.5$. We obtain W^4 by appending π^5.

$$
W^5 = \begin{bmatrix} 1 & -1 \\ -1 & -1 \\ -1 & 0 \\ 0 & -1 \\ -0.2 & -1 \\ -1 & 0 \\ 1 & -1 \\ 0 & -1 \end{bmatrix}.
$$

$$
\begin{aligned}
&\hat{\nu}_0^4(\omega^1) = (0, 0, 0, 0, 0, 0, 0)(-0.6, -1.2, -1, -1, -0.6, -3, -3)^\top = 0 \\
&\hat{\nu}_1^4(\omega^1) = (0, 0, 0, 0, 2.5, 0, 0)(-0.6, -1.2, -1, -1, -0.6, -3, -3)^\top + 1.5 = 0 \\
&\hat{\gamma}_0^4(\omega^1) = (0, 0, 0, 0, 0, 0, 0)(0, -0.5, 0, 0, 0, -3, -3)^\top = 0 \\
&\hat{\gamma}_1^4(\omega^1) = (0, 0, 0, 0, 2.5, 0, 0)(0, -0.5, 0, 0, 0, -3, -3)^\top = 0
\end{aligned}
$$

$\hat{\nu}_0^4(\omega^2) = (0, 0, 0, 0, 0, 0, 0)(-0.6, -1.2, -1, -1, -0.6, -5, -5)^\top = 0$
$\hat{\nu}_1^4(\omega^2) = (0, 0, 0, 0, 2.5, 0, 0)(-0.6, -1.2, -1, -1, -0.6, -5, -5)^\top + 1.5 = 0$
$\hat{\gamma}_0^4(\omega^2) = (0, 0, 0, 0, 0, 0, 0)(0, -0.5, 0, 0, 0, -5, -5)^\top = 0$
$\hat{\gamma}_1^4(\omega^2) = (0, 0, 0, 0, 2.5, 0, 0)(0, -0.5, 0, 0, 0, -5, -5)^\top = 0$

(a) Form and solve RHS-LP (9.35) for scenario ω^1:

$$
\begin{aligned}
\text{Max } & \delta(\omega^1) - \sigma_0(\omega^1) - (x^4)^\top \sigma_1(\omega^1) \\
\text{s.t. } & \sigma_0(\omega^1) - \tau_{00} \geq 0 \\
& \sigma_0(\omega^1) - \tau_{01} \geq 0 \\
& \tau_{00} + \tau_{01} = 1 \\
& \sigma_1(\omega^1) - (-1)\tau_0 - 0\tau_{00} \geq 0 \\
& \sigma_1(\omega^1) - (-1)\tau_1 - 0\tau_{01} \geq 0 \\
& \delta(\omega^1) - (-1)\tau_0 - 0\tau_{00} \geq 0 \\
& \delta(\omega^1) - (-1)\tau_1 - 0\tau_{01} \geq 0 \\
& \sigma_0(\omega^1), \delta(\omega^1), \sigma(\omega^1) \text{ free} \\
& \tau_0, \tau_1, \tau_{00}, \tau_{01} \geq 0.
\end{aligned}
$$

The optimal solution for scenario ω^1 is: $\sigma_0(\omega^1) = 0.5, \sigma_1(\omega^1) = 0, \delta(\omega^1) = 0, \tau_0 = 0, \tau_1 = 0, \tau_{00} = 0.5, \tau_{01} = 0.5$. Therefore, $\nu^4(\omega^1) = 0/0.5 = 0$ and $\gamma^4(\omega^1) = 0/0.5 = 0$. We obtain $r^5(\omega^1)$ by appending $\nu^4(\omega^1)$ to $r^4(\omega^1)$ and obtain $T^5(\omega^1)$ by appending $\gamma^4(\omega^1)$ to $T^4(\omega^1)$:

$$r^5(\omega^1) = (-0.6, -1.2, -1, -1, -0.6, -3, -3, 0)^\top$$

and

$$T^5(\omega^1) = (0, -0.5, 0, 0, 0, -3, -3, 0)^\top.$$

Form and solve RHS-LP (9.35) for scenario ω^2:

$$
\begin{aligned}
\text{Max } & \delta(\omega^2) - \sigma_0(\omega^2) - (x^4)^\top \sigma_1(\omega^2) \\
\text{s.t. } & \sigma_0(\omega^2) - \tau_{00} \geq 0 \\
& \sigma_0(\omega^2) - \tau_{01} \geq 0 \\
& \tau_{00} + \tau_{01} = 1 \\
& \sigma_1(\omega^2) - (-1)\tau_0 - 0\tau_{00} \geq 0 \\
& \sigma_1(\omega^2) - (-1)\tau_1 - 0\tau_{01} \geq 0
\end{aligned}
$$

$$\delta(\omega^2) - (-1)\tau_0 - 0\tau_{00} \geq 0$$

$$\delta(\omega^2) - (-1)\tau_1 - 0\tau_{01} \geq 0$$

$$\sigma_0(\omega^2), \delta(\omega^2), \sigma(\omega^2) \text{ free}$$

$$\tau_0, \tau_1, \tau_{00}, \tau_{01} \geq 0.$$

The optimal solution for scenario ω^2 is: $\sigma_0(\omega^2) = 0.5, \sigma_1(\omega^2) = 0, \delta(\omega^2) = 0, \tau_0 = 0, \tau_1 = 0, \tau_{00} = 0.5, \tau_{01} = 0.5$. Therefore, $\nu^4(\omega^2) = 0/0.5 = 0$ and $\gamma^4(\omega^2) = 0/0.5 = 0$. We obtain $r^5(\omega^2)$ by appending $\nu^4(\omega^2)$ to $r^4(\omega^2)$ and obtain $T^5(\omega^2)$ by appending γ^4 to $T^4(\omega^2)$:

$$r^5(\omega^2) = (-0.6, -1.4, -1, -1, -0.6, -5, -5, 0)^\top$$

and

$$T^5(\omega^2) = (0, -0.5, 0, 0, 0, -5, -5, 0)^\top.$$

Step 3. Update and Solve Subproblem LPs. Update and solve subproblems using

$$\begin{aligned}
\varphi_{LP}(x^4, \omega^1) := \text{Min} \quad & -2y_1 - 2y_2 \\
\text{s.t.} \quad & y_1 - y_2 \geq -0.6 \\
& -y_1 - y_2 \geq -1.2 \\
& -y_1 \geq -1 \\
& -y_2 \geq -1 \\
& -0.2y_1 - y_2 \geq -0.6 \\
& -y_1 \geq -3 \\
& y_1 - y_2 \geq -3 \\
& -y_2 \geq 0 \\
& y_1, \quad y_2 \geq 0.
\end{aligned}$$

Solving the subproblem, we get $y(\omega^1) = (1, 0)^\top$ (see Fig. 9.20a), and dual multipliers $d(\omega^1) = (0, 0, 2, 0, 0, 0, 0, 2)^\top$.

$$\begin{aligned}
\varphi_{LP}(x^4, \omega^2) := \text{Min} \quad & -2y_1 - 2y_2 \\
\text{s.t.} \quad & y_1 - y_2 \geq -0.6 \\
& -y_1 - y_2 \geq -1.4 \\
& -y_1 \geq -1 \\
& -y_2 \geq -1 \\
& -0.2y_1 - y_2 \geq -0.6 \\
& -y_1 \geq -5 \\
& y_1 - y_2 \geq -5 \\
& -y_2 \geq 0 \\
& y_1, \quad y_2 \geq 0.
\end{aligned}$$

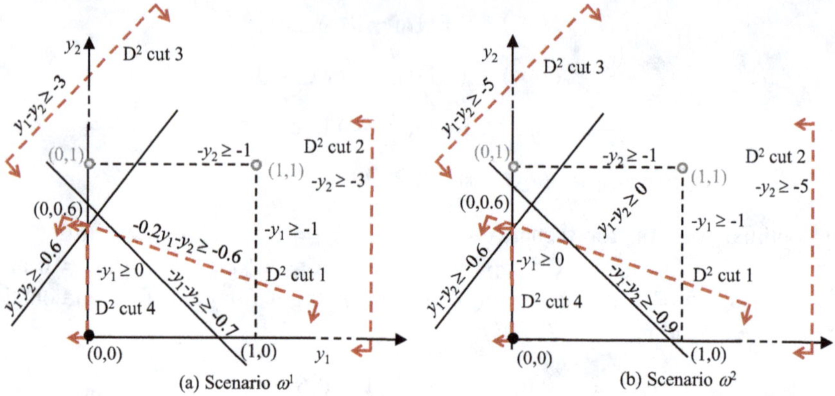

Fig. 9.20 Subproblem feasible region at iteration 4 after adding the D^2 cuts. (**a**) Scenario ω^1. (**b**) Scenario ω^2

Solving the subproblem, we get $y(\omega^2) = (1,0)^\top$ (see Fig. 9.20b), and dual multipliers $d(\omega^2) = (0,0,2,0,0,0,0,2)^\top$. Since the integer restrictions are satisfied, we update $U^5 \leftarrow \min\{c^\top x^4 + \mathbb{E}[\varphi_{LP}(x^4,\tilde{\omega})] + 4, U^4\} = \min\{-2 + 0 + 4, 2\} = 2$, which remains the same. This means that we have an alternative incumbent solution.

Step 4. Update and Solve Master Problem. Using the dual solution for each subproblem from Step 3, construct Benders optimality cuts of the form $\eta_s \geq d(\omega^s)^\top(r^2(\omega^s) - (T^2(\omega^s))^\top(\omega^s)x)$:

The resulting cuts are

$$\eta_1 \geq (0,0,2,0,0,0,0,2)(\begin{pmatrix} -0.6 \\ -1.2 \\ -1 \\ -1 \\ -0.6 \\ -3 \\ -3 \\ 0 \end{pmatrix} - \begin{bmatrix} 0 \\ -0.5 \\ 0 \\ 0 \\ 0 \\ -3 \\ -3 \\ 0 \end{bmatrix} x)$$

$$\Rightarrow \eta_1 \geq -2.$$

The resulting cuts are

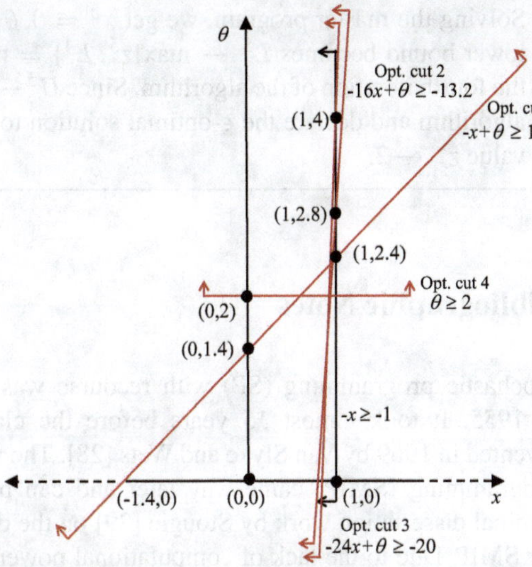

Fig. 9.21 D^2 algorithm master program feasible region after iteration 4

$$\eta_2 \geq (0, 0, 2, 0, 0, 0, 0, 2)\left(\begin{pmatrix} -0.6 \\ -1.4 \\ -1 \\ -1 \\ -0.6 \\ -5 \\ -5 \\ 0 \end{pmatrix} - \begin{bmatrix} 0 \\ -0.5 \\ 0 \\ 0 \\ 0 \\ -5 \\ -5 \\ 0 \end{bmatrix} x\right)$$

$$\Rightarrow \eta_2 \geq -2.$$

Since the two scenarios are equally likely, the expected values associated with the cut coefficients yield $\eta \geq -2$. Applying the translation $\theta = \eta + 4$, we get $\theta \geq 2$ as the optimality cut to add to the master program (see Fig. 9.21):

$$\begin{aligned} z^5 := \text{Min} \ & -2x + \theta \\ \text{s.t.} \ & -x \geq -1 \\ & -x + \theta \geq 1.4 \\ & -16x + \theta \geq 13.2 \\ & -24x + \theta \geq -20 \\ & \theta \geq 2 \\ & x \in \{0, 1\}, \theta \geq 0. \end{aligned}$$

Solving the master program, we get $x^5 = 0, \theta = 2$ and $z^5 = 0$. Therefore, the lower bound becomes $L^5 \leftarrow \max\{z^5, L^4\} = \max\{2, 1.4\} = 2$. This completes the fourth iteration of the algorithm. Since $U^5 - L^5 = 2-2 = 0$, we terminate the algorithm and declare the ϵ-optimal solution to be $x^*_\epsilon \leftarrow x^3 = 1$ with objective value $z^*_\epsilon \leftarrow 2$.

Bibliographic Notes

Stochastic programming (SP) with recourse was first proposed by Dantzig [13] in 1955. It took almost 15 years before the classical L-shaped algorithm was invented in 1969 by Van Slyke and Wets [28]. The work on stochastic mixed-integer programming (SMIP) came way later and can be traced back to 1985 with the seminal dissertation work by Stougie [29] on the design and analysis of algorithms for SMIP. Due to the lack of computational power in the early years of SMIP, it is interesting to note that the development of SP somehow parallels the developments in the semiconductor and computer industry. It is only later in the nineties that we see several works on SMIP. These include the establishment of continuity properties of expected resource function [20], properties of with simple integer recourse [14], properties and methods for integer recourse [30], and structure and stability of SMIP [21].

Early solution methods for SMIP include stochastic branch-and-bound [16], enumeration using Gröbner basis reduction [24], and dual decomposition [10–12]. Properties of SMIP with integer recourse were formalized in [22]. This was followed by a finite branch-and-bound method for SMIP with general mixed-integer variables [2] and the D^2 method [25]. The D^2 method was illustrated in [26] and applied in [19]. Extensions of these works include D^2 with branch-and-cut (D^2-BAC) [27] and later, D^2 for random recourse [17]. The FD method was introduced in [18] for SMIP with arbitrary first- and second-stage decision variables. Properties of risk-averse SP were formalized in [1] while those for risk-averse SMIP were proposed later in [23].

Problems

9.1 Properties

Consider a two-stage MR-SMIP of the form

$$\underset{x \in X \cap \mathbb{B}}{\text{Min}} \ \mathbb{E}[f(x, \tilde{\omega})] + \lambda \mathbb{D}[f(x, \tilde{\omega})],$$

where $f(x, \tilde{\omega}) := c^\top x + \varphi(x, \tilde{\omega})$ and for an outcome $\omega \in \Omega$ of $\tilde{\omega}$

$$\begin{aligned} \varphi(x, \omega) :=& \text{Min } q^\top y \\ & \text{s.t. } Wy \geq r(\omega) - T(\omega)x \\ & y \in \mathbb{Y}. \end{aligned}$$

Supposed that the problem data is given as follows:

First-stage:
$c = (1)^\top$, $A = [-1]$, $b = (-3)^\top$, and $X = \{x \in \mathbb{R}_+ : Ax \geq b\}$.
Second-stage:
$\Omega = \{\omega^1, \omega^2\}$ and $p(\omega^1) = p(\omega^2) = 0.5$.
$q = (2, 1)^\top$.
$T = \begin{bmatrix} -1 \\ 1 \end{bmatrix}$, $W = \begin{bmatrix} 1 & 0 \\ 0 & 1 \end{bmatrix}$, $r(\omega^1) = (-1, 1)^\top$, and $r(\omega^2) = (-2, 2)^\top$.
$\mathbb{D} := EE$ (expected excess) with target $\eta := 1.5$.
$\mathbb{Y} := \mathbb{Z}^2_+$.

(a) Write the formulation of the instance, specifying the first- and second-stage models, with explicit expressions for $f(x, \tilde{\omega})$ and $\varphi(x, \tilde{\omega})$.
(b) Derive closed-form expressions for $\varphi(x, \omega)$ and $f(x, \omega)$ for ω^1 and ω^2, respectively. Plot the two functions and describe them in terms of continuity and convexity. Is either function lower semicontinuous?
(c) Write closed-form expressions for $\mathbb{E}[\varphi(x, \tilde{\omega})]$ and $\mathbb{E}[f(x, \tilde{\omega})]$. Plot the two functions on the same x-axis and describe them in terms of continuity and convexity. Is either function lower semicontinuous?
(d) For $\lambda = 1$ (*risk-averse*), write a closed-form expression for $\mathbb{D}[f(x, \tilde{\omega})]$. Plot $\mathbb{D}[f(x, \tilde{\omega})]$ and $\mathbb{E}[f(x, \tilde{\omega})]$ on the same x-axis and describe the two functions in terms of continuity and convexity. Is either function lower semicontinuous?

9.2 Properties
An instance of a two-stage MR-SMIP of the form

$$\underset{x \in X \cap \mathbb{B}}{\text{Min}} \ \mathbb{E}[f(x, \tilde{\omega})] + \lambda \mathbb{D}[f(x, \tilde{\omega})],$$

where $f(x, \tilde{\omega}) := c^\top x + \varphi(x, \tilde{\omega})$ and for an outcome $\omega \in \Omega$ of $\tilde{\omega}$

$$\begin{aligned} \varphi(x, \omega) :=& \text{Min } q^\top y \\ & \text{s.t. } Wy \geq r(\omega) - T(\omega)x \\ & y \in \mathbb{B}^2 \end{aligned}$$

has the following problem data:

First-stage:
$c = (1)^\top$, $A = [-1]$, $b = (-1)^\top$, and $X = \{x \in \mathbb{R}_+ : Ax \geq b\}$.

Second-stage:

$\Omega = \{\omega^1, \omega^2\}$ and $p(\omega^1) = p(\omega^2) = 0.5$.
$q = (-2, -4)^\top$.

$$T(\omega^1) = T(\omega^2) = T = \begin{bmatrix} -1 \\ 1 \\ 0 \end{bmatrix}, \; W = \begin{bmatrix} 1 & 0 \\ 0 & 1 \\ -1 & -1 \end{bmatrix}, \; r(\omega^1) = (-1, 1, -1.5)^\top,$$

and $r(\omega^2) = (-0.5, 0.5, -1.5)^\top$.
$\mathbb{D} := EE$ (expected excess) with target $\eta := -3$.
$\mathbb{Y} := \mathbb{Z}^2_+$.

(a) Write the formulation of the instance, specifying the first- and second-stage models, with explicit expressions for $f(x, \tilde{\omega})$ and $\varphi(x, \tilde{\omega})$.
(b) Derive closed-form expressions for $\varphi(x, \omega)$ and $f(x, \omega)$ for ω^1 and ω^2, respectively, for $0 \le x \le 1$. Plot the two functions and describe them in terms of continuity and convexity. Is either function lower semicontinuous?
(c) Write closed-form expressions for $\mathbb{E}[\varphi(x, \tilde{\omega})]$ and $\mathbb{E}[f(x, \tilde{\omega})]$ for $0 \le x \le 1$. Plot the two functions on the same x-axis and describe them in terms of continuity and convexity. Is either function lower semicontinuous?
(d) For $\lambda = 1$ (*risk-averse*), write a closed-form expression for $\mathbb{D}[f(x, \tilde{\omega})]$ for $0 \le x \le 1$. Plot $\mathbb{D}[f(x, \tilde{\omega})]$ and $\mathbb{E}[f(x, \tilde{\omega})]$ on the same x-axis and describe the two functions in terms of continuity and convexity. Is either function lower semicontinuous?

9.3 Formulation
You are given a two-stage MR-SMIP of the form

$$\underset{x \in X \cap \mathbb{B}}{\text{Min}} \; \mathbb{E}[f(x, \tilde{\omega})] + \lambda \mathbb{D}[f(x, \tilde{\omega})],$$

where $f(x, \tilde{\omega}) := c^\top x + \varphi(x, \tilde{\omega})$ and for an outcome $\omega \in \Omega$ of $\tilde{\omega}$

$$\begin{aligned} \varphi(x, \omega) := &\text{Min } q^\top y \\ &\text{s.t. } Wy \ge r(\omega) - T(\omega)x \\ &\quad y \in \mathbb{B}^2. \end{aligned}$$

Consider an instance of this MR-SMIP with the following problem data:

First-stage:
$c = (-2)^\top$, $A = [-1]$, $b = (-1)^\top$, and $X = \{x \in \mathbb{R}_+ : Ax \ge b\}$.
Second-stage:
$\Omega = \{\omega^1, \omega^2\}$ and $p(\omega^1) = p(\omega^2) = 0.5$.
$q(\omega^1) = (-4, -4)^\top$ and $q(\omega^2) = (-2, -2)^\top$.

$$T = T(\omega^1) = T(\omega^2) = \begin{bmatrix} 0 \\ -0.5 \\ 0 \\ 0 \end{bmatrix}, W = \begin{bmatrix} 1 & -1 \\ -1 & -1 \\ -1 & 0 \\ 0 & -1 \end{bmatrix},$$

$r(\omega^1) = (-0.6, -1.2, -1, -1)^\top$, and $r(\omega^2) = (-0.6, -1.5, -1, -1)^\top$.

(a) Write the formulation of the instance, specifying the first- and second-stage models, with explicit expressions for $f(x, \tilde{\omega})$ and $\varphi(x, \tilde{\omega})$ for each realization ω^1 and ω^2 of $\tilde{\omega}$.
(b) Write the DEP formulation for $\mathbb{D} := EE$ (expected excess) with target $\eta := 4$ and weight $\lambda := 10$.
(c) Write the DEP formulation for $\mathbb{D}$ of your choice.

9.4 Formulation
Supposed that the problem data for the MR-SMIP in problem 9.3 is given as follows:
First-stage:
$c = (2)^\top$, $A = [-1]$, $b = (-1)^\top$, and $X = \{x \in \mathbb{R}_+ : Ax \geq b\}$.
Second-stage:
$\Omega = \{\omega^1, \omega^2\}$ and $p(\omega^1) = p(\omega^2) = 0.5$.
$q = (-60, 20)^\top$.

$$T(\omega^1) = \begin{bmatrix} 0 \\ 2 \\ 0 \\ 0 \end{bmatrix}, T(\omega^2) = \begin{bmatrix} 0 \\ -1 \\ 0 \\ 0 \end{bmatrix}, W = \begin{bmatrix} \frac{4}{3} & -1 \\ -\frac{5}{2} & -1 \\ -1 & 0 \\ 0 & -1 \end{bmatrix},$$

$r(\omega^1) = (-\frac{4}{3}, -2, -1, -1)^\top$, and $r(\omega^2) = (-\frac{1}{3}, -2, -1, -1)^\top$.

(a) Write the formulation of the instance, specifying the first- and second-stage models, with explicit expressions for $f(x, \tilde{\omega})$ and $\varphi(x, \tilde{\omega})$ for each realization ω^1 and ω^2 of $\tilde{\omega}$.
(b) Write the DEP formulation for $\mathbb{D} := EE$ (expected excess) with target $\eta := 0$ and weight $\lambda := 10$.
(c) Write the DEP formulation for $\mathbb{D}$ of your choice.

9.5 Formulation
Consider a two-stage MR-SMIP of the form

$$\underset{x \in X \cap \mathbb{B}^2}{\text{Min}} \ \mathbb{E}[f(x, \tilde{\omega})] + \lambda \mathbb{D}[f(x, \tilde{\omega})],$$

where $f(x, \tilde{\omega}) := c^\top x + \varphi(x, \tilde{\omega})$ and for an outcome $\omega \in \Omega$ of $\tilde{\omega}$

$$\begin{aligned} \varphi(x, \omega) :=& \text{Min } q(\omega)^\top y \\ & \text{s.t. } Wy \geq r(\omega) - T(\omega)x \\ & y \in \mathbb{B}^2. \end{aligned}$$

Supposed that the problem data for is given as follows:
First-stage:

$c = (-3, -8)^\top$, $A = \begin{bmatrix} -1 & 0 \\ 0 & -1 \end{bmatrix}$, $b = (-1, -1)^\top$, and $X = \{x \in \mathbb{R}^2_+ : Ax \geq b\}$.

Second-stage:

$\Omega = \{\omega^1, \omega^2\}$ and $p(\omega^1) = p(\omega^2) = 0.5$.
$q(\omega^1) = (-16, -20)^\top$ and $q(\omega^2) = (-32, -10)^\top$.

$$T = T(\omega^1) = T(\omega^2) = \begin{bmatrix} -1 & 0 \\ 0 & -1 \\ 0 & 0 \\ 0 & 0 \end{bmatrix}, W = \begin{bmatrix} -2 & -3 \\ -4 & -1 \\ -1 & 0 \\ 0 & -1 \end{bmatrix},$$

$r(\omega^1) = (-6, -5, -1, -1)^\top$, and $r(\omega^2) = (-4, -6, -1, -1)^\top$.

(a) Write the formulation of the instance, specifying the first- and second-stage models, with explicit expressions for $f(x, \tilde{\omega})$ and $\varphi(x, \tilde{\omega})$ for each realization ω^1 and ω^2 of $\tilde{\omega}$.
(b) Write the DEP formulation for $\mathbb{D} := EE$ (expected excess) with target $\eta := 50$ and weight $\lambda := 10$.
(c) Write the DEP formulation for $\mathbb{D}$ of your choice.

9.6 Formulation
Consider a two-stage MR-SMIP of the form

$$\underset{x \in X \cap \mathbb{B}^2}{\text{Min}} \ \mathbb{E}[f(x, \tilde{\omega})] + \lambda \mathbb{D}[f(x, \tilde{\omega})],$$

where $f(x, \tilde{\omega}) := c^\top x + \varphi(x, \tilde{\omega})$ and for an outcome $\omega \in \Omega$ of $\tilde{\omega}$

$$\begin{aligned} \varphi(x, \omega) :=& \text{Min } q^\top y \\ & \text{s.t. } Wy \geq r(\omega) - T(\omega)x \\ & y \in \mathbb{B}^4. \end{aligned}$$

Supposed that the problem data is given as follows:

First-stage:

$c = (-3, -3)^\top$, $A = \begin{bmatrix} -1 & 0 \\ 0 & -1 \end{bmatrix}$, $b = (-1, -1)^\top$, and $X = \{x \in \mathbb{R}^2_+ : Ax \geq b\}$.

Second-stage:

$\Omega = \{\omega^1, \omega^2\}$ and $p(\omega^1) = p(\omega^2) = 0.5$.
$q = (-16, -12, -22, -28)^\top$.

$$T(\omega^1) = T(\omega^2) = \begin{bmatrix} -1 & 0 \\ 0 & -1 \\ 0 & 0 \\ 0 & 0 \end{bmatrix}, \; W = \begin{bmatrix} -2 & -3 & -4 & -5 \\ -4 & -1 & -3 & -2 \\ -1 & 0 & 0 & 0 \\ 0 & -1 & 0 & 0 \\ 0 & 0 & -1 & 0 \\ 0 & 0 & 0 & -1 \end{bmatrix}, \; r(\omega^1) =$$

$(-6, -5, -1, -1, -1, -1)^\top$, and $r(\omega^2) = (-4, -6, -1, -1, -1, -1)^\top$.

(a) Write the formulation of the instance, specifying the first- and second-stage models, with explicit expressions for $f(x, \tilde{\omega})$ and $\varphi(x, \tilde{\omega})$ for each realization ω^1 and ω^2 of $\tilde{\omega}$.

(b) Write the DEP formulation for $\mathbb{D} := EE$ (expected excess) with target $\eta := 40$ and weight $\lambda := 10$.

(c) Write the DEP formulation for $\mathbb{D}$ of your choice.

9.7 ***BFS*** **Algorithm**

(a) Apply two iterations of the BFS algorithm towards solving the instance in Problem 9.4 for the risk-neutral case ($\lambda := 0$).

(b) Apply two iterations of the BFS algorithm towards solving the instance in Problem 9.5 for the risk-neutral case ($\lambda := 0$).

(c) Apply two iterations of the BFS algorithm towards solving the instance in Problem 9.6 for the risk-neutral case ($\lambda := 0$).

(d) Extend the BFS algorithm and apply two iterations towards solving the instance with $\mathbb{D} := EE$ (expected excess) in Problem 9.4 (b) or Problem 9.5(b) or Problem 9.6 (b).

9.8 ***FD*** **Algorithm**

(a) Apply two iterations of the FD algorithm towards solving the instance in Problem 9.4 for the risk-neutral case.

(b) Apply two iterations of the FD algorithm towards solving the instance in Problem 9.5 for the risk-neutral case.

(c) Apply two iterations of the FD algorithm towards solving the instance in Problem 9.6 for the risk-neutral case.

(d) Extend the FD algorithm and apply two iterations towards solving the instance with $\mathbb{D} := EE$ (expected excess) in Problem 9.4 (b) or Problem 9.5(b) or Problem 9.6 (b).

9.9 D^2 **algorithm**

(a) Apply two iterations of the D^2 algorithm towards solving the instance in Problem 9.4 for the risk-neutral case.

(b) Apply two iterations of the D^2 algorithm towards solving the instance in Problem 9.5 for the risk-neutral case.

(c) Apply two iterations of the D^2 algorithm towards solving the instance in Problem 9.6 for the risk-neutral case.

(d) Extend the D^2 algorithm and apply two iterations towards solving the instance with $\mathbb{D} := EE$ (expected excess) in Problem 9.4 (b) or Problem 9.5(b) or Problem 9.6 (b).

9.10 Algorithm Implementation: BFS Algorithm

(a) Implement (code) the BFS algorithm using your favorite programming language and LP/MIP solver. Test your code using the MR-SMIP instance given in Example 9.2 as well as your own instances to make sure that the code is working correctly.
(b) Use you implementation to solve instances from the literature and/or those available online.

9.11 Algorithm Implementation: FD Algorithm

(a) Implement (code) the FD algorithm using your favorite programming language and LP/MIP solver. Test your code using the SMIP instance given in Example 9.2 as well as your own instances to make sure that the code is working correctly.
(b) Use you implementation to solve instances from the literature and/or those available online.

9.12 Algorithm Implementation: D^2 Algorithm

(a) Implement (code) the D^2 algorithm using your favorite programming language and LP/MIP solver. Test your code using the SMIP instance given in Example 9.2 as well as your own instances to make sure that the code is working correctly.
(b) Use you implementation to solve instances from the literature and/or those available online.

References

1. S. Ahmed. Convexity and decomposition of mean-risk stochastic programs. *Mathematical Programming*, 106(3):433–446, 2006.
2. S. Ahmed, M. Tawarmalani, and N. V. Sahinidis. A finite branch and bound algorithm for two-stage stochastic integer programs. *Mathematical Programming*, 100:355–377, 2004.
3. E. Balas. Disjunctive programming: cutting planes from logical conditions. In O.L. Mangasarian, R.R. Meyer, and S.M. Robinson, editors, *Nonlinear Programming 2*. Academic Press, New York, 1975.
4. J. F. Benders. Partitioning procedures for solving mixed-variable programming problems. *Numerische Mathematik*, 4:238–252, 1962.
5. C. Blair and R. Jeroslow. A converse for disjunctive constraints. *Journal of Optimization Theory and Applications*, 25:195–206, 1978.
6. C.E. Blair and R.G. Jeroslow. The value function of a mixed-integer program: I. *Discrete Mathematics*, 19:121–138, 1977.
7. C.E. Blair and R.G. Jeroslow. The value function of an integer program. *Mathematical Programming*, 23:237–273, 1982.

8. E. A. Boyd. Fenchel cuts for integer programs. *Operations Research*, 42(1):53–64, 1994.
9. E. A. Boyd. On the convergence of Fenchel cutting planes in mixed-integer programming. *SIAM Journal on Optimization*, 5(2):421–435, 1995.
10. C. C. Carøe. *Decomposition in Stochastic Integer Programming*. Ph.D. thesis, Dept. of Operations Research, University of Copenhagen, Denmark, 1998.
11. C. C. Carøe and J. Tind. A cutting-plane approach to mixed 0–1 stochastic integer programs. *European Journal of Operational Research*, 101:306–316, 1997.
12. C.C. Carøe and R. Schultz. Dual decomposition in stochastic integer programming. *Operations Research Letters*, 24:37–45, 1999.
13. G.B. Dantzig. Linear programming under uncertainty. *Management Science*, 1(3–4):197–206, 1955. Republished in the 50th anniversary issue of Management Science 50(12):1764–1769, 2004.
14. W.K. Haneveld, W.K. Stougie, and M.H. van der Vlerk. On the convex hull of the simple integer recourse objective function. *Annals of Operations Research*, 56:209–224, 1995.
15. G. Laporte and F. V. Louveaux. The integer L-shaped method for stochastic integer programs with complete recourse. *Operations Research Letters*, 1:133–142, 1993.
16. V.I. Norkin, Y.M. Ermoliev, and A. Ruszczyński. On optimal allocation of indivisibles under uncertainty. *Operations Research*, 46(3):381–395, 1998.
17. L. Ntaimo. Disjunctive decomposition for two-stage stochastic mixed-binary programs with random recourse. *Operations Research*, 58(1):229–243, 2010.
18. L. Ntaimo. Fenchel decomposition for stochastic mixed-integer programming. *Journal of Global Optimization*, 55:141–163, 2013.
19. L. Ntaimo and S. Sen. The million-variable 'march' for stochastic combinatorial optimization. *Journal of Global Optimization*, 32(3):385–400, 2005.
20. R. Schultz. Continuity properties of expectation functions in stochastic integer programming. *Mathematics of Operations Research*, 18:578–589, 1993.
21. R. Schultz. On structure and stability in stochastic programs with random technology matrix and complete integer recourse. *Mathematical Programming*, 70:73–89, 1995.
22. R. Schultz. Stochastic programming with integer variables. *Mathematical Programming*, 97:285–309, 2003.
23. R. Schultz. Risk aversion in two-stage stochastic integer programming. In G. Infanger, editor, *Stochastic Programming: The State of the Art In Honor of George B. Dantzig*, International Series in Operations Research & Management Science 150, page 165–188. Springer, New York, USA, 2011.
24. R. Schultz, L. Stougie, and M. H. van der Vlerk. Solving stochastic programs with integer recourse by enumeration: a framework using Gröbner basis reduction. *Mathematical Programming*, 83(2):71–94, 1998.
25. S. Sen and J. L. Higle. The C3 theorem and a D2 algorithm for large scale stochastic mixed-integer programming: Set convexification. *Mathematical Programming*, 104(1):1–20, 2005.
26. S. Sen, J. L. Higle, and L. Ntaimo. A summary and illustration of disjunctive decomposition with set convexification. In D. L. Woodruff, editor, *Stochastic Integer Programming and Network Interdiction Models*, chapter 6, pages 105–123. Kluwer Academic Press, Dordrecht, The Netherlands, 2002.
27. S. Sen and H. D. Sherali. Decomposition with branch-and-cut approaches for two stage stochastic mixed-integer programming. *Mathematical Programming*, 106(2):203–223, 2006.
28. R. Van Slyke and R.-B. Wets. L-shaped linear programs with application to optimal control and stochastic programming. *SIAM Journal on Applied Mathematics*, 17:638–663, 1969.
29. L. Stougie. *Design and analysis of algorithms for stochastic integer programming*. Ph.D. Thesis, Center for Mathematics and Computer Science, Amsterdam, 1985.
30. M.H. van der Vlerk. Stochastic programming with integer recourse. Ph.d. thesis, Rijksuniversiteit Groningen, The Netherlands, 1995.

Part V
Computational Considerations

Chapter 10
Computational Experimentation

10.1 Introduction

The preceding chapters in this book cover mathematical modeling, optimization theory, derivation of decomposition algorithms, and application of the algorithms. It is, therefore, fitting to end the book with a chapter on a pragmatic aspect of optimization: computational experimentation. Decision-making regarding many interesting systems often involves modeling the system using mathematical optimization and running computational experiments to solve instances of the models to determine the optimal decisions. Real complex systems involve uncertain data, which has to be captured in the models and somehow read into the software platform to run the algorithms to solve the problem. Therefore, it becomes imperative to have a thorough understanding of the key issues associated with problem data and standard input data formats for two-stage stochastic programming (SP).

In this chapter, we provide an overview of fundamental issues associated with performing computational experiments in SP. In addition to understanding the theory, models, and algorithms, students in operations research and engineering should be able to implement (software coding/programming) and apply the models and algorithms. Several test problems have been created in the literature on SP for testing algorithms and are typically written in standard input formats. Therefore, we review some of the standard input formats that are useful for conducting computational experiments using standard test instances.

Let us restate the generic mean-risk SP (MR-SP) problem covered in the previous chapter, which we shall use as a basis for describing the different standard input formats:

$$\underset{x \in X \cap \mathbb{X}}{\text{Min}} \quad \mathbb{E}[f(x, \tilde{\omega})] + \lambda \mathbb{D}[f(x, \tilde{\omega})], \tag{10.1}$$

where $\mathbb{E} : \mathscr{F} \mapsto \mathbb{R}$ denotes the expected value, $\mathbb{D} : \mathscr{F} \mapsto \mathbb{R}$ is the *risk measure*, and $\lambda \geq 0$ is a suitable weight factor that quantifies the trade-off between expected

L. Ntaimo, *Computational Stochastic Programming*, Springer Optimization and Its Applications 774, https://doi.org/10.1007/978-3-031-52464-6_10

cost and risk. For ease of exposition, we focus on the risk-neutral case, i.e., when $\lambda := 0$. The set $X = \{x \in \mathbb{R}^{n_1}_+ : Ax \geq b\}$ is nonempty and $X \cap \mathbb{X}$ defines the first-stage feasible set, where $\mathbb{X}$ imposes integer restrictions on x. The matrix $A \in \mathbb{R}^{m_1 \times n_1}$ and vector $b \in \mathbb{R}^{m_1}$ are the first-stage matrix and right hand side (RHS) vector, respectively. The family of real random cost variables $\{f(x, \tilde{\omega})\}_{x \in X \cap \mathbb{X}} \subseteq \mathscr{F}$ is defined on $(\Omega, \mathscr{A}, \mathbb{P})$, where $\mathscr{F}$ is the space of all real random cost variables $f : \Omega \mapsto \mathbb{R}$ satisfying $\mathbb{E}[|f(\tilde{\omega})|] < \infty$. For a given $x \in X \cap \mathbb{X}$, the real random cost variable $f(x, \tilde{\omega})$ is given by

$$f(x, \tilde{\omega}) := c^\top x + \varphi(x, \tilde{\omega}). \tag{10.2}$$

By definition, if $x \notin X \cap \mathbb{X}$, $f(x, \tilde{\omega}) = \infty$. For a given outcome ω of $\tilde{\omega}$, the *recourse function* $\varphi(x, \omega)$ is given by

$$\begin{aligned} \varphi(x, \omega) := \text{Min } & q(\omega)^\top y(\omega) \\ \text{s.t. } & W y(\omega) \geq r(\omega) - T(\omega)x \\ & y(\omega) \in \mathbb{Y}, \end{aligned} \tag{10.3}$$

where $q(\omega) \in \mathbb{R}^{n_2}$ is the second-stage cost vector and $y(\omega)$ is the recourse decision. The matrix $W \in \mathbb{R}^{m_2 \times n_2}$ is the *recourse matrix*, $T(\omega) \in \mathbb{R}^{m_2 \times n_1}$ is the *technology matrix*, and $r(\omega) \in \mathbb{R}^{m_2}$ is the RHS vector. A *scenario* ω defines the outcome of the stochastic problem data, i.e., $\omega := (q(\omega), T(\omega), r(\omega))$. Next, we provide an overview of problem data standard input formats.

10.2 Problem Data Input Formats

In this subsection, we review the problem data standard input formats for mathematical programming. This is followed by an overview of the problem data standard input format for SP. The reader familiar with these standard input formats can skip this subsection without loss in the continuity of the foregoing discussion.

10.2.1 LP and MPS File Formats

Optimization software packages for mathematical programming are designed to read in problem data for a given instance and then optimize the instance using a specific algorithm. Therefore, instance data has to be placed in some format that the software package understands. To that end, standard input data formats have been developed over the years. The two standard formats for deterministic mathematical programs (LP, MIP, NLP, etc.) are *Linear Programming* (*LP*) file format and *Mathematical Programming System* (*MPS*) file format. The LP file

```
\Problem name: probname.lp

Minimize        obj1 y1   + obj2 y2   + ··· + objn yn
s.t.
  c1:           mat11 y1 + mat12 y2 + ··· + mat1n yn >= rhs1
  c2:           mat21 y1 + mat22 y2 + ··· + mat2n yn <= rhs2
  ···:             ···        ···             ···     ···  ···
  cm:           matm1 y1 + matm2 y2 + ··· + matmn yn  =  rhsm
Bounds
  0 <= y1 <= 1
Binaries
  y1
Generals
  y2
End
```

Fig. 10.1 A generic LP file

format is based on the 1970s FORTRAN programming language and is relatively easy to read as it captures the instance in a natural, algebraic LP formulation of how an LP is written. The input file is saved as "probname.lp," where "probname" is the instance name and the extension ".lp" indicates that the instance is in the LP file format.

Figure 10.1 shows an example of a generic LP file. As shown in the figure, in the LP file format an instance is characterized by its NAME, the type of optimization (`Minimization` or `Maximization`), `ROWS` (objective function and constraints), `COLUMNS` (decision variables), `RHS`, `BOUNDS` on the decision variables, and restrictions on the decision variables (`Binaries`, `Generals`, etc.). This instance is a minimization problem with objective function named `obj`, having decision variables `y1, y2, ···, yn` with corresponding objective function coefficients `obj1, obj2, ···, objn`. The constraints are named `c1, c2, ···, cm`, with each row $i = 1, \cdots, m$ having matrix coefficients `mati1, mati2, ···, matin` and RHS data `rhsi`. In this example LP file, the decision variable `y1` is bounded between zero and one and is binary, while the decision variable `y2` is integer. By default, not showing any restriction on the decision variable in an LP file means that it is nonnegative. Lastly, the keyword `End` indicates the end of the file.

A sample LP file is given in Fig. 10.2 for an example SP instance, spex1.lp. This instance has three binary decision variables (`x1`, `x2`, and `y2`) and one continuous variable bounded between 0 and 1 (`y2`). The instance has four constraints. We shall use this instance to illustrate the different SP files.

The MPS file format is based on sparse matrix storage to save on computer memory and to enable efficient computations. This format captures only nonzero elements of the instance in column format. Solvers such as CPLEX and GUROBI exploit this format for efficient computations and memory storage. In this case, the input file is saved as "probname.mps," where "probname" is the problem (instance) name and the extension ".mps" indicates that the instance is in the MPS file format.

```
\Problem name: spex1.lp

Minimize
  OBJ:              - 5 x1 - 10 x2 - 15 y1 - 20 y2
Subject To
  c1:           - 0.5 x1                        >= -0.4
  c2:                          - x2             >= -1
  c3:                 - x1          - 2 y1  - 3 y2 >= -5
  c4:                          - x2  - 4 y1  - 1 y2 >= -2
Bounds
  0 <= x1 <= 1
  0 <= x2 <= 1
  0 <= y1 <= 1
  0 <= y2 <= 1
Binaries
  x1 x2 y1
End
```

Fig. 10.2 A sample LP file `spex1.lp`

The generic LP file given above can be written in the MPS file format as shown in Fig. 10.3. In the shown sample MPS file, after the keyword `NAME` is the instance name followed by `ROWS`, which lists the rows as follows: `N` is the objective function row, `G` is for a "≥" constraint, `L` is for a "≤" constraint, and `E` is for the "=" constraint. The keyword `COLUMNS` lists the instance data by columns (decision variables), i.e., column name followed by row name and the corresponding data value in the instance.

Within the columns the phrase `MARK0000 'MARKER' 'INTORG'` marks the beginning of integer decision variables and `MARK0000 'MARKER' 'INTEND'` marks the end. The keyword `RHS` lists the row name and the corresponding RHS value. Under the keyword `BOUNDS` we have a list of the variable type and the corresponding bounds. In this example, `BV bnd` (binary variable bound) indicates that `x1` is binary, while `LI bnd` (lower integer bound) indicates that `x2` is integer. The keywords `LO bnd UP bnd` would indicate a lower and a upper bound, respectively, on a variable. Lastly, the keyword `ENDATA` indicates the end of the MPS file. A sample MPS file, corresponding to the sample LP file in Fig. 10.2, is given in Fig. 10.4.

10.2.2 SMPS File Format

Let us now turn to the standard input file formats for SP known as the *Stochastic Mathematical Programming System* (*SMPS*) file format. This is an extension of the MPS file format to the stochastic setting. The SMPS file format was proposed in [5] and comprises three separate files: CORE, TIME, and STOCH files. A full description of the format is given in [10] and [11].

Fig. 10.3 A generic MPS file

```
NAME              probname.mps
ROWS
 N obj
 G c1
 L c2
  ... ...
 E cm
COLUMNS
    MARK0000    'MARKER'        'INTORG'
    y1          obj              obj1
    y1          c1               mat11
    y1          c2               mat21
    ...         ...              ...
    y1          cm               matm1
    y2          obj              obj2
    y2          c1               mat12
    y2          c2               mat22
    ...         ...              ...
    y2          cm               matm2
    MARK0001    'MARKER'        'INTEND'
    y3          obj              obj3
    ...         ...              ...
    yn          obj              objn
    yn          c1               mat1n
    yn          c2               mat2n
    ...         ...              ...
    yn          cm               matmn
RHS
    rhs         c1               rhs1
    rhs         c2               rhs2
    ...         ...              ...
    rhs         cm               rhsm
BOUNDS
    BV bnd      x1
    LI bnd      x2                0
ENDATA
```

CORE File

The CORE file contains the "core" problem in the MPS file format, which is given as follows:

$$
\begin{aligned}
\text{Min}\ & c^\top x + q^\top y \\
\text{s.t.}\ \ & Ax \qquad\quad \geq b \\
& Tx + Wy \geq r \\
& \ \ x \in \mathbb{X},\ y \in \mathbb{Y}.
\end{aligned}
\tag{10.4}
$$

Fig. 10.4 A sample MPS file spex1.mps

```
NAME          spex1.mps
ROWS
 N OBJ
 G c1
 G c2
 G c3
 G c4
COLUMNS
   MARK0000 'MARKER'   'INTORG'
   x1       OBJ           -5
   x1       c1            -0.5
   x1       c3            -1
   x2       OBJ           -10
   x2       c2            -1
   x2       c4            -1
   y1       OBJ           -15
   y1       c3            -2
   y1       c4            -4
   MARK0001 'MARKER'   'INTEND'
   y2       OBJ           -20
   y2       c3            -3
   y2       c4            -1
RHS
   rhs      c1            -0.4
   rhs      c2            -1
   rhs      c3            -5
   rhs      c4            -2
BOUNDS
   BV bnd   x1
   BV bnd   x2
   BV bnd   y1
   UP bnd   y2            1
ENDATA
```

The CORE file is saved as "probname.cor," where "probname" is the instance name and the extension ".cor" indicates that this file is the CORE problem in the MPS file format. Note that the core problem contains only the deterministic data and captures the order of the decision variables (columns) x and y as well as the constraints (rows). A sample CORE file, spex1.cor, corresponding to the sample LP file in Fig. 10.2 is given in Fig. 10.5. Notice that the CORE file is the same as the MPS file shown in Fig. 10.4 except in this case the file is named "spex1.cor" instead of "spex1.mps."

TIME File

The TIME file contains the "timing" information which shows where the first- and the second-stage begin, respectively, in the CORE file as illustrated in Fig. 10.6. The TIME file is saved as "probname.tim," where "probname" is the instance name and the extension ".tim" indicates that this is the TIME file. A sample TIME file is shown

Fig. 10.5 A sample CORE file `spex1.cor`

```
NAME            spex1.cor
ROWS
 N OBJ
 G c1
 G c2
 G c3
 G c4
COLUMNS
    MARK0000 'MARKER'   'INTORG'
    x1        OBJ            -5
    x1        c1           -0.5
    x1        c3             -1
    x2        OBJ           -10
    x2        c2             -1
    x2        c4             -1
    y1        OBJ           -15
    y1        c3             -2
    y1        c4             -4
    MARK0001 'MARKER'   'INTEND'
    y2        OBJ           -20
    y2        c3             -3
    y2        c4             -1
RHS
    rhs       c1           -0.4
    rhs       c2             -1
    rhs       c3             -5
    rhs       c4             -2
BOUNDS
    BV bnd    x1
    BV bnd    x2
    BV bnd    y1
    UP bnd    y2              1
ENDATA
```

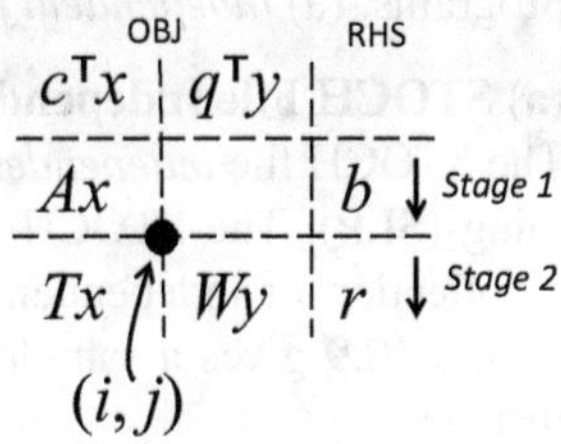

Fig. 10.6 The first- and second-stage data contained in the CORE file

in Fig. 10.7 for the spex1.cor instance given in Fig. 10.5. The first line contains the keyword `TIME` followed by the instance name. The second line contains the keyword `PERIODS` followed by the core problem type, i.e., `LP` in this case. The third line has the (column, row, and stage names) where the first-stage begins in the CORE file. Similarly, the fourth line contains the (column, row, and stage names) where the second-stage begins in the CORE file. The last line in the TIME file has the keyword `ENDATA` to indicate the end of the file. The purpose of the TIME file is to enable determining the coordinates (i, j) shown in Fig. 10.6. The coordinates

Fig. 10.7 A sample TIME file `spex1.tim`

```
TIME     spex1
PERIODS LP
   x1     c1     STAGE-1
   y1     c3     STAGE-2
ENDATA
```

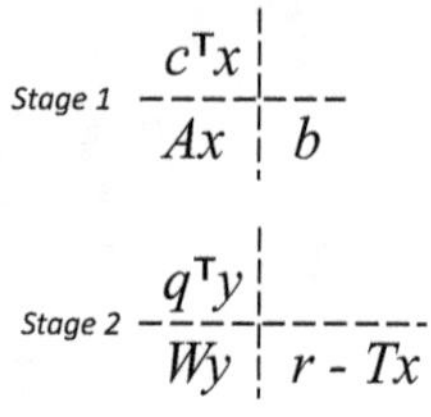

Fig. 10.8 The CORE file data split into first- and second-stage data

Fig. 10.9 A sample STOCH file `spex1.sto` in independent format

```
STOCH  spex1
INDEP  DISCRETE
       RHS        c3  -5  0.7
       RHS        c3  -6  0.3
       RHS        c4  -2  0.4
       RHS        c4  -4  0.6
ENDATA
```

allow to decompose the problem into separate first- and second-stage problems as shown in Fig. 10.8.

STOCH File

The third SMPS file is the STOCH file. As the name implies, this file contains the stochastic data of the instance. There are two basic formats for two-stage stochastic programs: (a) *independent format* and (b) *scenario format*.

(a) STOCH File Independent Format

The STOCH file *independent format* is typically used for stochastic linear programming (SLP). The STOCH file in this case contains the marginal distributions for a collection of independent random variables, thus the name *independent format.* Figure 10.9 gives a sample of a STOCH file in independent format for the spex1 instance.

The first line of the file starts with the keyword `STOCH` followed by the instance name. The next line has the keyword `INDEP` followed by `DISCRETE` to indicate that this stochastic data is in an independent format and the marginal distribution is discretely distributed. The next line in the file lists the (column, row) names of each random variable in the core problem from the CORE file followed by its outcome (value) and probability of occurrence. In the example STOCH file in Fig. 10.9, the column name `RHS` means the RHS of the instance. This means that this instance has two RHS random variables, (`RHS, c3`) and (`RHS, c4`). The random variable

(RHS, c3) has two possible outcomes, -5 and -6, with the probability of 0.7 and 0.3, respectively. The second random variable (RHS, c4) also has two possible outcomes, -2 and -4, with the probability of 0.4 and 0.6, respectively. Note that the probabilities for each of the random variables sum to one. The last line in the STOCH file has the keyword ENDATA to indicate the end of the file.

If an instance contains random objective function coefficients, the STOCH file will contain the (*column, row*) of the type (OBJ, row name). Similarly, if the instance has a random technology matrix, the (*column, row*) pair will take the form (column name, row name). The STOCH file in an independent format will have, say, N random variables with the i-th random variable having n_i outcomes each with a probability of occurrence $p_{i,j}$. This means that the instance has a total of $\prod_{i=1}^{N} n_i$ scenarios. Also, $\sum_j p_{i,j} = 1$ for each $i = 1, \cdots, N$. In the sample STOCH file in Fig. 10.9, we have $N = 2$ with $n_1 = 2$ and $n_2 = 2$. Therefore, the total number of scenarios is $\prod_{i=1}^{N} n_i = 2 \times 2 = 4$. The probabilities for each of the random variables must sum to one.

After reading the STOCH file data in an implementation (code) of a decomposition algorithm, the next step is to "create and process the scenarios on the fly." By this we mean creating and setting the stochastic data (OBJ, RHS, etc.) at runtime for a scenario subproblem, solving the subproblem, and then repeating the process until all scenarios are processed. The idea is to avoid creating and setting scenario data upfront for all scenarios because this can be computationally prohibitive and outright memory intensive when you have a large number of scenarios. For example, an instance with ten random variables each with ten possible outcomes results in $10^{10} = 10,000,000,000$ scenarios. This would mean creating and setting the stochastic data for these many scenarios. Depending on the instance size, this can be computationally inefficient.

Turning back to the sample STOCH file in Fig. 10.9, we would create "on the fly" the following RHS data for each scenario ω^s, $s = 1, \cdots, 4$:

$$r(\omega^1) = \begin{bmatrix} -5 \\ -2 \end{bmatrix}, \quad p(\omega^1) = 0.7 \times 0.4 = 0.28.$$

$$r(\omega^2) = \begin{bmatrix} -5 \\ -4 \end{bmatrix}, \quad p(\omega^2) = 0.7 \times 0.6 = 0.42.$$

$$r(\omega^3) = \begin{bmatrix} -6 \\ -2 \end{bmatrix}, \quad p(\omega^3) = 0.3 \times 0.4 = 0.12.$$

$$r(\omega^4) = \begin{bmatrix} -6 \\ -4 \end{bmatrix}, \quad p(\omega^4) = 0.3 \times 0.6 = 0.18.$$

Note that the scenario probabilities sum to one as expected, i.e., $0.28+0.42+0.12+0.18 = 1.0$

Fig. 10.10 A sample STOCH file `spex1.sto` in scenario format

```
STOCH      spex1
SCENARIOS       DISCRETE
  SC         Scen1 'ROOT'     0.28 STAGE-2
             RHS    c3        -5
             RHS    c4        -2
  SC         Scen2 'ROOT'     0.42 STAGE-2
             RHS    c3        -5
             RHS    c4        -4
  SC         Scen3 'ROOT'     0.12 STAGE-2
             RHS    c3        -6
             RHS    c4        -2
  SC         Scen4 'ROOT'     0.18 STAGE-2
             RHS    c3        -6
             RHS    c4        -4
ENDATA
```

(b) STOCH File Scenario Format

The STOCH file *scenario format* is usually used for SMIP instances as they typically have relatively fewer scenarios than SLP instances. The STOCH file in a scenario format contains the distribution of the stochastic data in terms of a collection of scenarios, each with a probability of occurrence. Figure 10.10 shows a sample STOCH file in scenario format for the spex1 instance. As with the independent format, the first line of the STOCH file starts with the keyword `STOCH` followed by the instance name. The next line has the keywords `SCENARIOS` and `DISCRETE`. This indicates that the probability distribution for the instance is given in scenario format and that it is discrete. The next line begins with `SC` (which indicates the beginning of a scenario) followed by the scenario name (`Scen1`), `'ROOT'` (which indicates that this scenario starts at the root node of the scenario tree), the probability of occurrence for this scenario (`0.28`), and the stage name (`STAGE-2`).

The next line lists the (*column, row*) names of each random variable in the core problem from the CORE file, followed by its outcome (value) under this scenario. In the example in Fig. 10.10, the column name `RHS` is the RHS of the instance, which has two random variables: (`RHS`, `c3`) with an outcome of −5 and (`RHS` and `c4`) with an outcome of −2. The next line lists the second scenario (`Scen2`), starting at the root node `'ROOT'` with a probability of occurrence of `0.42`, and stage name `STAGE-2`. The next two lines list the two random variables and their outcomes. The file lists the rest of the scenarios until the last line, which has the keyword `ENDATA` to indicate the end of the file. Notice that the STOCH file in Fig. 10.10 is simply a translation of STOCH in Fig. 10.9 from an *independent* format to a *scenario* format. Next, we illustrate the SMPS data formats using an example problem instance from Chap. 3.

Example 10.1 (The *abc*-Production Planning Problem—*abc*PPP) We recall the two-stage problem *abc*PPP from Chap. 3 and rewrite its formulation, which we shall refer to as *abc*PPPmodel, as follows:

$$\begin{array}{llllll}
\text{Min} & \mathbb{E}[f(x,\tilde{\omega})] & + \lambda\mathbb{D}[f(x,\tilde{\omega})] & & & \\
\text{s.t.} & -x_1 & & & & \geq -300 \\
& & -x_2 & & & \geq -700 \\
& & & -x_3 & & \geq -600 \\
& & & & -x_4 & \geq -500 \\
& & -x_2 & -x_3 & -x_4 & \geq -1600 \\
& x_1, & x_2, & x_3, & x_4 & \geq 0,
\end{array}$$

where

$$f(x,\tilde{\omega}) := 50x_1 + 30x_2 + 15x_3 + 10x_4 + \varphi(x,\tilde{\omega}),$$

and for a given scenario ω^s of $\tilde{\omega}$, the second-stage subproblem is given as follows:

$$\begin{array}{lllll}
\varphi(x,\omega^s) = \text{Min} & -1150y_1 & -1525y_2 & -1900y_3 & \\
\text{s.t.} & -6y_1 & -8y_2 & -10y_3 \geq & 0 \quad -x_1 \\
& -20y_1 & -25y_2 & -28y_3 \geq & 0 \quad -x_2 \\
& -12y_1 & -15y_2 & -18y_3 \geq & 0 \quad -x_3 \\
& -8y_1 & -10y_2 & -14y_3 \geq & 0 \quad -x_4 \\
& -y_1 & & \geq & -d_a(\omega^s) \\
& & -y_2 & \geq & -d_b(\omega^s) \\
& & & -y_3 \geq & -d_c(\omega^s) \\
& y_1, & y_2, & y_3 \geq & 0.
\end{array}$$

The problem data can be summarized in terms of the first- and second-stage problems as follows:

First-Stage

Object coefficients vector: $c = (50, 30, 15, 10)^\top$.

Constraint matrix: $A = \begin{bmatrix} -1 & 0 & 0 & 0 \\ 0 & -1 & 0 & 0 \\ 0 & 0 & -1 & 0 \\ 0 & 0 & 0 & -1 \\ 0 & -1 & -1 & -1 \end{bmatrix}$.

RHS: $b = (-300, -700, -600, -500, -1600)^\top$.

Second-Stage

Objective coefficients vector: $q = q(\omega^s) = (-1150, -1525, -1900)^\top$ for all s.

Table 10.1 Product demand scenarios for abcPPPmodel

	Scenarios				
	Low	*Moderate*	*High*	*Extreme*	*Rare*
	$s = 1$	$s = 2$	$s = 3$	$s = 4$	$s = 5$
$d_a(\omega^s)$	15	20	25	30	10
$d_b(\omega^s)$	10	15	20	25	10
$d_c(\omega^s)$	5	15	25	30	10
$p(\omega^s)$	0.15	0.30	0.30	0.20	0.05

$$\text{Recourse matrix: } W = \begin{bmatrix} -6 & -8 & -10 \\ -20 & -25 & -28 \\ -12 & -15 & -18 \\ -8 & -10 & -14 \\ -1 & 0 & 0 \\ 0 & -1 & 0 \\ 0 & 0 & -1 \end{bmatrix}.$$

$$\text{Technology matrix: } T = T(\omega^s) = \begin{bmatrix} 1 & 0 & 0 & 0 \\ 0 & 1 & 0 & 0 \\ 0 & 0 & 1 & 0 \\ 0 & 0 & 0 & 1 \\ 0 & 0 & 0 & 0 \\ 0 & 0 & 0 & 0 \\ 0 & 0 & 0 & 0 \end{bmatrix} \text{ for all } s.$$

RHS vector: $r(\omega^s) = (0, 0, 0, 0, d_a(\omega^s), d_b(\omega^s), d_c(\omega^s))^\top$ for all s.

Answer the following questions:

(a) Write the CORE problem for abcPPPmodel in the LP file format and create an LP file and save it as `abcPPPmodel.lp`.
(b) Create the CORE problem file for abcPPPmodel in the MPS file format and save it as `abcPPPmodel5.cor`.
(c) Write the TIME file for abcPPPmodel and save it as `abcPPPmodel5.tim`.
(d) Write the STOCH file abcPPPmodel in scenario format based on the scenario data given in Table 10.1 and save it as `abcPPPmodel5.sto`.
(e) Write the STOCH file for an instance of abcPPPmodel in independent format based on the scenario data given in Table 10.2 and save the file as `abcPPPmodel125.sto`. In the table, each of the three products has five *independent* demand outcomes: *Low, Moderate, High, Extreme, and Rare.* The collection of demand outcomes and the corresponding probabilities of occurrence forms a multivariate probability distribution for the instance. The outcomes are depicted in the scenario tree in Fig. 10.11. Notice that given the marginal distributions for each product, the total number of scenarios is $5 \times 5 \times 5 = 125$.

Table 10.2 Product marginal demand distribution for *abc*PPPmodel

Random variable	Demand	Probability	Level
Product-a (d_a)	15	0.15	Low
	20	0.30	Moderate
	25	0.30	High
	30	0.20	Extreme
	10	0.05	Rare
Product-b (d_b)	10	0.15	Low
	15	0.30	Moderate
	20	0.30	High
	25	0.20	Extreme
	10	0.05	Rare
Product-c (d_c)	5	0.15	Low
	15	0.30	Moderate
	25	0.30	High
	30	0.20	Extreme
	10	0.05	Rare

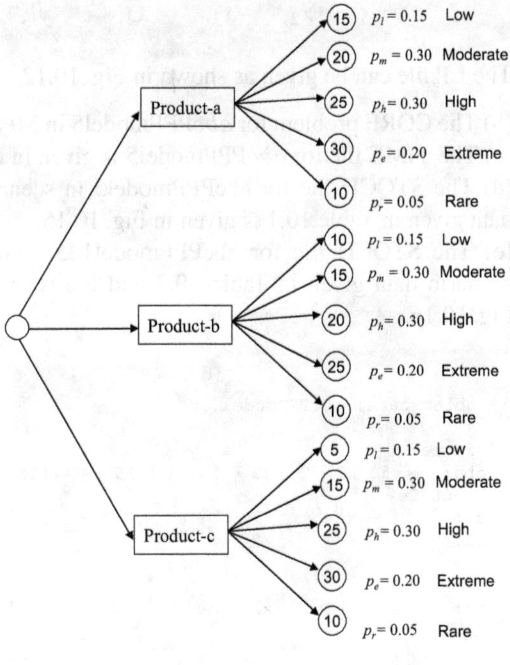

Fig. 10.11 Product demand scenario tree for *abc*PPPmodel

Note: In this case, create a copy of the CORE file `abcPPPmodel5.cor` and name it `abcPPPmodel125.cor`. Also, create a copy of the TIME file `abcPPPmodel5.tim` and name it `abcPPPmodel125.tim`.

Solution

(a) The CORE problem for *abc*PPPmodel can be written as follows:

$$
\begin{array}{llllllllll}
\text{Min} & 50x_1 & +30x_2 & +15x_3 & +10x_4 & -1150y_1 & -1525y_2 & -1900y_3 & & \\
\text{s.t.} & -x_1 & & & & & & & \geq & -300 \\
 & & -x_2 & & & & & & \geq & -700 \\
 & & & -x_3 & & & & & \geq & -600 \\
 & & & & -x_4 & & & & \geq & -500 \\
 & & -x_2 & -x_3 & -x_4 & & & & \geq & -1600 \\
 & & & & & -6y_1 & -8y_2 & -10y_3 & \geq & 0 \\
 & & & & & -20y_1 & -25y_2 & -28y_3 & \geq & 0 \\
 & & & & & -12y_1 & -15y_2 & -18y_3 & \geq & 0 \\
 & & & & & -8y_1 & -10y_2 & -14y_3 & \geq & 0 \\
 & & & & & -y_1 & & & \geq & -15 \\
 & & & & & & -y_2 & & \geq & -10 \\
 & & & & & & & -y_3 & \geq & -5 \\
 & x_1, & x_2, & x_3, & x_4 & y_1, & y_2, & y_3 & \geq & 0.
\end{array}
$$

The LP file can be given as shown in Fig. 10.12.

(b) The CORE problem for *abc*PPPmodel5 in MPS file format is given in Fig. 10.13.
(c) The TIME file for *abc*PPPmodel5 is given in Fig. 10.14.
(d) The STOCH file for abcPPPmodel5 in scenario format based on the scenario data given in Table 10.1 is given in Fig. 10.15.
(e) The STOCH file for abcPPPmodel125 in independent format based on the scenario data given in Table 10.2 and the scenario tree in Fig. 10.11 is given in Fig. 10.16.

```
\Problem name: abcPPPmodel.lp

Minimize
  obj:     50 x1 + 30 x2 + 15 x3 + 10 x4 - 1150 y1 - 1525 y2 - 1900 y3
Subject To
  c1:        - x1                                                  >= -300
  c2:                - x2                                          >= -700
  c3:                        - x3                                  >= -600
  c4:                                - x4                          >= -500
  c5:                - x2    - x3    - x4                          >= -1600
  c6:                                       - 6 y1   - 8 y2  - 10 y3 >=    0
  c7:                                      - 20 y1  - 25 y2  - 28 y3 >=    0
  c8:                                      - 12 y1  - 15 y2  - 18 y3 >=    0
  c9:                                       - 8 y1  - 10 y2  - 14 y3 >=    0
  c10:                                        - y1                 >=  -15
  c11:                                               - y2          >=  -10
  c12:                                                      - y3 >=    -5
End
```

Fig. 10.12 The CORE problem for *abc*PPPmodel in LP file format

Fig. 10.13 The CORE problem for *abc*PPPmodel5 in MPS file format

```
NAME      abcPPPmodel5.cor
ROWS
 N obj
 G c1
 G c2
 G c3
 G c4
 G c5
 G c6
 G c7
 G c8
 G c9
 G c10
 G c11
 G c12
COLUMNS
   x1   obj                   50
   x1   c1                    -1
   x2   obj                   30
   x2   c2                    -1
   x2   c5                    -1
   x3   obj                   15
   x3   c3                    -1
   x3   c5                    -1
   x4   obj                   10
   x4   c4                    -1
   x4   c5                    -1
   y1   obj                -1150
   y1   c6                    -6
   y1   c7                   -20
   y1   c8                   -12
   y1   c9                    -8
   y1   c10                   -1
   y2   obj                -1525
   y2   c6                    -8
   y2   c7                   -25
   y2   c8                   -15
   y2   c9                   -10
   y2   c11                   -1
   y3   obj                -1900
   y3   c6                   -10
   y3   c7                   -28
   y3   c8                   -18
   y3   c9                   -14
   y3   c12                   -1
RHS
   rhs  c1                  -300
   rhs  c2                  -700
   rhs  c3                  -600
   rhs  c4                  -500
   rhs  c5                 -1600
   rhs  c10                  -15
   rhs  c11                  -10
   rhs  c12                   -5
ENDATA
```

Fig. 10.14 The TIME file for abcPPPmodel5

```
TIME      abcPPPmodel5.tim
PERIODS LP
   x1   c1                  STAGE-1
   y1   c6                  STAGE-2
ENDATA
```

To enable efficient memory storage and computations, the data is typically stored in sparse matrix format. So in the next subsection we review sparse matrices and provide numerical examples.

Fig. 10.15 The STOCH file for abcPPPmodel5 in a scenario format

```
STOCH      abcPPPmodel5.sto
SCENARIOS        DISCRETE
  SC       Scen1 'ROOT'       0.15 STAGE-2
           RHS    c10         -15
           RHS    c11         -10
           RHS    c12         -5
  SC       Scen2 'ROOT'       0.30 STAGE-2
           RHS    c10         -20
           RHS    c11         -15
           RHS    c12         -15
  SC       Scen3 'ROOT'       0.30 STAGE-2
           RHS    c10         -25
           RHS    c11         -20
           RHS    c12         -25
  SC       Scen4 'ROOT'       0.20 STAGE-2
           RHS    c10         -30
           RHS    c11         -25
           RHS    c12         -30
  SC       Scen5 'ROOT'       0.05 STAGE-2
           RHS    c10         -10
           RHS    c11         -10
           RHS    c12         -10
ENDATA
```

Fig. 10.16 The STOCH file for abcPPPmodel125 in independent format

```
STOCH  abcPPPmodel125.stoc
INDEP  DISCRETE
       RHS        c10          -15    0.15
       RHS        c10          -20    0.30
       RHS        c10          -25    0.30
       RHS        c10          -30    0.20
       RHS        c10          -10    0.05
       RHS        c11          -10    0.15
       RHS        c11          -15    0.30
       RHS        c11          -20    0.30
       RHS        c11          -25    0.20
       RHS        c11          -10    0.05
       RHS        c12          -5     0.15
       RHS        c12          -15    0.30
       RHS        c12          -25    0.30
       RHS        c12          -30    0.20
       RHS        c12          -10    0.05
ENDATA
```

10.3 Sparse Matrices

When an SP problem is read into an optimization software package, data is stored in sparse matrix format to save memory due to a large number of zeros in the problem data in typical real-life instances. Therefore, the sparse matrix format is exploited for efficient computations. Specifically, *vector-matrix* and *matrix-vector* multiplications have to be handled using sparse matrices. A matrix can be written in either a *row* or

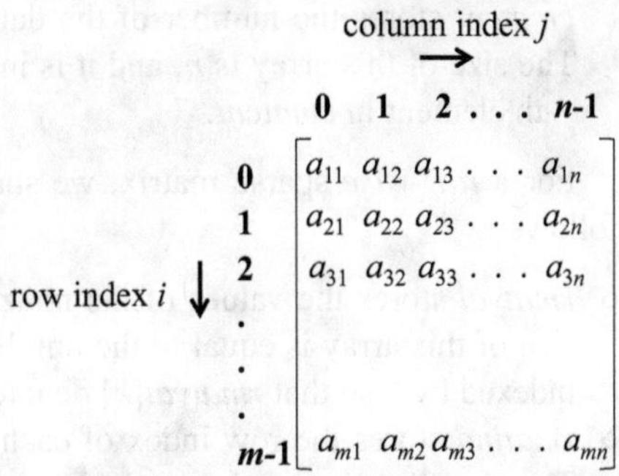

Fig. 10.17 Matrix showing the row and column indices

a *column* sparse format. The *row* sparse format stores the matrix nonzero elements (nonzeros) in the order of the rows (row-wise), while the *column* sparse format stores the matrix nonzeros in the order of the columns (column-wise).

To define a sparse matrix format, we need to first define how a matrix is structured. As shown in Fig. 10.17, for an $m \times n$ matrix, we shall have the row index i and the column index j starting at the top left corner of the matrix. As with many optimization software packages, we shall have the row indexed by $0, 1, 2, \cdots, m-1$ and the column indexed by $0, 1, 2, \cdots, n-1$.

In the optimization software packages, matrix data is stored as an *array*, which is a data structure consisting of a collection of elements each identified by an index or a key. A sparse matrix can be described using three arrays which we shall name *matbeg*, *matval*, and *matind*. These arrays are defined as follows:

matval stores the values of the nonzero elements in the matrix row-wise or column-wise.
matind stores the row or the column index of each nonzero element in the matrix.
matbeg stores the starting index of each matrix row or column in array *matval*.

For convenience, one can also include a fourth array, *matcnt*:

matcnt stores the number (count) of the nonzero data elements in the row or the column of the matrix.

To specify a *column-wise* sparse matrix, we shall prefix each array name with a "c" as follows:

◇ *cmatval* stores the values of the nonzero elements in the matrix column-wise. The size of this array is equal to the number of nonzeros in the matrix. We index this array by k such that *cmatval*$[k]$ denotes the k-th element in *cmatval*.
◇ *cmatind* stores the row index of each nonzero element in the matrix column-wise. The size of this array is equal to the number of nonzeros in the matrix, and we index it by k with *cmatind*$[k]$ denoting the k-th element in *cmatind*.
◇ *cmatbeg* stores the starting index of each matrix column nonzero element in array *cmatval*. This array is indexed by j so that *cmatbeg*$[j]$ denotes the j-th element in *cmatbeg*.

◇ *cmatcnt* stores the number of the data elements in each column of the matrix. The size of this array is n, and it is indexed by j so that *cmatcnt*$[j]$ denotes the j-th element in *cmatcnt*.

For a *row-wise* sparse matrix, we shall prefix each array name with an "r" as follows:

◇ *rmatval* stores the values of the nonzero elements in the matrix row-wise. The size of this array is equal to the number of nonzeros in the matrix. This array is indexed by k so that *rmatval*$[k]$ denotes the k-th element in *rmatval*.
◇ *rmatind* stores the row index of each nonzero element in the matrix row-wise. The size of this array is equal to the number of nonzeros in the matrix, and it is also indexed by k so that *rmatind*$[k]$ denotes the k-th element in *rmatind*.
◇ *rmatbeg* stores the starting index of each matrix row nonzero element in array *rmatval*. This array is indexed by j so that *rmatbeg*$[j]$ denotes the j-th element in *rmatbeg*.
◇ *rmatcnt* stores the number of the data elements in each row of the matrix. The size of this array is n, and it is indexed by j so that *rmatcnt*$[j]$ denotes the j-th element in *cmatcnt*.

Example 10.2 Consider the *abc*PPPmodel instance from the previous section and then answer the following questions:

(a) Write matrix A in sparse format: (i) column-wise and (ii) row-wise.
(b) Write matrix T in sparse format: (i) column-wise and (ii) row-wise.
(c) Write matrix W in sparse format: (i) column-wise and (ii) row-wise.

Solution

(a) We are given $A = \begin{bmatrix} -1 & 0 & 0 & 0 \\ 0 & -1 & 0 & 0 \\ 0 & 0 & -1 & 0 \\ 0 & 0 & 0 & -1 \\ 0 & -1 & -1 & -1 \end{bmatrix}$.

(i) Name the arrays for column-wise A sparse matrix as follows: *cmatvalA*, *cmatindA*, *cmatbegA*, and *cmatcntA*. Then we can define these arrays as follows:

- $cmatvalA = [-1, -1, -1, -1, -1, -1, -1]$.
- $cmatindA = [\ 0,\ 1,\ 4,\ 2,\ 4,\ 3,\ 4]$.
- $cmatbegA = [0,\ 1,\ 3,\ 5]$.
- $cmatcntA = [1,\ 2,\ 2,\ 2]$.

(ii) Name the arrays for row-wise A sparse matrix as follows: *rmatvalA*, *rmatindA*, *rmatbegA*, and *rmatcntA*. Then we can define these arrays as follows:

- $rmatvalA = [-1, -1, -1, -1, -1, -1, -1]$.

- $rmatindA = [\ 0,\ 1,\ \ 1,\ 2,\ \ 2,\ 3,\ \ 3]$.
- $rmatbegA = [0,\ 1,\ 2,\ 3,\ 4]$.
- $rmatcntA = [1,\ 1,\ 1,\ 1,\ 3]$.

(b) We are given $T = \begin{bmatrix} 1\,0\,0\,0 \\ 0\,1\,0\,0 \\ 0\,0\,1\,0 \\ 0\,0\,0\,1 \\ 0\,0\,0\,0 \\ 0\,0\,0\,0 \\ 0\,0\,0\,0 \end{bmatrix}$.

(i) Name the arrays for column-wise T sparse matrix as follows: *cmatbegT*, *cmatcntT*, *cmatindT*, and *cmatvalT*. Then we can define these arrays as follows:

- $cmatvalA = [1,\ 1,\ 1,\ 1]$.
- $cmatindA = [0,\ 1,\ 2,\ 3]$.
- $cmatbegA = [0,\ 1,\ 2,\ 3]$.
- $cmatcntA = [1,\ 1,\ 1,\ 1]$.

(ii) Name the arrays for row-wise T sparse matrix as follows: *rmatvalT*, *rmatindT*, *rmatbegT*, and *rmatcntT*. Then we can define these arrays as follows:

- $rmatvalA = [1,\ 1,\ 1,\ 1]$.
- $rmatindA = [0,\ 1,\ 2,\ 3]$.
- $rmatbegA = [0,\ 1,\ 2,\ 3]$.
- $rmatcntA = [1,\ 1,\ 1,\ 1]$.

(c) We are given $W = \begin{bmatrix} -6 & -8 & -10 \\ -20 & -25 & -28 \\ -12 & -15 & -18 \\ -8 & -10 & -14 \\ -1 & 0 & 0 \\ 0 & -1 & 0 \\ 0 & 0 & -1 \end{bmatrix}$.

(i) Name the arrays for column-wise W sparse matrix as follows: *cmatvalW*, *cmatindW*, *cmatbegW*, and *cmatcntW*. Then we can define these arrays as follows:

- $cmatvalA = [-6, -20, -12, -8, -1, -8, -25, -15, -10, -1, -10, -28, -18, -14, -1]$.
- $cmatindA = [\ 0,\ \ 1,\ \ 2,\ \ 3,\ \ 4,\ 0,\ \ 1,\ \ 2,\ \ 3,\ \ 5,\ \ 0,\ \ 1,\ \ 2,\ 3,\ \ 6]$.
- $cmatbegA = [0,\ 5,\ 10]$.
- $cmatcntA = [5,\ 5,\ 5]$.

(*ii*) Name the arrays for row-wise W sparse matrix as follows: *rmatvalW*, *rmatindW*, *rmatbegW*, and *rmatcntW*. Then we can define these arrays as follows:

- $rmatvalA = [-6, -8, -10, -20, -25, -28, -12, -15, -18, -8, -10, -14, -1, -1, -1]$.
- $rmatindA = [0, 0, 0, 1, 1, 1, 2, 2, 2, 3, 3, 3, 4, 5, 6]$.
- $rmatbegA = [0, 3, 6, 9, 12, 13, 14]$.
- $rmatcntA = [3, 3, 3, 3, 1, 1, 1]$.

Clearly, the sparse matrix format enables savings in terms of memory for data storage since only the nonzeros are stored. As we pointed out earlier, the sparse format is exploited for efficient computations involving *vector-matrix* and *matrix-vector* multiplications. For example, in a computer implementation of the Benders decomposition algorithm (Chap. 5) and the L-shaped algorithm (Chap. 6), computing the RHS $r - Tx^k$ and $r(\omega) - T(\omega)x^k$, respectively, would involve using sparse matrix-vector multiplication. Similarly, in computing an optimality cut the terms $\pi_k^\top T$ and $\pi_k^\top(\omega)T(\omega)$, respectively, would involve vector-matrix multiplication. Thus, ignoring zeros in matrices T and $T(\omega)$ reduces the number of elementary operations in computing matrix-vector and vector-matrix multiplication. Next, we give an overview of program design for SP algorithm implementation in SP.

10.4 Program Design for SP Algorithm Implementation

Understanding the standard input data formats (LP, MPS, and SMPS) and sparse matrix formats (row-wise and column-wise) and proficiency in a programming language (e.g., C/C++, Java, and Python) are essential for computer implementation of algorithms in SP. Without delving into specific programming languages and optimization software platforms (e.g., CPLEX, GUROBI, and Julia), in this subsection we highlight some essential elements for computer implementation of algorithms.

Many recent optimization software platforms follow the object-oriented programming (OOP) design paradigm, which allows for re-usability of codes. The software is typically implemented using classes following a hierarchical structure based on *inheritance*. Following this OOP paradigm, an algorithm implementation can follow the structure shown in Fig. 10.18, where `SuperClass` is an abstract class that contains all the *methods* (or *functions* in modular programming) and attributes that are common to `SpecialClassA` and `SpecialClassB`. `Algorithm` contains the *main* program that implements an optimization algorithm(s) of interest and uses the three classes above it.

Adapting the structure in Fig. 10.18, we show an example OOP design for software implementation of a suite of SP decomposition algorithms in Fig. 10.19. This design involves three major classes: `LPobjectClass`, `Reader`, `Master`, and `Sublp`. We assume that each of the classes has an interface that specifies the

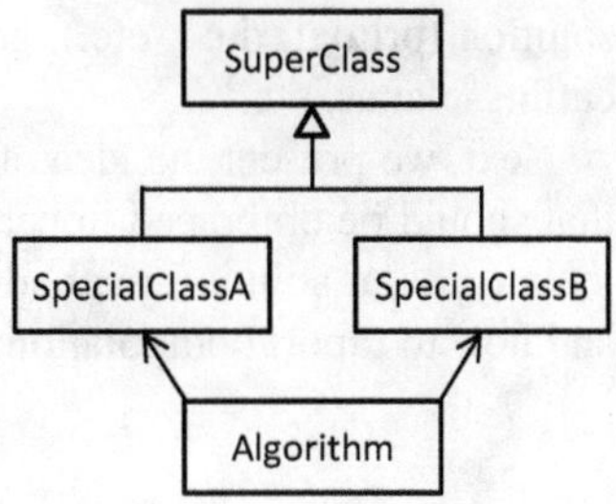

Fig. 10.18 An OOP program structure

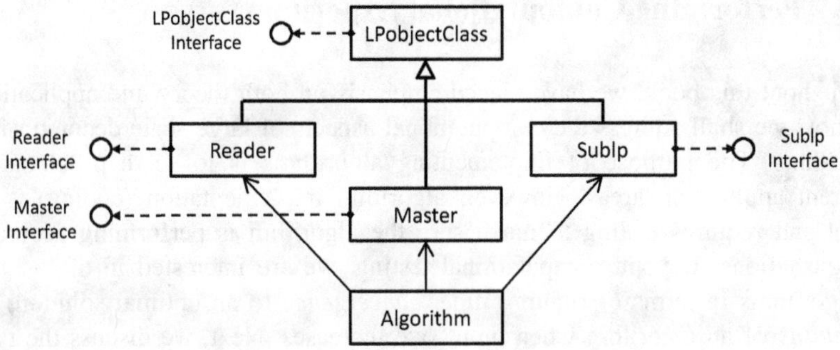

Fig. 10.19 An example OOP program design for SP decomposition algorithms

methods and constant parameters for the class. `LPobjectClass` is a superclass that includes all the methods and attributes that can be inherited in the classes `Reader`, `Master`, and `Sublp` and in `Algorithm`. The *main* program is written in `Algorithm`, where the algorithm(s) of interest (e.g., L-shaped algorithm) is coded. The program structure in Fig. 10.19 naturally follows typical decomposition schemes for large-scale optimization problems, which typically involves a master program and subproblems.

The `Reader` class is needed for reading problem instance data in standard data formats (LP, MPS, and SMPS) into the program. The `Reader` class can also be responsible for decomposition or splitting the core problem into *master* and *subproblem* LP objects. By *LP object* we mean an encapsulation of the problem data, which includes the following: *objective sense* (minimization or maximization), *objective function*, *constraints*, and *restrictions on the decision variables*. Recall that the instance data has to be stored in sparse matrix format for efficient memory storage and computations. This split data is passed to the `Master` and `Sublp` classes, respectively, and is coordinated in `Algorithm` based on the algorithm.

The `Master` class implements aspects of the algorithm pertaining to the master program, including the master program LP object, solving the master program and extracting the solution, adding optimality and feasibility cuts, etc. The `Sublp` class implements aspects of the algorithm pertaining to the (scenario) subproblem, including the subproblem LP object, solving the subproblem and extracting the solution (primal, dual, etc.), generating optimality and feasibility cuts, generating cutting-planes, etc.

Next, we present the idea of empirical analysis as part of the scientific approach that should be embraced in performing computational experiments. We discuss the main types of solution methods in SP, methods of analysis, types of test problems, and how to report computational results.

10.5 Performing Computational Experiments

Throughout this book, we have placed emphasis on both theory and applications, and now we shall address the computational aspects of large-scale decomposition algorithms. The purpose of implementing algorithms is to solve problems for different application areas. However, algorithm implementation (coding) is not trivial and requires testing to make sure the algorithm is performing according to expectations. Beyond computational testing, we are interested in the analysis of algorithms in terms of running time, *convergence* to an optimal solution, and *scalability* of an algorithm when input size increases. Next, we discuss the types of solution methods and algorithm analysis methods. Later, we then discuss test problems and reporting computational experiments.

10.5.1 Solution Methods and Analysis of Algorithms

The three main types of solution methods in large-scale optimization are *reduction* methods, *inductive* methods, and *decomposition* methods. Reduction methods involve reducing the size of the problem via some form of reformulation of the original problem and then applying an exact algorithm on the reformulated or reduced problem. Inductive methods involve solving smaller instances of the problem and then infer ring solution properties to the original problem. The focus in this book is on decomposition, which is a divide-and-conquer approach and involves breaking the problem down into smaller and manageable subproblems, solving each subproblem, and then coordinating the subproblem solutions to determine the solution to the original problem.

To determine or estimate how an algorithm would perform computationally, different methods of analysis are employed. These methods include *worst-case analysis*, *probabilistic analysis*, and *empirical analysis*. The analysis involves characterizing the *running time* of an algorithm, which is the number of primitive operations or steps executed based on the input data. *Worst-case analysis* deals with determining the longest running time for an algorithm for any input size. In other words, with worst-case analysis we want to know how "bad" an algorithm can perform. It provides an upper bound on the running time for any input size. For worst-case analysis, the Θ-notation ("Big-theta") is used to characterize the asymptotic lower and upper bounds on the running time. Asymptotic efficiency is concerned with how the running time of an algorithm increases with the input size in the limit as the size of the input increases without bound. As with the Θ-notation, the O-notation ("Big-O") and the Ω-notation ("Big-omega") are used for asymptotic analysis. The O-notation is used to characterize an asymptotic upper bound for an algorithm, while the Ω-notation is used for an asymptotic lower bound. For further details on this topic, we refer the reader to textbooks on analysis of algorithms (e.g., [6]).

Probabilistic analysis of an algorithm deals with using probability to analyze the running time of an algorithm in terms of the distribution of the input data to determine the average or expected running time. This is much more involved as one has to deal with deciding the appropriate distribution of the inputs. In stochastic optimization we often deal with randomized algorithms, i.e., algorithms whose behavior is determined by both the input and values produced by a pseudorandom number generator. Such algorithms require probability analysis to determine their expected performance.

Empirical analysis of algorithms, which is the focus of this chapter, deals with the design and running of computational experiments to analyze the performance of algorithms for different inputs or classes of problems. Empirical analysis of algorithms is useful in terms of providing insights into why an algorithm performs a particular way in theory versus in practice. In general, we gain insights from applying the *scientific method*, which involves empirical investigation. The basic steps of the scientific method can be summarized as follows:

(1) *State the purpose or question, or make an observation.*
(2) *Conduct background research.*
(3) *Propose a hypothesis.*
(4) *Design and conduct an experiment to test the hypothesis.*
(5) *Record observations and perform data analysis.*
(6) *Make conclusion whether to accept or reject the hypothesis.*

The fundamental aspect of the scientific method is to make sure that the experiment is *reproducible*. This requirement is needed for empirical analysis, and therefore, the scientific method should be applied in designing computational experiments or tests in SP. An experiment has *independent* and *dependent* variables. The independent variable is one that is controlled, while the dependent variable is measured during the experiment, i.e., it depends on the independent variable. The purpose of the experiment is to control and measure the variables so that the experiment is reproducible. The controlled variables are parts of the experiment that are kept constant throughout the experiment.

Computational experiments in SP should be reproducible, and care must be taken in selecting *test problems* to be used in the experiments. This is because randomly generated test problems generally do *not* resemble *real-life* problems, i.e., problems that will be encountered in the real setting on a regular basis. Benchmarking is necessary in the algorithm development and should be part of the empirical analysis. For two exemplars of empirical analysis studies in SP, we refer the readers to [16] who study the empirical behavior of sampling methods for SLP and to [7] who study the empirical behavior of decomposition algorithms for MR-SLP. We will discuss test problems in SP in the next subsection.

10.5.2 Empirical Analysis and Test Problems

Empirical analysis requires test problems as a basis for a computational study. The two basic types of test problems in SP are *real-life/practical* and *artificial/constructed* test problems. Practical problems are based on real-life cases, can be proprietary, and may not be readily available. On the other hand, constructed test problems can be created as needed and include pseudorandomly generated test instances. Nevertheless, constructed test problems can be prone to errors even though they provide a large pool of test instances. Another class of test problems, referred to as *standard* test problems, are those that have been used over time and have become "standard" for testing and benchmarking purposes. Standard test problems can be either practical or constructed problems.

Test problems have two fundamental purposes: (1) *to gain insight* and (2) *to find a best performer*. By using experimental design principles based on the scientific method as explained in the previous section, empirical test problems are used to gain insights. Competitive test problems, practical or constructed, are used to find the best performer among a set of algorithms. This requires ensuring a full range of problem types and knowing parameter settings and systematically varying them so that one can attribute variation to cause as with statistical tests. Therefore, one has to be aware of possible sources of such variation. The sources include variation among the algorithms, factor levels, among problem instances, measurement, and oversight. Variation due to problem instances is not desirable and should be avoided, while variation due to measurement (systematic) error and variation due to oversight are both very bad and should be prevented.

It is important in empirical analysis to carefully plan and explicitly state the experimental design. Doing this is crucial in making valid inferences about the performance of an algorithm. Both practical and constructed test problems are needed in empirical analysis. Each test problem should be justified by stating the *characteristics* that make it a good test problem and what will be *learned* as a result of running the problem. Testing more than one algorithm, the same set of problems should be solved by each algorithm under investigation.

Real-world problems usually have one or more characteristics found in problems which the algorithm under test will be solving on an ongoing basis. Therefore, practical test problems must be representative of the real-world applications to which the algorithm will be applied. Constructed test problems are needed in the testing process as well as for software validation. The advantage of constructed test problems is that they can be designed to test specific aspects of the algorithm which might be experienced infrequently by real-life problems. In addition, with constructed test problems, one can vary their size and data to test an algorithm at or near the boundary of solvable problems.

A few remarks regarding practical problems versus constructed pseudorandomly generated problems are now in order. Practical problems are usually representative of real-world behavior, while randomly generated problems are often not. However, practical problems can be expensive to collect and document. Randomly generated problems, on the other hand, are usually produced by a problem generator program. The characteristics of the population from which a randomly generated problem is drawn are known and can be controlled. However, this is not necessarily true for real-life problems as the population is typically unknown. Consequently, generalizations based on a sample are open to doubt. On the contrary, for randomly generated problems it is possible to use sampling methods and draw generalizations with a known degree of certainty.

10.5.3 *Standard Test Instances*

There are several test instances in SP that have been developed over the years from different applications and have become "standard" test instances. A selection of these instances is listed in Table 10.3. The table gives a summary of the characteristics of these test instances in terms of the instance name, SP class, application, year instance published, the number of scenarios (Scens.), and the number of first- and second-stage constraints (Cons.) and decision variables (Vars.). The test instances are listed in chronological order of their publication, and we specify the SP class the instance belongs to, i.e., SLP, SMIP, or PC-SP (probabilistically constrained SP).

Mathematical descriptions and test results of some of the SLP test instances are given in [12] and [16]. In particular, the empirical study by Linderoth et al. [16] uses several of these test instances, which can be accessed online at http://pages.cs.wisc.edu/~swright/stochastic/sampling/. The listed SMIP and PC-SP instances are available at the Stochastic Integer Programming Library (SIPLIB) website: https://www2.isye.gatech.edu/~sahmed/siplib/. There are also several instances that have been published in the literature for different applications (see Chap. 4). Next, we provide a brief description of each instance, noting the application and the nature of the decision variables and constraints involved.

gbd

Test instance *gbd* is derived from an aircraft allocation problem [9] that involves the optimal allocation of different types of aircraft to routes with uncertain demand. The objective is to maximize profit in the face of costs related to operating the aircraft and to bumping passengers when demand exceeds capacity. In this problem, the first-stage variables involve the number of aircraft (out of four types) to assign to a particular route, while the first-stage constraints bound the availability of the

Table 10.3 Example SP standard test instances

Name	SP class	Application	Year	Scens.	First-stage (Cons., vars.)	Second-stage (Cons., vars.)
gbd	SLP	Aircraft Allocation [9]	1963	6.5×10^5	(4, 17)	(5, 10)
pgp2	SLP	Power Generation Planning [12, 20]	1982	576	(2, 4)	(7, 12)
LandS	SLP	Electricity Planning [17]	1988	10^6	(2, 4)	(7, 12)
ssn	SLP	Telecom Network Design [24]	1994	10^{70}	(1, 89)	(175, 706)
storm	SLP	Cargo Flight Scheduling [19]	1995	6×10^{81}	(185, 121)	(528, 1259)
cep1	SLP	Capacity Expansion Planning [12]	1996	216	(9, 8)	(7, 15)
20term	SLP	Vehicle Assignment [18]	1999	1.1×10^{12}	(3, 64)	(4, 17)
SIZES	SMIP	Product Substitution [14, 15]	1995	[a]	[a]	[a]
SEMI	SMIP	Semiconductor Tool Purchasing [4]	2001	[a]	[a]	[a]
DCAP	SMIP	Dynamic Capacity Acquisition [1, 2]	2003	[a]	[a]	[a]
SSLP	SMIP	Telecommunications [21]	2005	[a]	[a]	[a]
MPTSP	SMIP	City Logistics [22, 25]	2017	[a]	[a]	[a]
SMKP	SMIP	Stochastic Multiple Knapsack [3]	2016	[a]	[a]	[a]
VACCINE	PC-SP	Vaccine Allocation [26, 27]	2008	[a]	[a]	[a]
PROBPORT	PC-SP	Portfolio Optimization [23]	2014	[a]	[a]	[a]

[a] Several instances

planes. The second-stage variables are the number of carried and bumped passengers on each of the (five) routes. The constraints in the second stage enforce balancing demand for each of the routes. This uncertain demand appears on the RHS of the constraints. The original model had 750 scenarios, but by increasing the number of demand states the number of scenarios was increased to 646, 425 by Linderoth et al. [16].

pgp2

Test instance *pgp2* is a power generation planning problem described by Murphy et al. [20], appears in [12], and deals with electrical capacity expansion with certain load forecasts to select the minimum cost strategy for investing in electricity generated from gas-fired, coal-fired, and nuclear generators. The first-stage variables model the annualized amount of power generation (kW) based on the specific type of generator acquired, while the second-stage decisions select a specific operational plan to satisfy the realized regional demand (based on satisfying kW of demand from a specific type of generator). The second-stage power generation costs and regional demands are the stochastic elements of this model.

LandS

Test instance *LandS* [17] is a model of a problem from electrical investment planning. In this problem, the first-stage decision variables are capacities of new technologies (four), and the second-stage decision variables represent the production of each of the three different modes of electricity from the four technologies. The first-stage constraints specify the minimum total capacity and budget constraints, while the second-stage constraints involve three demand constraints, each with random RHS demand.

ssn

The test instance *ssn* is a telecommunications network design problem under uncertain demand described by Sen et al. [24]. Specifically, the problem involves telephone-switching network expansion with the objective of adding capacity in terms of lines to a network of existing point-to-point connections subject to budget constraints. The goal is to minimize the number of unsatisfied service requests. The first-stage decision variables allocate capacity to routes before the service requests happen. The second-stage decision variables route the call requests to enable operation of the network. In this problem, the random demand is the number of known requests for connections at a given instant.

Storm

The test instance *storm* is a problem dealing with the allocation of US military aircraft routes during the Gulf War of 1991 and is described by Mulvey and Ruszcynski [19]. Flights are planed over a set of network routes with uncertain amounts of cargo across two periods. The first-stage variables involve flight route schedules with the objective of minimizing the cost of the scheduled fights and the uncertain cargo handling costs and a penalty cost. The second-stage decision variables allocate cargo delivery routes after demand has been realized, and the goal is to satisfy any unmet demand while minimizing holding and penalty cost.

cep1

Test instance *cep1* is a flexible manufacturing two-stage machine capacity expansion planning (CEP) problem described by Higle and Sen [12] with the objective of minimizing weekly amortized expansion cost plus expected weekly production costs. The first-stage variables represent the number of weekly hours of new capacity for each machine, and the second-stage decision variables specify the number of hours each machine is assigned to process a specific part. In this problem, the weekly demands are treated as independent and identically distributed (IID) random variables coming from a known distribution. A detailed formulation of the CEP problem is illustrated in Chap. 4.

20term

Test instance *20term* [18] comes from freight carrier operations planning and is a model of a freight carrier's operations. The first-stage decision variables specify the positions of fleet vehicles at the start of the day, while the second-stage decision variables specify the fleet's movements through a network to fulfil point-to-point demand for shipments. In this problem, the fleet must finish the day with the same fleet configuration as at the start of the day while penalizing unsatisfied demand.

SIZES

The SIZES test set consists of three instances of a two-stage multi-period SMIP arising in product substitution applications and is described in [14] and [15]. The problem is motivated by the manufacturing of different multiple-piece blind fasteners where substituting long-grip fasteners for short-grip fasteners to meet uncertain demand can result in savings. The problems have mixed-integer decision variables in both stages, where the first-stage decision variables represent whether a particular size product is produced. The second-stage decision variables specify the number of units of a particular size produced as well as the number of units of a particular size to cut to meet demand for a given scenario. The objective is to minimize the expected setup, production, and substitution cost.

SEMI

The SEMI test set has three instances of a two-stage multi-period SMIP problem arising in the planning of semiconductor tool purchases and is described in [4]. The first-stage decision variables are mixed-integer and specify new tool purchases, while the second-stage decision variables are continuous and represent the number of wavers per day that enter a production line for a given product and the number of wafers that require a given operation on a specific tool. The product demand is stochastic, and the objective is to minimize the expected value of the unmet demand subject to budget and capacity constraints.

DCAP

The DCAP test set is a collection of 12 two-stage SMIP problems arising in dynamic capacity acquisition and allocation under uncertainty and is described in [1]. The problem involves deciding the capacity expansion schedule for a set of resources over a given number of time periods to satisfy uncertain processing requirements of a given set of tasks. The first-stage decision variables are mixed-integer and represent the capacity acquisition of a given resource in a specific time period, while the second-stage decision variables are binary and indicate whether a resource is assigned to a given task in a specific time period under a given scenario. The objective is to minimize acquisition costs plus expected assignment costs. The instances have complete recourse and discrete distributions for the processing requirements and costs. The computational results for the DCAP test set were obtained using a decomposition based branch-and-bound (DBB) algorithm described in [2].

SSLP

The SSLP test set consists of 12 instances of a two-stage stochastic mixed-integer programming problems arising in server location under uncertainty described in [21]. The stochastic server location problems (SSLPs) have pure binary first-stage decision variables that represent whether or not a "server" is located at a potential site. The second-stage decision variables are mixed-binary and prescribe whether or not a given "client" is served by a server at a given location under a given scenario. The uncertainty in this application is in the availability of a client in the second stage after the servers have been located in the first stage. A detailed formulation of the problem is given in Chap. 4.

MPTSP

The MPTSP test set consists of five instances of two-stage SMIP problems arising in city logistics and is described in [22, 25]. Specifically, the instances are for the multi-path traveling salesman problem with stochastic travel costs. The problem is given over a graph with a set on nodes connected by arcs. There are multiple paths between every pair of nodes and each path has a random travel time. This travel time is a sum of a deterministic and a stochastic term, which represents the travel time oscillation due to path congestion. The objective is to find the shortest expected path. The first-stage decision variables are binary and prescribe whether or not a path is selected, and the second-stage decision variables are also pure binary and represent if a node is visited immediately after another node in the path. Each of the instances in the test set has 100 scenarios. Details of how the instances are generated are available in [25].

SMKP

The SMKP test set consists of 30 instances of two-stage stochastic multiple knapsack problems with a random second-stage vector and is described in [3]. This paper provides the details regarding formulation, data generation, and computational results. The problem has pure binary first-stage with 240 decision variables and 50 knapsack constraints and pure binary second-stage with 120 decision variables and 5 knapsack constraints. Each of the 30 instances has 20 scenarios.

VACCINE

The VACCINE test set has 30 instances of a chance constrained vaccine allocation problem in epidemiology described in [26]. The decision variables in this model prescribe the proportion of households of a given type to vaccinate under a given vaccination policy to prevent epidemics. Therefore, the model includes a chance constraint to require the post vaccination reproduction number to be below one. The objective is to minimize the probability that an epidemic will occur under resource constraints. Further computational results of the model are available in [27].

PROBPORT

The PROBPORT test set is a collection of 20 instances of a chance constrained portfolio optimization problem described in [23]. The decision variables are continuous and represent investment in a given asset with uncertain return which comes at a cost. The objective is to minimize the cost of investment such that the portfolio return exceeds the required return in at least a certain number of the scenarios.

10.5.4 Reporting Computational Experiments

In an empirical analysis study the next step after performing computational experiments is to report and interpret the results. Reporting computational experiments in SP should follow the standards expected for mathematical exposition. Fundamentally, computational tests of an algorithm must be *reproducible*. This means that the computational experiments can be replicated to obtain results that agree with the original experiments within a tolerance that may be reasonably attributed to changes in the available technology (software, hardware, and bioware) being used.

Table 10.4 Performance indicators for large-scale decomposition algorithms

Performance indicator	Meaning
Numerical accuracy	Ability to compute a correct solution in the face of numerical instability
Robustness or scope	Domain or class of problems that can be effectively solved by the algorithm
Scalability	This is an indication of how "well" the algorithm performs as the size of the test problems is increased
Observed rate of convergence	Some problems have an inherent "accuracy versus effort" the tabulation of this effort can be summarized as the empirical or observed rate of convergence
Number of iterations	Number of steps the algorithms take to solve the problem
Number of function evaluations	Number of times some subfunction is called during program execution
Basic operation count	Number of times a basic operation (e.g., addition and subtraction) is required during the execution of the algorithm
Storage requirements	As the size of solvable problems increases, the storage requirements for executing the algorithm become crucial
CPU time	Total central processing time for executing the procedure

Therefore, one has to carefully choose appropriate *performance indicators* for a computational experiment and consider all measures that are applicable to a given algorithm. Although often subjective, performance indicators should be ranked in order of importance and the experiment designed accordingly.

There are several performance indicators that are used for assessing large-scale decomposition algorithms in the literature and for evaluating the efficiency of competing computer codes. Table 10.4 provides examples of such indicators, and there are many others that one can use depending on the research question(s) and computational experiments. For example, for a software implementation performance measures include *user friendliness* and *portability* or ease of use of the software. Other subjective measurements of a computer program's versatility include the program user interface, program setup time, and time required to learn the program.

Finally, when reporting computational experiments, it is important to include certain items that would enable independent replication of the experiments to meet the scientific requirement. We give a list of such items in Table 10.5.

Bibliographic Notes

A full description of the SMPS file format can be found in [10] and [11]. An earlier version of the SMPS file format was published by Birge et al. [5]. Empirical analysis as a science for the analysis of algorithms can be traced back to [13] who

Table 10.5 Important items to include in a computational study report

Item	Description
Algorithm description	Name of the algorithm and detailed steps as exemplified in previous chapters, including the *initialization* and *termination* steps
Computer implementation	State the programming language used, software platform, and computer platform: processor name, speed (GHz), random access memory size (RAM), cores, threads, etc.
Experimental design	Description of the design of experiments including research problem or question(s), goals, important factors, performance indicators, number of replications, etc.
Results	Include tables and graphs with detailed descriptions, report in logical order of importance; report appropriate statistics
Discussion	Explain the results, findings, implications of empirical findings to both theory and practice, etc.
Conclusions	Summary of key findings or answers, insights, implications, and what remains to be answered (future work) if any

advocated for the empirical science of algorithms. The need for a scientific way of reporting computational experiments is not new and was pointed out by Crowder et al. [8]. This work remains relevant with the advance of computer technology and the development of a diverse number of algorithms. Examples of empirical studies in SP include a study on the empirical behavior of sampling methods for SLP [16] and another study on decomposition algorithms for MR-SLP [7]. In terms of analysis of algorithms in computer science, we refer the interested reader to textbooks on this topic such as [6].

Problems

Note to Students and Instructors The problems below classified as *Mini-Project* require significant effort to satisfactorily complete. Each problem is expected to take the student a few weeks to complete depending on their computer programming competence. The *Semester-Project*, as the name implies, requires a substantial amount of time (several weeks) and effort to complete. In terms of grading both the mini- and semester-projects, the grade should be based on running code and a well-written project report, following the guidelines for reporting computational results outlined in this chapter.

10.1 First- and Second-Stage Matrices and Sparse Matrices

Consider the following CORE LP for a two-stage SLP, where the first-stage decision variables are $x_i,\ i = 1, \cdots, 8$, and the second-stage decision variables are $y_j,\ j = 1, \cdots, 15$:

$$
\begin{aligned}
\text{Min} \quad & 2.5x_1 + 3.75x_2 + 5x_3 + 3x_4 \\
& + 2.6y_1 + 3.4y_2 + 3.4y_3 + 2.5y_4 + 1.5y_5 + 2.3y_6 + 2y_7 + 3.6y_8 \\
& + 4y_9 + 3.8y_{10} + 3.5y_{11} + 3.2y_{12} + 400y_{13} + 400y_{14} + 400y_{15} \\
\text{s.t.} \quad & -x_1 + x_5 <= 500 \\
& -x_2 + x_6 <= 500 \\
& -x_3 + x_7 <= 500 \\
& -x_4 + x_8 <= 500 \\
& 0.08x_5 + 0.04x_6 + 0.03x_7 + 0.01x_8 <= 100 \\
& x_5 <= 2000 \\
& x_6 <= 2000 \\
& x_7 <= 3000 \\
& x_8 <= 3000 \\
& 0.6y_1 + 0.6y_2 + 0.9y_3 + 0.8y_4 + y_{13} >= 1500 \\
& 0.1y_5 + 0.9y_6 + 0.6y_7 + 0.8y_8 + y_{14} >= 1500 \\
& 0.05y_9 + 0.2y_{10} + 0.5y_{11} + 0.8y_{12} + y_{15} >= 1500 \\
& -x_5 + y_1 + y_5 + y_9 <= 0 \\
& -x_6 + y_2 + y_6 + y_{10} <= 0 \\
& -x_7 + y_3 + y_7 + y_{11} <= 0 \\
& -x_8 + y_4 + y_8 + y_{12} <= 0 \\
& x_i, y_j \geq 0, \forall i = 1, \cdots, 8; \ j = 1, \cdots, 15.
\end{aligned}
$$

Answer the following questions:

(a) Write the *first-stage matrix* A. What is the number of rows (m_1), number of columns (n_1), and number of nonzeros of A? Write A in sparse format: (i) column-wise and (ii) row-wise.

(b) Write the *technology matrix* T. What is the number of rows (m_1), number of columns (n_1), and number of nonzeros of T? Write T in sparse format: (i) column-wise and (ii) row-wise.

(c) Write the *recourse matrix* W. What is the number of rows (m_1), number of columns (n_1), and number of nonzeros of W? Write W in sparse format: (i) column-wise and (ii) row-wise.

10.2 Modeling, SMPS Files, and Computational Study

(a) Choose an application area of interest, and then model a real-life problem from this application area using one of the SP approaches: SLP, MR-SLP

SMIP, MR-SMIP, or PC-SP. Define the decision variables and problem data for the first- and second-stages, respectively, and then write the mathematical formulation followed by the description of the objective function and constraints (see Chap. 4).

(b) What is the suitable algorithm(s) for solving the problem in part (a)?

(c) Conceptualize an empirical study of the algorithm in part (b) based on your problem in part (a). Your write-up should include a design of experiments to be conducted, including goals, important factors to measure, and performance indicators.

(d) Create test instances of your model for performing a computational study based on part (c) by randomly generating certain parameters (data) for your problem created in part (a). The data should include the following:

$$c,\ A,\ b,\ q,\ W,\ \{T(\omega), r(\omega)\}_{\omega\in\Omega}.$$

You should write code (a program) to generate several instances by varying the problem size (number of decision variables and constraints) and number of scenarios. For each instance, create SMPS files: CORE, TIME, and STOCH files.

(e) Perform a computational study based on parts (a)–(d), and write a report based on the guidelines for reporting computational experiments outlined in this chapter. This may require implementing part (b) if the algorithm is not readily available.

10.3 Mini-Project: Benders Decomposition Algorithm

Use your choice of programming language (e.g., C/C++, Java, and Python) and a software package (e.g., CPLEX Callable Library, CPLEX Concert Technology, Gurobi, and Julia) to code the *Benders decomposition algorithm* (Chap. 5). Your program should be designed to read SMPS data files, i.e., CORE, TIME, and STOCH files. This will enable your code to solve standard SP test instances. Conduct a computational study following the empirical analysis approach described in this chapter. Specifically, perform the following tasks:

(a) Test your Benders decomposition algorithm code on all the appropriate standard test instances (CORE files) in Table 10.3 that your implementation can solve.

(b) Test your code on *two* additional test instances of your choice from this book and/or from the literature.

(c) Report your results using appropriate tables and plots, including graphs showing a trace of the upper and lower bounds versus the algorithm iteration number until termination of the algorithm.

(d) Write a report including an introduction, problem statement, detailed step of the algorithm, your design of experiments, computational results, discussion of the results, and your conclusions. The results must include the following: *instance name, optimal objective value, number of iterations, CPU time of the algorithm,* and *CPU time for directly solving the instance using an LP solver.*

10.4 Mini-Project: Regularized Benders Decomposition Algorithm
Use your choice of programming language (e.g., C/C++, Java, and Python) and a software package (e.g., CPLEX Callable Library, CPLEX Concert Technology, Gurobi, and Julia) to code the *regularized Benders decomposition algorithm* (Chap. 5). The software of your software package must include a QP solver. Your program should be designed to read SMPS data files, i.e., CORE, TIME, and STOCH files. This will enable your code to solve standard SP test instances. Conduct a computational study following the empirical analysis approach described in this chapter. Specifically, perform the following tasks:

(a) Test your regularized Benders Decomposition algorithm code on all the appropriate standard test instances (CORE files) in Table 10.3 that your implementation can solve.
(b) Test your code on *two* additional test instances of your choice from this book and/or from the literature.
(c) Report your results using appropriate tables and plots, including graphs showing a trace of the upper and lower bounds versus the algorithm iteration number until termination of the algorithm.
(d) Write a report including an introduction, problem statement, detailed step of the algorithm, your design of experiments, computational results, discussion of the results, and your conclusions. The results must include the following: *instance name, optimal objective value, number of iterations, CPU time of the algorithm,* and *CPU time for directly solving the instance using an LP solver.*

10.5 Mini-Project: Dantzig–Wolfe Decomposition Algorithm
Use your choice of programming language (e.g., C/C++, Java, Python) and a software package (e.g., CPLEX Callable Library, CPLEX Concert Technology, Gurobi, Julia) to code the *Dantzig–Wolfe decomposition algorithm* (Chap. 5). Your program should be designed to read SMPS data files, i.e., CORE, TIME, and STOCH files. This will enable your code to solve standard SP test instances. Conduct a computational study following the empirical analysis approach described in this chapter. Specifically, perform the following tasks:

(a) Test your Dantzig–Wolfe decomposition algorithm code on all the appropriate standard test instances (CORE files) in Table 10.3 that your implementation can solve.
(b) Test your code on *two* additional test instances of your choice from this book and/or from the literature.
(c) Report your results using appropriate tables and plots, including graphs showing a trace of the upper and lower bounds versus the algorithm iteration number until termination of the algorithm.
(d) Write a report including an introduction, problem statement, detailed step of the algorithm, your design of experiments, computational results, discussion of the results, and your conclusions. The results must include the following: *instance name, optimal objective value, number of iterations, CPU time of the algorithm,* and *CPU time for directly solving the instance using an LP solver.*

10.6 Mini-Project: L-Shaped Algorithm

Use your choice of programming language (e.g., C/C++, Java, and Python) and a software package (e.g., CPLEX Callable Library, CPLEX Concert Technology, Gurobi, and Julia) to code the *L-shaped algorithm* (Chap. 6). You are encouraged to build your code from your Benders code from Mini-Project Problem 10.1. Your program should be designed to read SMPS data files, i.e., CORE, TIME, and STOCH files. This will enable your code to solve standard SP test instances. Conduct a computational study following the empirical analysis approach described in this chapter. Specifically, perform the following tasks:

(a) Test your L-shaped algorithm code on all the appropriate standard test instances in Table 10.3 that your implementation can solve.
(b) Test your code on *two* additional test instances of your choice from this book and/or from the literature.
(c) Report your results using appropriate tables and plots, including graphs showing a trace of the upper and lower bounds versus the algorithm iteration number until termination of the algorithm.
(d) Write a report including an introduction, problem statement, detailed step of the algorithm, your design of experiments, computational results, discussion of the results, and your conclusions. The results must include the following: *instance name, optimal objective value, number of iterations, CPU time of the algorithm,* and *CPU time for directly solving the DEP instance using an LP solver.*

10.7 Semester-Project: Risk-Neutral SLP

Use your choice of programming language (e.g., C/C++, Java, and Python) and a software package (e.g., CPLEX Callable Library, CPLEX Concert Technology, Gurobi, and Julia) to code one of the following algorithms from Chap. 6:

- *Multicut L-shaped Algorithm.*
- *Adaptive Multicut L-shaped Algorithm.*
- *PH Algorithm.*

You are encouraged to build your code from your L-shaped code from Mini-Project Problem 10.4. Your program should be designed to read SMPS data files, i.e., CORE, TIME, and STOCH files. This will enable your code to solve standard SP test instances. Conduct a computational study following the empirical analysis approach described in this chapter. Specifically, perform the following tasks:

(a) Test your code on all the appropriate standard test instances in Table 10.3 that your implementation can solve.
(b) Test your code on *two* additional test instances of your choice from this book and/or from the literature.
(c) Report your results using appropriate tables and plots, including graphs showing a trace of the upper and lower bounds versus the algorithm iteration number until termination of the algorithm.

(d) Write a report including an introduction, problem statement, detailed step of the algorithm, your design of experiments, computational results, discussion of the results, and your conclusions. The results must include the following: *instance name, optimal objective value, number of iterations, CPU time of the algorithm,* and *CPU time for directly solving the DEP instance using an LP solver.*

10.8 Semester-Project: Risk-Averse SLP

Use your choice of programming language (e.g., C/C++, Java, and Python) and a software package (e.g., CPLEX Callable Library, CPLEX Concert Technology, Gurobi, and Julia) to code one of the following algorithms from Chap. 7:

- $\mathbb{D}$-*AGG Algorithm* (for risk measures QDEV, CVaR, and EE)
- $\mathbb{D}$-*SEP Algorithm* (for risk measures QDEV, CVaR, and EE)
- *ASD-AGG Algorithm*
- *ASD-SEP Algorithm*

You are encouraged to build your code from your L-shaped code from Mini-Project Problem 10.4. Your program should be designed to read SMPS data files, i.e., CORE, TIME, and STOCH files. This will enable your code to solve standard SP test instances. Conduct a computational study following the empirical analysis approach described in this chapter. Specifically, perform the following tasks:

(a) Test your code on all the appropriate standard test instances in Table 10.3 that your implementation can solve.
(b) Test your code on *two* additional test instances of your choice from the literature not listed in Table 10.3.
(c) Report your results using appropriate tables and plots, including graphs showing a trace of the upper and lower bounds versus the algorithm iteration number until termination of the algorithm.
(d) Write a report including an introduction, problem statement, detailed step of the algorithm, your design of experiments, computational results, discussion of the results, and your conclusions. The results must include the following: *instance name, optimal objective value, number of iterations, CPU time of the algorithm,* and *CPU time for directly solving the DEP instance using an LP solver.* Recall that for MR-SP you have to trace the efficient frontier and report a solution for each value of the risk factor.

10.9 Semester-Project: Sampling-Based SLP

Use your choice of programming language (e.g., C/C++, Java, and Python) and a software package (e.g., CPLEX Callable Library, CPLEX Concert Technology, Gurobi, and Julia) to code one of the following methods from Chap. 8:

- *SAA with L-shaped Algorithm*
- *SD with L-shaped Algorithm*

You are encouraged to build your code from your L-shaped code from Mini-Project Problem 10.4. Your program should be designed to read SMPS data files, i.e.,

CORE, TIME, and STOCH files. This will enable your code to solve standard SP test instances. Conduct a computational study following the empirical analysis approach described in this chapter. Specifically, perform the following tasks:

(a) Test your code on all the appropriate standard test instances in Table 10.3 that your implementation can solve.
(b) Test your code on *two* additional test instances of your choice from the literature not listed in Table 10.3.
(c) Report your results using appropriate tables and plots, including graphs showing a trace of the upper and lower bounds versus the algorithm iteration number until termination of the algorithm.
(d) Write a report including an introduction, problem statement, detailed step of the algorithm, your design of experiments, computational results, discussion of the results, and your conclusions. The results must include the following: *instance name, optimal objective value, number of iterations, CPU time of the algorithm,* and *CPU time for directly solving the DEP instance using an LP solver.*

10.10 Semester-Project: SMIP

Use your choice of programming language (e.g., C/C++, Java, and Python) and a software package (e.g., CPLEX Callable Library, CPLEX Concert Technology, Gurobi, and Julia) to code one of the following algorithms from Chap. 9:

- *BFS Algorithm*
- *FD Algorithm*
- *D^2 Algorithm*

You are encouraged to build your code from your L-shaped code from Mini-Project Problem 10.4. Your program should be designed to read SMPS data files, i.e., CORE, TIME, and STOCH files. This will enable your code to solve standard SP test instances. Conduct a computational study following the empirical analysis approach described in this chapter. Specifically, perform the following tasks:

(a) Test your code on all the appropriate standard test instances in Table 10.3 that your implementation can solve.
(b) Test your code on *two* additional test instances of your choice from the literature not listed in Table 10.3.
(c) Report your results using appropriate tables and plots, including graphs showing a trace of the upper and lower bounds versus the algorithm iteration number until termination of the algorithm.
(d) Write a report including an introduction, problem statement, detailed step of the algorithm, your design of experiments, computational results, discussion of the results, and your conclusions. The results must include the following: *instance name, optimal objective value, number of iterations, CPU time of the algorithm,* and *CPU time for directly solving the DEP instance using an LP solver.*

References

1. S. Ahmed and R. Garcia. Dynamic capacity acquisition and assignment under uncertainty. *Annals of Operations Research*, 124:267–283, 2003.
2. S. Ahmed, M. Tawarmalani, and N. V. Sahinidis. A finite branch and bound algorithm for two-stage stochastic integer programs. *Mathematical Programming*, 100:355–377, 2004.
3. G. Angulo, S. Ahmed, and S. S. Dey. Improving the integer l-shaped method. *INFORMS Journal on Computing*, 28(3):483–499, 2016.
4. F. Barahona, S. Bermon, O. Gunluk, and S. Hood. Robust capacity planning in semiconductor manufacturing. Technical Report RC22196, IBM Research, 2001.
5. J.R. Birge, M.A.H. Dempster, H.I. Gassmann, E.A. Gunn, A.J. King, and S.W. Wallace. A standard input format for multiperiod stochastic linear programs. *COAL Newsletter*, 17:1–19, 1987.
6. T.H. Cormen, C.E. Leiserson, R.L. Rivest, and C. Stein. *Introduction to Algorithms*. MIT Press, Cambridge, MA, Fourth edition, 2022.
7. T.G. Cotton and L. Ntaimo. Computational study of decomposition algorithms for mean-risk stochastic linear programs. *Mathematical Programming Computation*, 7.4:471–499, 2015.
8. H. Crowder, R.S. Dembo, and J.M. Mulvey. On reporting computational experiments with mathematical software. *ACM Transactions on Mathematical Software*, 5(2):193–203, 1979.
9. G.B. Dantzig. *Linear Programming and Extensions*. Princeton University Press, Princeton, New Jersey, 1963.
10. H. I. Gassmann. The SMPS format for stochastic linear programs. In *Applications of Stochastic Programming*, chapter 2, pages 9–19. SIAM, 2005.
11. H. I. Gassmann and B. Kristjansson. The SMPS format explained. *IMA Journal of Management Mathematics*, 19:347–377, 2007.
12. Julie Higle and Suvrajeet Sen. *Stochastic Decomposition*. Kluwer Academic Publishers, 101 Phillip Drive, Norwell, MA 02061, 1996.
13. J.N. Hooker. An empirical science of algorithms. *Operations Research*, 42(2):201–212, 1994.
14. S. Jorjani, C. H. Scott, and D. L. Woodruff. Selection of an optimal subset of sizes. Technical report, University of California, Davis, CA, 1995.
15. S. Jorjani, C. H. Scott, and D. L. Woodruff. Selection of an optimal subset of sizes. *International Journal of Production Research*, 37(16):3697–3710, 1999.
16. J.T. Linderoth, A. Shapiro, and S.J. Wright. The empirical behavior of sampling methods for stochastic programming. *Annals of Operations Research*, 142:215–241, 2006.
17. F. Louveaux and Y. Smeers. Optimal investments for electricity generation: A stochastic model and a test problem. In Y Ermoliev and R Wets, editors, *Numerical techniques for stochastic optimization problems*, chapter 24, pages 445–452. Springer-Verlag, Berlin, 1988.
18. W.K. Mak, D.P. Morton, and R.K. Wood. Monte Carlo bounding techniques for determining solution quality in stochastic programs. *Operations Research Letters*, 24:47–56, 1999.
19. J.M Mulvey and A. Ruszcynski. A new scenario decomposition method for large scale stochastic optimization. *Operations Research*, 43:477–490, 1995.
20. F. H. Murphy, S. Sen, and A. L. Soyster. Electric utility capacity expansion with certain load forecasts. *IIE Transactions*, 14:52–59, 1982.
21. L. Ntaimo and S. Sen. The million-variable 'march' for stochastic combinatorial optimization. *Journal of Global Optimization*, 32(3):385–400, 2005.
22. G. Perboli, L. Gobbato, and F. Maggioni. A progressive hedging method for the multi-path travelling salesman problem with stochastic travel times. *IMA Journal of Management Mathematics*, 28(1):65–86, 2017.
23. F. Qiu, S. Ahmed, S.S. Dey, and L.A. Wolsey. Covering linear programming with violations. *INFORMS Journal on Computing*, 26(3):531–546, 2014.
24. S. Sen, R.D. Doverspike, and S. Cosares. Network planning with random demand. *Telecommunications Systems*, 3(11–30), 1994.

25. R. Tadei, G. Perboli, and F. Perfetti. The multi-path traveling salesman problem with stochastic travel costs. *EURO Journal on Transportation and Logistics*, 6(1):3–23, 2017.
26. M. T. Tanner, L. Sattenspiel, and L. Ntaimo. Finding optimal vaccination strategies under parameter uncertainty using stochastic programming. *Mathematical Biosciences*, 215:144–151, 2008.
27. M. W. Tanner and L. Ntaimo. IIS branch-and-cut for joint chance-constrained stochastic programs and application to optimal vaccine allocation. *European Journal of Operational Research*, 207(1):290–296, 2010.

Index

L. Ntaimo, *Computational Stochastic Programming*, Springer Optimization and Its Applications 774, https://doi.org/10.1007/978-3-031-52464-6

www.ingramcontent.com/pod-product-compliance
Lightning Source LLC
Chambersburg PA
CBHW072033260326
41796CB00047B/80
9783031524622